Random Processes for Image and Signal Processing

SPIE/IEEE SERIES ON IMAGING SCIENCE & ENGINEERING

The SPIE/IEEE Series on Imaging Science & Engineering publishes original books on theoretical and applied electronic and optical imaging.

Other titles in the series:
Fundamentals of Electronic Image Processing,
Arthur R. Weeks, Jr.

Random Processes for Image and Signal Processing

Edward R. Dougherty
Texas A&M University

SPIE Optical Engineering Press
A Publication of SPIE—The International Society for Optical Engineering
Bellingham, Washington USA

The Institute of Electrical and Electronics Engineers, Inc., New York

Library of Congress Cataloging-in-Publication Data

Dougherty, Edward R.
Random processes for image and signal processing / Edward R. Dougherty.
p. cm. – (SPIE/IEEE series on imaging science & engineering)
Includes bibliographical references and index.
ISBN 0-8194-2513-3 (hardcover)
1. Image processing–Statistical methods. 2. Stochastic processes.
3. Signal processing–Statistical methods. I. Title. II. Series.
TA1637.D685 1998
621.382'2'0151923—dc21 97-42909
CIP

Copublished by

SPIE—The International Society for Optical Engineering
P.O. Box 10
Bellingham, Washington 98227-0010
Phone: 360/676-3290
Fax: 360/647-1445
Email: spie@spie.org
WWW: http://www.spie.org
SPIE Press Volume PM44
ISBN 0-8194-2513-3

IEEE Press
445 Hoes Lane
P.O. Box 1331
Piscataway, NJ 08855-1331
Phone: 1-800/678-IEEE
Fax: 732/562-1746
Email: ieeepress@ieee.org
WWW: http://www.ieee.org
IEEE Order No. PC5747
ISBN 0-7803-3495-7

Printed in the United States of America.

To my mentor

Joanne Elliott

Contents

Preface

Science and engineering deal with temporal, spatial, and higher-dimensional processes that vary randomly from observation to observation. While one might study a specific observation for its algebraic or topological content, deterministic analysis does not provide a framework for understanding the ensemble of observations, nor does it provide a mechanism for prediction of future events. A system that functions over time needs to be designed in accordance with the manner in which input processes vary over time. System performance needs to be measured in terms of expected behavior and other statistical characteristics concerning operation on random inputs. There is an input random process, a transformation, and an output random process. System *analysis* begins with the input process and, based on the transformation, derives characteristics of the output process; system *synthesis* begins with an input process and a desired output process, and derives a transformation that estimates the desired output from the input.

Image and (one-dimensional) signal processing concern the analysis and synthesis of linear and nonlinear systems that operate on spatial and temporal random functions. As areas of applied science, image and signal processing mark off a region within the overall domain of random processes that suits the set of applications they encompass. Because our major concern is algorithm development, our major interest is operator synthesis. We focus on three basic problems: representation, filter design, and modeling. These are not independent; they form a unity that is the key to algorithm development in a stochastic framework. The end goal is design of a filter (operator). If the input process is given in terms of a representation that is compatible with the form of the desired filter, then design is enhanced. Ultimately, the filter is to be used on some class of real-world images or signals. Therefore we need models that fit real processes and whose mathematical structure facilitates design of filters to extract desired structural information.

My goal is not just to present the theory along with applications, but also to help students intuitively appreciate random functions. Were this a mathematics book, I would have taken the mathematical approach of stating general theorems and then giving corollaries for special situations. Instead, I have often begun with special cases in which probabilistic insight is more readily achievable. Moreover, I have not taken a theorem-proof approach. When provided, proofs are in the main body of the text and clearly delineated; sometimes they are either not provided or outlines of conceptual arguments are given. The intent is to state theorems carefully and to draw clear distinctions between rigorous mathematical arguments and heuristic explanations. When a proof can be given at a mathematical level commensurate with the text and when it enhances conceptual understanding, it is usually provided; in other cases, the effort is to explain subtleties of the definitions and properties concerning random functions, and to state conditions under which a proposition applies. Attention is drawn to the differences between deterministic concepts and their random counterparts, for instance, in the mean-square calculus, orthonormal representation, and linear filtering. Such differences are sometimes glossed over in method books; however, lack of differentiation between random and deterministic analysis can lead to misinterpretation of experimental results and misuse of techniques.

My motivation for the book comes from my experience in teaching graduate-level image processing and having to end up teaching random processes. Even students who have taken a course on random processes have often done so in the context of linear operators on signals. This approach is inadequate for image processing. Nonlinear operators play a widening role in image processing, and the spatial nature of imaging makes it significantly different from one-dimensional signal processing. Moreover, students who have some background in stochastic processes often lack a unified view in terms of canonical representation and orthogonal projections in inner product spaces.

The book can be used for courses in a number of ways. For students with a strong background in probability and statistics, it can form a one-semester course on random processes, with the obvious necessity of omitting a number of sections. Since the first chapter provides the essential probability and estimation theory that would normally be taught in a one-semester undergraduate course, the book can be used for a full-year course for students who lack a good undergraduate probability course. The book is structured with this use in mind. My experience shows me that very few engineering students come to graduate school having an adequate background in probability theory. Finally, owing to the large number of imaging applications, with the addition of some supplementary papers, the book can be used for a graduate course on image processing; indeed, I have taken such an approach here at Texas A&M. I suspect that a similar approach can be used for signal processing. For research-oriented departments, cookbook-style texts are totally inadequate and future researchers receive significant benefit from learning their specialty in the proper mathematical framework.

The first chapter covers basic probability theory, with attention paid to multivariate distributions and functions of several random variables. The probability theory concludes with a section on laws of large numbers. There follows a general section on parametric estimation. Maximum-likelihood estimators are covered and applied to a constant signal corrupted by various noise models. Estimation plays an important role throughout the book because it is not enough to know the probabilistic theory behind algorithm design; one also needs to be aware of the problems associated with estimating an optimal filter from sample signals. The chapter concludes with sections on entropy and coding.

The second chapter covers the basic properties of random functions typically found in general texts on engineering random processes. Differences are mainly in orientation, but this is important. The stochastic problems of image processing differ substantially from those of one-dimensional signal processing. Because the latter is more mature as a discipline, books on random processes tend to orient their image-signal processing applications toward signals. Historically, such an approach is natural; nevertheless, it is often not sufficient to give a definition, property, or application for signal processing and then say that, as presented, it is suitable for image processing. Many stochastic problems that are straightforward in one dimension become either very difficult or intractable in two dimensions. I try to take a balanced approach with the basic theory and continue to pay attention to both one- and two-dimensional processes throughout the text.

The third chapter treats canonical representation in the natural context of Fourier expansions in inner product spaces. The Karhunen-Loeve expansion is covered in detail: the meaning of the expansion, its relation to deterministic Fourier representation, and the role played by the eigenfunctions resulting from the covariance function. There follow sections on noncanonical representation, trigonometric representation of wide-sense stationary processes, and the role of canonical expansions as transforms. There is a substantial section on transform coding. It is placed in the context of the Karhunen-Loeve expansion, so that transform efficiency for other transforms, such as the discrete cosine and Walsh-Hadamard transforms, can be better understood. The text then goes into the general theory of discrete canonical expansions whose coefficients are generated by linear functionals. Properties of the coefficients and related coordinate functions are discussed. Because a canonical expansion of a random function can be derived from a canonical expansion of its covariance function, covariance expansions are discussed. Integral canonical expansions are theoretically more difficult and rely on the theory of generalized functions. Therefore the subject is treated formally with the understanding that we often consider random functions whose covariance functions are distributions. Integral canonical expansions provide an appropriate framework for discussion of the power spectral density and the Wiener-Khinchin theory. We are not merely concerned with the power spectral density as the Fourier transform of a covariance function; we are also concerned with white-noise representation of random functions. Representation is discussed in the context of necessary and sufficient conditions for an integral canonical expansion. The next section introduces vector random functions and their canonical representation. The chapter closes with an algorithm that produces a canonical representation over a discrete set that can be applied in very general circumstances.

The fourth chapter treats filter design. The basic theory of optimal mean-square-error filters is covered first. The next eight sections are committed to optimal linear filters. Coverage begins with application of the orthogonality principle to find the optimal linear filter for a finite number of observations. A number of examples are provided to explain both the algebraic and statistical aspects of the theory. The steepest-descent and LMS iterative algorithms are covered next, including convergence. Vector estimation begins with least-squares-estimation of a nonrandom vector, with the best estimator given in terms of the pseudoinverse of the design matrix. Because we assume that the columns of the design matrix are linearly independent, the pseudoinverse takes a simple form. The next section treats random vectors and requires the pseudoinverse of an autocorrelation matrix that may be singular. The main theorem follows directly from a basic theorem on finding projections into subspaces spanned by linearly dependent vectors in a Hilbert space. It is the main result for optimal finite-observation linear filters. The last section on finite-observation filters discusses recursive linear filters, concluding with the Kalman filter. Recursive linear filters are based on the fact that projections into direct sums of subspaces can be decomposed into sums of projections. This principle is introduced and both static and dynamic recursive filtering follow from it. The last three sections on optimal linear filtering involve infinite observations. The Kolmogorov theory places optimal linear filtering into the context of projections into subspaces generated by operators (in our case, integral operators). The Wiener-Hopf equation is derived and the Wiener filter for wide-sense stationary processes is covered. The next section considers optimal linear filtering in the context of a linear model. This approach can lead to an optimal filter being derived from a system of linear equations; in other cases, it leads to the solution of Wiener-Hopf–like equations that are simpler than the original equation. Optimal linear filtering culminates with the derivation of optimal filters in the framework of canonical expansions. Because it provides a general method for design, it may be considered the main section on optimal linear filtering. Having concluded coverage of linear filters, the chapter turns to nonlinear filters. There is extensive coverage of optimal binary filters, which are a key concern of digital image processing. For discrete binary images, optimization stays close to the probabilistic nature of the processes and optimal versus suboptimal design is appreciable at a very low level. Next, pattern classification is treated in the context of filter design and the Gaussian maximum-likelihood classifier is obtained under the appropriate model conditions. The chapter concludes with neural networks. Error back-propagation is discussed in the context of sum-of-squares error and adaptive network design. Overall, the chapter emphasizes the fundamental role of filter representation. This begins with the integral representation of linear filters and canonical decomposition of random functions, continues with the morphological representation of binary filters, and concludes with the representational power of neural-network filters.

The final chapter treats random models. A major portion is devoted to discrete- and continuous-time Markov chains. The range of application for these models is extensive, including engineering, computer science, and operations research. The key role of the Chapman-Kolmogorov equations is emphasized. Steady-state and stationary distributions are covered for both finite and infinite state spaces. Conditions for existence are given and methods for finding stationary distributions are explored. Special treatment is given to the birth-death model, random walks, and queues. Passing to two dimensions, Markov random fields and Gibbs distributions are discussed. Next comes the random Boolean model, which is fundamental to coverage processes, filtering, and texture analysis. Vacancy and hitting are discussed, and the simple linear Boolean model is discussed in some detail. Granulometries are treated in the following section, which describes granulometric classification and adaptive design of openings to filter clutter in the signal-union-noise model. The final section of the book provides elements of the theory of random closed sets and is at a higher mathematical level than the rest of the text. Its goal is twofold: to explain how the hit-or-miss topology arises naturally from the probabilistic requirements of a random closed set and to present the capacity theorem.

For those who wish to use the text without following the given order, I will outline the essential logical order of the book. Except for Sections 1.10 and 1.11, which are independent from the remainder of the book, one should be familiar with the first chapter before proceeding to random processes. Chapter 2 should be read in its entirety. The remaining three chapters are close to being independent. Except for Section 4.9, Chapter 4 does not depend on Chapter 3. In Chapter 4, one can go directly from Section 4.3 to Section 4.7; indeed, if desired, one can go from Section 4.1 to Section 4.10; however, I do not recommend reading Section 4.12 prior to Sections 4.2 and 4.3. Chapter 5 can be studied after completing Chapter 2. The last three sections of Chapter 5 can be read before the first four sections (exceptions being reference to Gaussian-maximum-likelihood classification in Section 5.6.2 and random walks in Section 5.6.3).

I hope readers find this book both instructive and enjoyable, and that it provides them with insight and knowledge useful to the pursuit of ground-breaking research. For me, a person trained in analysis, it represents a growing appreciation of the profound differences between deterministic and stochastic scientific epistemology.

Edward R. Dougherty
College Station, Texas
July 1998

Random Processes for Image and Signal Processing

Chapter 1
Probability Theory

1.1. Probability Space

Probability theory is concerned with measurements of random phenomena and the properties of such measurements. This opening section discusses the formulation of event structures, the axioms that need to be satisfied by a measurement to be a valid probability measure, and the basic set-theoretic properties of events and probability measures.

1.1.1. Events

At the outset we posit a set S, called the *sample space*, containing the possible experimental outcomes of interest. Mathematically, we simply postulate the existence of a set S to serve as a universe of discourse. Practically, all statements concerning the experiment must be framed in terms of elements in S and therefore S must be constrained relative to the experiment. Every physical outcome of the experiment should refer to a unique element of S. In effect, this practical constraint embodies two requirements: every physical outcome of the experiment must refer to some element in S and each physical outcome must refer to only one element in S. Elements of S are called *outcomes*.

Probability theory pertains to measures applied to subsets of a sample space. For mathematical reasons, subsets of interest must satisfy certain conditions. A collection $\mathcal{E}$ of subsets of a sample space S is called a *σ-algebra* if three conditions are satisfied:

(S1) $S \in \mathcal{E}$.

(S2) If $E \in \mathcal{E}$, then $E^c \in \mathcal{E}$, where E^c is the complement of E.

(S3) If the countable (possibly finite) collection $E_1, E_2, \ldots \in \mathcal{E}$, then the union $E_1 \cup E_2 \cup \cdots \in \mathcal{E}$.

Subsets of S that are elements of $\mathcal{E}$ are called *events*. If S is finite, then we usually take the set of all subsets of S to be the σ-algebra of events. However, when S is infinite, the problem is more delicate. An in-depth discussion of σ-algebras properly belongs to a course on measure theory and here we will restrict ourselves to a few basic points.

Since S is an event and the complement of an event is an event, the null set $\varnothing$ is an event. If $E_1, E_2, \ldots$ are events, then, according to De Morgan's law,

$$\bigcap_{i=1}^{\infty} E_i = \left(\bigcup_{i=1}^{\infty} E_i^c\right)^c \tag{1.1}$$

Since E_1^c, E_2^c,... are events, their union is an event, as is the complement of their union. Thus, a countable intersection of events is an event. In particular, set subtraction, $E_2 - E_1 = E_2 \cap E_1^c$, is an event.

On the real line $\Re$ there exists a smallest σ-algebra containing all open intervals (a, b), where $-\infty \leq a < b \leq \infty$. This σ-algebra, called the *Borel σ-algebra*, contains all intervals (open, closed, half-open-half-closed). It also contains all countable unions and intersections of intervals. For applications, we are only concerned with these countable unions and intersections. Sets in the Borel σ-algebra are called *Borel sets*.

Given a sample space S and a σ-algebra $\mathcal{E}$, a *probability measure* is a real-valued function P defined on the events in $\mathcal{E}$ such that the following three axioms are satisfied:

(P1) $P(E) \geq 0$ for any $E \in \mathcal{E}$.

(P2) $P(S) = 1$.

(P2) If E_1, E_2,... is a disjoint (mutually exclusive) countable collection of events, then

$$P\left(\bigcup_{n=1}^{\infty} E_n\right) = \sum_{n=1}^{\infty} P(E_n) \tag{1.2}$$

The triple $(S, \mathcal{E}, P)$ is called a *probability space* and Eq. 1.2 is called the *countable additivity* property. In the case of two disjoint events, the additivity property reduces to

$$P(E_1 \cup E_2) = P(E_1) + P(E_2) \tag{1.3}$$

Example 1.1. If $S = \{a_1, a_2,..., a_n\}$ is a finite set of cardinality n, then a σ-algebra $\mathcal{E}$ is defined by the *power set* (set of all subsets) of S. For any singleton event $\{a_i\} \in \mathcal{E}$, assign a nonnegative value $P(\{a_i\})$ such that

$$\sum_{i=1}^{n} P(\{a_i\}) = 1$$

For any event $E = \{e_1, e_2,..., e_m\} \subset S$, define

$$P(E) = \sum_{j=1}^{m} P(\{e_j\})$$

and define $P(\varnothing) = 0$. Then P is a probability measure on $\mathcal{E}$. For convenience, the outcome probabilities $P(\{a_i\})$ are typically denoted by $P(a_i)$. In the special case where the outcomes are *equiprobable*, $P(a_i) = 1/n$ for $i = 1, 2,..., n$, event probabilities reduce to $P(E) = m/n$, where m is the cardinality of E. When a probability space is equiprobable, outcomes are said to occur *uniformly randomly*. ■

Example **1.2.** An *urn model* consists of a set $U = \{1, 2,..., n\}$ representing n numbered balls in an urn and a protocol for randomly selecting balls from the urn. Probabilities are determined for events resulting from the protocol. We consider three protocols.

Ordered selection with replacement involves selecting $k > 0$ balls, one at a time, returning a selected ball to the urn, and recording in order the numbers of selected balls. A sample space for this protocol is given by the Cartesian product $U^k = U \times U \times \cdots \times U$, with each outcome being a k-vector. With uniformly random selection, the probability of any outcome $(b_1, b_2,..., b_k)$ is n^{-k} and event probabilities are thereby determined.

Ordered selection without replacement involves selecting $k \leq n$ balls without returning selected balls to the urn and recording in order the numbers of selected balls. Each outcome is a k-vector in which no component is repeated. Each such vector is known as a *permutation* of n objects taken k at a time. The number of such permutations is

$$P_{n,k} = n(n-1)\cdots(n-k+1) = \frac{n!}{(n-k)!}$$

When selecting k balls uniformly randomly with replacement, the probability of choosing a permutation (the event E composed of permutations) is

$$P(E) = \frac{n!}{n^k(n-k)!}$$

Unordered selection without replacement involves selecting $k \leq n$ balls without returning selected balls to the urn and recording without respect to order the numbers of selected balls. Each outcome is a subset of U and, in this context, is known as a *combination* of n objects taken k at a time. The number of such combinations is the number of subsets of U containing k elements and is given by

$$C_{n,k} = \binom{n}{k} = \frac{n!}{k!(n-k)!}$$

The expression for $C_{n,k}$ follows from $P_{n,k}$ because for each subset containing k elements there are $k!$ permutations of the subset, so that $P_{n,k} = k!C_{n,k}$. ■

Both the counting of elements in a product set and of permutations are particular instances of a more general principle, namely, counting k-vectors formed according to the following scheme: (1) the first component of the vector can be occupied by any one of r_1 elements; (2) no matter which element is chosen for the first component, any one of r_2 elements can occupy the second component; (3) proceeding recursively, no matter which elements are chosen for the first $j - 1$ components, any one of r_j elements can occupy the jth component. According to the *fundamental principle of counting*, there are $r_1 r_2 \cdots r_k$ possible vectors that can result from application of the selection scheme.

***Example* 1.3.** Let $f(x)$ be a nonnegative integrable function defined on the real line $\mathfrak{R}$ whose integral over $\mathfrak{R}$ is unity. For any Borel set $B \subset \mathfrak{R}$, a probability measure on the Borel σ-algebra is defined by

$$P(B) = \int_B f(x)\,dx$$

The first two axioms of a probability measure are satisfied owing to the assumptions on f and the third is a fundamental property of integration. For instance, the two-event additivity property of Eq. 1.3 states that, if B_1 and B_2 are disjoint Borel sets, then

$$\int_{B_1 \cup B_2} f(x)\,dx = \int_{B_1} f(x)\,dx + \int_{B_2} f(x)\,dx$$ ■

A number of basic properties of probability measures follow immediately from the axioms for a probability space $(S, \mathcal{E}, P)$. For any $E \in \mathcal{E}$, $E \cup E^c = S$. Applying additivity together with $P(S) = 1$ yields

$$P(E^c) = 1 - P(E) \tag{1.4}$$

It follows at once that

$$P(\varnothing) = P(S^c) = 1 - P(S) = 0 \tag{1.5}$$

If $E_1, E_2 \in \mathcal{E}$ and $E_1 \subset E_2$, then $E_2 = E_1 \cup (E_2 - E_1)$ and additivity implies

$$P(E_2 - E_1) = P(E_2) - P(E_1) \tag{1.6}$$

Since $P(E_2 - E_1)$ is nonnegative, $P(E_1) \leq P(E_2)$.

For any events $E_1, E_2 \in \mathcal{E}$, additivity implies

$$P(E_1 \cup E_2) = P(E_1 - E_2) + P(E_2 - E_1) + P(E_1 \cap E_2) \tag{1.7}$$

Using

$$P(E_1 - E_2) = P(E_1) - P(E_1 \cap E_2) \tag{1.8}$$

and an analogous expression for $P(E_2 - E_1)$, Eq. 1.7 becomes

$$P(E_1 \cup E_2) = P(E_1) + P(E_2) - P(E_1 \cap E_2) \tag{1.9}$$

Mathematical induction yields the *probability addition theorem*.

Theorem 1.1. If $(S, \mathcal{E}, P)$ is a probability space and $E_1, E_2, \ldots, E_n \in \mathcal{E}$, then

$$P\left(\bigcup_{k=1}^{n} E_k\right) = \sum_{j=1}^{n} (-1)^{j+1} \sum_{1 \leq i_1 < i_2 < \cdots < i_j \leq n} P\left(\bigcap_{k=1}^{j} E_{i_k}\right) \quad \blacksquare \tag{1.10}$$

Theorem 1.2. If $(S, \mathcal{E}, P)$ is a probability space, $E_1, E_2, \ldots \in \mathcal{E}$, and $E_1 \subset E_2 \subset E_3 \subset \cdots$, then there is *continuity from below*:

$$P\left(\bigcup_{n=1}^{\infty} E_n\right) = \lim_{n \to \infty} P(E_n) \tag{1.11}$$

If $E_1 \supset E_2 \supset E_3 \supset \cdots$, then there is *continuity from above*:

$$P\left(\bigcap_{n=1}^{\infty} E_n\right) = \lim_{n \to \infty} P(E_n) \quad \blacksquare \tag{1.12}$$

To show continuity from below, let $F_k = E_k - E_{k-1}$ for $k = 1, 2, \ldots$, where $E_0 = \varnothing$, and let E denote the union in Eq. 1.11. Then

$$E_n = \bigcup_{k=1}^{n} F_k \tag{1.13}$$

$$E = \bigcup_{k=1}^{\infty} F_k \tag{1.14}$$

where both unions are disjoint. Countable additivity applied to Eq. 1.13 yields

$$P(E_n) = \sum_{k=1}^{n} P(F_k) \tag{1.15}$$

Countable additivity applied to Eq. 1.14 yields

$$P(E) = \sum_{k=1}^{\infty} P(F_k) = \lim_{n\to\infty} \sum_{k=1}^{n} P(F_k) = \lim_{n\to\infty} P(E_n) \tag{1.16}$$

which verifies Eq. 1.11.

As for continuity from above, with $E_1 \supset E_2 \supset E_3 \supset \cdots$ and E denoting the intersection in Eq. 1.12, De Morgan's law together with continuity from below yields

$$\begin{aligned} P(E_1) - P(E) &= P\left(E_1 - \bigcap_{n=1}^{\infty} E_n\right) \\ &= P\left(\bigcup_{n=1}^{\infty} (E_1 - E_n)\right) \\ &= \lim_{n\to\infty} P(E_1 - E_n) \\ &= P(E_1) - \lim_{n\to\infty} P(E_n) \end{aligned} \tag{1.17}$$

Countable additivity requires disjointness of the events $E_1, E_2, \ldots \in \mathcal{E}$. In the absence of disjointness, one can still conclude *Boole's inequality*,

$$P\left(\bigcup_{n=1}^{\infty} E_n\right) \leq \sum_{n=1}^{\infty} P(E_n) \tag{1.18}$$

The inequality is demonstrated by expressing the union as a disjoint union and applying countable additivity:

$$P\left(\bigcup_{n=1}^{\infty} E_n\right) = P\left(E_1 \cup \bigcup_{n=2}^{\infty} (E_1^c \cap E_2^c \cap \cdots \cap E_{n-1}^c) \cap E_n\right)$$

$$= P(E_1) + \sum_{n=2}^{\infty} P((E_1^c \cap E_2^c \cap \cdots E_{n-1}^c) \cap E_n) \qquad (1.19)$$

Equation 1.18 follows because each set composing the sum is a subset of E_n.

1.1.2. Conditional Probability

Rather than simply asking the probability of an event E occurring, one might ask the probability of E occurring given that some other event F is known to have occurred. The question arises because one wishes to predict the outcome of one measurement given knowledge of one or more other measurements. If E and F are two events, then the probability of observing event E conditioned by prior knowledge that event F has occurred is defined in the following manner: if $(S, \mathcal{E}, P)$ is a probability space and $P(F) > 0$, then the *conditional probability measure* relative to F is defined by

$$P(E|F) = \frac{P(E \cap F)}{P(F)} \qquad (1.20)$$

The definition can be motivated by the following considerations. Suppose a point is to be randomly chosen in a region R of volume $\nu(R) = 1$ and, for any subregion $E \subset R$, the probability of the point falling in E is given by its volume $\nu(E)$. If one is asked the probability of the point falling in E conditioned by the prior knowledge that it has fallen in subregion F, then it is geometrically reasonable to choose this new conditioned probability to be $\nu(E \cap F)/\nu(F)$.

Theorem 1.3. If $(S, \mathcal{E}, P)$ is a probability space and $P(F) > 0$, then $P(\cdot\ |F)$ is a probability measure on the σ-algebra $\mathcal{E}$. ■

The theorem means that $P(\cdot\ |F)$ satisfies the probability axioms. It is immediate from Eq. 1.20 that $P(E|F) \geq 0$ and, since $S \cap F = F$, that $P(S|F) = 1$. As for countable additivity, if events $E_1, E_2, \ldots$ are mutually disjoint, then the definition of conditional probability and the countable additivity of P yield

$$P\left(\bigcup_{n=1}^{\infty} E_n \middle| F\right) = \frac{1}{P(F)} P\left(\left(\bigcup_{n=1}^{\infty} E_n\right) \cap F\right)$$

$$= \frac{1}{P(F)} P\left(\bigcup_{n=1}^{\infty} (E_n \cap F)\right)$$

$$= \sum_{n=1}^{\infty} \frac{P(E_n \cap F)}{P(F)}$$

$$= \sum_{n=1}^{\infty} P(E_n | F) \tag{1.21}$$

Cross multiplication in Eq. 1.20 yields the *multiplication principle*:

$$P(E \cap F) = P(F)P(E|F) \tag{1.22}$$

The multiplication principle extends to $n > 2$ events: if

$$P(E_1 \cap E_2 \cap \cdots \cap E_n) > 0 \tag{1.23}$$

then

$$P(E_1 \cap E_2 \cap \cdots \cap E_n) = P(E_1)P(E_2|E_1)P(E_3|E_1, E_2) \cdots P(E_n|E_1, E_2, \ldots, E_{n-1}) \tag{1.24}$$

where $P(E_3 \mid E_1, E_2)$ denotes $P(E_3 \mid E_1 \cap E_2)$.

On many occasions one is interested in the conditional probability $P(F|E)$ but only knows $P(E|F)$. In such a situation, the following *Bayes' rule* can be applied:

$$P(F|E) = \frac{P(F \cap E)}{P(E)} = \frac{P(E \cap F)}{P(E)} = \frac{P(F)P(E|F)}{P(E)} \tag{1.25}$$

Now suppose events $F_1, F_2, \ldots, F_n$ form a *partition* of the sample space S, meaning that the collection is disjoint and S equals the union of $F_1, F_2, \ldots, F_n$. If event $E \subset S$, then

$$P(E) = P\left(\bigcup_{k=1}^{n} (E \cap F_k)\right) = \sum_{k=1}^{n} P(E \cap F_k) \tag{1.26}$$

Putting Eqs. 1.25 and 1.26 together yields *Bayes' theorem*: if events $F_1, F_2, \ldots, F_n$ form a partition of the sample space S, event $E \subset S$, and $E, F_1, F_2, \ldots, F_n$ have positive probabilities, then

$$P(F_k|E) = \frac{P(F_k)P(E|F_k)}{\sum_{i=1}^{n} P(F_i)P(E|F_i)} \tag{1.27}$$

for $k = 1, 2, \ldots, n$. The theorem is applied when the *prior probabilities* $P(F_i)$ and $P(E|F_i)$ composing the sum in the denominator can be obtained experimentally or from a model and we desire the *posterior probabilities* $P(F_k|E)$.

***Example* 1.4.** A basic classification paradigm is to decide whether an object belongs to a particular class based on a measurement (or measurements) pertaining to the object. Consider observing a set of objects, say geometric shapes, and classifying an object as belonging to class C_0 or C_1 based on a real-valued measurement X. For instance X might be the perimeter, area, or number of holes of a shape. Let E be the event that $X \geq t$, where t is a fixed value, and F be the event that a randomly selected object A belongs to class C_0. Suppose we know the conditional probabilities $P(E|F)$, the probability that $X \geq t$ given $A \in C_0$, and $P(E|F^c)$, the probability that $X \geq t$ given $A \in C_1$, and we also know the probability $P(F)$ that a randomly selected object belongs to C_0. For deciding whether or not a selected object belongs to C_0, we would like to know $P(F|E)$, the probability $A \in C_0$ given $X \geq t$. $P(F|E)$ is given by Bayes' theorem:

$$P(F|E) = \frac{P(F)P(E|F)}{P(F)P(E|F) + P(F^c)P(E|F^c)}$$

■

Events E and F are said to be *independent* if

$$P(E \cap F) = P(E)P(F) \tag{1.28}$$

Otherwise they are *dependent*. If $P(F) > 0$, then E and F are independent if and only if $P(E|F) = P(E)$. If E and F are independent, then so too are E and F^c, E^c and F, and E^c and F^c. More generally, events $E_1, E_2, \ldots, E_n$ are independent if, for any subclass $\{E_{i_1}, E_{i_2}, \ldots, E_{i_m}\} \subset \{E_1, E_2, \ldots, E_n\}$,

$$P\left(\bigcap_{j=1}^{m} E_{i_j}\right) = \prod_{j=1}^{m} P(E_{i_j}) \tag{1.29}$$

Note that pairwise independence of $E_1, E_2, \ldots, E_n$, namely, that each pair within the class satisfies Eq. 1.28, does not imply independence of the full class.

***Example* 1.5.** Suppose m components $C_1, C_2, \ldots, C_m$ compose a system, F_k is the event that component C_k fails during some stated period of operation, F is the event that the system fails during the operational period, and component failures are independent. The components are said to be arranged in *series* if the system fails if any component fails and to be arranged in *parallel* if the system fails if and only if all components fail. If the system is arranged in series, then

$$F = \bigcup_{k=1}^{m} F_k$$

$$P(F) = 1 - P(F^c) = 1 - P\left(\bigcap_{k=1}^{m} F_k^c\right) = 1 - \prod_{k=1}^{m} (1 - P(F_k))$$

If the series is arranged in parallel, then

$$P(F) = P\left(\bigcap_{k=1}^{m} F_k\right) = \prod_{k=1}^{m} P(F_k)$$ ■

1.2. Random Variables

Measurement randomness results from both the inherent randomness of phenomena and variability within observation and measurement systems. Quantitative description of random measurements is embodied in the concept of a random variable. The theory of random processes concerns random variables defined at points in time or space, as well as interaction between random variables.

1.2.1. Probability Distributions

Given a probability space $(S, \mathcal{E}, P)$, a *random variable* is a mapping X: $S \rightarrow \Re$, the space of real numbers, such that

$$X^{-1}((-\infty, x]) = \{z \in S: X(z) \leq x\} \tag{1.30}$$

is an element of $\mathcal{E}$ (an event) for any $x \in \Re$. If $X^{-1}((-\infty, x]) \in \mathcal{E}$ for any $x \in \Re$ (if X is a random variable), then it can be shown that $X^{-1}(B) \in \mathcal{E}$ for any Borel set $B \subset \Re$, which means in particular that $X^{-1}(B) \in \mathcal{E}$ if B is an open set, a closed set, an intersection of open sets, or a union of closed sets.

Theorem 1.4. A random variable X on a probability space $(S, \mathcal{E}, P)$ induces a probability measure P_X on the Borel σ-algebra $\mathcal{B}$ in $\mathfrak{R}$ by

$$P_X(B) = P(X^{-1}(B)) = P(\{z \in S: X(z) \in B\}) \qquad \blacksquare \qquad (1.31)$$

To prove the theorem, first note that $P_X(B)$ is defined for any $B \in \mathcal{B}$ because $X^{-1}(B) \in \mathcal{E}$ for any $B \in \mathcal{B}$. The first two probability axioms are easily verified:

$$P_X(\mathfrak{R}) = P(X^{-1}(\mathfrak{R})) = P(S) = 1 \qquad (1.32)$$

$$P_X(B^c) = P(X^{-1}(B^c)) = P([X^{-1}(B)]^c) = 1 - P(X^{-1}(B)) = 1 - P_X(B) \qquad (1.33)$$

If $B_1, B_2,...$ form a disjoint countable collection of Borel sets, then $X^{-1}(B_1)$, $X^{-1}(B_2)$,... form a disjoint countable collection of events in $\mathcal{E}$. Hence,

$$\begin{aligned} P_X\left(\bigcup_{i=1}^{\infty} B_i\right) &= P\left(X^{-1}\left(\bigcup_{i=1}^{\infty} B_i\right)\right) \\ &= P\left(\bigcup_{i=1}^{\infty} X^{-1}(B_i)\right) \qquad (1.34) \\ &= \sum_{i=1}^{\infty} P(X^{-1}(B_i)) \\ &= \sum_{i=1}^{\infty} P_X(B_i) \end{aligned}$$

Hence, the three probability axioms are satisfied by the induced probability measure.

The induced probability measure of a random variable X defines the inclusion probability $P(X \in B)$ by

$$P(X \in B) = P_X(B) = P(X^{-1}(B)) \qquad (1.35)$$

Example **1.6.** Let $(S, \mathcal{E}, P)$ be a probability space, E_0 and E_1 partition S, and random variable X be defined by $X(a) = 0$ if $a \in E_0$ and $X(a) = 1$ if $a \in E_1$. For any Borel set B, $P_X(B) = 0$, if $\{0, 1\} \cap B = \varnothing$, $P_X(B) = P(E_0)$, if $\{0, 1\} \cap B = \{0\}$, $P_X(B) = P(E_1)$, if $\{0, 1\} \cap B = \{1\}$, $P_X(B) = 1$, if $\{0, 1\} \cap B = \{0, 1\}$. $\blacksquare$

As a consequence of its inducing a probability measure on the Borel σ-algebra, the random variable X induces a probability space $(\mathfrak{R}, \mathcal{B}, P_X)$ on the real line. If we are only concerned with X and its inclusion probabilities $P(X \in B)$ for Borel sets, once we have a representation for P_X we need not concern ourselves with the original sample space. This observation is the key to probabilistic modeling: when measuring random phenomena we need only model the distribution of probability mass over the real line associated with the random variable.

For a random variable X defined on the probability space $(S, \mathcal{E}, P)$, define its *probability distribution function* $F_X: \mathfrak{R} \to [0, 1]$ by

$$F_X(x) = P(X \leq x) = P_X((-\infty, x]) \tag{1.36}$$

Interval probabilities are expressed via the probability distribution function. If $a < b$, then

$$\begin{aligned} P(a < X \leq b) &= F_X(b) - F_X(a) \\ P(a \leq X \leq b) &= F_X(b) - F_X(a) + P(X = a) \\ P(a < X < b) &= F_X(b) - F_X(a) - P(X = b) \\ P(a \leq X < b) &= F_X(b) - F_X(a) + P(X = a) - P(X = b) \end{aligned} \tag{1.37}$$

Theorem 1.5. If F_X is the probability distribution function for a random variable X, then

(i) F_X is increasing.
(ii) F_X is continuous from the right.
(iii) $\lim_{x \to -\infty} F_X(x) = 0$.
(iv) $\lim_{x \to \infty} F_X(x) = 1$.

Conversely, if F is any function satisfying the four properties, then there exists a probability space and a random variable X on that space such that the probability distribution function for X is given by F. ■

We demonstrate that the four conditions stated in Theorem 1.5 hold whenever F_X is a probability distribution function. First, if $x_1 \leq x_2$, then $(-\infty, x_1] \subset (-\infty, x_2]$, implying that

$$F_X(x_1) = P_X((-\infty, x_1]) \leq P_X((-\infty, x_2]) = F_X(x_2) \tag{1.38}$$

so that F_X is increasing. To show continuity from the right, suppose $\{x_n\}$ is a decreasing sequence and $x_n \to x$. Then, owing to continuity from above,

$$\lim_{n\to\infty} F_X(x_n) = \lim_{n\to\infty} P_X((-\infty, x_n])$$

$$= P_X\left(\bigcap_{n=1}^{\infty}(-\infty, x_n]\right) \tag{1.39}$$

$$= P_X((-\infty, x])$$

which is $F_X(x)$, thereby demonstrating continuity from the right. For property (iii), suppose $\{x_n\}$ is a decreasing sequence with $x_n \to -\infty$. Then, as in Eq. 1.39,

$$\lim_{n\to\infty} F_X(x_n) = P_X\left(\bigcap_{n=1}^{\infty}(-\infty, x_n]\right) = P_X(\varnothing) = 0 \tag{1.40}$$

For property (iv), suppose $\{x_n\}$ is an increasing sequence with $x_n \to \infty$. Then, owing to continuity from below,

$$\lim_{n\to\infty} F_X(x_n) = \lim_{n\to\infty} P_X((-\infty, x_n])$$

$$= P_X\left(\bigcup_{n=1}^{\infty}(-\infty, x_n]\right) \tag{1.41}$$

$$= P_X(\Re)$$

which is 1. We will not prove the converse of the theorem because it requires the theory of Lebesgue-Stieltjes measure.

Owing to the converse of Theorem 1.5, once a probability distribution function is defined, *ipso facto* there exists a random variable whose behavior is described by the function. The probability distribution function is then said to be the *law* of the random variable and it is chosen to model some phenomena. Unless there is potential confusion, we will drop the subscript when denoting a probability distribution function. The terminology *probability distribution* is applied to either the probability distribution function or to a random variable having the probability distribution function. Two random variables possessing the same probability distribution function are said to be *identically distributed* and are probabilistically indistinguishable.

1.2.2. Probability Densities

A probability distribution is commonly specified by a nonnegative function $f(x)$ for which

$$\int_{-\infty}^{\infty} f(x)\,dx = 1 \tag{1.42}$$

Such a function is called a *probability density* and yields a probability distribution function (and therefore a corresponding random variable) by

$$F(x) = \int_{-\infty}^{x} f(t)\,dt \tag{1.43}$$

$F(x)$ is a continuous function and models the probability mass of a random variable X taking values in a continuous range. $F(x)$, $f(x)$, and X are said to constitute a *continuous probability distribution*. According to the fundamental theorem of calculus,

$$F'(x) = \frac{d}{dx}F(x) = f(x) \tag{1.44}$$

at any point x where the density f is continuous.

According to Eq. 1.37, for $a < b$,

$$P(X = b) = F(b) - F(a) - P(a < X < b) \tag{1.45}$$

Owing to continuity from above, $P(a < X < b) \rightarrow 0$ as $a \rightarrow b$ from the left. Because F is continuous, $F(b) - F(a) \rightarrow 0$ as $a \rightarrow b$ from the left. Hence, $P(X = b) = 0$. Since the point b was chosen arbitrarily, we conclude that all point probabilities of a continuous probability distribution are zero. Consequently, for a continuous distribution, all interval-probability expressions of Eq. 1.37 reduce to the first, which now becomes

$$P(a < X \leq b) = \int_{a}^{b} f(t)\,dt \tag{1.46}$$

Example **1.7.** For $a < b$, the *uniform distribution* over interval $[a, b]$ is characterized by the probability density $f(x) = 1/(b - a)$ for $a \leq x \leq b$ and $f(x) = 0$ otherwise. The corresponding probability distribution function is defined by $f(x) = 0$ for $x < a$,

$$F(x) = \frac{x - a}{b - a}$$

for $a \le x \le b$, and $f(x) = 1$ for $x > b$. As determined by the density, the mass of the distribution is spread uniformly over $[a, b]$ and the probability distribution function ramps from 0 to 1 across $[a, b]$. ■

A discrete random variable X is modeled by specifying a countable *range* of points $\Omega_X = \{x_1, x_2,...\}$ and a nonnegative *probability mass function* (*discrete density*) $f(x)$ such that $f(x) > 0$ if and only if $x \in \Omega_X$ and

$$\sum_{k=1}^{\infty} f(x_k) = 1 \tag{1.47}$$

The probability mass function generates a probability distribution function by

$$F(x) = \sum_{\{k: x_k \le x\}} f(x_k) \tag{1.48}$$

$F(x)$ is a step function with jump $f(x_k)$ at x_k and, assuming $x_1 < x_2 < ...$, $F(x)$ is constant on the interval $[x_k, x_{k+1})$. A random variable X with probability distribution function $F(x)$ has point probabilities $P(X = x_k) = f(x_k)$. The first interval probability of Eq. 1.37 becomes

$$P(a < X \le b) = \sum_{\{k: a < x_k \le b\}} f(x_k) \tag{1.49}$$

Other interval probabilities differ depending on whether $f(a) > 0$ or $f(b) > 0$. If the point set Ω_X is finite, then the sum of Eq. 1.48 is finite and $F(x) = 1$ for sufficiently large x. A probability mass function and its random variable constitute a *discrete distribution*.

To unify the representation of continuous and discrete distributions, we employ delta functions to represent discrete distributions. As a generalized function, a delta function is an operator on functions; nevertheless, we take a view often adopted in engineering and treat a delta function $\delta(x)$ as though it were a function with the integration property

$$\int_{-\infty}^{\infty} g(x)\delta(x-a)\,dx = g(a) \tag{1.50}$$

Under this convention we apply Eq. 1.44 in a generalized sense, and the probability mass function $f(x)$ corresponding to the point set Ω_X can be represented as

$$f(x)=\sum_{k=1}^{\infty} f(x_k)\delta(x-x_k) \tag{1.51}$$

Use of generalized functions for discrete densities is theoretically justified. For applications, one need only recognize that use of delta functions according to Eq. 1.51 is appropriate so long as Ω_X does not have any limit points.

There exist probability distributions that are neither continuous nor discrete. These *mixed* distributions have contributions from both continuous and discrete parts. Although we will not explicitly employ them as models in the text, they constitute valid distributions and need to be considered in the general theory. The salient point is that Theorem 1.5 determines the central role of probability distribution functions and their properties. We will take a dual approach in this regard. We will utilize Eq. 1.44 in a generalized sense, so that the theory is unified in the framework of probability densities, but we will often prove theorems assuming a continuous distribution, thereby avoiding theoretical questions concerning generalized functions. A more direct and unified approach would be to remain in the context of probability distribution functions; however, this would require the use of Lebesgue-Stieltjes integration and force us into measure-theoretic questions outside the scope of the text. For those with a background in measure theory, recall that according to the Lebesgue decomposition of a function F satisfying the conditions of Theorem 1.5, F is differentiable almost everywhere and F is decomposed as $F = F_s + F_a$, where F_a is absolutely continuous, F_s is singular [$F_s' = 0$ almost everywhere], and F_s can be continuous and not be a constant.

1.2.3. Functions of a Random Variable

We are rarely concerned with a random variable in isolation, but rather with its relations to other random variables. The terminology "variable" applies because we usually consider a system with one or more random inputs and one or more random outputs. The simplest case involves a function of a single random variable.

For a discrete random variable X, the density of a function $Y = g(X)$ is given by

$$f_Y(y)=P(Y=y)=\sum_{\{x:g(x)=y\}} f_X(x) \tag{1.52}$$

In particular, if g is one-to-one, then $\{x: g(x) = y\}$ consists of the single element $g^{-1}(y)$ and

$$f_Y(y)=f_X[g^{-1}(y)] \tag{1.53}$$

Finding output distributions of continuous random variables is more difficult. One approach is to find the probability distribution function of the output random variable and then differentiate to obtain the output density.

***Example* 1.8.** Consider the affine transformation $Y = aX + b$, $a \neq 0$. For $a > 0$,

$$F_Y(y) = P(aX + b \leq y) = P\left(X \leq \frac{y-b}{a}\right) = F_X\left(\frac{y-b}{a}\right)$$

For $a < 0$,

$$F_Y(y) = P\left(X \geq \frac{y-b}{a}\right) = 1 - F_X\left(\frac{y-b}{a}\right)$$

Differentiation for $a > 0$ yields

$$f_Y(y) = \frac{d}{dx} F_X\left(\frac{y-b}{a}\right) = \frac{1}{a} f_X\left(\frac{y-b}{a}\right)$$

Except for a minus sign in front, the same expression is obtained for $a < 0$. Combining the two results yields

$$f_Y(y) = \frac{1}{|a|} f_X\left(\frac{y-b}{a}\right)$$ ■

Suppose $y = g(x)$ is differentiable for all x and has a derivative that is either strictly greater than or strictly less than 0. Then $x = g^{-1}(y)$ has a derivative known as the *Jacobian* of the transformation and denoted by $J(x; y)$. Moreover, there exist values y_1 and y_2, $y_1 < y_2$, either or both possibly infinite, such that for any y between y_1 and y_2, there exists exactly one value x such that $y = g(x)$.

Theorem 1.6. If X is a continuous random variable and $y = g(x)$ has a derivative that is either strictly greater than or strictly less than 0, then $Y = g(X)$ is a continuous random variable with density

$$f_Y(y) = \begin{cases} f_X[g^{-1}(y)]\,|J(x;y)|, & \text{if } y_1 < y < y_2 \\ 0, & \text{otherwise} \end{cases} \tag{1.54}$$

where y_1 and y_2 are the limiting values described prior to the theorem. ■

To prove the theorem, suppose $y_1 < y < y_2$ and $g' > 0$. Then

$$F_Y(y) = P(g(X) \le y) = P(X \le g^{-1}(y)) = F_X(g^{-1}(y)) \tag{1.55}$$

Differentiation with respect to y yields

$$f_Y(y) = f_X(g^{-1}(y))\frac{d}{dy}g^{-1}(y) = f_X(g^{-1}(y))J(x;y) \tag{1.56}$$

Now suppose $g' < 0$. Then

$$F_Y(y) = P(g(X) \le y) = P(X \ge g^{-1}(y)) = 1 - F_X(g^{-1}(y)) \tag{1.57}$$

Differentiation yields

$$f_Y(y) = -f_X(g^{-1}(y))\frac{d}{dy}g^{-1}(y) = f_X(g^{-1}(y))[-J(x;y)] \tag{1.58}$$

Since $g' < 0$, $J(x;y) < 0$, so that $-J(x;y) = |J(x;y)|$. Combining Eqs. 1.56 and 1.58 yields the result for $y_1 < y < y_2$. It is straightforward to show that $f_Y(y) = 0$ for $y \notin (y_1, y_2)$.

***Example* 1.9.** The exponential function $y = g(x) = e^{tx}$, $t > 0$, satisfies the preceding conditions. For $y > 0$,

$$g^{-1}(y) = \frac{\log y}{t}$$

$$J(x;y) = \frac{d}{dt}\left(\frac{\log y}{t}\right) = \frac{1}{ty}$$

$$f_Y(y) = \frac{1}{ty}f_X\left(\frac{\log y}{t}\right)$$

For $y < 0$, $f_Y(y) = 0$. For $y = 0$, $f_Y(0)$ can be defined arbitrarily. ■

1.3. Moments

Full description of a random variable requires characterization of its probability distribution function; however, most applications involve only partial description of a random variable via its moments. We state definitions and properties in terms of densities, for which moments have geometric intuition relative to probability mass. Definitions involving integrals can be interpreted directly for continuous random variables; for discrete random variables they can be interpreted using delta functions or can be re-expressed as sums.

1.3.1. Expectation and Variance

The *expected value* (*expectation*) of a random variable X with density $f(x)$ is defined by

$$E[X] = \int_{-\infty}^{\infty} x f(x)\,dx \tag{1.59}$$

as long as the defining integral is absolutely convergent, meaning

$$\int_{-\infty}^{\infty} |x| f(x)\,dx < \infty \tag{1.60}$$

Whenever we write $E[X]$, it is implicitly assumed that the defining integral is absolutely convergent. The expected value is the center of mass for the probability distribution (density). It is also called the *mean* and denoted by μ_X (or simply μ if X is clear from the context).

To apply Eq. 1.59 directly to a function $g(X)$ of a random variable would require first finding the density of $g(X)$. Fortunately, this need not be done.

Theorem 1.7. If $g(x)$ is any piecewise continuous, real-valued function and X is a random variable possessing density $f(x)$, then

$$E[g(X)] = \int_{-\infty}^{\infty} g(x) f(x)\,dx \quad \blacksquare \tag{1.61}$$

We demonstrate the theorem for the special case in which g is differentiable and has a derivative that is either strictly greater than or strictly less than 0. By Theorem 1.6,

$$E[g(X)] = \int_{y_1}^{y_2} y f_X(g^{-1}(y)) \frac{d}{dy} g^{-1}(y)\,dy \tag{1.62}$$

For the substitution $y = g(x)$, $dy = g'(x)dx$ and $x = g^{-1}(y)$. According to the definition of y_1 and y_2, as y varies from y_1 to y_2, x goes from $-\infty$ to ∞. Thus, the substitution yields

$$E[g(X)] = \int_{-\infty}^{\infty} g(x) f_X(x) \frac{d}{dy} g^{-1}(y)\, g'(x)dx \tag{1.63}$$

Equation 1.61 follows because

$$\frac{d}{dy} g^{-1}(y) = \frac{1}{g'(x)} \tag{1.64}$$

Assuming the defining integral is absolutely convergent, for integer $k \geq 1$, the kth *moment* about the origin of a random variable X possessing density $f(x)$ is defined by

$$\mu_k' = E[X^k] = \int_{-\infty}^{\infty} x^k f(x)\, dx \tag{1.65}$$

$E[X]$ is the first moment of X. Assuming the defining integral is absolutely convergent, the kth *central moment* is defined by

$$\mu_k = E[(X-\mu)^k] = \int_{-\infty}^{\infty} (x-\mu)^k f(x)\, dx \tag{1.66}$$

where μ is the mean of X. Note that Theorem 1.7 was invoked in defining both moments and central moments. The kth central moment is the kth moment of the *centered* random variable $X - \mu$, whose mean is 0.

The second central moment is called the *variance* and is defined by

$$\sigma^2 = \mu_2 = E[(X-\mu)^2] \tag{1.67}$$

The variance measures the spread of the probability mass about the mean: the more dispersed the mass, the greater the variance. To specify the random variable X when writing the variance, we write Var[X] or σ_X^2. The square root of the variance is called the *standard deviation*.

Rather than compute the variance by its defining integral, it is usually more easily computed via the relation

$$\sigma^2 = \mu_2' - \mu^2 \tag{1.68}$$

which is obtained by expanding the integral defining $E[(X - \mu)^2]$. It is straightforward to show that, for any constants a and b,

$$\mathrm{Var}[aX + b] = a^2\mathrm{Var}[X] \tag{1.69}$$

***Example* 1.10.** Consider the uniform distribution over the interval $[a, b]$ defined in Example 1.7. For $k = 1, 2, \ldots$, the kth moment is

$$\mu_k' = E[X^k] = \frac{1}{b-a}\int_a^b x^k\,dx = \frac{b^{k+1} - a^{k+1}}{(k+1)(b-a)}$$

The mean and second moment are found by substituting $k = 1$ and $k = 2$, respectively, into μ_k': $\mu = (a + b)/2$, $\mu_2' = (b^2 + ab + a^2)/3$. Equation 1.68 yields $\sigma^2 = (b - a)^2/12$. ■

The next theorem provides two inequalities bounding the probability mass of a random variable over a range of values. The first (the *generalized Chebyshev inequality*) bounds the probability mass of a nonnegative random variable above a given value in terms of its expectation; the second (*Chebyshev's inequality*) quantifies the manner in which the variance measures the absolute deviation of a random variable from its mean. Chebyshev's inequality plays important roles in statistical estimation and the convergence of sequences of random variables.

Theorem 1.8. If random variable X is nonnegative and has mean μ, then, for any $t > 0$,

$$P(X \geq t) \leq \frac{\mu}{t} \tag{1.70}$$

If X (not necessarily nonnegative) possesses mean μ and variance σ^2, then, for any $t > 0$,

$$P(|X - \mu| \geq t) \leq \frac{\sigma^2}{t^2} \quad ■ \tag{1.71}$$

For a continuous random variable, the generalized Chebyshev inequality results from the following inequality upon division by t:

$$\mu = \int_0^\infty xf(x)\,dx$$

$$\geq \int_t^\infty xf(x)\,dx$$

$$\geq t\int_t^\infty f(x)\,dx \tag{1.72}$$

$$= tP(X \geq t)$$

Applying the generalized Chebyshev inequality to $|X - \mu|^2$ and t^2 yields the ordinary form in the following way:

$$P(|X-\mu| \geq t) = P(|X-\mu|^2 \geq t^2) \leq \frac{E[|X-\mu|^2]}{t^2} = \frac{\sigma^2}{t^2} \tag{1.73}$$

Letting $t = k\sigma$, $k > 0$, in Eq. 1.71 expresses Chebyshev's inequality in terms of the number of standard deviations from the mean:

$$P(|X-\mu| \geq k\sigma) \leq \frac{1}{k^2} \tag{1.74}$$

A different form results from complementation in Eq. 1.71,

$$P(|X-\mu| < t) \geq 1 - \frac{\sigma^2}{t^2} \tag{1.75}$$

Thus, the probability mass over the interval $(\mu - t, \mu + t)$ is bounded from below. For small variance, the mass is tightly concentrated about the mean. The Chebyshev inequality does not take into account the actual distribution and therefore it is often rather loose; however, without a strengthened hypothesis it cannot be improved.

1.3.2. Moment-Generating Function

Using an exponential transform can help with some tasks involving probability densities. The moment-generating function of a random variable X having density $f(x)$ is defined by

$$M_X(t) = E[e^{tX}] = \int_{-\infty}^{\infty} e^{tx} f(x)\, dx \tag{1.76}$$

for all t for which the integral is finite. For any constants a and b, straightforward use of the properties of the exponential shows that

$$M_{aX+b}(t) = e^{bt} M_X(at) \tag{1.77}$$

To be useful, a transform requires unique inversion. The next theorem is the *uniqueness* theorem for the moment-generating function.

Theorem 1.9. If $M_X(t) = M_Y(t)$ for all t in some open interval containing $t = 0$, then the random variables X and Y are identically distributed. ■

The moment-generating function can be used to find moments. Suppose X has density $f(x)$ and the kth moment of X exists. Taking the derivative of $M_X(t)$ with respect to t yields

$$\begin{aligned} \frac{d}{dt} M_X(t) &= \frac{d}{dt} \int_{-\infty}^{\infty} e^{tx} f(x)\, dx \\ &= \int_{-\infty}^{\infty} \frac{\partial}{\partial t}[e^{tx} f(x)]\, dx \\ &= \int_{-\infty}^{\infty} x e^{tx} f(x)\, dx \end{aligned} \tag{1.78}$$

where we have assumed in the second equality that the derivative can be brought inside the integral (which it can be for all distributions in which the moment-generating function will be invoked in this text). Proceeding by recursive differentiation yields

$$M_X^{(k)}(t) = \int_{-\infty}^{\infty} x^k e^{tx} f(x)\, dx \tag{1.79}$$

Letting $t = 0$ gives

$$M_X^{(k)}(0) = \int_{-\infty}^{\infty} x^k f(x)\, dx = \mu_k' \tag{1.80}$$

where $M_X^{(k)}(0)$ is the kth derivative of $M_X(t)$ with respect to t, evaluated at $t = 0$.

Example 1.11. For $b > 0$, the *exponential distribution* is characterized by the density $f(x) = be^{-bx}$ for $x \geq 0$ and $f(x) = 0$ for $x < 0$, where $b > 0$. Its moment-generating function is

$$M_X(t) = \int_0^{\infty} e^{tx} be^{-bx}\, dx = b \int_0^{\infty} e^{-(b-t)x}\, dx = \frac{b}{b-t}$$

for $t < b$ (so that the integral is finite). Successive differentiations with respect to t yield

$$M_X^{(k)}(t) = \frac{k!b}{(b-t)^{k+1}}$$

for $k = 1, 2, \ldots$. Letting $t = 0$ yields $\mu_k' = k!/b^k$. Hence, the mean, second moment, and variance are $\mu = 1/b$, $\mu_2' = 2/b^2$, and $\sigma^2 = 1/b^2$, respectively. ■

1.4. Important Probability Distributions

This section provides definitions and properties of some commonly employed probability distributions. The binomial distribution describes probabilities associated with repeated, independent binary trials; the Poisson distribution models a fundamental class of random point processes; the normal distribution is used extensively in statistics, serves as a model for noise in image and signal filtering, and is a limiting distribution in many key circumstances; the gamma distribution governs a family of many useful distributions and is useful for modeling grain sizing and interarrival times for queues; the beta distribution takes on many types of shapes for different combinations of its parameters and is therefore useful for modeling various kinds of phenomena.

1.4.1. Binomial Distribution

Suppose an experiment consists of n trials, $n > 0$. The trials are called *Bernoulli trials* if three conditions are satisfied: (1) each trial is defined by a sample space $\{s, f\}$ having two outcomes, called *success* and *failure*; (2) there is a number p, $0 < p < 1$, such that, for each trial, $P(s) = p$ and $P(f) = q = 1 - p$;

and (3) the trials are independent. Bernoulli trials are realized by selecting n balls randomly, in succession, and with replacement after each selection from an urn containing k black balls and m white balls. If selecting a black ball constitutes a success, then $p = k/(k + m)$. The appropriate sample space for an experiment composed of n Bernoulli trials is $\{s, f\}^n$, the Cartesian product of $\{s, f\}$ with itself n times. Let the random variable X count the number of successes in the n trials. Its probability mass function is nonzero for $x = 0$, $1, \ldots, n$. The probability of any outcome $(u_1, u_2, \ldots, u_n) \in \{s, f\}^n$ for which exactly x of the components equal s is $p^x q^{n-x}$. There are

$$C_{n,x} = \binom{n}{x} = \frac{n!}{x!(n-x)!} \tag{1.81}$$

such outcomes. Hence, the probability mass function for X is

$$f(x) = P(X = x) = \binom{n}{x} p^x q^{n-x} \tag{1.82}$$

for $x = 0, 1, \ldots, n$. Any random variable having this density is said to possess a *binomial distribution* and is said to *binomially distributed.* The binomial probability distribution function is

$$F(x) = \sum_{k \le x} \binom{n}{k} p^k q^{n-k} \tag{1.83}$$

$F(x) = 0$ for $x < 0$, $F(x) = 1$ for $x \ge n$, $F(x)$ has jumps at $x = 0, 1, \ldots, n$, and $F(x)$ is constant on all intervals $[x, x + 1)$ for $x = 0, 1, \ldots, n - 1$.

The moment-generating function for the binomial distribution is given by

$$\begin{aligned} M_X(t) &= \sum_{x=0}^{n} e^{tx} \binom{n}{x} p^x q^{n-x} \\ &= \sum_{x=0}^{n} \binom{n}{x} (pe^t)^x q^{n-x} \\ &= (pe^t + q)^n \end{aligned} \tag{1.84}$$

where the last equality follows from the recognition that the preceding sum is the binomial expansion of the last expression. Taking the first and second derivative of the moment-generating function with respect to t and then setting $t = 0$ yields the mean and second moment according to Eq. 1.80, and application of Eq. 1.68 then gives the variance: $\mu = np$, $\mu_2' = np(1 + np - p)$, $\sigma^2 = npq$.

1.4.2. Poisson Distribution

Whereas the binomial distribution can be arrived at by means of an experimental protocol involving a finite sample space, the Poisson distribution cannot be so developed. Later in the text we will show how the Poisson distribution results from an arrival process in time; for now we define it and give some basic properties. A discrete random variable X is said to possess a *Poisson distribution* if it has probability mass function

$$f(x) = \frac{e^{-\lambda}\lambda^x}{x!} \tag{1.85}$$

for $x = 0, 1, 2, \ldots$, where $\lambda > 0$. The Poisson probability distribution function is

$$F(x) = \sum_{k \le x} \frac{e^{-\lambda}\lambda^k}{k!} \tag{1.86}$$

$F(x) = 0$ for $x < 0$ and $F(x)$ has jumps at $x = 0, 1, 2, \ldots$.

The moment-generating function for the Poisson distribution is given by

$$\begin{aligned} M_X(t) &= \sum_{x=0}^{\infty} \frac{e^{tx}e^{-\lambda}\lambda^x}{x!} \\ &= e^{-\lambda}\sum_{x=0}^{\infty} \frac{(\lambda e^t)^x}{x!} \\ &= \exp[\lambda(e^t - 1)] \end{aligned} \tag{1.87}$$

the last equality following from the fact that the series is the Taylor series for the exponential function. The mean, second moment, and variance are found from the moment-generating function: $\mu = \lambda$, $\mu_2' = \lambda + \lambda^2$, $\sigma^2 = \lambda$.

The binomial and Poisson distributions are asymptotically related. Let $b(x; n, p)$ and $\pi(x; \lambda)$ denote the binomial and Poisson densities, respectively. Then, for $x = 0, 1, 2, \ldots$,

$$\lim_{n \to \infty} b\left(x; n, \frac{\lambda}{n}\right) = \pi(x; \lambda) \tag{1.88}$$

This asymptotic relation can be used to approximate the binomial distribution by the Poisson distribution. Replacing λ/n by p in Eq. 1.88 yields

$$\lim_{n\to\infty} b(x;n,p) = \pi(x;np) \tag{1.89}$$

so that for large n, $b(x;\, n,\, p) \approx \pi(x;\, np)$. There is a caveat in applying this approximation: p must be sufficiently small so that np is not too large, even though n is large. The reason for this caveat has to do with the rate of convergence to the limit.

To obtain the limit in Eq. 1.88, first note that

$$\lim_{n\to\infty} \frac{n(n-1)\cdots(n-x+1)}{n^x} = \prod_{k=0}^{x-1} \lim_{n\to\infty}\left(1-\frac{k}{n}\right) = 1 \tag{1.90}$$

$$\lim_{n\to\infty}\left(1-\frac{\lambda}{n}\right)^{-x} = 1 \tag{1.91}$$

$$\lim_{n\to\infty}\left(1-\frac{\lambda}{n}\right)^{n} = e^{-\lambda} \tag{1.92}$$

Writing out $b(x;\, n,\, \lambda/n)$, rearranging terms, and taking the product of the limits yields the desired limit relation:

$$\begin{aligned}
\lim_{n\to\infty} b\left(x;n,\frac{\lambda}{n}\right) &= \lim_{n\to\infty}\binom{n}{x}\left(\frac{\lambda}{n}\right)^x\left(1-\frac{\lambda}{n}\right)^{n-x} \\
&= \lim_{n\to\infty}\frac{n(n-1)\cdots(n-x+1)}{x!}\left(\frac{\lambda}{n}\right)^x\left(1-\frac{\lambda}{n}\right)^{n-x} \\
&= \lim_{n\to\infty}\frac{n(n-1)\cdots(n-x+1)}{n^x}\frac{\lambda^x}{x!}\left(1-\frac{\lambda}{n}\right)^{n}\left(1-\frac{\lambda}{n}\right)^{-x} \\
&= \frac{\lambda^x e^{-\lambda}}{x!}
\end{aligned} \tag{1.93}$$

***Example* 1.12.** Suppose that for a particular communication channel the error rate is 1 incorrect transmission per 100 messages and transmissions are independent. If n messages are sent, then, in essence, there are n Bernoulli trials and the probability of a success, which is actually an erroneous transmission, is $p = 0.01$. Letting X denote the number of erroneous transmissions in the n messages, the Poisson approximation is

$$P(X=x) \approx \frac{e^{-np}(np)^x}{x!}$$

For $n = 200$, $P(X \geq 3) = 0.3233$ by both methods. ■

1.4.3. Normal Distribution

A continuous random variable X is said to possess a *normal* (*Gaussian*) *distribution* if it has the density

$$f(x) = \frac{1}{\sqrt{2\pi}\sigma} e^{-\frac{1}{2}\left(\frac{x-\mu}{\sigma}\right)^2} \tag{1.94}$$

where $-\infty < x < \infty$, $-\infty < \mu < \infty$, and $\sigma > 0$. Its probability distribution function is

$$F(x) = \frac{1}{\sqrt{2\pi}\sigma} \int_{-\infty}^{x} e^{-\frac{1}{2}\left(\frac{y-\mu}{\sigma}\right)^2} dy \tag{1.95}$$

In the parameterization of Eq. 1.94, the mean and variance of the normal distribution are μ and σ^2, respectively, as we will soon show.

Setting $\mu = 0$ and $\sigma = 1$ yields the *standard normal distribution*, whose density and probability distribution function are given by

$$\phi(z) = \frac{1}{\sqrt{2\pi}} e^{-\frac{z^2}{2}} \tag{1.96}$$

$$\Phi(z) = \frac{1}{\sqrt{2\pi}} \int_{-\infty}^{z} e^{-\frac{y^2}{2}} dy \tag{1.97}$$

respectively. $\phi(z)$ is a legitimate density because

$$\frac{1}{\sqrt{2\pi}} \int_{-\infty}^{\infty} e^{-\frac{z^2}{2}} dz = 1 \tag{1.98}$$

The integral from $-\infty$ to ∞ of the normal density of Eq. 1.94 can be shown to equal 1 by reducing it to the integral of Eq. 1.98 via the substitution $z =$

$(x - \mu)/\sigma$. In fact, the transformation $Z = (X - \mu)/\sigma$ transforms a normally distributed random variable X into a standard normal random variable. This is demonstrated by applying the results of Example 1.8 to the normal density of Eq. 1.94 with $a = 1/\sigma$ and $b = -\mu/\sigma$. Moreover, for any c and d,

$$P(c < X < d) = \Phi\left(\frac{d-\mu}{\sigma}\right) - \Phi\left(\frac{c-\mu}{\sigma}\right) \tag{1.99}$$

Hence, all interval probabilities for a normally distributed random variable can be found from the standard-normal probability distribution function, whose values can be found in standard-normal statistical tables.

The moment-generating function of the normal distribution is derived in the following manner:

$$\begin{aligned} M_X(t) &= \frac{1}{\sqrt{2\pi}\sigma}\int_{-\infty}^{\infty} e^{tx} e^{-\frac{1}{2}\left(\frac{x-\mu}{\sigma}\right)^2} dx \\ &= \frac{1}{\sqrt{2\pi}\sigma}\int_{-\infty}^{\infty} \exp\left[-\frac{1}{2}\left(\frac{x^2 - 2\mu x + \mu^2 - 2\sigma^2 tx}{\sigma^2}\right)\right] dx \\ &= \exp\left[\mu t + \frac{t^2\sigma^2}{2}\right]\frac{1}{\sqrt{2\pi}\sigma}\int_{-\infty}^{\infty} \exp\left[-\frac{1}{2}\left(\frac{x-(\mu+t\sigma^2)}{\sigma}\right)^2\right] dx \\ &= \exp\left[\mu t + \frac{t^2\sigma^2}{2}\right]\frac{1}{\sqrt{2\pi}}\int_{-\infty}^{\infty} e^{-\frac{y^2}{2}} dy \\ &= \exp\left[\mu t + \frac{t^2\sigma^2}{2}\right] \end{aligned} \tag{1.100}$$

where the third equality results from completing the square in the exponential, the fourth results from the substitution $y = [x - (\mu + t\sigma^2)]/\sigma$, and the last from Eq. 1.98. Obtaining the derivatives of the moment-generating function at $t = 0$ and applying Eqs. 1.68 and 1.80 shows that the mean and variance of the normal distribution are μ and σ^2, respectively.

1.4.4. Gamma Distribution

The gamma distribution involves the *gamma function*, which is defined for $x > 0$ by

$$\Gamma(x) = \int_0^\infty t^{x-1} e^{-t}\, dt \tag{1.101}$$

For $x > 0$, $\Gamma(x + 1) = x\Gamma(x)$. Thus, the gamma function is a generalization of the factorial function. In fact, if x is an integer, then $\Gamma(x + 1) = x!$. A random variable X is said to possess a *gamma distribution* with parameters $\alpha > 0$ and $\beta > 0$ if it has the density

$$f(x) = \frac{\beta^{-\alpha}}{\Gamma(\alpha)} x^{\alpha-1} e^{-x/\beta} \tag{1.102}$$

for $x \geq 0$ and $f(x) = 0$ for $x < 0$. Its probability distribution function is given by $F(x) = 0$ for $x \leq 0$ and

$$F(x) = \frac{\beta^{-\alpha}}{\Gamma(\alpha)} \int_0^x t^{\alpha-1} e^{-t/\beta}\, dt \tag{1.103}$$

for $x > 0$. Some gamma densities are shown in Fig. 1.1.

For $t < 1/\beta$, the moment-generating function is given by

$$\begin{aligned} M_X(t) &= \frac{\beta^{-\alpha}}{\Gamma(\alpha)} \int_{-\infty}^{\infty} x^{\alpha-1} e^{-[(1/\beta)-t]x}\, dx \\ &= \frac{\beta^{-\alpha}}{\Gamma(\alpha)} \left(\frac{1}{\beta} - t\right)^{-\alpha} \int_{-\infty}^{\infty} u^{\alpha-1} e^{-u}\, du \\ &= (1 - \beta t)^{-\alpha} \end{aligned} \tag{1.104}$$

where the first integral is finite for $t < 1/\beta$, the second follows from the substitution $u = [(1/\beta) - t]x$, and the final equality follows from the definition of $\Gamma(\alpha)$.

Differentiating the moment-generating function k times in succession yields

$$M_X^{(k)}(t) = \beta^k(\alpha + k - 1)\cdots(\alpha + 1)\alpha(1 - \beta t)^{-\alpha-k} \tag{1.105}$$

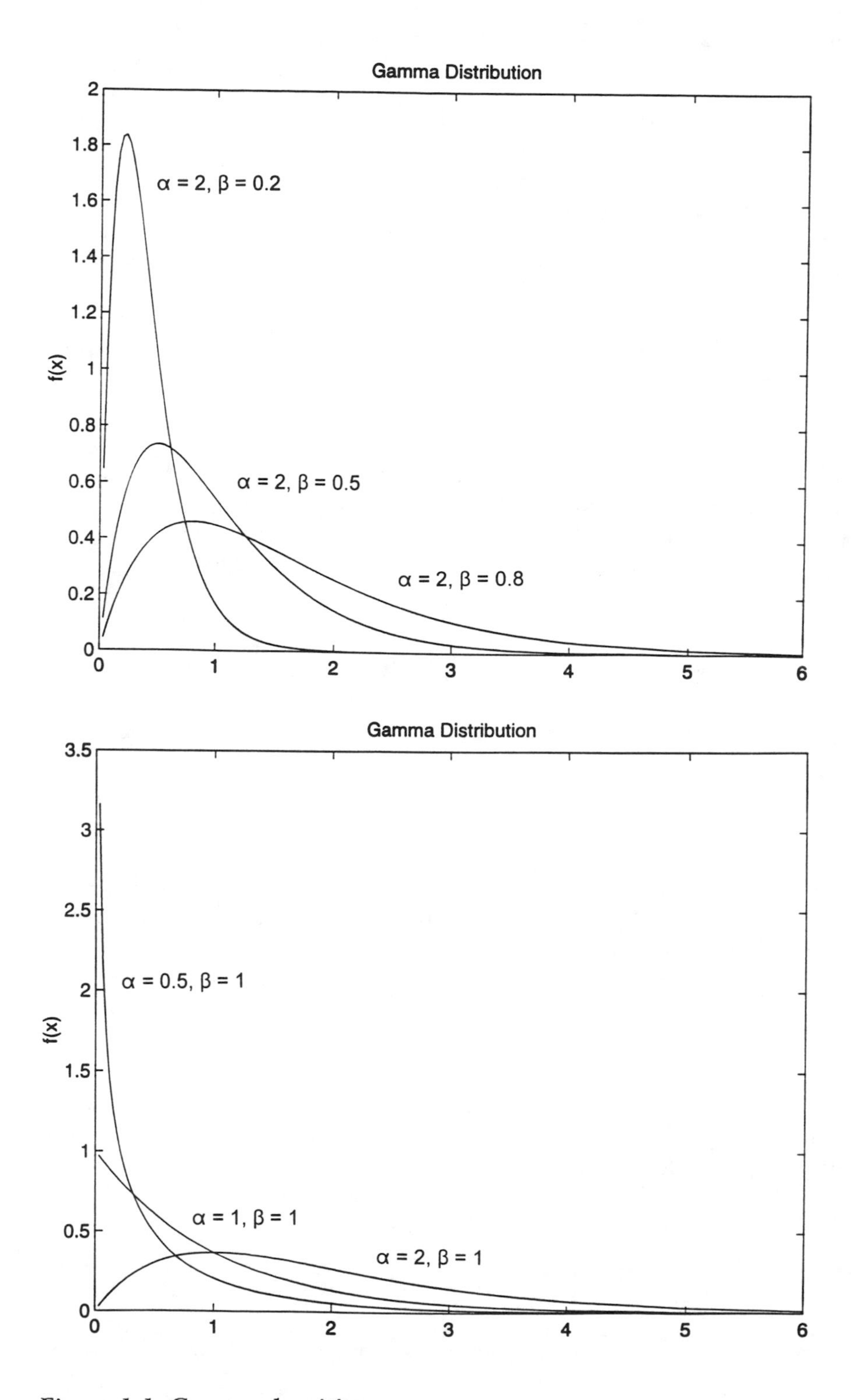

Figure 1.1 Gamma densities.

Letting $t = 0$ yields

$$\mu_k' = \beta^k(\alpha + k - 1)\cdots(\alpha + 1)\alpha = \frac{\Gamma(\alpha + k)}{\Gamma(\alpha)}\beta^k \tag{1.106}$$

Hence, $\mu = \alpha\beta$, $\mu_2' = \beta^2(\alpha + 1)\alpha$, $\sigma^2 = \alpha\beta^2$.

When $\alpha = k$ is an integer, the gamma distribution is sometimes called a *k-Erlang distribution* and it then takes the form

$$f(x) = \frac{\beta^{-k}}{(k-1)!}x^{k-1}e^{-x/\beta} \tag{1.107}$$

for $x \geq 0$ and $f(x) = 0$ for $x < 0$.

The exponential density, introduced in Example 1.11, results from the gamma density by letting $\alpha = 1$ and $\beta = 1/b$. As shown in the example, the exponential density has mean $\mu = 1/b$ and variance $\sigma^2 = 1/b^2$. A salient property of the exponential distribution is that it is memoryless: a random variable X is *memoryless* if for all nonnegative x and y,

$$P(X > x + y \mid X > y) = P(X > x) \tag{1.108}$$

A continuous random variable is memoryless if and only if it is exponentially distributed.

***Example* 1.13.** The time-to-failure distribution of any system is the probability distribution of the random variable T measuring the time the system runs prior to failure. Given T is a continuous random variable with density $f(t)$ and probability distribution function $F(t)$, a system's reliability is characterized by its *reliability function*, which is defined by

$$R(t) = P(T > t) = 1 - F(t) = \int_t^\infty f(u)\,du$$

$R(t)$ is monotonically decreasing, $R(0) = 1$, and $\lim_{t\to\infty} R(t) = 0$. System reliability is often judged by *mean time to failure* (*MTTF*),

$$E[T] = \int_0^\infty tf(t)\,dt = \int_0^\infty R(t)\,dt$$

The *hazard function* h of a system gives the instantaneous failure rate of the system,

$$h(t) = \lim_{\Delta t \to 0} \frac{P(t < T < t + \Delta t | T > t)}{\Delta t}$$

$$= \lim_{\Delta t \to 0} \frac{P(t < T < t + \Delta t, T > t)}{P(T > t)\Delta t}$$

$$= \lim_{\Delta t \to 0} \frac{P(t < T < t + \Delta t)}{P(T > t)\Delta t}$$

$$= \lim_{\Delta t \to 0} \frac{F(t + \Delta t) - F(t)}{R(t)\Delta t}$$

At all points at which $f(t)$ is continuous, $F'(t) = f(t)$. Therefore,

$$h(t) = \frac{f(t)}{R(t)} = -\frac{R'(t)}{R(t)}$$

Consequently,

$$R(t) = \exp\left[-\int_0^t h(u)\,du\right]$$

$$f(t) = h(t)\exp\left[-\int_0^t h(u)\,du\right]$$

From a modeling perspective, a constant hazard function, $h(t) = q$, corresponds to a situation in which wear-in failures have been eliminated and the system has not reached the wear-out stage. Such an assumption is often appropriate for the kind of electronic components used in digital signal processing. From the preceding representations of $R(t)$ and $f(t)$, for $h(t) = q$ the time-to-failure distribution is exponential with $f(t) = qe^{-qt}$, $R(t) = e^{-qt}$, and MTTF $= 1/q$. Since the exponential distribution is memoryless, the probability that the system will function longer than time $t + v$ given that it has functioned for time v is the same as the probability that it will function for time t from the outset: the system's reliability is not modified even though it has been in service for some length of time v. The MTTF is independent of the time the system has already been in operation. ■

1.4.5. Beta Distribution

For $\alpha > 0$, $\beta > 0$, the *beta function* is defined by

$$B(\alpha, \beta) = \int_0^1 t^{\alpha-1}(1-t)^{\beta-1}\,dt \tag{1.109}$$

It can be shown that

$$B(\alpha, \beta) = \frac{\Gamma(\alpha)\Gamma(\beta)}{\Gamma(\alpha+\beta)} \tag{1.110}$$

A random variable X is said to possess a *beta distribution* if it has density

$$f(x) = \frac{1}{B(\alpha,\beta)} x^{\alpha-1}(1-x)^{\beta-1} \tag{1.111}$$

for $0 < x < 1$ and $f(x) = 0$ elsewhere. The beta density takes on various shapes and is therefore useful for modeling many kinds of data distributions. If $\alpha < 1$ and $\beta < 1$, the beta density is U-shaped; if $\alpha < 1$ and $\beta \geq 1$, it is reverse J-shaped; if $\alpha \geq 1$ and $\beta < 1$, it is J-shaped; and if $\alpha > 1$ and $\beta > 1$, it possesses a single maximum. If $\alpha = \beta$, then the graph of the density is symmetric. Some beta densities are shown in Fig. 1.2.

The kth moment of the beta distribution is obtained by applying the moment definition directly to obtain

$$\mu_k' = \frac{1}{B(\alpha,\beta)} \int_0^1 x^{\alpha+k-1}(1-x)^{\beta-1}\,dx$$

$$= \frac{B(\alpha+k,\beta)}{B(\alpha,\beta)}$$

$$= \frac{\Gamma(\alpha+\beta)\Gamma(\alpha+k)}{\Gamma(\alpha)\Gamma(\alpha+\beta+k)} \tag{1.112}$$

Therefore, $\mu = \alpha(\alpha + \beta)^{-1}$ and $\sigma^2 = \alpha\beta(\alpha + \beta)^{-2}(\alpha + \beta + 1)^{-1}$.

The beta distribution is generalized so as to cover the interval (a, b). The *generalized beta distribution* has density

$$f(x) = \frac{(x-a)^{\alpha-1}(b-x)^{\beta-1}}{(b-a)^{\alpha+\beta-1}B(\alpha,\beta)} \tag{1.113}$$

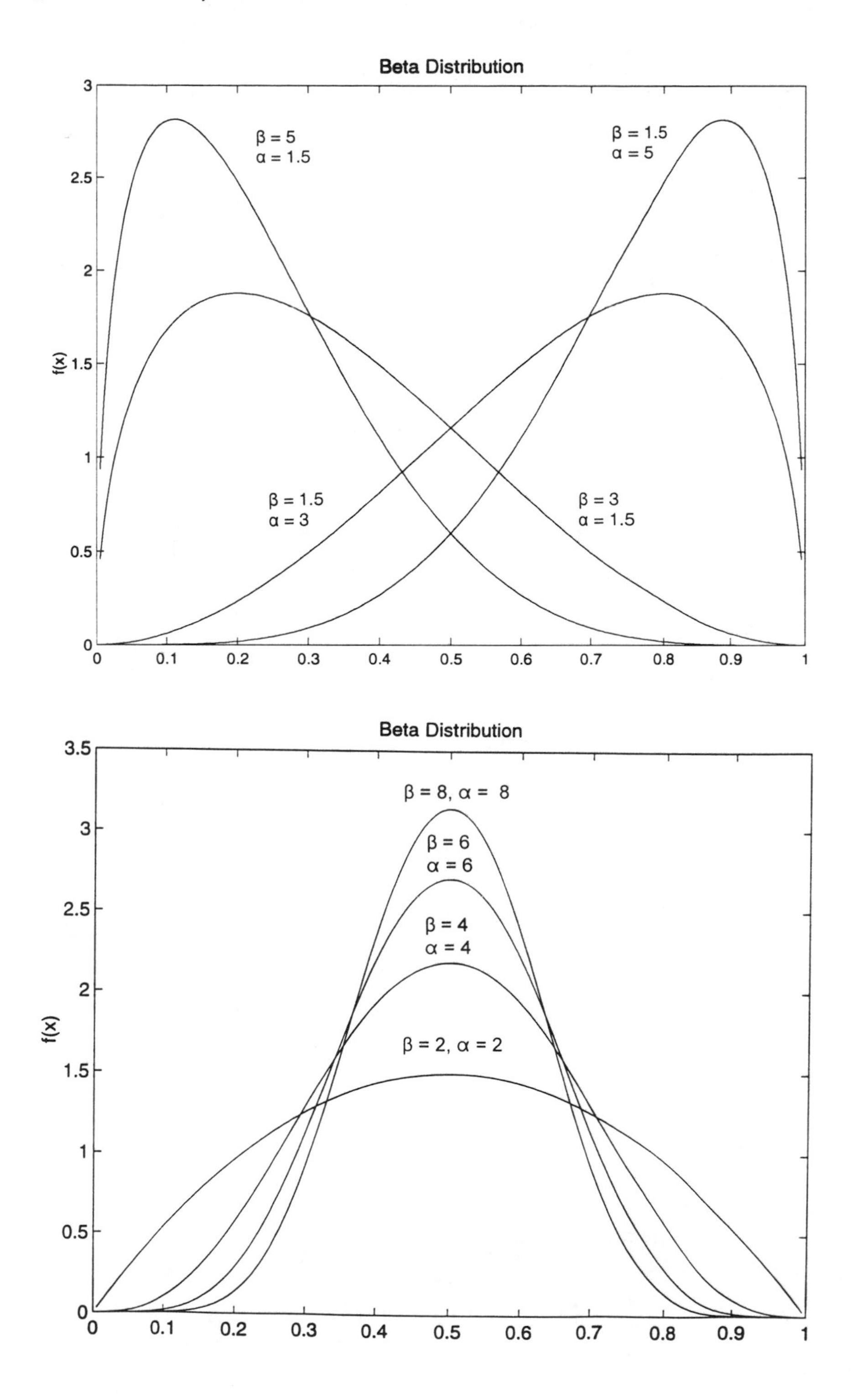

Figure 1.2 Beta densities.

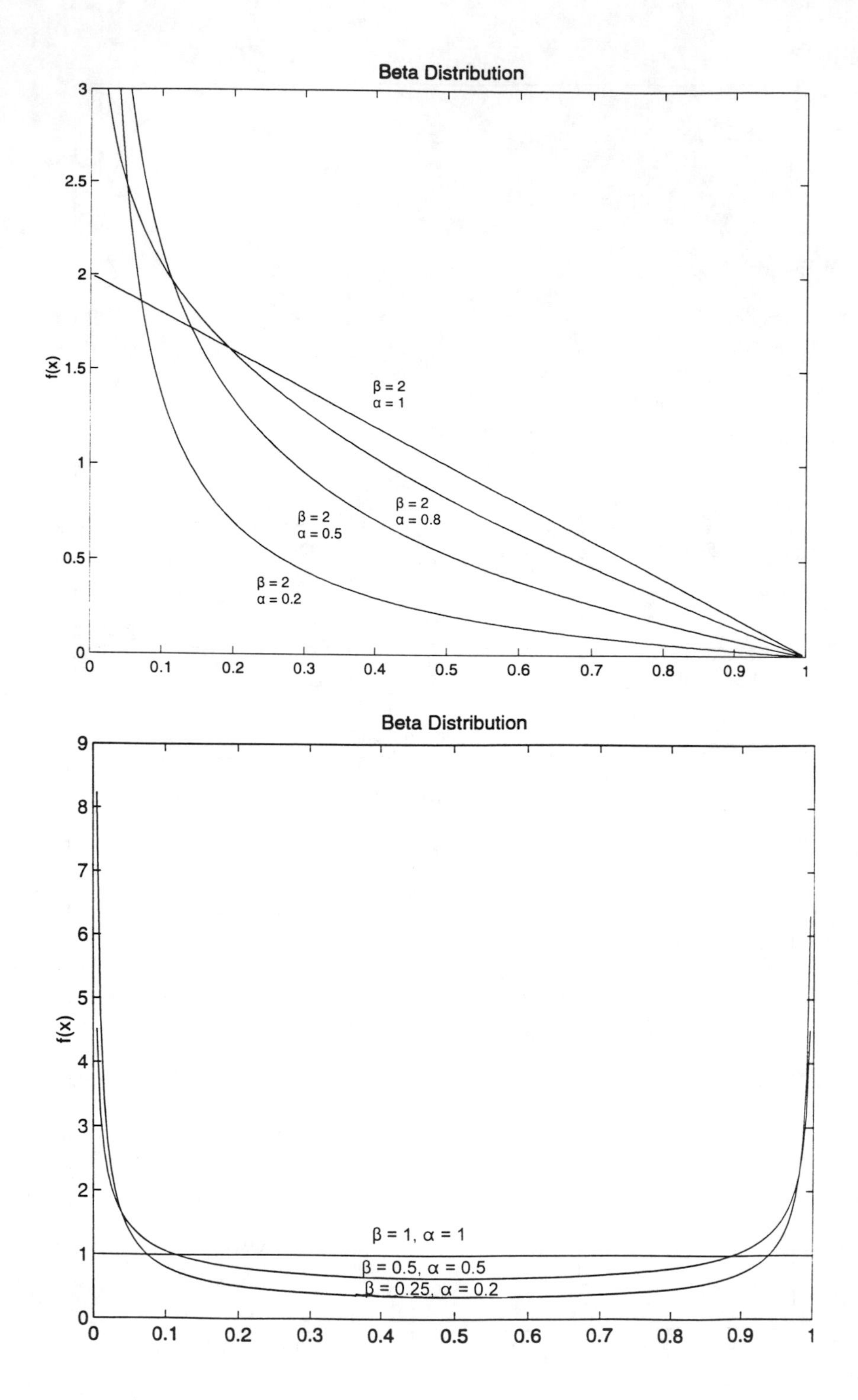

Figure 1.2 (cont): Beta densities.

for $a < x < b$, and $f(x) = 0$ elsewhere. Its mean and variance are

$$\mu = a + \frac{(b-a)\alpha}{\alpha+\beta} \tag{1.114}$$

$$\sigma^2 = \frac{(b-a)^2\alpha\beta}{(\alpha+\beta)^2(\alpha+\beta+1)} \tag{1.115}$$

Setting $a = 0$ and $b = 1$ gives the mean and variance of the original beta distribution. The uniform distribution introduced in Example 1.7 is a generalized beta distribution with $\alpha = \beta = 1$. It has mean $\mu = (a + b)/2$ and variance $\sigma^2 = (b - a)^2/12$.

1.4.6. Computer Simulation

Processing images and signals involves transformations of random variables. These transformations can be very complex, involving compositions of several highly involved mappings of numerous random variables. Even when there is only one function of a single random variable, it can be difficult to analytically describe the distribution of the output variable in terms of the input distribution. More generally, analytic description is impossible in the majority of cases encountered. Sometimes it is possible to describe some output moments in terms of input moments; however, such descriptions are rarely known for nonlinear operators. Moreover, even if some output moments are known, they may not provide sufficient description of the output distribution. Owing to the intractability of analytic description, it is common to simulate input distributions, operate on the resulting *synthetic* data, and statistically analyze the output data arising from the synthetic input.

The key to the entire procedure is the ability to simulate data whose empirical distribution fits well the theoretical distribution of the random variable of interest. Suppose X is a random variable and $x_1, x_2,\ldots, x_m$ are computer-generated data meant to simulate the behavior of X. If the empirical distribution composed of $x_1, x_2,\ldots, x_m$ is a good approximation of the theoretical distribution, then interval probabilities of the form $P(a < X < b)$ should be well approximated by the proportion of synthetic data in the interval (a, b).

The basis of computer generation of *random values* (outcomes of a random variable) is simulation of the uniformly distributed random variable U over (0, 1). From the perspective that an outcome of U is a nonterminating decimal, generation of U values involves uniformly random selection of digits between 0 and 9, inclusively, and concatenation to form a string of such digits. The result will be a finite decimal expansion and we will take the view that such finite expansions constitute the outcomes of U. Since the strings are finite, say of length r, the procedure does not actually generate all possible outcomes of U; nonetheless, we must be satisfied with a given degree of approximation. A typical generated value of U takes the form $u = 0.d_1d_2\cdots d_r$, where, for $i = 1, 2,..., r$, d_i is an outcome of a random variable possessing equally likely outcomes between 0 and 9.

The digits $d_1, d_2,..., d_r$ can be generated by uniformly randomly selecting, with replacement, balls numbered 0 through 9 from an urn. In practice, however, the digits are generated by a nonrandom process whose outcomes, called *pseudorandom values*, simulate actual randomness. There are various schemes, called *random-number generators*, for producing pseudorandom values. Not only do they vary in their ability to simulate randomness, but they also require experience to be used effectively.

Besides having routines to generate random values for the uniform distribution, many computer systems also generate random values for the standard normal distribution. Other commonly employed distributions can be simulated by using the random values generated for the uniform distribution.

If F is a continuous probability distribution function that is strictly increasing, then

$$X = F^{-1}(U) \tag{1.116}$$

is a random variable possessing the probability distribution function F. Indeed,

$$F_X(x) = P(F^{-1}(U) \leq x) = P(U \leq F(x)) = F(x) \tag{1.117}$$

the last equality holding because U is uniformly distributed over (0, 1) and $0 \leq F(x) \leq 1$.

***Example* 1.14.** To simulate an exponentially distributed random variable X with parameter b, consider its probability distribution function

$$u = F(x) = 1 - e^{-bx}$$

Solving for x in terms of u gives $x = -b^{-1}\log(1 - u)$. According to Eq. 1.116,

$$X = -b^{-1}\log(1 - U)$$

has an exponential distribution with mean $1/b$. To generate exponentially distributed random values, generate uniformly distributed random values and apply the expression of X in terms of U. Since $1 - U$ is uniformly distributed, the representation of X can be simplified to

$$X = -b^{-1}\log U$$ ■

1.5. Multivariate Distributions

To the extent that phenomena are related, so too are measurements of those phenomena. Since measurements are mathematically treated as random variables, we need to study properties of random variables taken as collections; indeed, in the most general sense the theory of random processes concerns collections of random variables. If $X_1, X_2,..., X_n$ are n random variables, a *random vector* is defined by

$$\mathbf{X} = \begin{pmatrix} X_1 \\ X_2 \\ \vdots \\ X_n \end{pmatrix} \tag{1.118}$$

Each random variable X_k is a mapping from a sample space into $\Re$; $\mathbf{X}$ is a mapping from the sample space into n-dimensional Euclidean space $\Re^n$. Whereas the probability distributions of $X_1, X_2,..., X_n$ can be determined from the distribution of $\mathbf{X}$, except in special circumstances the distribution of $\mathbf{X}$ cannot be determined from the individual distributions of $X_1, X_2,..., X_n$. To conserve space and to keep matrix-vector algebraic operations consistent, we will write $\mathbf{X} = (X_1, X_2,..., X_n)'$, the prime denoting transpose, so that $\mathbf{X}$ will always be a column vector.

For single random variables we have considered the manner in which a random variable induces a probability measure on the Borel σ-algebra in $\Re$ and how this probability is interpreted as the inclusion probability $P(X \in B)$ for a Borel set $B \subset \Re$. Although we will not go into detail, we note that the *Borel σ-algebra* in $\Re^n$ is the smallest σ-algebra containing all open sets in $\Re^n$, that sets in the Borel σ-algebra are again called *Borel sets*, that all unions and intersections of open and closed sets in $\Re^n$ are Borel sets, and that $\mathbf{X}$ induces a probability measure on the Borel σ-algebra by defining the inclusion probabilities $P(\mathbf{X} \in B)$ for any Borel set $B \subset \Re^n$. We leave the details of product measures and their related σ-algebras to a text on measure theory.

1.5.1. Jointly Distributed Random Variables

If $X_1, X_2,..., X_n$ are n discrete random variables, then their *joint* (*multivariate*) *distribution* is defined by the *joint probability mass function*

$$f(x_1, x_2,..., x_n) = P(X_1 = x_1, X_2 = x_2,..., X_n = x_n) \tag{1.119}$$

The joint mass function is nonnegative, there exists a countable set of points $(x_1, x_2,..., x_n) \in \Re^n$ such that $f(x_1, x_2,..., x_n) > 0$, and

$$\sum_{\{(x_1,x_2,\ldots,x_n):f(x_1,x_2,\ldots,x_n)>0\}} f(x_1,x_2,\ldots,x_n) = 1 \tag{1.120}$$

Moreover, for any Borel set $B \subset \Re^n$,

$$P((X_1, X_2,..., X_n)' \in B) = \sum_{\{(x_1,x_2,\ldots,x_n)\in B:f(x_1,x_2,\ldots,x_n)>0\}} f(x_1,x_2,\ldots,x_n) \tag{1.121}$$

The continuous random variables $X_1, X_2,..., X_n$ are said to possess a *multivariate distribution* defined by the *joint density* $f(x_1, x_2,..., x_n) \geq 0$ if, for any Borel set $B \subset \Re^n$,

$$P((X_1, X_2,..., X_n)' \in B) = \iint_B \cdots \int f(x_1,x_2,\ldots,x_n)\, dx_1 dx_2 \cdots dx_n \tag{1.122}$$

If $f(x_1, x_2,..., x_n)$ is a function of n variables such that $f(x_1, x_2,..., x_n) \geq 0$ and

$$\int_{-\infty}^{\infty}\int_{-\infty}^{\infty} \cdots \int_{-\infty}^{\infty} f(x_1,x_2,\ldots,x_n)\, dx_1 dx_2 \cdots dx_n = 1 \tag{1.123}$$

then there exist continuous random variables $X_1, X_2,..., X_n$ possessing multivariate density $f(x_1, x_2,..., x_n)$. As in the univariate setting, employing delta functions allows us to use continuous-variable notation to represent both continuous and discrete random variables.

***Example* 1.15.** This example treats a discrete multivariate distribution that generalizes the binomial distribution. Consider an experiment satisfying the following three conditions: (1) the experiment consists of n independent trials; (2) for each trial there are r possible outcomes, $w_1, w_2,..., w_r$; (3) there exist numbers $p_1, p_2,..., p_r$ such that, for $j = 1, 2,..., r$, on each trial the probability of outcome w_j is p_j. For $j = 1, 2,..., r$, the random variable X_j counts the number of times that outcome w_j occurs during the n trials. The sample space for the experiment is the set of n-vectors whose components are chosen from the set $\{w_1, w_2,..., w_r\}$ of trial outcomes. Assuming equiprobability, a specific n-vector having x_j components with w_j for $j = 1, 2,..., r$ has probability $p_1^{x_1} p_2^{x_2} \cdots p_r^{x_r}$ of occurring. To obtain the probability of obtaining $X_1 = x_1, X_2 = x_2,..., X_r = x_r$, we need to count the number of such vectors (as in the case of the binomial distribution). Although we will not prove it, the number of n-vectors having x_1 components with w_1, x_2

components with $w_2,\ldots, x_r$ components with w_r is given by the *multinomial coefficient*

$$\binom{n}{x_1, x_2, \ldots, x_r} = \frac{n!}{x_1!x_2!\cdots x_r!}$$

Hence, the joint density of $X_1, X_2,\ldots, X_r$ is

$$f(x_1, x_2,\ldots, x_r) = \frac{n!}{x_1!x_2!\cdots x_r!} p_1^{x_1} p_2^{x_2} \cdots p_r^{x_r}$$

The distribution is known as the *multinomial distribution*. ■

Consider two jointly distributed random variables X and Y. Each has its own univariate distribution and in the context of the joint density $f(x, y)$ the corresponding densities $f_X(x)$ and $f_Y(y)$ are called *marginal densities*. The marginal density for X is derived from the joint density by

$$f_X(x) = \int_{-\infty}^{\infty} f(x, y)\, dy \tag{1.124}$$

To see this, note that, for any Borel set $B \subset \Re$,

$$\int_B f_X(x)\, dx = P(X \in B)$$

$$= P((X, Y)' \in B \times \Re)$$

$$= \int_B \int_{-\infty}^{\infty} f(x, y)\, dy dx \tag{1.125}$$

The marginal density for Y is similarly derived by

$$f_Y(y) = \int_{-\infty}^{\infty} f(x, y)\, dx \tag{1.126}$$

More generally, if $X_1, X_2,\ldots, X_n$ possess joint density $f(x_1, x_2,\ldots, x_n)$, then the marginal density for X_k, $k = 1, 2,\ldots, n$, is obtained by

$$f_{X_k}(x_k) = \int_{-\infty}^{\infty} \cdots \int_{-\infty}^{\infty} f(x_1, x_2, \ldots, x_n)\, dx_n \cdots dx_{k+1} dx_{k-1} \cdots dx_1 \tag{1.127}$$

the integral being $(n - 1)$-fold over all variables, excluding x_k. Joint marginal densities can be obtained by integrating over subsets of the variables. For instance, if X, Y, U, and V are jointly distributed, then the joint marginal density for X and Y is given by

$$f_{X,Y}(x, y) = \int_{-\infty}^{\infty} \int_{-\infty}^{\infty} f_{X,Y,U,V}(x, y, u, v)\, dv du \tag{1.128}$$

Should the variables be discrete, integrals become sums over the appropriate variables.

The *joint probability distribution function* for random variables X and Y is defined by

$$F(x, y) = P(X \le x, Y \le y) \tag{1.129}$$

If X and Y possess the joint density $f(x, y)$, then

$$F(x, y) = \int_{-\infty}^{x} \int_{-\infty}^{y} f(r, s)\, dr ds \tag{1.130}$$

If the random variables are continuous, then partial differentiation with respect to x and y yields

$$\frac{\partial^2}{\partial x \partial y} F(x, y) = f(x, y) \tag{1.131}$$

at points of continuity of $f(x, y)$. As with univariate distributions, probability distribution functions characterize random variables up to identical distribution.

Theorem 1.10. If X and Y possess the joint probability distribution function $F(x, y)$, then

(i) $\lim_{\min\{x,y\} \to \infty} F(x, y) = 1$

(ii) $\lim_{x \to -\infty} F(x, y) = 0$, $\lim_{y \to -\infty} F(x, y) = 0$

(iii) $F(x, y)$ is continuous from the right in each variable.

(iv) If $a < b$ and $c < d$, then

$$F(b,d) - F(a,d) - F(b,c) + F(a,c) \geq 0 \tag{1.132}$$

Moreover, if $F(x, y)$ is any function satisfying the four conditions, then there exist random variables X and Y having joint probability distribution function $F(x, y)$. ■

In the bivariate continuous case,

$$P(a < X < b, c < Y < d) = F(b,d) - F(a,d) - F(b,c) + F(a,c) \tag{1.133}$$

Thus, we see the necessity of the condition in Eq. 1.132. Theorem 1.10 can be extended to any finite number of random variables.

1.5.2. Conditioning

Conditioning is a key probabilistic concept for signal processing because it lies at the foundation of filtering systems. If X and Y are discrete and possess joint density $f(x, y)$, then a natural way to define the probability that $Y = y$ given that $X = x$ is by

$$P(Y = y | X = x) = \frac{P(Y = y, X = x)}{P(X = x)} = \frac{f(x,y)}{f_X(x)} \tag{1.134}$$

The middle expression is undefined for continuous distributions because the denominator must be zero; however, the last is defined when $f_X(x) > 0$. It is taken as the definition.

If X and Y possess joint density $f(x, y)$, then for all x such that $f_X(x) > 0$, the *conditional density* of Y given $X = x$ is defined by

$$f(y|x) = \frac{f(x,y)}{f_X(x)} \tag{1.135}$$

For a given x, $f(y \mid x)$ is a function of y and is a legitimate density because

$$\int_{-\infty}^{\infty} f(y|x)\,dy = \frac{1}{f_X(x)} \int_{-\infty}^{\infty} f(x,y)\,dy = 1 \tag{1.136}$$

The random variable associated with the conditional density is called the *conditional random variable* Y given x and is denoted by $Y|x$. It has an expectation called the *conditional expectation* (*conditional mean*), denoted by $E[Y \mid x]$ or $\mu_{Y|x}$, and defined by

$$E[Y|x] = \int_{-\infty}^{\infty} y f(y|x)\, dy \tag{1.137}$$

It also has a conditional variance,

$$\mathrm{Var}[Y \mid x] = E[(Y|x - \mu_{Y|x})^2] \tag{1.138}$$

Conditioning can be extended to $n + 1$ random variables $X_1, X_2, \ldots, X_n, Y$. If the joint densities for $X_1, X_2, \ldots, X_n, Y$ and $X_1, X_2, \ldots, X_n$ are $f(x_1, x_2, \ldots, x_n, y)$ and $f(x_1, x_2, \ldots, x_n)$, respectively, then the conditional density of Y given $X_1, X_2, \ldots, X_n$ is defined by

$$f(y|x_1, x_2, \ldots, x_n) = \frac{f(x_1, x_2, \ldots, x_n, y)}{f(x_1, x_2, \ldots, x_n)} \tag{1.139}$$

for all $(x_1, x_2, \ldots, x_n)$ such that $f(x_1, x_2, \ldots, x_n) > 0$. The conditional expectation and variance of Y given $x_1, x_2, \ldots, x_n$ are defined via the conditional density. In particular,

$$E[Y|x_1, x_2, \ldots, x_n] = \int_{-\infty}^{\infty} y f(y|x_1, x_2, \ldots, x_n)\, dy \tag{1.140}$$

The conditional expectation plays a major role in filter optimization.

***Example* 1.16.** Random variables X and Y are said to possess a *joint uniform distribution* over region $R \subset \Re^2$ if their joint density $f(x, y)$ is defined by $f(x, y) = 1/v[R]$ for $(x, y) \in R$ and $f(x, y) = 0$ for $(x, y) \notin R$, where $v[R]$ is the area of R (assuming $v[R] > 0$). Let X and Y be jointly uniformly distributed over the triangular region R consisting of the portion of the plane bounded by the x axis, the line $x = 1$, and the line $y = x$. For $0 \le x \le 1$,

$$f_X(x) = \int_0^x 2\, dy = 2x$$

and $f_X(x) = 0$ for $x \notin [0, 1]$. For $0 \le y \le 1$,

$$f_Y(y) = \int_y^1 2\,dx = 2(1-y)$$

and $f_Y(y) = 0$ for $y \notin [0, 1]$. The conditional density for $Y|x$ is defined for $0 < x \le 1$ and is given by

$$f(y|x) = \frac{f(x,y)}{f_X(x)} = \frac{2}{2x} = \frac{1}{x}$$

for $0 \le y \le x$ and $f(y \mid x) = 0$ for $y \notin [0, x]$. The conditional density for $X|y$ is defined for $0 \le y < 1$ and is given by

$$f(x|y) = \frac{f(x,y)}{f_Y(y)} = \frac{2}{2(1-y)} = \frac{1}{1-y}$$

for $y \le x \le 1$ and $f(x \mid y) = 0$ for $x \notin [y, 1]$. Thus, $Y|x$ and $X|y$ are uniformly distributed over $[0, x]$ and $[y, 1]$, respectively. Consequently, $E[Y \mid x] = x/2$ and $E[X \mid y] = (1 + y)/2$. These conditional expectations can be obtained by integration (Eq. 1.137). ■

1.5.3. Independence

Cross multiplication in Eq. 1.135 yields

$$f(x, y) = f_X(x) f(y \mid x) \tag{1.141}$$

If

$$f(x, y) = f_X(x) f_Y(y) \tag{1.142}$$

then Eq. 1.141 reduces to $f(y \mid x) = f_Y(y)$, so that conditioning by x does not affect the probability distribution of Y. If Eq. 1.142 holds, then X and Y are said to be *independent*. In general, random variables $X_1, X_2, \ldots, X_n$ possessing multivariate density $f(x_1, x_2, \ldots, x_n)$ are said to be *independent* if

$$f(x_1, x_2, \ldots, x_n) = f_{X_1}(x_1) f_{X_2}(x_2) \cdots f_{X_n}(x_n) \tag{1.143}$$

Otherwise they are *dependent*.

If $X_1, X_2, \ldots, X_n$ are independent, then so too is any subset of $X_1, X_2, \ldots, X_n$. To see this for three jointly distributed continuous random variables X, Y, Z, note that

$$
\begin{aligned}
f_{X,Y}(x,y) &= \int_{-\infty}^{\infty} f_{X,Y,Z}(x,y,z)\,dz \\
&= \int_{-\infty}^{\infty} f_X(x) f_Y(y) f_Z(z)\,dz \\
&= f_X(x) f_Y(y) \int_{-\infty}^{\infty} f_Z(z)\,dz \qquad (1.144) \\
&= f_X(x) f_Y(y)
\end{aligned}
$$

If $X_1, X_2,\ldots, X_n$ are independent, then for any Borel sets $B_1, B_2,\ldots, B_n$,

$$
\begin{aligned}
P\left(\bigcap_{i=1}^{n} (X_i \in B_i)\right) &= P((X_1, X_2, \ldots, X_n) \in B_1 \times B_2 \times \cdots \times B_n) \\
&= \int_{B_1}\int_{B_2}\cdots\int_{B_n} f(x_1, x_2, \ldots, x_n)\,dx_n dx_{n-1} \cdots dx_1 \\
&= \int_{B_1}\int_{B_2}\cdots\int_{B_n} f_{X_1}(x_1) f_{X_2}(x_2) \cdots f_{X_n}(x_n)\,dx_n dx_{n-1} \cdots dx_1 \\
&= \prod_{i=1}^{n} \int_{B_i} f_{X_i}(x_i)\,dx_i \\
&= \prod_{i=1}^{n} P(X_i \in B_i) \qquad (1.145)
\end{aligned}
$$

In general, given the multivariate density for a finite collection of random variables, it is possible to derive the marginal densities via integration according to Eq. 1.127. On the contrary, it is not generally possible to obtain the multivariate density from the marginal densities; however, if the random variables are independent, then the multivariate density can be so derived. Many experimental designs postulate independence just so the joint density can be expressed as a product of the marginal densities.

***Example* 1.17.** If $X_1, X_2,\ldots, X_n$ are independent normally distributed random variables with means $\mu_1, \mu_2,\ldots, \mu_n$ and standard deviations $\sigma_1, \sigma_2,\ldots, \sigma_n$, respectively, then the multivariate density for $X_1, X_2,\ldots, X_n$ is given by

$$f(x_1, x_2,\ldots, x_n) = \prod_{k=1}^{n} \frac{1}{\sqrt{2\pi}\sigma_k} e^{-\frac{1}{2}\left(\frac{x_k-\mu_k}{\sigma_k}\right)^2}$$

$$= \frac{1}{(2\pi)^{n/2}}\left(\prod_{k=1}^{n} \sigma_k^{-1}\right)\exp\left[-\frac{1}{2}\left(\sum_{k=1}^{n}\left(\frac{x_k-\mu_k}{\sigma_k}\right)^2\right)\right]$$

$$= \frac{1}{\sqrt{(2\pi)^n \det[\mathbf{K}]}}\exp\left[-\frac{1}{2}(\mathbf{x}-\boldsymbol{\mu})'\mathbf{K}^{-1}(\mathbf{x}-\boldsymbol{\mu})\right]$$

where $\mathbf{x} = (x_1, x_2,\ldots, x_n)'$, $\boldsymbol{\mu} = (\mu_1, \mu_2,\ldots, \mu_n)'$, and $\mathbf{K}$ is the diagonal matrix whose diagonal contains the variances of $X_1, X_2,\ldots, X_n$,

$$\mathbf{K} = \begin{pmatrix} \sigma_1^2 & 0 & \cdots & 0 \\ 0 & \sigma_2^2 & \cdots & 0 \\ \vdots & \vdots & \ddots & \vdots \\ 0 & 0 & \cdots & \sigma_n^2 \end{pmatrix}$$ ■

1.6. Functions of Several Random Variables

When dealing with random processes we are rarely concerned with a function of a single random variable; for most systems there are several random inputs and one or more random outputs. Given n input random variables $X_1, X_2,\ldots, X_n$ to a system, if the system outputs a single numerical value, then the output is a function of the form

$$Y = g(X_1, X_2, \ldots, X_n) \tag{1.146}$$

We would like to describe the probability distribution of the output given the joint distribution of the inputs, or at least describe some of the output moments. Typically, this problem is very difficult.

1.6.1. Basic Arithmetic Functions of Two Random Variables

For some basic arithmetic functions of two continuous random variables X and Y having joint density $f(x, y)$, output densities can be found by differentiation of the probability distribution functions. The probability distribution function of the sum $X + Y$ is given by

$$
\begin{aligned}
F_{X+Y}(z) &= P(X+Y \leq z) \\
&= \iint\limits_{\{(x,y):x+y\leq z\}} f(x,y)\,dxdy \\
&= \int_{-\infty}^{\infty} dx \int_{-\infty}^{z-x} f(x,y)\,dy \\
&= \int_{-\infty}^{\infty} dx \int_{-\infty}^{z} f(x,u-x)\,du
\end{aligned}
\tag{1.147}
$$

the last integral resulting from the substitution $u = x + y$. Differentiation yields the density

$$
f_{X+Y}(z) = \int_{-\infty}^{\infty} f(x,z-x)\,dx = \int_{-\infty}^{\infty} f(z-y,y)\,dy \tag{1.148}
$$

where the second integral follows by symmetry. If X and Y are independent, then

$$
f_{X+Y}(z) = \int_{-\infty}^{\infty} f_X(x) f_Y(z-x)\,dx = \int_{-\infty}^{\infty} f_X(z-x) f_Y(x)\,dx \tag{1.149}
$$

which is the convolution of the densities.

The probability distribution function of the product XY is given by

$$
\begin{aligned}
F_{XY}(z) &= \int_{-\infty}^{\infty} dx \int_{-\infty}^{z/x} f(x,y)\,dy \\
&= \int_{-\infty}^{\infty} \frac{dx}{|x|} \int_{-\infty}^{z} f\left(x, \frac{u}{x}\right) du
\end{aligned}
\tag{1.150}
$$

the last integral resulting from the substitution $u = xy$. Differentiation yields the density

$$f_{X/Y}(z) = \int_{-\infty}^{\infty} f\left(x, \frac{z}{x}\right) \frac{dx}{|x|} = \int_{-\infty}^{\infty} f\left(\frac{z}{y}, y\right) \frac{dy}{|y|} \tag{1.151}$$

The probability distribution function of the quotient Y/X is given by

$$F_{Y/X}(z) = \int_{-\infty}^{\infty} dx \int_{-\infty}^{xz} f(x,y)\, dy$$

$$= \int_{-\infty}^{\infty} |x|\, dx \int_{-\infty}^{z} f(x, ux)\, du \tag{1.152}$$

the last integral resulting from the substitution $u = y/x$. Differentiation yields the density

$$f_{Y/X}(z) = \int_{-\infty}^{\infty} f(x, zx)|x|\, dx = \int_{-\infty}^{\infty} f(zy, y)|y|\, dy \tag{1.153}$$

Example **1.18.** For a quotient $Z = Y/X$ of two independent standard normal random variables, Eq. 1.153 yields

$$f_Z(z) = \int_{-\infty}^{\infty} f_X(x) f_Y(zx)|x|\, dx$$

$$= \frac{1}{2\pi} \int_{-\infty}^{\infty} e^{-x^2/2} e^{-(zx)^2/2} |x|\, dx$$

$$= \frac{1}{\pi} \int_{0}^{\infty} e^{-(1+z^2)x^2/2} x\, dx$$

$$= \frac{1}{\pi(1+z^2)}$$

which is known as the *Cauchy density*. This density does not possess an expectation since

$$\int_{-\infty}^{\infty} |z| f_Z(z)\, dz = \int_{-\infty}^{\infty} \frac{|z|}{\pi(1+z^2)}\, dz$$

does not converge when evaluated by improper integration. ■

The probability distribution function for $Z = (X^2 + Y^2)^{1/2}$, the Euclidean norm of the random vector $(X, Y)'$, is given by $F_Z(z) = 0$ for $z < 0$ and, for $z \geq 0$,

$$F_Z(z) = \iint_{\{(x,y):x^2+y^2 \leq z^2\}} f(x,y)\, dxdy$$

$$= \int_0^{2\pi} d\theta \int_0^z f(r\cos\theta, r\sin\theta) r\, dr \tag{1.154}$$

where we have transformed to polar coordinates. Differentiation yields

$$f_Z(z) = z \int_0^{2\pi} f(z\cos\theta, z\sin\theta)\, d\theta \tag{1.155}$$

for $z \geq 0$, and $f_Z(z) = 0$ for $z < 0$.

Example **1.19.** If X and Y are independent zero-mean, normal random variables possessing a common variance σ^2, then, according to Eq. 1.155, the density of the Euclidean norm Z of $(X, Y)'$ is given by

$$f_Z(z) = z \int_0^{2\pi} f_X(z\cos\theta) f_Y(z\sin\theta)\, d\theta$$

$$= \frac{z}{2\pi\sigma^2} \int_0^{2\pi} e^{-\frac{z^2}{2\sigma^2}}\, d\theta$$

$$= \frac{z}{\sigma^2} e^{-\frac{z^2}{2\sigma^2}}$$

for $z \geq 0$ and $f_Z(z) = 0$ for $z < 0$, which is known as a *Rayleigh density*. ■

1.6.2. Distributions of Sums of Independent Random Variables

An important special case of a function of several random variables X_1, X_2,..., X_n occurs when the function of the random variables is linear,

$$Y = a_1 X_1 + a_2 X_2 + \cdots + a_n X_n \tag{1.156}$$

where a_1, a_2,..., a_n are constants. If X_1, X_2,..., X_n are independent, then we can employ moment-generating functions to discover output distributions for linear functions.

If X_1, X_2,..., X_n are independent random variables possessing joint density $f(x_1, x_2,..., x_n)$ and their moment-generating functions exist for $t < t_0$, then the moment-generating function of

$$Y = X_1 + X_2 + \cdots + X_n \tag{1.157}$$

exists for $t < t_0$ and is given by

$$\begin{aligned} M_Y(t) &= \int_{-\infty}^{\infty} \cdots \int_{-\infty}^{\infty} \exp\left[t \sum_{k=1}^{n} x_k \right] f(x_1, x_2, \ldots, x_n)\, dx_1 dx_2 \cdots dx_n \\ &= \int_{-\infty}^{\infty} \cdots \int_{-\infty}^{\infty} \prod_{k=1}^{n} e^{tx_k} f(x_k)\, dx_1 dx_2 \cdots dx_n \\ &= \prod_{k=1}^{n} M_{X_k}(t) \end{aligned} \tag{1.158}$$

where the second equality follows from the properties of exponentials and the independence of the random variables. The relation of Eq. 1.158 has been demonstrated for continuous random variables; the demonstration is similar for discrete random variables, except that integrals are replaced by sums.

Example 1.20. Suppose X_1, X_2,..., X_n are independent gamma-distributed random variables with X_k having parameters α_k and β, for $k = 1, 2,..., n$. Since the moment-generating function of X_k is $(1-\beta t)^{-\alpha_k}$, Eq. 1.158 shows the moment-generating function of the sum of X_1, X_2,..., X_n to be

$$M_Y(t) = \prod_{k=1}^{n} (1-\beta t)^{-\alpha_k} = (1-\beta t)^{-(\alpha_1+\alpha_2+\cdots+\alpha_n)}$$

which is the moment-generating function of a gamma-distributed random variable with parameters $\alpha_1 + \alpha_2 + \cdots + \alpha_n$ and β. By uniqueness, the sum Y is such a variable. As a special case, the sum of n identically distributed exponential random variables with parameter b is gamma distributed with parameters $\alpha = n$ and $\beta = 1/b$. Such a random variable is also said to be n-Erlang with parameter $1/b$.

If $U_1, U_2,..., U_n$ are independent uniform random variables on (0, 1), then, from Example 1.14, we know that $-b^{-1}\log U_k$ is exponentially distributed with parameter b, for $k = 1, 2,..., n$. Hence

$$X = -b^{-1}\sum_{k=1}^{n} \log U_k$$

is gamma distributed with $\alpha = n$ and $\beta = 1/b$. X can be simulated via computer generation of independent uniform random variables. ■

***Example* 1.21.** Suppose $X_1, X_2,..., X_n$ are independent Poisson-distributed random variables with X_k having mean λ_k, for $k = 1, 2,..., n$. Since the moment-generating function of X_k is $\exp[\lambda_k(e^t - 1)]$, Eq. 1.158 shows the moment-generating function of the sum of $X_1, X_2,..., X_n$ to be

$$M_Y(t) = \prod_{k=1}^{n} \exp[\lambda_k(e^t - 1)] = \exp\left[(e^t - 1)\sum_{k=1}^{n} \lambda_k\right]$$

which is the moment-generating function for a Poisson-distributed random variable with mean $\lambda_1 + \lambda_2 + \cdots + \lambda_n$. By uniqueness, the sum Y is such a variable. ■

***Example* 1.22.** Suppose $X_1, X_2,..., X_n$ are independent normally distributed random variables such that, for $k = 1, 2,..., n$, X_k has mean μ_k and variance σ_k^2. Since the moment-generating function of X_k is $\exp[\mu_k t + \sigma_k^2 t^2/2]$, Eq. 1.158 shows the moment-generating function of the linear combination of Eq. 1.156 to be

$$M_Y(t) = \prod_{k=1}^{n} M_{X_k}(a_k t)$$

$$= \prod_{k=1}^{n} \exp\left[a_k \mu_k t + \frac{a_k^2 \sigma_k^2 t^2}{2}\right]$$

$$= \exp\left[\left(\sum_{k=1}^{n} a_k \mu_k\right)t + \left(\sum_{k=1}^{n} a_k^2 \sigma_k^2\right)\frac{t^2}{2}\right]$$

Hence, Y is normally distributed with mean and variance

$$\mu_Y = \sum_{k=1}^{n} a_k \mu_k$$

$$\sigma_Y^2 = \sum_{k=1}^{n} a_k^2 \sigma_k^2$$

■

1.6.3. Joint Distributions of Output Random Variables

Many systems have several output random variables as well as several input variables and for these it is desirable (if possible) to have the joint distribution of the output random variables in terms of the distribution of the input variables.

For the case of two discrete input random variables, X and Y, and two discrete output random variables, U and V, there exist functions g and h such that

$$U = g(X, Y) \tag{1.159}$$

$$V = h(X, Y)$$

and the output probability mass function is

$$f_{U,V}(u, v) = P(g(X, Y) = u, h(X, Y) = v)$$

$$= \sum_{\{(x,y):g(x,y)=u,h(x,y)=v\}} f_{X,Y}(x, y) \tag{1.160}$$

Now suppose the vector mapping

$$\begin{pmatrix} u \\ v \end{pmatrix} = \begin{pmatrix} g(x, y) \\ h(x, y) \end{pmatrix} \tag{1.161}$$

is one-to-one and has the inverse vector mapping

$$\begin{pmatrix} x \\ y \end{pmatrix} = \begin{pmatrix} r(u,v) \\ s(u,v) \end{pmatrix} \tag{1.162}$$

Then

$$\{(x, y): g(x, y) = u,\ h(x, y) = v\} = \{(r(u, v), s(u, v))\} \tag{1.163}$$

and Eq. 1.160 reduces to

$$f_{U,V}(u, v) = f_{X,Y}(r(u, v), s(u, v)) \tag{1.164}$$

The analysis extends to any finite-dimensional vector mapping.

***Example* 1.23.** Suppose X and Y are independent binomially distributed random variables, X having parameters n and p, and Y having parameters m and d. Their joint density is the product of their individual densities,

$$f_{X,Y}(x, y) = \binom{n}{x}\binom{m}{y} p^x d^y (1-p)^{n-x}(1-d)^{m-y}$$

Suppose the output random variables are defined by $U = X + Y$ and $V = X - Y$. The vector mapping

$$\begin{pmatrix} u \\ v \end{pmatrix} = \begin{pmatrix} g(x,y) \\ h(x,y) \end{pmatrix} = \begin{pmatrix} x+y \\ x-y \end{pmatrix}$$

possesses the unique solution

$$\begin{pmatrix} x \\ y \end{pmatrix} = \begin{pmatrix} r(u,v) \\ s(u,v) \end{pmatrix} = \begin{pmatrix} (u+v)/2 \\ (u-v)/2 \end{pmatrix}$$

From Eq. 1.164,

$$f_{U,V}(u, v) = \binom{n}{(u+v)/2}\binom{m}{(u-v)/2} p^{(u+v)/2} d^{(u-v)/2} (1-p)^{n-(u+v)/2}(1-d)^{m-(u-v)/2}$$

Constraints on the variables u and v are determined by the constraints on x and y. ■

Theorem 1.11. Suppose the continuous random vector $\mathbf{X} = (X_1, X_2, \ldots, X_n)'$ possesses multivariate density $f(x_1, x_2, \ldots, x_n)$ and the vector mapping $\mathbf{x} \to \mathbf{u}$ is defined by

$$\mathbf{u} = \begin{pmatrix} u_1 \\ u_2 \\ \vdots \\ u_n \end{pmatrix} = \begin{pmatrix} g_1(x_1, x_2, \ldots, x_n) \\ g_2(x_1, x_2, \ldots, x_n) \\ \vdots \\ g_n(x_1, x_2, \ldots, x_n) \end{pmatrix} \tag{1.165}$$

where $g_1, g_2, \ldots, g_n$ have continuous partial derivatives and the mapping is one-to-one on the set $A_\mathbf{x}$ of all vectors $\mathbf{x}$ such that $f(\mathbf{x}) > 0$. If $A_\mathbf{u}$ denotes the set of all vectors corresponding to vectors in $A_\mathbf{x}$ and if the inverse vector mapping $\mathbf{u} \to \mathbf{x}$ is defined by

$$\mathbf{x} = \begin{pmatrix} x_1 \\ x_2 \\ \vdots \\ x_n \end{pmatrix} = \begin{pmatrix} r_1(u_1, u_2, \ldots, u_n) \\ r_2(u_1, u_2, \ldots, u_n) \\ \vdots \\ r_n(u_1, u_2, \ldots, u_n) \end{pmatrix} \tag{1.166}$$

then the joint density of the random vector

$$\mathbf{U} = \begin{pmatrix} U_1 \\ U_2 \\ \vdots \\ U_n \end{pmatrix} = \begin{pmatrix} g_1(X_1, X_2, \ldots, X_n) \\ g_2(X_1, X_2, \ldots, X_n) \\ \vdots \\ g_n(X_1, X_2, \ldots, X_n) \end{pmatrix} \tag{1.167}$$

is given by

$$f_\mathbf{U}(\mathbf{u}) = \begin{cases} f_\mathbf{X}(r_1(\mathbf{u}), r_2(\mathbf{u}), \ldots, r_n(\mathbf{u}))|J(\mathbf{x};\mathbf{u})|, & \text{if } \mathbf{u} \in A_\mathbf{u} \\ 0, & \text{otherwise} \end{cases} \tag{1.168}$$

where $J(\mathbf{x}; \mathbf{u})$, the *Jacobian* of the mapping $\mathbf{u} \to \mathbf{x}$, is defined by the determinant

$$J(\mathbf{x};\mathbf{u}) = \det\begin{bmatrix} \partial x_1/\partial u_1 & \partial x_1/\partial u_2 & \cdots & \partial x_1/\partial u_n \\ \partial x_2/\partial u_1 & \partial x_2/\partial u_2 & \cdots & \partial x_2/\partial u_n \\ \vdots & \vdots & \ddots & \vdots \\ \partial x_n/\partial u_1 & \partial x_n/\partial u_2 & \cdots & \partial x_n/\partial u_n \end{bmatrix} \quad \blacksquare \tag{1.169}$$

Example **1.24.** As in Example 1.23, consider the vector mapping $U = X + Y$, $V = X - Y$, but now let X and Y be jointly uniformly distributed over the unit square $A_{\mathbf{x}} = (0, 1)^2$. $A_{\mathbf{u}}$ is determined by solving the variable constraint pair

$$0 < \frac{u+v}{2} < 1, \qquad 0 < \frac{u-v}{2} < 1$$

Since $\partial x/\partial u = \partial x/\partial v = \partial y/\partial u = 1/2$ and $\partial y/\partial v = -1/2$, the Jacobian is

$$J(\mathbf{x}; \mathbf{u}) = \det\begin{bmatrix} 1/2 & 1/2 \\ 1/2 & -1/2 \end{bmatrix} = -\frac{1}{2}$$

By Theorem 1.11, $f_{U,V}(u, v) = 1/2$ for $(u, v) \in A_{\mathbf{u}}$ and $f_{U,V}(u, v) = 0$ otherwise. ■

1.6.4. Expectation of a Function of Several Random Variables

For taking the expectation of a function of several random variables, there exists the following extension of Theorem 1.7. An important consequence of the new theorem is that the expected-value operator is a linear operator.

Theorem 1.12. Suppose $X_1, X_2,\ldots, X_n$ have joint density $f(x_1, x_2,\ldots, x_n)$ and $g(x_1, x_2,\ldots, x_n)$ is a piecewise continuous function of $x_1, x_2,\ldots, x_n$. Then

$$E[g(X_1, X_2,\ldots, X_n)] = \int_{-\infty}^{\infty}\cdots\int_{-\infty}^{\infty} g(x_1, x_2,\ldots, x_n) f(x_1, x_2,\ldots, x_n)\, dx_1 dx_2 \cdots dx_n \tag{1.170}$$

where the integral is n-fold. ■

For discrete random variables there is no restriction on $g(x_1, x_2,\ldots, x_n)$ and the integral representation of Eq. 1.170 becomes

$$E[g(X_1, X_2,\ldots, X_n)] = \sum_{\{(x_1,x_2,\ldots,x_n): f(x_1,x_2,\ldots,x_n)>0\}} g(x_1, x_2,\ldots, x_n) f(x_1, x_2 \ldots, x_n) \tag{1.171}$$

Theorem 1.13. For any random variables $X_1, X_2,\ldots, X_n$ possessing expectations and constants $a_1, a_2,\ldots, a_n$,

$$E\left[\sum_{k=1}^{n} a_k X_k\right] = \sum_{k=1}^{n} a_k E[X_k] \quad ■ \tag{1.172}$$

Linearity is demonstrated by

$$E\left[\sum_{k=1}^{n} a_k X_k\right] = \int_{-\infty}^{\infty}\cdots\int_{-\infty}^{\infty}\left(\sum_{k=1}^{n} a_k x_k\right) f(x_1, x_2, \ldots, x_n)\, dx_1 dx_2 \cdots dx_n$$

$$= \sum_{k=1}^{n} \int_{-\infty}^{\infty}\cdots\int_{-\infty}^{\infty} a_k x_k f(x_1, x_2, \ldots, x_n)\, dx_1 dx_2 \cdots dx_n$$

$$= \sum_{k=1}^{n} a_k \int_{-\infty}^{\infty} x_k f_{X_k}(x_k)\, dx_k \tag{1.173}$$

$$= \sum_{k=1}^{n} a_k E[X_k]$$

1.6.5. Covariance

Moments can be generalized to collections of jointly distributed random variables. Key to the analysis of linear systems and random processes is the covariance. We are mainly concerned with bivariate moments.

Given two random variables X and Y and integers $p \geq 0$ and $q \geq 0$, the $(p + q)$-order *product moment* is defined by

$$\mu'_{pq} = E[X^p Y^q] = \int_{-\infty}^{\infty}\int_{-\infty}^{\infty} x^p y^q f(x, y)\, dydx \tag{1.174}$$

The $(p + q)$-order *central moment* is

$$\mu_{pq} = E[(X - \mu_X)^p (Y - \mu_Y)^q]$$

$$= \int_{-\infty}^{\infty}\int_{-\infty}^{\infty} (x - \mu_X)^p (y - \mu_Y)^q f(x, y)\, dydx \tag{1.175}$$

In both cases we assume that the defining integral is absolutely convergent.

The second-order central product moment μ_{11} is called the *covariance* and is given by

$$\mathrm{Cov}[X, Y] = E[(X - \mu_X)(Y - \mu_Y)] \tag{1.176}$$

We sometimes denote the covariance of X and Y by σ_{XY}^2. If it exists, the covariance is conveniently expressed as

$$\operatorname{Cov}[X, Y] = E[XY] - \mu_X\mu_Y \tag{1.177}$$

which can be seen by expanding the integral expression for the covariance.

If X and Y are independent, then

$$\begin{aligned} E[XY] &= \int_{-\infty}^{\infty}\int_{-\infty}^{\infty} xyf(x,y)\,dydx \\ &= \int_{-\infty}^{\infty}\int_{-\infty}^{\infty} xyf_X(x)f_Y(y)\,dydx \\ &= \int_{-\infty}^{\infty} xf_X(x)\,dx \int_{-\infty}^{\infty} yf_Y(y)\,dy \qquad (1.178) \\ &= E[X]E[Y] \end{aligned}$$

In particular, from Eq. 1.177, if X and Y are independent, then $\operatorname{Cov}[X, Y] = 0$.

The covariance provides a measure of the linear relationship between random variables; however, the deviations $X - \mu_X$ and $Y - \mu_Y$ are dependent on the units in which X and Y are measured. A normalized measure is given by the *correlation coefficient*

$$\rho_{XY} = \frac{\operatorname{Cov}[X,Y]}{\sigma_X\sigma_Y} \tag{1.179}$$

If $\rho_{XY} = 0$, then the random variables are said to be *uncorrelated*. If the variables are independent, then according to Eq. 1.178 they are uncorrelated. The converse, however, is not valid: uncorrelated variables need not be independent. The next theorem specifies the manner in which the correlation coefficient provides a measure of linearity between random variables. It is simply a statement of the Schwarz inequality in terms of the correlation coefficient.

Theorem 1.14. For any random variables X and Y,

$$-1 \le \rho_{XY} \le 1 \tag{1.180}$$

and $|\rho_{XY}| = 1$ if and only if there exist constants $a \neq 0$ and b such that

$$P(Y = aX + b) = 1 \qquad \blacksquare \tag{1.181}$$

According to the theorem, the probability mass of the joint distribution lies on a straight line if and only if $|\rho_{XY}| = 1$. The correlation coefficient lies between −1 and +1. The closer it is to either extreme, the more the mass is linearly concentrated. For $|\rho_{XY}| = 1$, Y is (up to possibly a set of probability zero) a linear function of X.

Theorem 1.15. If Y is given by the linear combination of the random variables $X_1, X_2, \ldots, X_n$ in Eq. 1.156, then

$$\mathrm{Var}[Y] = \sum_{j=1}^{n}\sum_{k=1}^{n} a_j a_k \mathrm{Cov}[X_j, X_k] \tag{1.182}$$

If $X_1, X_2, \ldots, X_n$ are uncorrelated, then all covariance terms vanish except for $j = k$ and

$$\mathrm{Var}[Y] = \sum_{k=1}^{n} a_k^2 \mathrm{Var}[X_k] \qquad \blacksquare \tag{1.183}$$

Equation 1.182 is demonstrated in the following manner:

$$\begin{aligned}
\mathrm{Var}[Y] &= E[Y^2] - E[Y]^2 \\
&= E\left[\left(\sum_{k=1}^{n} a_k X_k\right)^2\right] - \left(\sum_{k=1}^{n} a_k E[X_k]\right)^2 \\
&= E\left[\sum_{k=1}^{n}\sum_{j=1}^{n} a_k a_j X_k X_j\right] - \sum_{k=1}^{n}\sum_{j=1}^{n} a_k a_j E[X_k]E[X_j] \\
&= \sum_{k=1}^{n}\sum_{j=1}^{n} a_k a_j (E[X_k X_j] - E[X_k]E[X_j]) \qquad (1.184) \\
&= \sum_{j=1}^{n}\sum_{k=1}^{n} a_j a_k \mathrm{Cov}[X_j, X_k]
\end{aligned}$$

the last equality following from Eq. 1.177.

The *mean vector* and *covariance matrix* for the random vector $\mathbf{X} = (X_1, X_2,..., X_n)'$ are defined by

$$\boldsymbol{\mu} = \begin{pmatrix} E[X_1] \\ E[X_2] \\ \vdots \\ E[X_n] \end{pmatrix} \tag{1.185}$$

and

$$\mathbf{K} = E[(\mathbf{X} - \boldsymbol{\mu})(\mathbf{X} - \boldsymbol{\mu})'] = \begin{pmatrix} \sigma_{11}^2 & \sigma_{12}^2 & \cdots & \sigma_{1n}^2 \\ \sigma_{21}^2 & \sigma_{22}^2 & \cdots & \sigma_{2n}^2 \\ \vdots & \vdots & \ddots & \vdots \\ \sigma_{n1}^2 & \sigma_{n2}^2 & \cdots & \sigma_{nn}^2 \end{pmatrix} \tag{1.186}$$

where $\sigma_{ij}^2 = \mathrm{Cov}[X_i, X_j]$. The diagonal elements of $\mathbf{K}$ are the variances of $X_1, X_2,..., X_n$.

The covariance matrix $\mathbf{K}$ is real and symmetric. It is *nonnegative definite*, meaning that $\mathbf{v}'\mathbf{K}\mathbf{v} \geq 0$ for any n-vector $\mathbf{v}$. This is shown by the inequality

$$\begin{aligned} 0 &\leq E[|\mathbf{v}'(\mathbf{X} - \boldsymbol{\mu})|^2] \\ &= \mathbf{v}'E[(\mathbf{X} - \boldsymbol{\mu})(\mathbf{X} - \boldsymbol{\mu})']\mathbf{v} \\ &= \mathbf{v}'\mathbf{K}\mathbf{v} \end{aligned} \tag{1.187}$$

If $\mathbf{K}$ has eigenvalues $\lambda_1, \lambda_2,..., \lambda_n$, since $\mathbf{K}$ is real and symmetric, it has mutually orthogonal unit-length eigenvectors $\mathbf{e}_1, \mathbf{e}_2,..., \mathbf{e}_n$ corresponding to $\lambda_1, \lambda_2,..., \lambda_n$. If

$$\mathbf{E} = [\mathbf{e}_1 \ \mathbf{e}_2 \ \cdots \ \mathbf{e}_n] \tag{1.188}$$

is the matrix whose columns are $\mathbf{e}_1, \mathbf{e}_2,..., \mathbf{e}_n$, then E is *unitary*, meaning $\mathbf{E}^{-1} = \mathbf{E}'$, and

$$\mathbf{E}^{-1}\mathbf{K}\mathbf{E} = \begin{pmatrix} \lambda_1 & 0 & \cdots & 0 \\ 0 & \lambda_2 & \cdots & 0 \\ \vdots & \vdots & \ddots & \vdots \\ 0 & 0 & \cdots & \lambda_n \end{pmatrix} \tag{1.189}$$

meaning that K is *similar* to the diagonal matrix composed of the eigenvalues of K (in order down the diagonal). Finally, since K is real and symmetric, it is *positive definite*, meaning $\mathbf{v}'\mathbf{K}\mathbf{v} > 0$ for any nonzero n-vector $\mathbf{v}$, if and only if all eigenvalues are positive.

Covariance functions of random functions play important roles in the theory and application of random processes. We will have much to say about the covariance matrix in the context of finite random processes; however, we have introduced it here and discussed a few of its linear-algebraic properties for completeness, in particular, because we wish to introduce the multivariate normal distribution.

1.6.6. Multivariate Normal Distribution

To define the multivariate normal distribution, let K be a positive definite, real symmetric matrix and $\boldsymbol{\mu} = (\mu_1, \mu_2,..., \mu_n)'$ be an arbitrary vector. A random vector $\mathbf{X} = (X_1, X_2,..., X_n)'$ is said to have a *multivariate normal (Gaussian) distribution* if it possesses the multivariate density

$$f(\mathbf{x}) = \frac{1}{\sqrt{(2\pi)^n \det[\mathbf{K}]}} \exp\left[-\frac{1}{2}(\mathbf{x}-\boldsymbol{\mu})'\mathbf{K}^{-1}(\mathbf{x}-\boldsymbol{\mu})\right] \tag{1.190}$$

where $\mathbf{x} = (x_1, x_2,..., x_n)'$. When $n = 1$, $\mathbf{K} = (\sigma^2)$, $\det[\mathbf{K}] = \sigma^2$, and $\boldsymbol{\mu}$ is the mean.

The following properties hold for the multivariate normal distribution: (1) $\boldsymbol{\mu}$ is the mean vector; (2) K is the covariance matrix; (3) $X_1, X_2,..., X_n$ are independent if and only if K is diagonal (which is the case considered in Example 1.17); and (4) the marginal densities are normally distributed.

In the special case where $n = 2$,

$$\mathbf{K} = \begin{pmatrix} \sigma_X^2 & \rho\sigma_X\sigma_Y \\ \rho\sigma_X\sigma_Y & \sigma_Y^2 \end{pmatrix} \tag{1.191}$$

and the density is given by

$$f(x,y) = \frac{\exp\left\{-\frac{1}{2(1-\rho^2)}\left[\left(\frac{x-\mu_X}{\sigma_X}\right)^2 - 2\rho\left(\frac{x-\mu_X}{\sigma_X}\right)\left(\frac{y-\mu_Y}{\sigma_Y}\right) + \left(\frac{y-\mu_Y}{\sigma_Y}\right)^2\right]\right\}}{2\pi\sigma_X\sigma_Y\sqrt{1-\rho^2}} \tag{1.192}$$

where the marginal random variables X and Y are normally distributed and μ_X, μ_Y, σ_X^2, σ_Y^2, and ρ are the mean of X, mean of Y, variance of X, variance of

Y, and correlation coefficient, respectively. These properties can be demonstrated by performing the appropriate integrations. It follows immediately from the form of the joint density that X and Y are independent if and only if they are uncorrelated (whereas, generally, uncorrelatedness does not imply independence).

In Example 1.22 we have seen that a linear combination of independent normally distributed random variables is normally distributed and we have found the mean and variance of the linear combination. Now consider the more general situation where $\mathbf{X}$ is a random vector possessing an n-dimensional multivariate normal distribution with mean vector $\boldsymbol{\mu}$ and covariance matrix $\mathbf{K}$, $\mathbf{A}$ is a nonsingular $n \times n$ matrix, and $\mathbf{U} = \mathbf{AX}$ is the output random vector. Using Theorem 1.11, we demonstrate that $\mathbf{U}$ possesses a multivariate normal distribution with mean vector $\mathbf{A}\boldsymbol{\mu}$ and covariance matrix $\mathbf{AKA}'$. The transformation $\mathbf{u} = \mathbf{Ax}$ is inverted by $\mathbf{x} = \mathbf{A}^{-1}\mathbf{u}$. Hence, the Jacobian of the mapping $\mathbf{u} \to \mathbf{x}$ is

$$J(\mathbf{x}; \mathbf{u}) = \det[\mathbf{A}^{-1}] = \det[\mathbf{A}]^{-1} \tag{1.193}$$

and, according to Eqs. 1.168 and 1.190,

$$\begin{aligned} f_{\mathbf{U}}(\mathbf{u}) &= \frac{1}{|\det[\mathbf{A}]|\sqrt{(2\pi)^n \det[\mathbf{K}]}} \exp\left[-\frac{1}{2}(\mathbf{A}^{-1}\mathbf{u}-\boldsymbol{\mu})'\mathbf{K}^{-1}(\mathbf{A}^{-1}\mathbf{u}-\boldsymbol{\mu})\right] \\ &= \frac{1}{\sqrt{(2\pi)^n \det[\mathbf{AKA}']}} \exp\left[-\frac{1}{2}(\mathbf{u}-\mathbf{A}\boldsymbol{\mu})'(\mathbf{AKA}')^{-1}(\mathbf{u}-\mathbf{A}\boldsymbol{\mu})\right] \end{aligned} \tag{1.194}$$

where the second equality follows from the matrix relations

$$|\det[\mathbf{A}]|\det[\mathbf{K}]^{1/2} = \det[\mathbf{AKA}']^{1/2} \tag{1.195}$$

$$(\mathbf{A}^{-1}\mathbf{u} - \boldsymbol{\mu})'\mathbf{K}^{-1}(\mathbf{A}^{-1}\mathbf{u} - \boldsymbol{\mu}) = (\mathbf{u} - \mathbf{A}\boldsymbol{\mu})'(\mathbf{AKA}')^{-1}(\mathbf{u} - \mathbf{A}\boldsymbol{\mu}) \tag{1.196}$$

1.7. Laws of Large Numbers

Some of the most fundamental theorems of probability concern limiting properties for sums of random variables. This section introduces various types of convergence used in probability theory and discusses laws of large numbers and the central limit theorem. The weak law of large numbers is proven and the strong law is stated without proof. Three forms of the central limit theorem for sequences of independent random variables are discussed, all without proof. We state the form for identically distributed random variables that is typically stated in statistics books, and also discuss Liapounov's and Lindberg's conditions for the central limit theorem to apply to independent random variables that are not necessarily identically distributed. While these may be outside the ordinary sphere of application, when interpreted for sequences of uniformly bounded random variables they help to explain the naturalness of the central limit theorem.

1.7.1. Weak Law of Large Numbers

Averages play an important role in both probability and statistics. From an empirical perspective, if a random variable X is observed n times (meaning rigorously that n identically distributed random variables are observed) and the numerical average of the observations is taken, we would like to quantify the degree to which that average can be taken as an estimate of the mean of X. More generally, what is the relationship between an average of random variables and the average of their means?

Let $X_1, X_2,...$ be random variables defined on a common sample space and possessing finite second moments. Let their means be $\mu_1, \mu_2,...$, respectively, and let

$$Y_n = \frac{1}{n}\sum_{k=1}^{n} X_k \tag{1.197}$$

be the arithmetic mean of $X_1, X_2,..., X_n$. Owing to the linearity of expectation,

$$E[Y_n] = \frac{1}{n}\sum_{k=1}^{n} \mu_k \tag{1.198}$$

and, according to Theorem 1.15,

$$\mathrm{Var}[Y_n] = \frac{1}{n^2}\sum_{j=1}^{n}\sum_{k=1}^{n} \mathrm{Cov}[X_j, X_k] \tag{1.199}$$

For $\varepsilon > 0$, applying Chebyshev's inequality to Y_n yields

$$P(|Y_n - E[Y_n]| \geq \varepsilon) \leq \frac{1}{\varepsilon^2 n^2}\sum_{j=1}^{n}\sum_{k=1}^{n} \mathrm{Cov}[X_j, X_k] \tag{1.200}$$

The probability on the left will converge to 0 as $n \to \infty$ if the sum of the covariances divided by n^2 converges to 0 as $n \to \infty$. This observation gives rise to Markov's form of the *weak law of large numbers*.

Theorem 1.16. Suppose X_1, X_2,... are random variables defined on a common sample space, possessing finite second moments, and having means μ_1, μ_2,..., respectively. If

$$\lim_{n\to\infty} \frac{1}{n^2} \sum_{j=1}^{n} \sum_{k=1}^{n} \mathrm{Cov}[X_j, X_k] = 0 \tag{1.201}$$

then, for any $\varepsilon > 0$,

$$\lim_{n\to\infty} P\left(\left|\frac{1}{n}\sum_{k=1}^{n} X_k - \frac{1}{n}\sum_{k=1}^{n} \mu_k\right| \geq \varepsilon\right) = 0 \qquad \blacksquare \tag{1.202}$$

The theorem asserts that, given an arbitrarily small quantity $\varepsilon > 0$, the probability that the difference between the average of the random variables and the average of their means exceeds ε can be made arbitrarily small by averaging a sufficient number of random variables.

The weak law of large numbers represents a type of convergence involving a probability measure. In general, a sequence of random variables Z_1, Z_2,... is said to *converge in probability* to random variable Z if, for any $\varepsilon > 0$,

$$\lim_{n\to\infty} P\left(|Z_n - Z| \geq \varepsilon\right) = 0 \tag{1.203}$$

The weak law of large numbers asserts that the difference between the average of a sequence of random variables and the average of their means converges to 0 in probability if the limit of Eq. 1.201 holds.

The weak law of large numbers is usually employed for specific cases. One particular case occurs when X_1, X_2,... are uncorrelated. If σ_1^2, σ_2^2,... are the variances of X_1, X_2,..., respectively, and there exists a bound M such that $\sigma_k^2 \leq M$ for $k = 1, 2,...$, then

$$\frac{1}{n^2} \sum_{j=1}^{n} \sum_{k=1}^{n} \mathrm{Cov}[X_j, X_k] = \frac{1}{n^2} \sum_{k=1}^{n} \sigma_k^2 \leq \frac{M}{n} \tag{1.204}$$

Consequently, the limit of Eq. 1.201 is satisfied and the weak law holds for X_1, X_2,.... If X_1, X_2,... are uncorrelated and there exists a bound M such that $\sigma_k^2 \leq M$, and, moreover, X_1, X_2,... possess a common mean μ, then the weak law states that

$$\lim_{n\to\infty} P\left(\left|\frac{1}{n}\sum_{k=1}^{n} X_k - \mu\right| \geq \varepsilon\right) = 0 \tag{1.205}$$

This is its most commonly employed form. If $X_1, X_2, \ldots$ are independent and identically distributed with common mean μ and finite variance σ^2, then they are uncorrelated, Eq. 1.204 holds with $M = \sigma^2$, and, as stated in Eq. 1.205, the arithmetic mean of the random variables converges in probability to their common mean.

The type of convergence that will play the dominant role in our study of random processes is defined via the second moments of the differences between the random variables of the sequence and the limiting random variable. The sequence of random variables $Z_1, Z_2, \ldots$ is said to *converge in the mean-square* to random variable Z if

$$\lim_{n\to\infty} E[|Z_n - Z|^2] = 0 \tag{1.206}$$

It is an immediate consequence of Chebyshev's inequality that, if Z_n converges to Z in the mean-square, then Z_n converges to Z in probability. The converse, however, is not valid: it is possible for a sequence of random variables to converge in probability but not in the mean-square.

***Example* 1.25.** Consider an infinite sequence of independent trials and assume that a certain event A occurs with probability p_n on trial n. If the random variable X_n is defined by $X_n = 1$ if event A occurs on trial n and $X_n = 0$ if event A does not occur on trial n, then X_n is a binomial random variable with parameters 1 and p_n. Therefore, $E[X_n] = p_n$ and $\mathrm{Var}[X_n] = p_n(1 - p_n)$. In this context, the random variable Y_n of Eq. 1.197 is the relative frequency of event A on the first n trials. Since the random variables constituting the sequence are uncorrelated and $\mathrm{Var}[X_n] \leq 1$ for all n, the weak law of large numbers applies and the relative frequency converges in probability to the arithmetic mean of the trial probabilities of event A. This proposition was proved by Poisson. Now, if the trial probabilities are constant, $p_n = p$, then for any $\varepsilon > 0$,

$$\lim_{n\to\infty} P(|Y_n - p| \geq \varepsilon) = 0$$

and the relative frequency converges in probability to the common probability of event A. This simplified result was proved by Jacob Bernoulli and represents the first law of large numbers in probability theory. ■

1.7.2. Strong Law of Large Numbers

A random variable is a real-valued function defined on a sample space. Therefore convergence of a sequence of random variables can be considered from the perspective of ordinary function convergence. If $(S, \mathcal{E}, P)$ is a probability space, $Z_n: S \to \Re$ is a random variable for $n = 1, 2, \ldots,$ and $Z: S \to \Re$, then, as functions, Z_n converges to Z if, for any $w \in S$,

$$\lim_{n\to\infty} Z_n(w) = Z(w) \tag{1.207}$$

If two random variables differ only on a set of probability zero, then they are identically distributed. Hence, it is sufficient to have the limit of Eq. 1.207 hold everywhere in S except for an event of probability zero. We make the following definition: Z_n *converges almost surely* to Z if there exists an event G such that $P(G) = 0$ and the limit of Eq. 1.207 holds for all $w \in S - G$. This means that, for any $w \in S - G$ and any $\varepsilon > 0$, there exists a positive integer $N_{w,\varepsilon}$ such that, for $n \geq N_{w,\varepsilon}$,

$$|Z_n(w) - Z(w)| < \varepsilon \tag{1.208}$$

Almost-sure convergence is often expressed by

$$P\left(\lim_{n\to\infty} Z_n = Z\right) = 1 \tag{1.209}$$

To investigate the relationship between convergence in probability and almost-sure convergence, we introduce another type of convergence, which generalizes the classical concept of uniform convergence for functions: Z_n *converges almost uniformly* to Z if, for any $\delta > 0$, there exists an event F_δ such that $P(F_\delta) < \delta$ and Z_n converges uniformly to Z on $S - F_\delta$. Uniform convergence on $S - F_\delta$ means that, for any $\varepsilon > 0$, there exists a positive integer $N_{\delta,\varepsilon}$ such that, for $n \geq N_{\delta,\varepsilon}$, Eq. 1.208 holds for all $w \in S - F_\delta$. It is a fundamental property (*Egoroff's theorem*) of random variables that almost-sure and almost-uniform convergence are equivalent (under the assumption that we restrict our attention to random variables that are finite, except perhaps on sets of zero probability). Consequently, to show that almost-sure convergence implies convergence in probability, we need only show that almost-uniform convergence implies convergence in probability. To do so, consider arbitrary $\delta > 0$ and $\varepsilon > 0$, and choose F_δ and $N_{\delta,\varepsilon}$ so that Eq. 1.208 holds for $n \geq N_{\delta,\varepsilon}$ and $w \in S - F_\delta$. Then, for $n \geq N_{\delta,\varepsilon}$,

$$\begin{aligned} P(|Z_n - Z| \geq \varepsilon) &= P(\{w \in S: |Z_n(w) - Z(w)| \geq \varepsilon\}) \\ &= 1 - P(\{w \in S: |Z_n(w) - Z(w)| < \varepsilon\}) \end{aligned} \tag{1.210}$$

$$\leq 1 - P(S - F_\delta)$$

$$= P(F_\delta)$$

which is less than δ. Since δ has been chosen arbitrarily, the limit of Eq. 1.203 holds for arbitrary ε and Z_n converges to Z in probability. The converse is not true: convergence in probability does not imply almost-sure convergence. Consequently, almost-sure convergence is a stronger form of convergence than convergence in probability. We now state Kolmogorov's *strong law of large numbers*.

Theorem 1.17. If $X_1, X_2,...$ are independent and identically distributed random variables possessing finite mean μ, then

$$\lim_{n\to\infty} \frac{1}{n}\sum_{k=1}^{n} X_k = \mu \quad \text{(almost surely)} \qquad \blacksquare \qquad (1.211)$$

The strong law is equivalently written as

$$P\left(\lim_{n\to\infty} \frac{1}{n}\sum_{k=1}^{n} X_k = \mu\right) = 1 \qquad (1.212)$$

The conclusion of Theorem 1.17 is stronger than the conclusion of Theorem 1.16, but so too is the hypothesis.

1.7.3. Central Limit Theorem

Inclusion probabilities of a random variable are determined by its probability distribution function (or density). If we are interested in approximating probabilities for a random variable X from probabilities of a sequence of random variables $X_1, X_2,...$ converging (in some sense) to X, then we need to be concerned about the relationship between the probability distribution functions of $X_1, X_2,...$ and the probability distribution function of X. X_n is said to *converge in law* (*converge in distribution*) to X if, for any point a at which the probability distribution function F_X of X is continuous,

$$\lim_{n\to\infty} F_{X_n}(a) = F_X(a) \qquad (1.213)$$

Equivalently, if F_X is continuous at a and b, $a < b$, then

$$\lim_{n\to\infty}\left(F_{X_n}(b)-F_{X_n}(a)\right)=F_X(b)-F_X(a) \tag{1.214}$$

which, in terms of interval probabilities, means that

$$\lim_{n\to\infty} P(a < X_n \le b) = P(a < X \le b) \tag{1.215}$$

If X is a continuous random variable, then its probability distribution function is continuous and the preceding limit can be expressed in terms of densities as

$$\lim_{n\to\infty}\int_a^b f_{X_n}(x)\,dx = \int_a^b f_X(x)\,dx \tag{1.216}$$

Relative to our original point concerning approximation, if X_n converges to X in law, then, for large n,

$$P(a < X_n \le b) \approx P(a < X \le b) \tag{1.217}$$

and probabilities of X can be used to approximate probabilities of X_n.

As with laws of large numbers, our concern is with limits of averages, or sums, of sequences of independent random variables. Here, however, our interest lies in the limiting distributions of the averages. The manner in which distributions of averages converge to the standard normal distribution (that is, how averages of nonnormal distributions converge in law to the standard normal distribution) explains the special role of the normal distribution in probability and statistics, a role recognized by Gauss. We first state the most commonly employed form of the *central limit theorem* and subsequently discuss more general formulations. The first form of the theorem is stated in terms of standardized random variables.

For a random variable X with mean μ and variance σ^2, define the *standardized* random variable $(X - \mu)/\sigma$. The standardized variable has zero mean and unit variance. If $X_1, X_2,...$ are independent, identically distributed random variables possessing common mean μ and variance σ^2, then the average Y_n of Eq. 1.197 has mean μ and variance σ^2/n. Hence, the standardized variable corresponding to Y_n is $(Y_n - \mu)\sqrt{n}/\sigma$. The central limit theorem asserts that this standardized average converges in law to the standard normal random variable.

Theorem 1.18. If $X_1, X_2,...$ are independent, identically distributed random variables possessing common mean μ and variance σ^2, then, for any z,

$$\lim_{n\to\infty} P\left(\frac{\frac{1}{n}\sum_{k=1}^{n} X_k - \mu}{\frac{\sigma}{\sqrt{n}}} \le z\right) = \frac{1}{\sqrt{2\pi}}\int_{-\infty}^{z} e^{-y^2/2}\,dy \quad \blacksquare \tag{1.218}$$

If we denote the standardized average by

$$Z_n = \frac{\frac{1}{n}\sum_{k=1}^{n} X_k - \mu}{\sigma/\sqrt{n}} \tag{1.219}$$

the standard normal random variable by Z, and the probability distribution function of the standard normal variable by $\Phi(z)$, then the central limit theorem can be written simply as

$$\lim_{n\to\infty} Z_n = Z \qquad \text{(in law)} \tag{1.220}$$

or, in terms of interval probabilities, as

$$\lim_{n\to\infty} P(a < Z_n \le b) = \Phi(b) - \Phi(a) \tag{1.221}$$

Example **1.26.** The individual trials of the binomial distribution with success probability p are independent and identically distributed. Let $X_k = 1$ if there is a success on trial k and $X_k = 0$ if there is a failure on trial k. The binomial random variable for n trials is given by

$$X^n = \sum_{k=1}^{n} X_k$$

$X_1, X_2, \ldots$ possess common mean p and variance $p(1-p)$, X^n has mean np and variance $np(1-p)$, the average X^n/n has mean p and variance $p(1-p)/n$, and the central limit theorem states that

$$\lim_{n\to\infty} \frac{X^n/n - p}{\sqrt{p(1-p)/n}} = Z \quad \text{(in law)}$$

In terms of interval probabilities,

$$\lim_{n\to\infty} P\left(a < \frac{X^n - np}{\sqrt{np(1-p)}} \le b \right) = \Phi(b) - \Phi(a)$$

This is the classical *DeMoivre-Laplace* theorem. It can be used to estimate binomial probabilities via the standard normal distribution. Writing the preceding limit as an approximation for large n and changing variables yields

$$P(a < X^n \le b) \approx \Phi\left(\frac{b - np}{\sqrt{np(1-p)}} \right) - \Phi\left(\frac{a - np}{\sqrt{np(1-p)}} \right) \qquad \blacksquare$$

There are more general conditions under which the central limit theorem applies to a sequence of independent random variables X_1, X_2,... that need not be identically distributed. The most celebrated of these, which we state next, is due to Liapounov and for this reason the central limit theorem is often called *Liapounov's theorem*. To simplify expressions, we employ the following notation for the remainder of the section, under the assumption that X_1, X_2,... are independent random variables with means μ_1, μ_2,... and variances σ_1^2, σ_2^2,..., respectively. Let

$$S_n = \sum_{k=1}^{n} X_k \tag{1.222}$$

$$m_n = \sum_{k=1}^{n} \mu_k \tag{1.223}$$

$$s_n^2 = \sum_{k=1}^{n} \sigma_k^2 = \sum_{k=1}^{n} E[|X_k - \mu_k|^2] \tag{1.224}$$

S_n is the sum of the first n random variables in the sequence, m_n is the mean of S_n, and, because the random variables are independent, s_n^2 is the variance of S_n.

Theorem 1.19. Suppose X_1, X_2,... are independent random variables with means μ_1, μ_2,... and variances σ_1^2, σ_2^2,..., respectively. If there exists $\delta > 0$ such that $E[|X_n - \mu_n|^{2+\delta}]$ is finite for $n = 1, 2,...$ and

$$\lim_{n\to\infty} \frac{1}{s_n^{2+\delta}} \sum_{k=1}^{n} E[|X_k - \mu_k|^{2+\delta}] = 0 \tag{1.225}$$

then

$$\lim_{n\to\infty} \frac{S_n - m_n}{s_n} = Z \qquad \text{(in law)} \qquad \blacksquare \tag{1.226}$$

The limiting condition of Eq. 1.225, known as *Liapounov's condition*, represents the profound part of the theorem. It is not easily apprehended; however, from a practical perspective, the theorem has an easily applied corollary: if there exists a fixed bound C such that $|X_n| \le C$ for all n and

$$\lim_{n\to\infty} s_n^2 = \sum_{k=1}^{\infty} \sigma_k^2 = \infty \tag{1.227}$$

then the limiting conclusion of the central limit theorem applies. In fact, the uniform boundedness of the random variables implies

$$\begin{aligned} E[|X_k - \mu_k|^{2+\delta}] &= E[|X_k - \mu_k|^{\delta}\,|X_k - \mu_k|^2] \\ &\le 2^{\delta} C^{\delta} E[|X_k - \mu_k|^2] \end{aligned} \tag{1.228}$$

Hence,

$$\frac{1}{s_n^{2+\delta}} \sum_{k=1}^{n} E[|X_k - \mu_k|^{2+\delta}] \le \frac{2^{\delta} C^{\delta}}{s_n^{\delta} s_n^2} \sum_{k=1}^{n} E[|X_k - \mu_k|^2] = \frac{2^{\delta} C^{\delta}}{s_n^{\delta}} \tag{1.229}$$

which, according to the assumption of Eq. 1.227, converges to 0 as $n \to \infty$. While we might model random variables by unbounded distributions, real-world data are uniformly bounded. Moreover, divergence of the sum of the variances can be expected from real-world random phenomena since, for the sum of the variances to converge, it would be necessary that $\mathrm{Var}[X_n] \to 0$ as $n \to \infty$. Consequently, convergence of the distribution of $(S_n - m_n)/s_n$ to the standard normal distribution can be expected for many natural phenomena.

There is a more general sufficient condition than that of Liapounov for the central limit theorem. *Lindberg's condition* states that the central limit theorem (Eq. 1.226) holds for a sequence of independent random variables if, for any $\varepsilon > 0$,

$$\lim_{n\to\infty} \frac{1}{s_n^2} \sum_{k=1}^{n} \int_{\{x:|x-\mu_k|>\varepsilon s_n\}} (x - \mu_k)^2 f_k(x)\, dx = 0 \tag{1.230}$$

where $f_k(x)$ is the density for X_k. Not only is Lindberg's condition sufficient, with a slight modification in the statement of the theorem, it is also necessary. Specifically, we have the following restatement of the central limit theorem: for a sequence of independent random variables, there is convergence in law according to Eq. 1.226 and

$$\lim_{n\to\infty} \frac{\max_{k\leq n} \sigma_k}{s_n} = 0 \tag{1.231}$$

if and only if Lindberg's condition holds. The condition of Eq. 1.231, which when added to convergence in law of $(S_n - m_n)/s_n$ to Z, makes Lindberg's condition necessary and sufficient, is due to Feller and states that the maximum variance among the random variables in the sequence up to n becomes negligible relative to the variance of the sum as $n \to \infty$.

In conclusion, one should recognize the long history of convergence theorems in the theory of probability. The great scientific names associated with the various theorems and the degree to which they have remained a central theme are testimony to their importance. The central limit theorem alone shows the profound relationship between theoretical mathematics, applied science, and the philosophy of nature.

1.8. Parametric Estimation via Random Samples

If we model a random variable by some probability distribution (normal, gamma, etc.), there remains the task of specifying the distributional parameters in such a way that the probability mass reflects the distribution of the phenomena described by the random variable. For instance, suppose X is assumed to be gamma distributed. Then X has a density $f(x; \alpha, \beta)$, where the notation is meant to imply that the parameters α and β are unknown and need to be particularized to X. From observations of random variables related to X, we wish to estimate the relevant values of the parameters.

1.8.1. Random-Sample Estimators

A common way of proceeding is to observe a set of random variables X_1, $X_2, \ldots, X_n$ that are both independent and identically distributed to X. Such a set of random variables is called a *random sample* for X. If X has density $f(x)$, then, owing to independence and identical distribution, the joint density of $X_1, X_2, \ldots, X_n$ is given by

$$f(x_1, x_2, \ldots, x_n) = \prod_{k=1}^{n} f(x_k) \tag{1.232}$$

Now suppose X has density $f(x; \theta)$ with unknown parameter θ. A function of a random sample $X_1, X_2, \ldots, X_n$, say,

$$\hat{\theta} = \hat{\theta}(X_1, X_2, \ldots, X_n) \tag{1.233}$$

is needed to provide estimation of θ based on $X_1, X_2, \ldots, X_n$. $\hat{\theta}(X_1, X_2, \ldots, X_n)$ is itself a random variable and is called an *estimator* of θ. Given sample values $x_1, x_2, \ldots, x_n$ from observations of $X_1, X_2, \ldots, X_n$, the functional value $\hat{\theta}(x_1, x_2, \ldots, x_n)$ provides an *estimate* of θ for the particular sample. The function rule defining $\hat{\theta}(X_1, X_2, \ldots, X_n)$ is called the *estimation rule* for the estimator. $\hat{\theta}(X_1, X_2, \ldots, X_n)$ is called a *statistic* if the estimation rule defining it is free of unknown parameters (which does not mean that the distribution of $\hat{\theta}$ is free of unknown parameters). We assume that estimators are statistics.

Two sets of sample values, $\{x_1, x_2, \ldots, x_n\}$ and $\{z_1, z_2, \ldots, z_n\}$, obtained from observation of a random sample $X_1, X_2, \ldots, X_n$ give rise to two, almost certainly distinct, estimates of θ, $\hat{\theta}(x_1, x_2, \ldots, x_n)$ and $\hat{\theta}(z_1, z_2, \ldots, z_n)$. Is one of the estimates better than the other? Is either a good estimate? As posed, these questions are not meaningful. Goodness must be posed in terms of the estimator (estimation rule). Whether an estimator provides "good" estimates depends on positing some measure of goodness and then examining the estimator relative to the measure of goodness.

An estimator $\hat{\theta}$ of θ is said to be *unbiased* if $E[\hat{\theta}] = \theta$. Implicit in the definition is that $E[\hat{\theta}] = \theta$ regardless of the value of θ. $E[\hat{\theta}] - \theta$ is called the *bias* of $\hat{\theta}$. Since $\hat{\theta} = \hat{\theta}(X_1, X_2, \ldots, X_n)$, $\hat{\theta}$ depends on the sample size n. In many circumstances an estimator is biased but the bias diminishes as the sample size increases. Estimator $\hat{\theta}$ is said to be *asymptotically unbiased* if $E[\hat{\theta}] \to \theta$ as $n \to \infty$. Unbiasedness is a desirable property because it means that on average the estimator provides the desired parameter. However, this may be of little practical value if the variance of the estimator is too great.

To desire *precision* in an estimator is to wish the estimator to be within some tolerance of the parameter. Since the estimator is random, this cannot be guaranteed; however, for $r > 0$, we can consider the probability that the estimator is within r of the parameter, namely, $P(|\hat{\theta} - \theta| < r)$. $\hat{\theta}$ is said to be a *consistent* estimator of θ if, for any $r > 0$,

$$\lim_{n \to \infty} P(|\hat{\theta} - \theta| < r) = 1 \tag{1.234}$$

An equivalent way of stating the matter is that $\hat{\theta}$ is a consistent estimator of θ if and only if $\hat{\theta}$ converges in probability to θ.

If $\hat{\theta}$ is unbiased, then, according to Chebyshev's inequality,

$$P(|\hat{\theta}-\theta|<r)\geq 1-\frac{\mathrm{Var}[\hat{\theta}]}{r^2} \tag{1.235}$$

Consequently, if $\hat{\theta}$ is unbiased and $\mathrm{Var}[\hat{\theta}] \to 0$ as $n \to \infty$, then $\hat{\theta}$ is a consistent estimator. This proposition can be strengthened: $\hat{\theta}$ is a consistent estimator of θ if $\hat{\theta}$ is an asymptotically unbiased estimator of θ and $\mathrm{Var}[\hat{\theta}] \to 0$ as $n \to \infty$. Indeed, asymptotic unbiasedness implies that $|E[\hat{\theta}] - \theta| < r/2$ for sufficiently large n, so that

$$\begin{aligned}
P(|\hat{\theta}-\theta|\geq r) &\leq P(|\hat{\theta}-E[\hat{\theta}]|+|E[\hat{\theta}]-\theta|\geq r)\\
&\leq P\left(\left(|\hat{\theta}-E[\hat{\theta}]|\geq\frac{r}{2}\right)\cup\left(|\theta-E[\hat{\theta}]|\geq\frac{r}{2}\right)\right)\\
&= P\left(|\hat{\theta}-E[\hat{\theta}]|\geq\frac{r}{2}\right)\\
&\leq \frac{4\mathrm{Var}[\hat{\theta}]}{r^2}
\end{aligned} \tag{1.236}$$

for sufficiently large n (the last inequality resulting from Chebyshev's inequality).

1.8.2. Sample Mean and Sample Variance

Given a random sample $X_1, X_2,\ldots, X_n$ for the random variable X possessing mean μ, a commonly employed estimator of μ is the *sample mean*

$$\overline{X}=\frac{X_1+X_2+\cdots+X_n}{n} \tag{1.237}$$

The distribution of the sample mean (as a random variable) is called the *sampling distribution of the mean*. Generically, the sample-mean estimation rule is the function

$$\hat{\theta}(\xi_1,\xi_2,\ldots,\xi_n)=\frac{\xi_1+\xi_2+\cdots+\xi_n}{n} \tag{1.238}$$

Applied to the random sample $X_1, X_2,\ldots, X_n$, this estimation rule gives the sample-mean estimator. If sample values $x_1, x_2,\ldots, x_n$ are observed and the sample-mean estimator is used, then the estimation rule applied to $x_1, x_2, \ldots, x_n$ yields an estimate of μ,

$$\bar{x} = \frac{x_1 + x_2 + \cdots + x_n}{n} \tag{1.239}$$

called an *empirical mean*.

The sample mean is unbiased, $E[\bar{X}] = \mu$, and, by Theorem 1.15, $\mathrm{Var}[\bar{X}] = \sigma^2/n$. Since the sample mean is unbiased and its variance tends to 0 as $n \to \infty$, it is a consistent estimator of the mean. Consistency of the sample mean is a form of the weak law of large numbers for the special case of a random sample. In fact, the complementary form of Eq. 1.200 applied to the special case of an average of a random sample is given by

$$P(|\bar{X} - \mu| < r) \geq 1 - \frac{\sigma^2}{nr^2} \tag{1.240}$$

which is itself the complementary form of Chebyshev's inequality given in Eq. 1.75 as applied to the sample mean.

Not only does the sample mean converge to the mean in probability, but according to the strong law of large numbers, it converges to the mean almost surely. Furthermore, according to the central limit theorem, the standardized version of the sample mean converges in law to the standard normal random variable. Consequently, for large n,

$$P\left(a < \frac{\bar{X} - \mu}{\sigma/\sqrt{n}} < b\right) \approx \Phi(b) - \Phi(a) \tag{1.241}$$

where Φ(z) is the probability distribution function for the standard normal variable. This approximation is used to compute approximate probabilities concerning the sample mean via the standard normal probability distribution.

The most common variance estimator is the *sample variance*, which for a random sample $X_1, X_2, \ldots, X_n$ arising from the random variable X, is defined by

$$S^2 = \frac{1}{n-1}\sum_{k=1}^{n}(X_k - \bar{X})^2 \tag{1.242}$$

The variance is an unbiased estimator of the variance:

$$E[S^2] = \frac{1}{n-1} E\left[\sum_{k=1}^{n}(X_k - \bar{X})^2\right]$$

$$= \frac{1}{n-1} E\left[\sum_{k=1}^{n} (X_k - \mu)^2 - n(\overline{X} - \mu)^2\right]$$

$$= \frac{1}{n-1}\left(\sum_{k=1}^{n} E[(X_k - \mu)^2] - nE[(\overline{X} - \mu)^2]\right) \tag{1.243}$$

$$= \frac{1}{n-1}\left(\sum_{k=1}^{n} \mathrm{Var}[X_k] - n\mathrm{Var}[\overline{X}]\right)$$

$$= \mathrm{Var}[X]$$

where the second equality results from some algebraic manipulation and the last from the identical distribution of $X_1, X_2, \ldots, X_n$ and the fact that $\mathrm{Var}[\overline{X}] = \mathrm{Var}[X]/n$.

A basic theorem of statistics states that, for a random sample $X_1, X_2, \ldots, X_n$ arising from a normally distributed random variable X with mean μ and variance σ^2, $\overline{X}$ and S^2 are independent and $(n - 1)S^2/\sigma^2$ possesses a gamma distribution with $\alpha = (n - 1)/2$ and $\beta = 2$. Since the variance of the gamma distribution is $\alpha\beta^2$,

$$\mathrm{Var}[S^2] = \mathrm{Var}\left[\frac{\sigma^2}{n-1}\left(\frac{(n-1)S^2}{\sigma^2}\right)\right] = \frac{2\sigma^4}{n-1} \tag{1.244}$$

Hence, $\mathrm{Var}[S^2] \to 0$ as $n \to \infty$. Since S^2 is an unbiased estimator, for a normal random variable the sample variance is a consistent estimator of the variance. Practically, however, there is a difficulty. If the variance of X is large, then the variance of the sample variance will be large because the original variance is squared and then doubled in the numerator of $\mathrm{Var}[S^2]$. The sample size may have to be prohibitively large to obtain a sufficiently small variance of the sample variance. Specifically, the complementary form of Chebyshev's inequality takes the form

$$P(|S^2 - \sigma^2| < r) \geq 1 - \frac{2\sigma^4}{(n-1)r^2} \tag{1.245}$$

The precision of the sample variance compared with the precision of the sample mean can be seen by comparing Eqs. 1.245 and 1.240.

1.8.3. Minimum-Variance Unbiased Estimators

Application of Chebyshev's inequality in Eq. 1.235 yields a lower bound on the precision $P(|\hat{\theta} - \theta| < r)$ in terms of the variance of $\hat{\theta}$ under the assumption that $\hat{\theta}$ is an unbiased estimator. From a limiting perspective, if $\text{Var}[\hat{\theta}] \to 0$ as $n \to \infty$, then $\hat{\theta}$ is a consistent estimator, meaning $\hat{\theta} \to \theta$ in probability as $n \to \infty$. From the standpoint of comparing two unbiased estimators of θ, the one with smaller variance gives a greater lower bound.

Another approach is to consider the mean-square error, $E[|\hat{\theta} - \theta|^2]$, of $\hat{\theta}$ as an estimator of θ. Instead of consistency, we can ask whether $\hat{\theta}$ converges to θ in the mean-square; that is, does

$$\lim_{n\to\infty} E[|\hat{\theta} - \theta|^2] = 0 \tag{1.246}$$

The mean-square error can be expanded in terms of the variance and bias as

$$\begin{aligned} E[|\hat{\theta} - \theta|^2] &= E[\hat{\theta}^2] - 2E[\hat{\theta}\,\theta] + E[\theta^2] \\ &= E[\hat{\theta}^2] - E[\hat{\theta}]^2 + E[\hat{\theta}]^2 - 2\theta E[\hat{\theta}] + \theta^2 \\ &= \text{Var}[\hat{\theta}] + (E[\hat{\theta}] - \theta)^2 \end{aligned} \tag{1.247}$$

If $\hat{\theta}$ is not asymptotically unbiased, then $(E[\hat{\theta}] - \theta)^2$ does not converge to 0 and $\hat{\theta}$ cannot converge to θ in the mean-square: if $\hat{\theta}$ is asymptotically unbiased, then

$$\lim_{n\to\infty} E[|\hat{\theta} - \theta|^2] = \lim_{n\to\infty} \text{Var}[\hat{\theta}] \tag{1.248}$$

and $\hat{\theta} \to \theta$ in the mean-square as $n \to \infty$ if and only if $\text{Var}[\hat{\theta}] \to 0$ as $n \to \infty$. Recall that mean-square convergence implies convergence in probability (consistency).

If $\hat{\theta}$ is unbiased, then

$$E[|\hat{\theta} - \theta|^2] = \text{Var}[\hat{\theta}] \tag{1.249}$$

Given unbiased estimators $\hat{\theta}_1$ and $\hat{\theta}_2$ of θ, the one with smaller variance has a smaller mean-square error. Hence, we call $\hat{\theta}_1$ a *better* estimator than $\hat{\theta}_2$ if $\text{Var}[\hat{\theta}_1] < \text{Var}[\hat{\theta}_2]$, where it is implicit that the estimators are being compared for the same sample size. More generally, an unbiased estimator $\hat{\theta}$ is said to be a *minimum-variance unbiased estimator* (*MVUE*) of θ if for any other unbiased estimator $\hat{\theta}_0$ of θ, $\text{Var}[\hat{\theta}] \le \text{Var}[\hat{\theta}_0]$. $\hat{\theta}$ is also called a *best unbiased estimator* of θ. Even if an MVUE cannot be found, it may be possible to find an unbiased estimator of θ that has minimum variance among all estimators of θ in some restricted class C of estimators. Such an estimator is a best estimator relative to C.

***Example* 1.27.** Let $X_1, X_2,..., X_n$ be a random sample from a random variable X with mean μ. Let C be the class of all linear unbiased estimators of μ, these being of the form

$$\hat{\mu} = \sum_{k=1}^{n} a_k X_k$$

with $E[\hat{\mu}] = \mu$. By linearity of the expectation,

$$E[\hat{\mu}] = \left(\sum_{k=1}^{n} a_k \right) \mu$$

Therefore it must be that $a_1 + a_2 + \cdots + a_n = 1$. The variance of $\hat{\mu}$ is

$$\sigma_{\hat{\mu}}^2 = \sum_{k=1}^{n} a_k^2 \sigma^2 = \left(\sum_{k=1}^{n-1} a_k^2 + \left(1 - \sum_{k=1}^{n-1} a_k \right)^2 \right) \sigma^2$$

For $j = 1, 2,..., n - 1$,

$$\frac{\partial}{\partial a_j} \sigma_{\hat{\mu}}^2 = \left(a_j - 1 + \sum_{k=1}^{n-1} a_k \right) 2\sigma^2$$

Setting the derivative equal to 0 yields

$$a_1 + \cdots + a_{j-1} + 2a_j + a_{j+1} + \cdots + a_{n-1} = 1$$

Setting all derivatives, $j = 1, 2,..., n - 1$, to 0 yields the system

$$\begin{pmatrix} 2 & 1 & 1 & \cdots & 1 \\ 1 & 2 & 1 & \cdots & 1 \\ 1 & 1 & 2 & \cdots & 1 \\ \vdots & \vdots & \vdots & \ddots & \vdots \\ 1 & 1 & 1 & \cdots & 2 \end{pmatrix} \begin{pmatrix} a_1 \\ a_2 \\ a_3 \\ \vdots \\ a_{n-1} \end{pmatrix} = \begin{pmatrix} 1 \\ 1 \\ 1 \\ \vdots \\ 1 \end{pmatrix}$$

The system is solved by $a_j = 1/n$ for $j = 1, 2, \ldots, n-1$. Since the system matrix is nonsingular, this solution is unique. Moreover, since the sum of the coefficients is 1, $a_n = 1/n$ and the best linear unbiased estimator is the sample mean. ■

Finding a minimum-variance unbiased estimator is generally difficult; however, checking whether a particular unbiased estimator has minimum variance can often be accomplished with the aid of the next theorem, known as the *Cramer-Rao inequality*. If an estimator has a variance equal to the lower bound stated in the theorem, then the estimator must be an MVUE. There are a number of regularity conditions on the density whose parameter is being estimated. These have to do with existence of the partial derivative in the Cramer-Rao lower bound, finiteness and positivity of the expectation in the bound denominator, and sufficient regularity of the density to bring partial derivatives inside expectation-defining integrals in the proof of the theorem. We omit them from the statement of the theorem, noting that they are satisfied for commonly employed densities.

Theorem 1.20. Under certain regularity conditions, if $X_1, X_2, \ldots, X_n$ comprise a random sample arising from a random variable X and $\hat{\theta}$ is an unbiased estimator of the parameter θ in the density $f(x; \theta)$ of X, then

$$\operatorname{Var}[\hat{\theta}] \geq \frac{1}{nE\left[\left(\frac{\partial}{\partial\theta}\log f(X;\theta)\right)^2\right]} \qquad ■ \qquad (1.250)$$

***Example* 1.28.** We demonstrate that the sample mean is an MVUE for the mean of a normally distributed random variable. If X has mean μ and variance σ^2, then

$$f(x;\mu) = \frac{1}{\sqrt{2\pi}\sigma} e^{-\frac{1}{2}\left(\frac{x-\mu}{\sigma}\right)^2}$$

$$\log f(x;\mu) = -\log(\sigma\sqrt{2\pi}) - \frac{1}{2}\left(\frac{x-\mu}{\sigma}\right)^2$$

$$\frac{\partial}{\partial\mu}\log f(x;\mu) = \frac{x-\mu}{\sigma^2}$$

$$E\left[\left(\frac{\partial}{\partial\mu}\log f(X;\mu)\right)^2\right]=\frac{E[(X-\mu)^2]}{\sigma^4}=\frac{1}{\sigma^2}$$

Hence, the Cramer-Rao lower bound is σ^2/n, the variance of the sample mean. ■

There is crucial difference between the results of Examples 1.27 and 1.28. The former shows that the sample mean is the best estimator of the mean among the class of linear unbiased estimators, regardless of the distribution; the latter shows that the sample mean is the best estimator among all unbiased estimators of the mean of a normal distribution. In fact, the sample mean is not always the MVUE for the mean of a distribution.

1.8.4. Method of Moments

Just because an estimator is best for one distribution does not imply it is best for another distribution. In many cases one cannot find a best estimator and therefore settles for one that performs satisfactorily. There are a number of techniques for finding estimation rules. The same technique applied to different distributions often produces different estimation rules. Bestness depends on both the distribution and the measure of goodness. More interesting, different techniques can yield different estimators for the same distribution. In addition, the properties of the estimator produced by a technique for finding estimation rules will vary, depending on the distribution to which it is applied. Two commonly used techniques for finding estimators are maximum-likelihood and the method of moments. Generally, the properties of maximum-likelihood estimators are preferable to method-of-moment estimators, but the method of moments may be applied in many circumstances where maximum likelihood is mathematically intractable. Here we consider the method of moments; the next section is devoted to maximum-likelihood estimation.

If $X_1, X_2,\dots, X_n$ comprise a random sample arising from the random variable X, then the rth *sample moment* of $X_1, X_2,\dots, X_n$ is

$$M_r' = \frac{1}{n}\sum_{k=1}^{n} X_k^r \tag{1.251}$$

For $r = 1$, the sample moment is the sample mean. For $r = 2$,

$$M_2' = \frac{n-1}{n}S^2 + \overline{X}^2 \tag{1.252}$$

For $r = 1, 2,..., M_r'$ is the arithmetic mean of $X_1^r, X_2^r, ..., X_n^r$, which themselves constitute a random sample arising from the random variable X^r. Consequently, if $E[X^r]$, the rth moment about the origin of X, exists, then M_r' is an unbiased estimator of $E[X^r]$, so that

$$E[M_r'] = E[X^r] \tag{1.253}$$

$$\text{Var}[M_r'] = \frac{\text{Var}[X^r]}{n} \tag{1.254}$$

and M_r' is a consistent estimator of $E[X^r]$.

If X has density $f(x; \theta_1, \theta_2,..., \theta_p)$, where parameters $\theta_1, \theta_2,..., \theta_p$ need to be estimated, then $E[X^r]$ is a function of $\theta_1, \theta_2,..., \theta_p$, meaning there exists a function h_r such that

$$E[X^r] = h_r(\theta_1, \theta_2, ..., \theta_p) \tag{1.255}$$

Method-of-moments estimation is done by setting $E[X^r] = M_r'$, which is reasonable since M_r' is a consistent estimator of $E[X^r]$. This leads to the system of equations

$$\begin{aligned} h_1(\theta_1, \theta_2, ..., \theta_p) &= M_1' \\ h_2(\theta_1, \theta_2, ..., \theta_p) &= M_2' \\ &\vdots \\ h_N(\theta_1, \theta_2, ..., \theta_p) &= M_N' \end{aligned} \tag{1.256}$$

where N, the number of moments employed, is chosen so that a unique solution for $\theta_1, \theta_2,..., \theta_p$ can be found. The solutions $\hat{\theta}_1, \hat{\theta}_2, ..., \hat{\theta}_p$ are the *method-of-moments estimators* for $\theta_1, \theta_2,..., \theta_p$. Properties of the estimators depend on the distribution and often very little is known about method-of-moment estimators. Once found, they are usually tested on data (often synthetic) to see how they perform.

***Example* 1.29.** Consider the gamma distribution with parameters α and β. Since

$$E[X] = \alpha\beta$$

$$E[X^2] = (\alpha + 1)\alpha\beta^2$$

the method-of-moments system is

$$\alpha\beta = M_1' = \overline{X}$$

$$(\alpha+1)\alpha\beta^2 = M_2' = \frac{n-1}{n}S^2 + \overline{X}^2$$

Solving for α and β yields

$$\hat{\alpha} = \frac{\overline{X}^2}{\frac{n-1}{n}S^2}$$

$$\hat{\beta} = \frac{\frac{n-1}{n}S^2}{\overline{X}}$$ ■

The system of Eq. 1.256 reveals that the key to the method is finding estimators of functions of the unknown parameters and then setting the estimators equal to the functions (as if the estimators had zero variance). As demonstrated, the method employs sample moments; however, the functions and estimators can come from other sources. In image processing the functions can provide geometric characteristics in terms of the unknown parameters and the estimators can be for these geometric characteristics.

1.8.5. Order Statistics

Sample values are often ordered from least to greatest, with some value in the ordering being of interest. If $X_1, X_2,..., X_n$ comprise a random sample arising from the random variable X, then the n *order statistics* for the sample are the n random variables $Y_1 \leq Y_2 \leq ... \leq Y_n$ resulting from ordering the values of the sample variables from lowest to highest. Each order statistic can be expressed as a function of the sample variables. For instance,

$$Y_1 = \min\{X_1, X_2,..., X_n\} \tag{1.257}$$

$$Y_n = \max\{X_1, X_2,..., X_n\} \tag{1.258}$$

A number of statistics are defined in terms of the order statistics of a random sample. Perhaps the most important is the *sample median*, which for odd n is defined by $\tilde{X} = Y_{(n+1)/2}$, the middle value of the observations. The sample median is often used to estimate the mean of symmetric distributions, especially when the distribution of the underlying random variable is heavy-tailed, meaning that values far from the mean possess relatively high probabilities, say in comparison to the tails of a normal distribution.

Theorem 1.21. If $X_1, X_2,..., X_n$ constitute a random sample arising from the continuous random variable X and $Y_1 \leq Y_2 \leq ... \leq Y_n$ are the n order statistics resulting from the sample, then, for $k = 1, 2,..., n$, the density of the kth order statistic is

$$f_{Y_k}(y) = \frac{n!}{(n-k)!(k-1)!} F_X(y)^{k-1}[1-F_X(y)]^{n-k} f_X(y) \tag{1.259}$$

In particular,

$$f_{Y_1}(y) = n[1-F_X(y)]^{n-1} f_X(y) \tag{1.260}$$

$$f_{Y_n}(y) = nF_X(y)^{n-1} f_X(y) \tag{1.261}$$

and, if n is odd, the density of the sample median is

$$f_{\tilde{X}}(y) = \frac{n!}{[((n-1)/2)!]^2} F_X(y)^{(n-1)/2}[1-F_X(y)]^{(n-1)/2} f_X(y) \quad \blacksquare \tag{1.262}$$

We demonstrate the cases for the minimum and maximum order statistics, Eqs. 1.260 and 1.261, respectively. The minimum Y_1 has probability distribution function

$$\begin{aligned} F_{Y_1}(y) &= P(\min\{X_1, X_2, \ldots, X_n\} \leq y) \\ &= P\left(\bigcup_{k=1}^{n}(X_k \leq y)\right) \\ &= 1 - P\left(\bigcap_{k=1}^{n}(X_k > y)\right) \qquad (1.263) \\ &= 1 - P(X > y)^n \\ &= 1 - (1 - F_X(y))^n \end{aligned}$$

Differentiation with respect to y gives Eq. 1.260. For the maximum Y_n,

$$F_{Y_n}(y) = P(\max\{X_1, X_2, \ldots, X_n\} \le y)$$

$$= P\left(\bigcap_{k=1}^{n} (X_k \le y)\right) \tag{1.264}$$

$$= F_X(y)^n$$

Differentiation with respect to y gives Eq. 1.261.

1.9. Maximum-Likelihood Estimation

Estimator properties are distribution dependent; nevertheless, maximum-likelihood estimators tend to perform quite well for commonly encountered distributions. A number of standard image/signal processing operators arise as maximum-likelihood estimators, in particular, the mean, the median, weighted medians, and flat morphological filters.

1.9.1. Maximum-Likelihood Estimators

Suppose X is a random variable with density $f(x; \theta)$, θ is a parameter to be estimated, and $X_1, X_2, \ldots, X_n$ are independent random variables identically distributed to X, so that $X_1, X_2, \ldots, X_n$ compose a random sample for X. The joint density of $X_1, X_2, \ldots, X_n$ is called the *likelihood function* of the random sample $X_1, X_2, \ldots, X_n$. Owing to independence and identical distribution with X, the likelihood function is given by

$$L(x_1, x_2, \ldots, x_n; \theta) = f(x_1; \theta) f(x_2; \theta) \cdots f(x_n; \theta) \tag{1.265}$$

To simplify notation, we may write the likelihood function as $L(\theta)$.

If a value of θ can be found to maximize the likelihood function for a set of sample outcomes $x_1, x_2, \ldots, x_n$, then that value is called the *maximum-likelihood estimate* with respect to $x_1, x_2, \ldots, x_n$. If the same functional relationship between the estimate and $x_1, x_2, \ldots, x_n$ holds for all possible choices of $x_1, x_2, \ldots, x_n$, then the functional relationship is taken as an estimation rule and provides an estimator $\hat{\theta} = \hat{\theta}(X_1, X_2, \ldots, X_n)$ called the *maximum-likelihood estimator* (*filter*) for θ.

For intuitive appreciation, suppose X is discrete. Then

$$L(x_1, x_2, \ldots, x_n; \theta) = \prod_{k=1}^{n} P(X = x_k; \theta) \tag{1.266}$$

Suppose a set of sample values $x_1, x_2, \ldots, x_n$ is observed. Given these n observations, what value of θ would be a reasonable choice? If there exists a value θ' such that

$$L(x_1, x_2, \ldots, x_n; \theta') \geq L(x_1, x_2, \ldots, x_n; \theta) \tag{1.267}$$

for all θ, then θ' maximizes $P(X_1 = x_1, X_2 = x_2, \ldots, X_n = x_n)$, which is the probability of obtaining the observations that were, in fact, obtained.

If the density of X has more than a single unknown parameter, say $f(x; \theta_1, \theta_2, \ldots, \theta_m)$, then the definitions of likelihood function and maximum-likelihood estimator remain the same, with the vector $\boldsymbol{\theta} = (\theta_1, \theta_2, \ldots, \theta_m)'$ taking the place of the single parameter θ. The maximum-likelihood estimator is now a vector estimator

$$\hat{\boldsymbol{\theta}} = \hat{\boldsymbol{\theta}}(X_1, X_2, \ldots, X_n) = \begin{pmatrix} \hat{\theta}_1(X_1, X_2, \ldots, X_n) \\ \hat{\theta}_2(X_1, X_2, \ldots, X_n) \\ \vdots \\ \hat{\theta}_m(X_1, X_2, \ldots, X_n) \end{pmatrix} \tag{1.268}$$

Maximum-likelihood estimators possess the *invariance* property: if $\hat{\theta}$ is a maximum-likelihood estimator of θ, g is a one-to-one function, and $\phi = g(\theta)$, then $\hat{\phi} = g(\hat{\theta})$ is a maximum-likelihood estimator of ϕ. In many cases maximum-likelihood estimators possess desirable properties, especially for large samples. For instance, the maximum- likelihood estimator is often (but not always) a minimum-variance-unbiased estimator. Indeed, there are instances in which the maximum-likelihood estimator is quite poor; however, for many commonly applied distributions, the maximum-likelihood estimator is either best or close to best. When the theoretical properties of the estimator are not known, it is prudent to perform simulations and estimate the bias and variance of the maximum-likelihood estimator for processes of interest.

***Example* 1.30.** For X normally distributed with unknown mean μ and known variance σ^2, the likelihood function is

$$L(x_1, x_2, \ldots, x_n; \mu) = \frac{1}{(2\pi\sigma^2)^{n/2}} \exp\left[-\frac{1}{2} \sum_{i=1}^{n} \left(\frac{x_i - \mu}{\sigma} \right)^2 \right]$$

$L(x_1, x_2, \ldots, x_n; \mu)$ is maximized if and only if its logarithm,

$$\log L(\mu) = -\frac{n}{2}\log(2\pi\sigma^2) - \frac{1}{2}\sum_{k=1}^{n}\left(\frac{x_k - \mu}{\sigma}\right)^2$$

is maximized. Differentiating with respect to μ gives

$$\frac{d}{d\mu}\log L(\mu) = \frac{1}{\sigma^2}\sum_{k=1}^{n}(x_k - \mu)$$

Setting the derivative equal to 0 shows the maximum to be the mean of x_1, $x_2,\ldots, x_n$. Since this estimation rule holds for any choice of $x_1, x_2,\ldots, x_n$, it provides the maximum-likelihood estimator $\hat{\mu} = \overline{X}$, the sample mean of the observations.

Continue to assume that X is normally distributed with mean μ and variance σ^2, but now assume that the variance is also unknown. By treating $(\mu, \sigma^2)'$ as a parameter vector, maximum-likelihood estimation can be used to find estimators of both μ and σ^2. The likelihood function remains the same, except now it is a function $L(\mu, \sigma^2)$ of both μ and σ^2. To maximize it, take its partial derivatives with respect to μ and σ^2 to obtain

$$\frac{\partial}{\partial\mu}\log L(\mu) = \frac{1}{\sigma^2}\sum_{k=1}^{n}(x_k - \mu)$$

$$\frac{\partial}{\partial\sigma^2}\log L(\mu,\sigma^2) = -\frac{n}{2\sigma^2} + \frac{1}{2\sigma^4}\sum_{k=1}^{n}(x_k - \mu)^2$$

Simultaneous solution yields the maximum-likelihood estimators $\hat{\mu} = \overline{X}$ and

$$\hat{\sigma}^2 = \frac{1}{n}\sum_{k=1}^{n}(X_k - \overline{X})^2 = \frac{n-1}{n}S^2$$

where S^2 is the sample variance. Maximum-likelihood estimation has produced an unbiased estimator of the mean and an asymptotically unbiased estimator of the variance. If the mean μ is known, then maximum-likelihood estimation applied to σ^2 alone yields the unbiased variance estimator

$$\hat{\sigma}^2 = \frac{1}{n}\sum_{k=1}^{n}(X_k - \mu)^2$$

■

1.9.2. Additive Noise

A substantial portion of the text concerns estimation of signals from noise-corrupted observations. A very simple model involving additive noise has a constant (pure) discrete signal corrupted by independent, identically distributed, zero-mean additive noise. The observed signal then takes the form

$$X(i) = \theta + N(i) \tag{1.269}$$

$i = 1, 2, \ldots$, where θ is a constant to be estimated, the noise random variables $N(i)$ possess a common distribution with variance σ^2, and any finite collection of noise random variables is independent. The observed signal $X(i)$ has mean θ and variance σ^2. We use maximum-likelihood estimation for θ, employing a finite number of observations, which for notational consistency we denote by $X_1, X_2, \ldots, X_n$. From this perspective the maximum-likelihood estimator is a filter operating on observed signal values to estimate the pure signal θ. For instance, relative to Example 1.30, the noise is normally distributed at each point with mean 0 and variance σ^2, the observed signal values are normally distributed with mean θ, and the maximum-likelihood filter is an estimator of θ.

Equation 1.269 can be viewed as either signal plus noise or $X(i)$ expressed as its mean θ plus its random displacement

$$N(i) = X(i) - \theta \tag{1.270}$$

from its mean. Under the latter interpretation, a set of observations is simply a random sample arising from an underlying random variable X, with each observation being identically distributed and maximum-likelihood estimation of θ being parametric estimation of the mean of X, as just discussed. Mathematically, nothing has changed; however, the interpretation has changed. Because our interest is in processing images and signals, we adopt the signal-model perspective for the remainder of this section.

In practice, a filter is applied to a signal or image by placing a *window* at a point and applying the filter to the observations in the window to estimate the signal value at the point. From this perspective, the signal of Eq. 1.269 is filtered by choosing a window containing some fixed number of points and applying the filter (estimator) to the values in the window at each location as the window is moved across the signal. Because the noise is the same across the entire signal in the model of Eq. 1.269, the same estimator is applied at each location. In this context, the sample mean is known as the *moving mean*.

***Example* 1.31.** The *Laplace distribution* is defined by the density

$$f(x) = \frac{\alpha}{2} e^{-\alpha|x-\mu|}$$

for all $x \in \Re$, and for parameters $\alpha > 0$ and $-\infty < \mu < \infty$. The Laplace distribution has mean μ and variance $2/\alpha^2$. Suppose $N(i)$ has a Laplace density with mean 0 and variance $2/\alpha^2$ in Eq. 1.269. Then the observations $X_1, X_2, \ldots, X_n$ form a random sample from a Laplace distribution with mean θ and variance $2/\alpha^2$. The likelihood function is

$$L(x_1, x_2, \ldots, x_n; \theta) = \left(\frac{\alpha}{2}\right)^n \exp\left[-\alpha \sum_{i=1}^{n} |x_i - \theta|\right]$$

Maximization occurs from minimizing the sum inside the exponential. Assuming the number of observations to be odd, this sum is minimized by letting $\hat{\theta}$ be the median of the observations; the resulting signal filter is called the *moving median*.

The maximum-likelihood estimator for additive Gaussian noise is the sample mean; the maximum-likelihood estimator for additive Laplacian noise is the sample median. Whatever the distribution, if $\mathrm{Var}[X] = \sigma^2$, then $\mathrm{Var}[\bar{X}] = \sigma^2/n$. For the sample median the matter is more complicated; however, for the Gaussian and Laplacian distributions, the sample median has asymptotic variances (as $n \to \infty$) $\pi\sigma^2/2n$ and $\sigma^2/2n$, respectively. Hence, for large n, the asymptotic variance of the sample median is approximately $\pi/2$ times as great as that for the sample mean when the underlying distribution is Gaussian, but the asymptotic variance of the sample median is approximately half as large as that for the sample mean when the underlying distribution is Laplacian. ■

***Example* 1.32.** Now assume in Eq. 1.269 that the noise is uniformly distributed over the interval $[-\beta, 0]$, where $\beta > 0$. The likelihood function is

$$L(x_1, x_2, \ldots, x_n; \theta) = \frac{1}{\beta^n} \prod_{k=1}^{n} I_{[\theta-\beta,\theta]}(x_k)$$

where $I_{[\theta-\beta,\theta]}$ is the indicator function for the interval $[\theta - \beta, \theta]$. The likelihood function is

$$L(x_1, x_2, \ldots, x_n; \theta) = \begin{cases} \beta^{-n}, & \text{if } \theta - \beta \le x_k \le \theta \text{ for } k = 1, 2, \ldots, n \\ 0, & \text{otherwise} \end{cases}$$

$L(\theta)$ is maximized for a set of sample values $x_1, x_2,..., x_n$, if and only if $\theta - \beta \leq x_k \leq \theta$ for all k, else $L(\theta) = 0$. Hence, $L(\theta)$ is maximized by having θ satisfy the inequality

$$\max\{x_1, x_2,..., x_n\} \leq \theta \leq \min\{x_1, x_2,..., x_n\} + \beta$$

If $\hat{\theta}$ is chosen so that

$$\max\{X_1, X_2,..., X_n\} \leq \hat{\theta} \leq \min\{X_1, X_2,..., X_n\} + \beta$$

then $\hat{\theta}$ is a maximum-likelihood estimator. If we assume β is unknown, then to be certain that the preceding double inequality holds, we must choose

$$\hat{\theta} = \max\{X_1, X_2,..., X_n\}$$

Applied across the entire signal, the maximum-likelihood filter is a *moving maximum*, which in morphological image processing is known as *flat dilation*.

If the model is changed with the noise uniformly distributed over the interval $[0, \beta]$, then the same basic analysis applies, except now $\hat{\theta}$ must be chosen so that

$$\max\{X_1, X_2,..., X_n\} - \beta \leq \hat{\theta} \leq \min\{X_1, X_2,..., X_n\}$$

If β is unknown, then we must choose

$$\hat{\theta} = \min\{X_1, X_2,..., X_n\}$$

which yields a *moving minimum* and is known as *flat erosion*. ■

***Example* 1.33.** Weighted medians are commonly employed in image processing. They give flexibility to the median approach by allowing certain observations to provide greater contributions to filter output. For the observations $x_1, x_2,..., x_n$ and the integer weights $\gamma_1, \gamma_2,..., \gamma_n$ (summing to an odd number), the *weighted median* for $x_1, x_2,..., x_n$ with weights $\gamma_1, \gamma_2,..., \gamma_n$ is defined to be the median with x_i repeated γ_i times for $i = 1, 2,..., n$. The weighted median can be viewed as a maximum-likelihood filter under the model of Eq. 1.269 if we drop the assumption that the noise observations are identically distributed and assume the noise for X_i is Laplacian distributed with mean 0 and variance $2/\gamma_i^2$. Then the likelihood function becomes

$$L(x_1, x_2,..., x_n; \theta) = \frac{\gamma_1\gamma_2\cdots\gamma_n}{2^n}\exp\left[-\sum_{i=1}^{n}\gamma_i|x_i-\theta|\right]$$

Maximization occurs with minimization of the sum inside the exponential. This sum can be expressed in the form of the sum in the exponent for the likelihood function for identically distributed Laplacian noise in Example 1.31. Just repeat each $|x_i - \theta|$ in that sum γ_i times. Hence the sum is minimized, and the likelihood function maximized, by the weighted median with weights $\gamma_1, \gamma_2,..., \gamma_n$. Since it has not been assumed that the noise is identically distributed, moving-window application is point dependent. ■

1.9.3. Minimum Noise

Rather than consider a constant signal corrupted by additive noise, as in Eq. 1.269, we can instead assume that the corrupted signal results from minimum noise; that is,

$$X(i) = \theta \wedge N(i) \tag{1.271}$$

where each $N(i)$ is identically distributed to a continuous random variable N. We assume that $F_N(\theta) < 1$; otherwise, we would have $N(i) \leq \theta$ with probability one. The probability distribution function for $X(i)$ is given by

$$F_{X(i)}(x) = P(X(i) \leq x) = \begin{cases} 1, & \text{if } x \geq \theta \\ F_N(x), & \text{if } x < \theta \end{cases} \tag{1.272}$$

Taking the derivative in a generalized sense yields $f_{X(i)}(x)$ and the likelihood function

$$L(x_1, x_2,..., x_n; \theta) = \prod_{k=1}^{n}[f_N(x_k)I_{(-\infty,\theta)}(x_k) + (1 - F_N(\theta))\delta(x_k - \theta)] \tag{1.273}$$

To find the maximum-likelihood filter, assume (without loss of generality) that $x_1, x_2,\ldots,x_q$ equal the maximum observation and $x_{q+1}, x_{q+2},\ldots,x_n$ are strictly less than the maximum. If

$$\theta = \max\{x_1, x_2,..., x_n\} \tag{1.274}$$

then

$$L(x_1, x_2,..., x_n; \theta) = \prod_{k=1}^{q}(1-F_N(\theta))\delta(x_k-\theta)\prod_{k=q+1}^{n} f_N(x_k) \tag{1.275}$$

If $\theta < \max\{x_1, x_2,..., x_n\}$, then

$$f_N(x_1)I_{(-\infty,\theta)}(x_1)+(1-F_N(\theta))\delta(x_1-\theta)=0 \tag{1.276}$$

so that $L(x_1, x_2,..., x_n; \theta) = 0$. If $\theta > \max\{x_1, x_2,..., x_n\}$, then

$$L(x_1, x_2,..., x_n; \theta) = \prod_{k=1}^{n} f_N(x_k) \tag{1.277}$$

Therefore, the maximum-likelihood filter is the maximum of the observations and, as a signal filter, is the flat dilation. A dual argument applies to maximum noise.

1.10. Entropy

Although we always lack certainty in predicting the outcome of a random variable, the degree of uncertainty is not the same in all cases. This section treats entropy, which is the quantification of uncertainty. Entropy plays a key role in coding theory, the subject of the next section.

1.10.1. Uncertainty

Consider a random variable X that can take on two values, 0 and 1, with probabilities $p = P(X = 1)$ and $q = P(X = 0)$. If $p = 0.99$ and $q = 0.01$, then the observer feels less uncertain than if the probabilities were $p = 0.6$ and $q = 0.4$. Intuitively, the observer's uncertainty is greatest when the probabilities are equal and least when one of the probabilities is 1. We now quantify uncertainty.

Suppose a discrete random variable X having probability mass function $f(x)$ can take on the n values $x_1, x_2,..., x_n$, with $f(x_i) = p_i$ for $i = 1, 2,..., n$. Because uncertainty should be increased when the probabilities p_i are more alike than when one or two carry most of the probability mass, the following criteria appear reasonable for a measure of uncertainty: (1) uncertainty is nonnegative and equal to zero if and only if there exists i such that $p_i = 1$; (2) uncertainty is maximum when the outcomes of X are equally likely; (3) if random variables X and Y have n and m equally likely outcomes, respectively, with $n < m$, then the uncertainty of X is less than the uncertainty of Y; and (4) uncertainty is a continuous function of $p_1, p_2,..., p_n$. These four conditions are met by defining the *uncertainty* of X by

$$H[X] = -\sum_{i=1}^{n} p_i \log_2 p_i \tag{1.278}$$

where the convention is adopted that $p_i \log_2 p_i = 0$ if $p_i = 0$. $H[X]$ is called the *entropy* of X; it is measured in *bits* and can be expressed in terms of expectation by

$$H[X] = -E[\log_2 f(X)] \tag{1.279}$$

The definition of entropy involves only probability masses, not their distribution on the x axis. Consequently, two random variables with the same outcome probabilities are indistinguishable from the perspective of entropy. Because total probability is 1, p_n can be considered to be a function of p_1, $p_2, \ldots, p_{n-1}$, so that it is appropriate to write entropy as

$$H = H(p_1, p_2, \ldots, p_{n-1}) \tag{1.280}$$

We have claimed that $H[X]$ satisfies the four criteria listed for uncertainty. First, since $0 \le p_i \le 1$ for all i, it is immediately clear from the definition that $H[X] \ge 0$ and $H[X] = 0$ if and only if there exists i such that $p_i = 1$. Because $\log_2$ is a continuous function, H is also. If the outcomes of X are equally likely, then $H[X] = \log_2 n$ and entropy is an increasing function of n. To complete verification of the four criteria, we need to demonstrate that $H[X]$ is maximized in the case of equally likely outcomes. We treat the case of two outcomes having probabilities p and $1 - p$. Writing entropy as a function of p, assuming $0 < p < 1$, employing the natural logarithm, and differentiating yields

$$H'(p) = -\frac{1}{\log 2}\left[\log p - \log(1-p)\right] \tag{1.281}$$

Solving $H'(p) = 0$ yields $p = 0.5$. Since $H''(p) < 0$, a maximum occurs at $p = 0.5$.

Given two random variables X and Y, observing X can affect our uncertainty regarding Y. If X and Y can take on the values $x_1, x_2, \ldots, x_n$ and $y_1, y_2, \ldots, y_m$, respectively, then, for $i = 1, 2, \ldots, n$, the *conditional entropy* of Y given x_i is defined via conditional densities by

$$H[Y|x_i] = -\sum_{j=1}^{m} f(y_j|x_i) \log_2 f(y_j|x_i) \tag{1.282}$$

This restates the entropy definition in terms of the conditional random variable $Y|x_i$. By letting x_i vary, conditional entropy becomes a function of X and is then written $H[Y|X]$.

As a function of X, $H[Y|X]$ has an expected value, called the *mean conditional entropy* of Y relative to X, and is defined by

$$\bar{H}[Y|X] = E[H[Y|X]]$$

$$= \sum_{i=1}^{n} H[Y|x_i] f_X(x_i) \tag{1.283}$$

Mean conditional entropy can be expressed via the conditional density of Y given X:

$$\bar{H}[Y|X] = -\sum_{i=1}^{n}\sum_{j=1}^{m} f_X(x_i) f(y_j|x_i) \log_2 f(y_j|x_i)$$

$$= -\sum_{i=1}^{n}\sum_{j=1}^{m} f(x_i, y_j) \log_2 f(y_j|x_i)$$

$$= -E[\log_2 f(Y|X)] \tag{1.284}$$

In particular, if X and Y are independent, then $\bar{H}[Y|X] = H[Y]$.

Generalizing Eq. 1.279 gives the definition of entropy for a random vector $(X, Y)'$:

$$H[X,Y] = -E[\log_2 f(X,Y)] \tag{1.285}$$

where $f(x, y)$ is the joint density of X and Y. A fundamental property is that the joint uncertainty of X and Y can be decomposed into a sum of the uncertainty of X plus the expected uncertainty remaining in Y following an observation of X, namely,

$$H[X,Y] = H[X] + \bar{H}[Y|X] \tag{1.286}$$

In particular, if X and Y are independent, then the joint uncertainty is the sum of the marginal uncertainties. Equation 1.286 is simply a restatement of the relation

$$-E[\log_2 f(Y|X) f_X(X)] = -E[\log_2 f(Y|X)] - E[\log_2 f_X(X)] \tag{1.287}$$

1.10.2. Information

Intuitively, if X and Y are dependent and X is observed, then information is obtained regarding Y because observation of X alters our uncertainty with respect to Y. The *expected amount of information* given for Y by observing X is defined by

$$I_X[Y] = H[Y] - \overline{H}[Y|X] \tag{1.288}$$

$I_X[Y]$ is the expected amount of information obtained, since, for a specific observation x_i of X, $H[Y] - H[Y|x_i]$ is the difference between the unconditional uncertainty of Y and the conditional uncertainty of Y given x_i, and $I_X[Y]$ results from taking the expected value of $H[Y] - H[Y|X]$ relative to X. If X and Y are independent, then $I_X[Y] = 0$.

Since the entropy of any random variable whose full probability mass is carried by a single outcome is 0 and since the conditional random variable $X|x_i$ is such a variable for any x_i, $H[X|x_i] = 0$. Hence, the mean conditional entropy of X relative to X is 0 and $I_X[X] = H[X]$, so that the entropy of X is the expected amount of information in X.

$I_X[Y]$ can be expressed via the joint and marginal densities of X and Y. From Eq. 1.288,

$$
\begin{aligned}
I_X[Y] &= -E[\log_2 f_Y(Y)] + E[\log_2 f(Y|X)] \\
&= -\sum_{i=1}^{n}\sum_{j=1}^{m} f(x_i, y_j)\log_2 f(y_j) + \sum_{i=1}^{n}\sum_{j=1}^{m} f(x_i, y_j)\log_2 f(y_j|x_i) \\
&= \sum_{i=1}^{n}\sum_{j=1}^{m} f(x_i, y_j)\left[\log_2 f(y_j|x_i) - \log_2 f(y_j)\right] \\
&= E\left[\log_2 \frac{f(Y|X)}{f_Y(Y)}\right] \\
&= E\left[\log_2 \frac{f(X,Y)}{f_X(X)f_Y(Y)}\right]
\end{aligned} \tag{1.289}
$$

Owing to the symmetry of the expression,

$$I_X[Y] = I_Y[X] \tag{1.290}$$

Using Eq. 1.289 and the fact that, for $a > 0$, $\log a \le a - 1$, we show that $I_X[Y] \ge 0$:

$$I_X[Y] = \sum_{i=1}^{n}\sum_{j=1}^{m} f(x_i, y_j)\log_2 \frac{f(x_i, y_j)}{f_X(x_i)f_Y(y_j)}$$

$$= -\frac{1}{\log 2}\sum_{i=1}^{n}\sum_{j=1}^{m} f(x_i, y_j)\log \frac{f_X(x_i)f_Y(y_j)}{f(x_i, y_j)}$$

$$\ge -\frac{1}{\log 2}\sum_{i=1}^{n}\sum_{j=1}^{m} f(x_i, y_j)\left[\frac{f_X(x_i)f_Y(y_j)}{f(x_i, y_j)} - 1\right]$$

$$= -\frac{1}{\log 2}\left[\sum_{i=1}^{n}\sum_{j=1}^{m} f_X(x_i)f_Y(y_j) - \sum_{i=1}^{n}\sum_{j=1}^{m} f(x_i, y_j)\right] \tag{1.291}$$

Since both double sums in the preceding expression sum to 1, $I_X[Y] \ge 0$.

From the definition of $I_X[Y]$, it follows from Eq. 1.291 that

$$\overline{H}[Y|X] \le H[Y] \tag{1.292}$$

Applying this inequality to the decomposition of Eq. 1.286 yields

$$H[X, Y] \le H[X] + H[Y] \tag{1.293}$$

1.10.3. Entropy of a Random Vector

The entropy of an arbitrary random vector $(X_1, X_2,..., X_r)'$ is defined by

$$H[X_1, X_2,..., X_r] = -E[\log_2 f(X_1, X_2,..., X_r)]$$

$$= -\sum_{x_1, x_2,...,x_r} f(x_1, x_2, \ldots, x_r)\log_2 f(x_1, x_2, \ldots, x_r) \tag{1.294}$$

where $f(x_1, x_2,..., x_r)$ is the joint density of $X_1, X_2,..., X_r$. $H[X_1, X_2,..., X_r]$ gives the expected information gained upon observation of $X_1, X_2,..., X_r$. The basic properties holding for two random variables generalize to $r > 2$ random variables. In general,

$$H[X_1, X_2,..., X_r] \le H[X_1] + H[X_2] + \cdots + H[X_r] \tag{1.295}$$

and, if $X_1, X_2, \ldots, X_r$ are independent, then

$$H[X_1, X_2, \ldots, X_r] = H[X_1] + H[X_2] + \cdots + H[X_r] \tag{1.296}$$

The mean conditional entropy for X_r given $X_1, X_2, \ldots, X_{r-1}$ is defined by

$$\overline{H}[X_r \mid X_1, X_2, \ldots, X_{r-1}] = -E[\log_2 f(X_r \mid X_1, X_2, \ldots, X_{r-1})] \tag{1.297}$$

Corresponding to the decomposition of Eq. 1.286 is the decomposition

$$H[X_1, X_2, \ldots, X_r] = H[X_1] + \sum_{k=2}^{r} \overline{H}[X_k \mid X_1, X_2, \ldots, X_{k-1}] \tag{1.298}$$

The inequality of Eq. 1.292 also extends: if $\{Z_1, Z_2, \ldots, Z_s\} \subset \{X_1, X_2, \ldots, X_{r-1}\}$, then

$$\overline{H}[X_r \mid X_1, X_2, \ldots, X_{r-1}] \leq \overline{H}[X_r \mid Z_1, Z_2, \ldots, Z_s] \tag{1.299}$$

For a sequence $\ldots, X_{-1}, X_0, X_1, \ldots$ of random variables, $H[X_{r-k}, X_{r-k+1}, \ldots, X_r]$ gives the uncertainty present in the $k + 1$ consecutive random variables terminating at time r. It follows from Eq. 1.299 that, for $k = 1, 2, \ldots,$

$$\overline{H}[X_r \mid X_{r-1}, \ldots, X_{r-k}] \leq \overline{H}[X_r \mid X_{r-1}, \ldots, X_{r-k-1}] \tag{1.300}$$

Thus, $\overline{H}[X_r \mid X_{r-1}, \ldots, X_{r-k}]$ decreases as $k \to \infty$. Since $\overline{H}[X_r \mid X_{r-1}, \ldots, X_{r-k}]$ is bounded below by 0, it has a limit as $k \to \infty$. This limit,

$$\overline{H}_c[X_r] = \lim_{k \to \infty} \overline{H}[X_r \mid X_{r-1}, X_{r-2}, \ldots, X_{r-k}] \tag{1.301}$$

is called the (*mean*) *conditional entropy* of the sequence at time r and is a measure of the uncertainty concerning the present, X_r, given observation of the entire past.

1.11. Source Coding

For digital communication and storage, symbols have to be encoded into binary form. These symbols can be alphabetic, numeric, or a combination thereof; they can be image data, image features, or descriptive words or phrases pertaining to images. They are coded at a source and decoded at some destination. Source coding involves a finite set Σ of symbols, called the *source*, and the symbols (*source words*) need to be placed into one-to-one correspondence with strings (*code words*) of 0s and 1s. Efficient coding requires that the expected number of bits in a randomly selected code word be kept small. To measure the efficiency of a particular encoding, we assume that the probability of each source word is known. If B is the random variable counting the number of bits transmitted for a randomly chosen source word, then the smaller the expected number of bits, $E[B]$, the more efficient the code. $E[B]$, which gives the expected length of a code word, does not depend on the actual symbols in Σ, but only on their codes and probabilities.

1.11.1. Prefix Codes

We consider only *prefix codes*. For these, no symbol code word can be obtained from the code word of a distinct symbol by adjoining 0s and 1s. For instance, if 001 is a code word, then 00110 cannot be the code word of a different symbol. A prefix code is uniquely decodable, meaning that a string of 0s and 1s forming a *message* (sequence of code words) can be unambiguously decoded by recognizing each code word in turn when reading from left to right without spacing between code words. For illustration, let $\Sigma = \{a, b, c, d\}$ and the code be given by the following correspondences: $a \leftrightarrow 01$, $b \leftrightarrow 001$, $c \leftrightarrow 10$, $d \leftrightarrow 110$. The message 0101101100010010110 is uniquely decoded as *aacdbbad*.

Since the actual source symbols are unimportant, we can assume without loss of generality that we are encoding the integers 1, 2,..., n. Thus, if p_1, p_2,..., p_n are the symbol probabilities, then in effect we have a random variable X taking on values 1, 2,..., n and having probability mass function $f(k) = p_k$ for $k = 1, 2,..., n$. If k is encoded using m_k bits, then the probability mass function for the code-word length B is defined, for $j = 1, 2,...$, by

$$f_B(j) = \sum_{\{k:m_k=j\}} p_k \tag{1.302}$$

Efficient coding requires using shorter code words for higher-probability symbols. Given n symbols, the problem is to design a prefix code for which $E[B]$ is minimal.

If each symbol probability is a negative power of 2, then the following coding algorithm yields a code possessing minimal expected code-word length:

C1. List the symbol probabilities in order from highest to lowest and label them from top to bottom as $p_1, p_2,..., p_n$. Ties can be broken arbitrarily.

C2. Beginning at the top of the list, select $p_1, p_2,..., p_k$ so that

$$p_1 + p_2 + \cdots + p_k = 1/2$$

Let 1 be the first bit in the encoding of symbols 1 through k and 0 be the first bit in the encoding of symbols $k + 1$ through n.

C3. Out of the first k probabilities, choose the first r of them, p_1, $p_2, \ldots, p_r$, so that

$$p_1 + p_2 + \cdots + p_r = 1/4$$

Let 1 be the second bit in the encoding of symbols 1 through r, so that their first two bits are 11, and let 0 be the second bit in the encoding of symbols $r + 1$ through k, so that their first two bits are 10. Similarly, choose the first s of the probabilities $k + 1$ through n, p_{k+1} through p_{k+s}, so that

$$p_{k+1} + p_{k+2} + \cdots + p_{k+s} = 1/4$$

Let 1 be the second bit in the encoding of symbols $k + 1$ through $k + s$, so that their first two bits are 01, and let 0 be the second bit in the encoding of symbols $k + s + 1$ through n, so that their first two bits are 00.

C4. Proceed recursively for the third and following bits, ceasing the process for any subcollection containing only a single symbol or for any subcollection containing two symbols after assigning 1 as the last bit of the encoding of the symbol higher on the list and 0 as the last bit of the encoding of the symbol lower on the list.

The coding algorithm generates a prefix code: if two bit strings are identical after generation of the rth bit, then they must represent symbols in the same subcollection at the rth stage of the algorithm, so that each string must get at least one more bit.

***Example* 1.34.** Consider the source Σ consisting of the first ten characters in the alphabet and the associated probabilities given in the following table:

Symbol	a	b	c	d	e	f	g	h	i	j
Probability	1/16	1/32	1/4	1/8	1/16	1/32	1/8	1/32	1/4	1/32

Application of the coding algorithm to Σ is illustrated in Table 1.1, together with the resulting code. According to the probabilities, $E[B] = 23/8$. Suppose we number the character rows of Table 1.1 and let X be the random variable that, for a selected character, gives its row number. Then a direct computation shows that $H[X] = E[B]$. ■

Table 1.1 Application of coding algorithm for Example 1.34.

Symbol	Prob	Bit Number					Code
		1	2	3	5	6	
c	1/4	1	1				11
i	1/4	1	0				10
d	1/8	0	1	1			011
g	1/8	0	1	0			010
a	1/16	0	0	1	1		0011
e	1/16	0	0	1	0		0010
b	1/32	0	0	0	1	1	00011
f	1/32	0	0	0	1	0	00010
h	1/32	0	0	0	0	1	00001
j	1/32	0	0	0	0	0	00000

The fact that $H[X] = E[B]$ in the preceding example is not exceptional; indeed, so long as the symbols are labeled by natural numbers in decreasing order of probability and all probabilities are negative powers of 2, the expected number of bits must equal the entropy of X when the coding algorithm is applied. To see this, suppose there are n symbols (n rows) and the probability corresponding to the kth row is $p_k = 2^{-m_k}$. According to the coding algorithm, the number of bits in the code of the symbol in the kth row is m_k. Thus,

$$E[B] = \sum_{k=1}^{n} m_k p_k$$

$$= \sum_{k=1}^{n} - p_k \log_2 p_k \tag{1.303}$$

Two questions arise. What happens when the symbol probabilities are not all negative powers of 2? And is it possible to find a more efficient coding scheme, one for which $E[B] < H[X]$? The coding algorithm provides a prefix code by assigning longer bit strings to symbols having lower probabilities. It is not optimal among all coding schemes, but it is optimal among prefix codes since, for these, $E[B] \geq H[X]$. When the probabilities are all powers of 2, the coding algorithm provides an optimal prefix code in which the lower bound $H[X]$ is achieved. When the probabilities are not all powers of 2, the lower bound need not be achieved; however, there always exists a prefix code for which $E[B] - H[X] \leq 1$. Before demonstrating these statements, we examine possible code-word lengths in a prefix code, specifically, the possible encodings that result in r_j code words of length j.

For $j = 1$, a coding scheme can have 0, 1, or 2 code words of length 1. If $r_1 = 0$, then the final code has no encodings of the form $k \leftrightarrow 0$ or $k \leftrightarrow 1$; if $r_1 = 1$, then there exists a single encoding of the preceding types; and if $r_1 = 2$, then both $k \leftrightarrow 0$ and $l \leftrightarrow 1$ are in the code and, by the prefix requirement, they constitute the entire code. In all cases, $r_1 \leq 2$.

The number r_2 of code words of length $j = 2$ depends on r_1. If $r_1 = 2$, then the code is complete and $r_2 = 0$. If $r_1 = 1$, then r_2 can be 0, 1, or 2. To see this, suppose without loss of generality that the single code word of unit length is 1. Then $r_2 = 0$ if the code contains the code word 1 and other code words of the form $0b_1b_2\cdots b_m$ where $m \geq 2$; $r_2 = 1$ if the code contains the code words 1, 01, and other code words of the form $00b_1b_2\cdots b_m$ where $m \geq 1$; and $r_2 = 2$ if the full code consists of the code words 1, 01, and 00. Finally, if $r_1 = 0$, then r_2 can take on the values 0, 1, 2, 3, or 4. The case $r_2 = 4$ occurs when the full code is 11, 10, 01, and 00. The case $r_2 = 3$ occurs when the code words compose a set of the form 11, 10, 01, and further code words of the form $00b_1b_2\cdots b_m$, where $m \geq 1$. The case $r_2 = 2$ occurs when the code words compose a set of the form 11, 01, and further code words of the forms $00b_1b_2\cdots b_m$ and $10c_1c_2\cdots c_q$, where m, $q \geq 1$. The case $r_2 = 1$ occurs when the code words compose a set of the form 11 and further code words of the forms $00b_1b_2\cdots b_m$, $10c_1c_2\cdots c_q$, and $01d_1d_2\cdots d_s$ where m, q, $s \geq 1$. The case $r_2 = 0$ occurs when all code words are of the form $b_1b_2\cdots b_m$ with $m \geq 3$. To summarize, if $r_1 = 0$, then $r_2 \leq 4$; if $r_1 = 1$, then $r_2 \leq 2$; and if $r_1 = 2$, then $r_2 = 0$. These expressions may be combined to yield the single inequality $r_2 \leq 4 - 2r_1$.

Were we to continue, the number r_3 of code words of length $j = 3$ depends on r_1 and r_2. Indeed, it can be shown by mathematical induction that, for $j = 1, 2, \ldots, \nu$, where ν is the maximum of the code-word lengths,

$$r_j \leq 2^j - 2^{j-1}r_1 - 2^{j-2}r_2 - \cdots - 2r_{j-1} \tag{1.304}$$

Conversely, since the entire argument is constructive, if $r_1, r_2, \ldots, r_\nu$ is any sequence of nonnegative integers satisfying the inequality for $j = 1, 2, \ldots, \nu$, then there exists a prefix code having r_j code words of length $j = 1, 2, \ldots, \nu$.

***Example* 1.35.** To illustrate Eq. 1.304, let $r_1 = 1$, $r_2 = 0$, $r_3 = 3$, and $r_4 = 2$. Since $r_1 \leq 2$,

$$r_2 \le 4 - 2r_1 = 2$$
$$r_3 \le 8 - 4r_1 - 2r_2 = 4$$
$$r_4 \le 16 - 8r_1 - 4r_2 - 2r_3 = 2$$

there exists a prefix code having 1, 0, 3, and 2 code words of lengths 1, 2, 3, and 4, respectively. One such code is composed of the code words 1, 011, 010, 001, 0001, and 0000. Another would be 1, 000, 011, 010, 0010, and 0011. ■

Theorem 1.22. For the source $\Sigma = \{1, 2, \ldots, n\}$, there exists a prefix code for which the code word for k has m_k bits, $k = 1, 2, \ldots, n$, if and only if

$$\sum_{k=1}^{n} 2^{-m_k} \le 1 \quad ■ \tag{1.305}$$

Equation 1.305 provides a lower bound on the sizes of the code-word lengths in a prefix code; however, since it is an equivalence, it has a valid converse. To demonstrate the meaning of the converse, suppose we are given code-word lengths $m_1 = 3$, $m_2 = 4$, $m_3 = 3$, $m_4 = 3$, and $m_5 = 2$. Since the sum in Eq. 1.305 yields 15/16, there exists a prefix code for the integers 1 through 5 with the given-word code lengths. One such code is given by $1 \leftrightarrow 101$, $2 \leftrightarrow 0001$, $3 \leftrightarrow 100$, $4 \leftrightarrow 011$, $5 \leftrightarrow 11$.

Theorem 1.22 is a consequence of Eq. 1.304, where in the context of the theorem, r_j gives the number of code-word lengths m_k equal to j. Owing to the relationship between $r_1, r_2, \ldots, r_v$ and $m_1, m_2, \ldots, m_n$,

$$\sum_{k=1}^{n} 2^{-m_k} = \sum_{j=1}^{v} r_j 2^{-j} \tag{1.306}$$

Furthermore, collecting all the r terms on one side of Eq. 1.304 and dividing by 2^j yields the equivalent inequality

$$\sum_{i=1}^{j} r_i 2^{-i} \le 1 \tag{1.307}$$

Thus, the desired prefix code exists if and only if Eq. 1.307 holds for all j, which from Eq. 1.306 is true if and only if Eq. 1.305 holds.

Theorem 1.23. Let X be a random variable taking on the values $1, 2, \ldots, n$. If B counts the number of bits in a prefix encoding of $\{1, 2, \ldots, n\}$, then

$$E[B] \geq H[X] \tag{1.308}$$

Moreover, there exists at least one prefix code such that

$$E[B] \leq H[X] + 1 \quad \blacksquare \tag{1.309}$$

To demonstrate that $H[X]$ is a lower bound for $E[B]$, let A denote the sum in Eq. 1.305, $p_k = P(X = k)$, and $t_k = 2^{-m_k}/A$. Then

$$\begin{aligned} -\sum_{k=1}^{n} p_k \log_2 \frac{p_k}{t_k} &= \log_2 e \sum_{k=1}^{n} p_k \log\left(\frac{t_k}{p_k}\right) \\ &\leq \log_2 e \sum_{k=1}^{n} p_k \left(\frac{t_k}{p_k} - 1\right) \\ &= \log_2 e \left(\sum_{k=1}^{n} t_k - \sum_{k=1}^{n} p_k\right) \end{aligned} \tag{1.310}$$

Since both sums in the last expression sum to 1, the sum on the left is bounded above by 0. Using this bound, expressing the logarithm of a quotient as the difference of the logarithms, and using the fact that $A \leq 1$, we demonstrate Eq. 1.308:

$$\begin{aligned} H[X] &= -\sum_{k=1}^{n} p_k \log_2 p_k \\ &\leq -\sum_{k=1}^{n} p_k \log_2 t_k \\ &= \sum_{k=1}^{n} p_k (m_k + \log_2 A) \\ &\leq \sum_{k=1}^{n} p_k m_k \end{aligned} \tag{1.311}$$

To fully prove Theorem 1.23, it remains to show that there exists at least one prefix code satisfying Eq. 1.309. For $k = 1, 2, \ldots, n$, let m_k be an integer for which

$$-\log_2 p_k \le m_k \le -\log_2 p_k + 1 \tag{1.312}$$

Then

$$\sum_{k=1}^{n} 2^{-m_k} \le \sum_{k=1}^{n} 2^{-\log_2 p_k} = 1 \tag{1.313}$$

and Theorem 1.22 ensures the existence of a prefix code with bit lengths m_k. Moreover,

$$\begin{aligned} E[B] &= \sum_{k=1}^{n} p_k m_k \\ &\le \sum_{k=1}^{n} p_k(-\log_2 p_k + 1) \\ &= H[X] + 1 \end{aligned} \tag{1.314}$$

***Example* 1.36.** This example elucidates the bit interpretation of entropy, mean conditional entropy, and expected amount of information in the context of prefix coding. Consider coding the sixteen numerals of the hexadecimal system assuming that their occurrences are equally likely. In effect, we have a random variable Y that can take on the values 0, 1, 2,..., 15 with equal probabilities. To obtain an optimal prefix code, we can apply the coding algorithm with the numerals listed ordinarily. Applying the coding algorithm and then interchanging the roles of 0 and 1 yields a code in which the code word of each hexadecimal integer is its binary equivalent: $0 \leftrightarrow 000$, $1 \leftrightarrow 0001$, $2 \leftrightarrow 0010$, $3 \leftrightarrow 0011$, $4 \leftrightarrow 0100$,..., $14 \leftrightarrow 1110$, $15 \leftrightarrow 1111$. Y has entropy $H[Y] = \log_2 16 = 4$. From an information perspective, the expected amount of information in Y is 4 bits, which is the number of bits required for the binary encoding of the hexadecimal system. Physically, transmission of a binary-encoded hexadecimal numeral requires 4 bits (so long as the probability mass is assumed to be equally distributed). Now suppose the numerals are grouped according to the first three bits of their binary codes, thereby creating eight classes numbered 0 through 7: $C_0 = \{0, 1\}$, $C_1 = \{2, 3\}$,..., $C_7 = \{14, 15\}$. If X gives the class number of an arbitrarily selected numeral, then X has eight equally likely outcomes and $H[X] = 3$. Given $X = i$, the equally likely outcomes of $Y|i$ mean that the conditional entropy of Y given $X = i$ is $H[Y|i] = \log_2 2 = 1$. Hence, the mean conditional entropy of Y relative to X is $\overline{H}[Y|X] = 1$ and the amount of information obtained for Y by observing X is $I_X[Y] = 3$ bits, which agrees with X physically supplying 3 bits. In general, for $0 < s < r$, if Y has 2^r equally likely outcomes and X denotes the number of one of the 2^s classes resulting from equally dividing up the outcomes of Y, then $H[Y] = \log_2 2^r = r$, $\overline{H}[Y|X] = \log_2 2^{r-s} = r - s$, and $I_X[Y] = s$. The physical interpretation is that X has supplied s bits. The clarity of the interpretation results from the assumption that there are 2^r equally likely outcomes. In other circumstances, $I_X[Y]$ will not possess such a striking interpretation; nevertheless, the example illustrates the meaning of entropy and its relationship to information, as well as the underlying meanings of $H[Y]$, $\overline{H}[Y|X]$, and $I_X[Y]$. ■

1.11.2. Optimal Coding

Optimal coding involves finding a coding scheme for which the expected code-word length is minimized over all possible acceptable codes, in our case, binary prefix codes.

A prefix code can be viewed via a *coding tree*. Consider a tree, each branch being labeled 0 or 1, emanating from a root node and having a set of n terminal nodes. A prefix code for an n-symbol source is obtained by tracing out the branches from the root to each terminal node, associating a symbol with each terminal node, and assigning to each symbol the code word formed by the string of 0s and 1s running along the branches from the root to its corresponding terminal node. Moreover, every prefix code for a finite source can be so viewed.

A subclass of prefix codes is defined by trees for which each node is either terminal or has two branches (0 and 1) emanating from it. From the perspective of coding efficiency, nothing is lost by restricting ourselves to these *binary trees*. If d is a nonterminal node in a nonbinary tree from which there is generated only a single branch, then there must be a terminal node a subsequent to d in the tree so that the bit-strings to nodes d and a are of the forms $b_1b_2\cdots b_m$ and $b_1b_2\cdots b_mb_{m+1}\cdots b_q$, respectively. The code using

$$a \leftrightarrow b_1b_2\cdots b_mb_{m+1}\cdots b_q \tag{1.315}$$

can be replaced by a code that is exactly the same except that $a \leftrightarrow b_1b_2\cdots b_m$. The new code is a prefix code with $E[B]$ reduced. Subsequent to Theorem 1.22 we considered the code: $1 \leftrightarrow 101$, $2 \leftrightarrow 0001$, $3 \leftrightarrow 100$, $4 \leftrightarrow 011$, $5 \leftrightarrow 11$. A more efficient code results from "pruning" the coding tree back to obtain the code: $1 \leftrightarrow 101$, $2 \leftrightarrow 00$, $3 \leftrightarrow 100$, $4 \leftrightarrow 01$, $5 \leftrightarrow 11$, which is described by a binary coding tree. Since we are now concerned with efficiency, we henceforth assume that all codes correspond to binary coding trees.

When all symbol probabilities are negative powers of 2, an optimal code is achieved by the previously given coding algorithm with the lower bound

$H[X]$ being attained. More generally, Theorem 1.23 guarantees there exists a code whose expected bit length is within 1 of $H[X]$ but it does not provide a procedure for constructing an optimal code. The *Huffman code*, whose encoding algorithm we now describe, is optimal in the class of codes under consideration. To construct the Huffman code, list the source symbols according to their probabilities of occurrence, the highest probability down to the lowest. The coding tree is generated backward with the starting (terminal) nodes being the symbol-probability pairs $(a, p(a))$. Begin by combining two nodes of lowest probability, say $(a, p(a))$ and $(b, p(b))$, to form a new node $([a, b], p(a) + p(b))$. Branches of the coding tree run from node $([a, b], p(a) + p(b))$ to nodes $(a, p(a))$ and $(b, p(b))$. Label the top branch 0 and the bottom branch 1. The next node is generated by again combining two nodes of lowest probability, where the recently created node and its probability are part of the listing, and adding their probabilities. Branches connect the new node to the two chosen nodes and are labeled 0 and 1. Node creation is continued recursively until only a single node exists. The final node is the root and the original symbols are its terminal nodes. The Huffman code associates with each symbol the string of 0s and 1s going along the branches from the root node to the symbol. The choice of labeling the upper branch 0 and the lower branch 1 is arbitrary; at each stage, either can be labeled 0 and the other 1. Figure 1.3 shows a Huffman coding tree and the resulting code words.

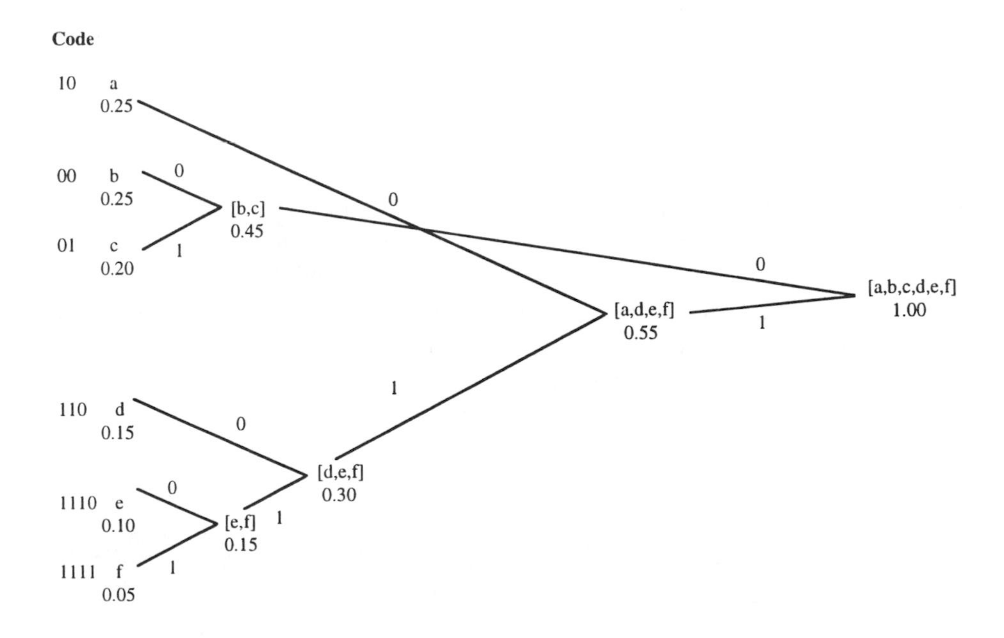

Figure 1.3 Huffman code.

Theorem 1.24. The Huffman code is optimal among prefix codes. ■

The theorem is demonstrated by mathematical induction on the number n of source words. If $n = 2$, then the Huffman code contains only two code words, 0 and 1, and is optimal. For the induction hypothesis, suppose the Huffman code is optimal for every source containing $m \leq n - 1$ symbols. To prove the theorem, we assume a source with n symbols, $\Sigma_n = \{a_1, a_2,..., a_n\}$ with associated probabilities $p_1, p_2,..., p_n$, and show that, based on the induction hypothesis, the Huffman code is optimal. Let β_n be the expected number of bits for the Huffman code and suppose there exists a different code, C_n, whose expected number of bits, α_n, is strictly less than β_n. For an optimal code, if the symbols a and b have probabilities $p(a)$ and $p(b)$ with $p(a) < p(b)$, then the length of the code word for a must be greater than or equal to the length of the code word for b. Consequently, if the symbols of Σ_n are listed in order of decreasing probability, it can be assumed without loss of generality that a_{n-1} and a_n appear as a node pair whose branches emanate from the same node at the last stage of the coding tree for C_n (for otherwise we could change the coding tree by interchanging symbols to produce a coding tree with all symbols having the same code-word lengths and with a_{n-1} and a_n appearing as a node pair whose branches emanate from the same node at the last stage of the tree). If the terminal nodes for symbols a_{n-1} and a_n are removed from the tree and a symbol c_{n-1} is placed at the new terminal node created, then the resulting tree provides a code C_{n-1} for the source $\Sigma_{n-1} = \{a_1, a_2,..., a_{n-2}, c_{n-1}\}$. If the probability associated with c_{n-1} in Σ_{n-1} is $p_{n-1} + p_n$, then the expected number of bits for the new code C_{n-1} is

$$\alpha_{n-1} = \alpha_n - p_n - p_{n-1} \tag{1.316}$$

On the other hand, by construction of the Huffman code $\mathcal{H}_n$ for the n-symbol source Σ_n, a_{n-1} and a_n appear as a node pair whose branches emanate from the same node at the last stage of the coding tree for $\mathcal{H}_n$. If they are removed and replaced as in the case for C_n to form the source $\Sigma_{n-1} = \{a_1, a_2,..., a_{n-2}, c_{n-1}\}$ with c_{n-1} having probability $p_{n-1} + p_n$, then the reduced tree is the coding tree for the Huffman code $\mathcal{H}_{n-1}$ corresponding to Σ_{n-1} and the expected number of bits for the Huffman code $\mathcal{H}_{n-1}$ is

$$\beta_{n-1} = \beta_n - p_n - p_{n-1} \tag{1.317}$$

Since, by supposition, $\alpha_n < \beta_n$, Eqs. 1.316 and 1.317 imply that $\alpha_{n-1} < \beta_{n-1}$; however, this contradicts the induction hypothesis and therefore we conclude that the Huffman code on n symbols is optimal.

Exercises for Chapter 1

1. Prove the probability addition theorem for $n = 3$ without using mathematical induction.
2. Show that, for $n \geq 1$ and $1 \leq k \leq n$, $C_{n,k} = C_{n-1,k-1} + C_{n-1,k}$.
3. Suppose there are m_k white balls and n_k black balls in urn A_k for $k = 1, 2, \ldots, m$. Suppose an urn is randomly selected, with p_k being the probability of selecting urn A_k, and a ball is uniformly randomly selected from the chosen urn. Given that a black ball is selected, what is the probability that it was chosen from urn A_1?
4. Prove that, if E and F are independent, then so are E and F^c, E^c and F, and E^c and F^c.
5. Suppose an experiment has two possible outcomes, a and b, with the probability of a being $p > 0$ and the probability of b being $q > 0$. The experiment is independently repeated until the outcome a occurs and the random variable X counts the number of times b occurs before the outcome a. Find the probability density for X.
6. Find the constant c so the function $f(x)$ defined by $f(x) = cxe^{-bx}$ for $x \geq 0$ and $f(x) = 0$ for $x < 0$ is a legitimate density.
7. Demonstrate all the relations of Eq. 1.37.
8. Derive the density for $Y = |X|$ in terms of the density for X. Apply the result to the uniform density over $[-1, 2]$.
9. Derive the density for $Y = X^2$ in terms of the density for X.
10. Assuming $X \geq 0$, find the density for $Y = X^{1/2}$ in terms of the density for X. Apply the result to the case where X is uniformly distributed over $[4, 9]$.
11. Apply the result of Example 1.9 when X has an exponential density with parameter b.
12. The *Pareto distribution* has positive parameters r and a, and is defined by the density $f(x) = ra^r x^{-(r+1)}$ for $x \geq a$ and $f(x) = 0$ for $x < a$. Show that the Pareto distribution possesses a kth moment if and only if $k < r$. Assuming they exist, show that the mean and variance of the Pareto distribution are given by $\mu = ra/(r-1)$ and $\sigma^2 = ra^2/(r-1)^2(r-2)$.
13. Derive the mean and variance of the exponential distribution directly from the density without using the moment-generating function.
14. Show that $\mathrm{Var}[X] = E[X^2] - E[X]^2$.
15. Show that $\mathrm{Var}[aX + b] = a^2\mathrm{Var}[X]$ for constants a and b.
16. Show that, for constants a and b, $M_{aX+b}(t) = e^{bt}M_X(at)$.
17. Compare the bound given by Chebyshev's inequality of Eq. 1.74 for the exponential distribution with the actual probability.
18. For the discrete random variable characterized by the density $f(-2) = f(2) = 1/8$ and $f(0) = 3/4$, show that Chebyshev's inequality is tight (cannot be improved).

19. If $f(x)$ is a continuous density, then its *median* is the value $\tilde{\mu}$ for which $P(X \leq \tilde{\mu}) = P(X \geq \tilde{\mu})$. Find the median of the exponential and uniform densities.
20. For a continuous distribution, the *mean deviation* is defined by

$$MD(X) = \int_{-\infty}^{\infty} |x - \mu_X| f_X(x)\, dx$$

Find the mean deviation for the exponential and uniform distributions.
21. Find $M_X'(t)$ and $M_X''(t)$ for the binomial distribution. From these, derive μ, μ_2', and σ^2.
22. The *negative binomial distribution* is defined by the discrete density

$$f(x) = \binom{x+k-1}{k-1} p^k q^x$$

for $x = 0, 1, 2, \ldots$, where k is a positive integer, $0 < p < 1$ and $p + q = 1$. Show that

$$M_X(t) = p^k(1 - qe^t)^{-k}$$

Using the moment-generating function, derive μ, μ_2', and σ^2.
23. For the Poisson density $\pi(x; \lambda)$ with parameter λ, show $\pi(0; \lambda) + \pi(1; \lambda) + \cdots = 1$.
24. Show that the Poisson density $\pi(x; \lambda)$ with parameter λ satisfies the recursion relation

$$\pi(x+1; \lambda) = \frac{\lambda}{x+1} \pi(x; \lambda)$$

25. Find $M_X'(t)$ and $M_X''(t)$ for the Poisson distribution. From these, derive μ, μ_2', and σ^2.
26. Using integration by parts, show that the probability distribution function of a Poisson random variable with parameter λ is given by

$$\Pi(x; \lambda) = \frac{1}{x!} \int_{\lambda}^{\infty} t^x e^{-t}\, dt$$

27. A discrete random variable X with density $f(x) = p(1 - p)^x$ for $x = 0, 1, 2, \ldots$, where $0 < p < 1$, is called a *geometric* density (see Exercise 1.5). Show

$$M_X(t) = \frac{p}{1-(1-p)e^t}$$

Use the moment-generating function to show that $\mu = (1 - p)/p$ and $\sigma^2 = (1 - p)/p^2$.

28. Assuming the order of integration and summation can be interchanged, show that, for a continuous random variable X possessing finite moments of all orders,

$$M_X(t) = \sum_{k=0}^{\infty} E[X^k]\frac{t^k}{k!}$$

Assuming the order of summation can be interchanged, show the same result for a discrete random variable. Hint: In both instances, expand e^{tx} as a Maclaurin series.

29. Show that the total integral of the standard normal density is 1 (Eq. 1.98). Hint: Let I denote the integral, so that

$$I^2 = \left(\frac{1}{\sqrt{2\pi}}\int_{-\infty}^{\infty} e^{-x^2/2}\,dx\right)\left(\frac{1}{\sqrt{2\pi}}\int_{-\infty}^{\infty} e^{-y^2/2}\,dy\right) = \frac{1}{2\pi}\int_{-\infty}^{\infty}\int_{-\infty}^{\infty} e^{-(x^2+y^2)/2}\,dydx$$

and change to polar coordinates.

30. Find $M_X'(t)$, $M_X''(t)$, and $M_X'''(t)$ for the normal distribution. From these, derive μ, μ_2', μ_3', and σ^2.
31. In the text it was stated that the transformation $Z = (X - \mu)/\sigma$ transforms a normally distributed random variable X with mean μ and variance σ^2 into a standard normal variable and this is demonstrated by applying the result of Example 1.8. Show the details.
32. Use standard calculus maximum-minimum techniques to prove that the normal density has a maximum at μ and points of inflection at $\mu \pm \sigma$.
33. Find $M_X'''(t)$ for the gamma distribution and use it to find μ_3'.
34. For $\alpha = \nu/2$ and $\beta = 2$, the gamma distribution is known as the *chi-square distribution*. Write out the probability density, moment-generating function, mean, and variance for the chi-square distribution.
35. The *Weibull distribution* has density $f(x) = \alpha\beta^{-\alpha}x^{\alpha-1}e^{-(x/\beta)^{\alpha}}$ for $x > 0$ and $f(x) = 0$ for $x \le 0$, where $\alpha > 0$ and $\beta > 0$. Show that the kth moment of the Weibull distribution is $\mu_k' = \beta^k\Gamma((\alpha + \beta)/\alpha)$. Hint: Use the substitution $u = (x/\beta)^{\alpha}$ in the integral giving $E[X^k]$. From μ_k' obtain the mean and variance.
36. Letting $\alpha = 2$ and $\beta = \sqrt{2}\,\eta$ in the Weibull distribution gives the *Rayleigh distribution* with parameter η (also see Example 1.19). From the

results of Exercise 1.35, find the mean and variance of the Rayleigh distribution.

37. Find the moment-generating function of the Laplace distribution and show that its mean and variance are given by μ and $2/\alpha^2$, respectively.
38. A basic operation in image processing is *thresholding*. For the input random variable X, the output $Y = g(X)$ is defined by $Y = 1$ if $X \geq \tau$ and $Y = 0$ if $X < \tau$, where τ is a fixed parameter. Show that, for X with a continuous density, the probability distribution function for Y is given by $F_Y(y) = 0$ for $y < 0$, $F_Y(y) = F_X(\tau)$ for $0 \leq y < 1$, and $F_Y(y) = 1$ for $y \geq 1$. Apply this result when X has a Laplace density with parameters μ and α.
39. Show that an exponentially distributed random variable is memoryless.
40. Consider the joint discrete densities

$$f(x,y) = \frac{3!}{x!\,y!(3-x-y)!\,2^x 3^y 6^{3-x-y}}$$

defined for all $x, y = 0, 1, 2, 3$ such that $x + y \leq 3$, and

$$g(x,y) = \binom{3}{x}\binom{3}{y}\frac{2^{3-y}}{216}$$

defined for all $x, y = 0, 1, 2, 3$. Numerically compute the outcome probabilities for both densities. Show that $g(x, y)$ is the product of two binomial densities, one with $n = 3$ and $p = 1/2$, and the other with $n = 3$ and $p = 1/3$. Show that both have the same marginal densities, these being the densities whose product gives $g(x, y)$.

41. Show parts (i) and (ii) of Theorem 1.10.
42. Repeat Example 1.16 with X and Y uniformly distributed over the region bounded by the curves $y = x^2$ and $y = x^{1/2}$, $x > 0$.
43. Suppose X and Y are uniformly distributed over the region bounded by the x axis, y axis, and the line $y = x/2 + 1$. Find $P(X < 1/2)$, $P(Y < X)$, and $P(Y > X^2)$.
44. Find the constant c so that the function $f(x, y) = ce^{-(2x+4y)}$ for $x, y \geq 0$ and $f(x, y) = 0$ otherwise is a bivariate density. Find $P(X > Y)$ and the marginal densities.
45. Find the constant c so that the function $f(x, y) = cxe^{-(x+y)}$ for $x, y \geq 0$ and $f(x, y) = 0$ otherwise is a bivariate density. Find the marginal densities.
46. Let Z_1, Z_2, and Z_3 be independent standard normal random variables, $Y_1 = Z_1 + Z_2 + Z_3$, $Y_2 = Z_2 - Z_1$, and $Y_3 = Z_3 - Z_1$. Find the joint density of Y_1, Y_2, and Y_3.
47. Let X_1 and X_2 be independently distributed exponential random variables and find the density of $Y = X_1X_2$.

48. For $\kappa > 1$, find the covariance and correlation coefficient for the random variables X and Y possessing a uniform distribution over the region bounded by the curves $y = x^{\kappa}$ and $y = x^{1/\kappa}$, $x > 0$. Find the limits of the correlation coefficient as $\kappa \to 1$ and $\kappa \to \infty$.
49. Look up the Schwarz inequality and show how it relates to Theorem 1.14.
50. Show that if $Y = aX + b$, then $\mathrm{Cov}[X, Y] = a\mathrm{Var}[X]$.
51. Show that $\mathrm{Cov}[X, Y] = E[XY] - \mu_X\mu_Y$.
52. Show that the marginal distribution for X in the bivariate normal distribution is normally distributed with mean μ_X and variance σ_X^2 by integrating the joint normal density with respect to dy. Hint: let $u = (x - \mu_X)/\sigma_X$ and $v = (y - \mu_Y)/\sigma_Y$, complete the square in the quadratic in the resulting exponential, and make the change of variables $z = (1 - \rho^2)^{-1/2}(v - \rho u)$.
53. Suppose X possesses a Poisson distribution with parameter λ. Show that the limit as $\lambda \to \infty$ of the moment-generating function of the standardized Poisson random variable $(X - \lambda)/\lambda^{1/2}$ converges to the moment-generating function of the standard normal variable.
54. Rewrite Chebyshev's inequality in Eq. 1.200 for the case when $X_1, X_2,\ldots, X_n$ possess a multivariate normal distribution according to Eq. 1.190.
55. Suppose X is binomially distributed with parameters n and p. Based on the central limit theorem, write out an approximate expression for $P(a < X < b)$ in terms of the standard normal density under the supposition that n is large. Note that a rule-of-thumb is that the approximation is acceptable if $np \geq 5$ and $n(1 - p) \geq 5$.
56. Suppose X is Poisson distributed with mean λ. Based on the central limit theorem, write out an approximate expression for $P(a < X < b)$ in terms of the standard normal density under the supposition that n is large.
57. Apply the weak law of large numbers to the sequence of independent random variables $X_1, X_2,\ldots$ where X_n is Poisson distributed with mean λ when n is odd and binomially distributed with fixed parameters m and p when n is even.
58. Consider a sequence of uncorrelated random variables $X_1, X_2,\ldots,$ where X_k has mean $k\alpha$, α fixed, and all variances are bounded by a common bound. Show that the weak law of large numbers states that, for any $\varepsilon > 0$,

$$\lim_{n\to\infty} P\left(\left|\frac{1}{n}\sum_{k=1}^{n} X_k - \frac{(n+1)\alpha}{2}\right| \geq \varepsilon\right) = 0$$

59. Show that the sample mean is an MVUE for the mean of the exponential distribution.

60. Show that the sample mean is an MVUE for the mean of the Poisson distribution.
61. Show that the sample mean is an MVUE for the mean of the binomial distribution with parameters $n = 1$ and p.
62. Find the method-of-moments estimator for the mean of the Poisson distribution.
63. Find the method-of-moments estimator for the mean of the geometric distribution.
64. Find the method-of-moments estimator for the parameter α of the beta distribution, given $\beta = 1$.
65. For X having an exponential distribution and $X_1, X_2,..., X_n$ being a random sample arising from X, find the densities of the maximum and minimum order statistics.
66. Show that the median of a random sample of size $2n + 1$ from a uniform distribution over (0, 1) has a beta distribution with parameters $\alpha = \beta = n + 1$.
67. Suppose $X_1, X_2,..., X_n, X_{n+1},..., X_m$ are independent identically distributed continuous random variables. Find

$$P(\min\{X_{n+1},\ldots,X_m\} \geq \max\{X_1,X_2,\ldots,X_n\})$$

68. Using 1,000 random values generated by a uniform random-value generator, simulate a gamma distribution with parameters $\alpha = 20$ and $\beta = 0.5$. Construct the histogram of the data, use the sample mean and sample variance to obtain estimates of the mean and variance, and compare the estimates with the theoretical values.
69. Verify the last statement in Example 1.30, including the unbiasedness assertion.
70. Show that the sample mean is the maximum-likelihood estimator for the mean of the Poisson distribution.
71. Show that the sample mean is the maximum-likelihood estimator for the mean of the exponential distribution.
72. For a random sample of size n arising from a gamma distribution with α known and β unknown, show that $\hat{\beta} = \overline{X}/\alpha$ is the maximum-likelihood estimator for β.
73. Reconsider Example 1.32 with the noise uniformly distributed over $[0, \beta]$.
74. Reconsider Section 1.9.3 for maximum noise and show that the maximum-likelihood filter is a moving minimum (flat erosion).
75. Let X possess a uniform distribution over the interval $[\mu - 1/2, \mu + 1/2]$ and $Y_1, Y_2,..., Y_n$ be the n order statistics corresponding to a random sample of odd size n. Find the expected values of the minimum Y_1 and the maximum Y_n. Suppose μ is unknown. Compare the sample-mean

estimator for μ with the estimator $\hat{\mu} = (Y_1 + Y_n)/2$. Specifically, check the bias and variance of $\hat{\mu}$.

76. In the text we showed that entropy is maximum for equally likely outcomes when there are two outcomes. Show the corresponding statement when there is an arbitrary number n of outcomes.
77. Demonstrate Eq. 1.295.
78. Demonstrate Eq. 1.298.
79. Demonstrate Eq. 1.299.
80. If the random variables ..., X_{-1}, X_0, X_1,... are independent and identically distributed, what can be said about the conditional entropy (Eq. 1.301)?
81. The entropy of a continuous random variable X possessing density $f(x)$ is defined by

$$H[X] = -E[\log_2 f(X)] = -\int_{-\infty}^{\infty} f(x) \log_2 f(x) dx$$

where the integration is assumed to be over $\{x: f(x) > 0\}$. Show that the entropy of a random variable uniformly distributed over the interval $(0, a)$ is $\log_2 a$.

82. Referring to the definition of Exercise 1.81, find the entropy of a normally distributed random variable possessing mean μ and variance σ^2.
83. Apply the coding algorithm of Section 1.11.1 to the source in the following table.

symbol	a	b	c	d	e	f	g	h	i
probability	1/8	1/32	1/8	1/8	1/8	1/32	1/8	1/16	1/4

Show that the lower entropy bound is achieved by the expected bit length.

84. The case for $j = 2$ in Eq. 1.304 is explained in detail in the text. Give a similar detailed explanation for $j = 3$.
85. Show that the code-word lengths $m_1 = 2$, $m_2 = 2$, $m_3 = 3$, $m_4 = 3$, and $m_5 = 3$ satisfy the condition of Theorem 1.22. Find all codes possessing these code-word lengths.
86. Find the Huffman code, $H[X]$, and $E[B]$ for the source in the following table:

symbol	a	b	c	d	e	f	g	h	i
probability	1/10	1/10	1/20	1/5	1/5	1/20	1/20	3/20	1/10

87. Constructing the Huffman code can be complicated when there is a large number of symbols. To avoid cumbersome coding, with the loss of some efficiency, one can employ a *truncated Huffman code*. For truncation, the symbols with the smallest probabilities are grouped into a single group symbol for the purpose of the Huffman code tree. When the tree is completed and coding accomplished, symbols forming the group symbol are individually coded by appending a fixed-bit-length binary code to the code word generated by the coding tree. If there are between 5 and 8 grouped symbols, then 3 bits are appended; if there are between 9 and 16 grouped symbols, then 4 bits are appended; etc. Apply truncated Huffman coding to the following source by combining the symbols with the six smallest probabilities into a group:

symbol	a	b	c	d	e	f	g	h
probability	0.025	0.015	0.050	0.050	0.075	0.010	0.020	0.025

symbol	i	j	k	l	m	n	o	p
probability	0.075	0.125	0.075	0.175	0.005	0.050	0.100	0.125

88. Truncate the Huffman code for the source of Exercise 1.86 by grouping the symbols with the three lowest probabilities. Compare the expected bit lengths for the Huffman and truncated Huffman codes for this source.

89. Suppose a_1 and a_2 are two source symbols with probabilities p_1 and p_2, respectively, the source is coded optimally, and the code words for a_1 and a_2 are of lengths m_1 and m_2, respectively. Show that, if $p_1 < p_2$, then $m_1 \geq m_2$.

Chapter 2
Random Processes

2.1. Random Functions

Consider three commonplace imaging scenarios: (1) a document image is sent through various digital operations, say, scanning, printing, copying, and then faxing; (2) an image is compressed to reduce the number of bytes for storage or transmission; and (3) geometric features are computed from an image to characterize the degree to which an industrial process is in control. In each scenario, characterization of the image processing cannot be based on the effects of processing a single image, or on the effects of processing any finite number of images. For the document image, if one wishes to design a filter that will *restore* it, then that filter needs to be designed in accordance with how the various stages of image processing affect the class of images to be filtered, in particular, how the processing affects the probabilistic distribution of the image class. In the case of the compressed image, if one wishes to measure the degree of compression or to design a decompression filter, then both the compression and goodness of the restoration filter must be evaluated relative to the class of images to be compressed and decompressed. Any particular image will likely occur very rarely and the system must be designed and evaluated probabilistically. Finally, for feature generation, image observations will vary, features will be random variables, and classification accuracy will depend on the joint distribution of the features. At their root, image and signal processing are applied disciplines within the domain of random processes.

For a more quantitative example, consider an ordinary function, say,

$$x(t) = a \cos bt$$

Suppose there is variability in the system generating the signal, so that both amplitude and frequency are subject to variation. Suppose also that the signal is transmitted and additive noise is thereby imposed on the signal. From the standpoint of the receiver, the signal is not $a \cos bt$; rather, it is some variant of the intended cosine wave, say,

$$x_1(t) = a_1 \cos b_1 t + n_1(t)$$

where a_1, b_1, and $n_1(t)$ correspond to the amplitude, frequency, and additive noise of the actual signal received. However, these correspond only to the currently received signal; for the next transmission they will be different. Hence, rather than modeling the signal deterministically as $x(t) = a \cos bt$, it is more realistic to model it as

$$X(t) = A \cos Bt + N(t)$$

where A, B, and $N(t)$ (for each t) are random variables, thereby making $X(t)$ a random variable for each t. The properties of the signal can then be studied as one might study the properties of a deterministic function; however, now all properties are relevant to the distribution of the random variables A, B, and $N(t)$, and $X(t)$ is now a random function.

The theory of random functions permeates contemporary science and has its roots in modern physics. It applies whenever there is sufficient indeterminism in the functions under study that a deterministic model would prove too crude.

A *random function*, or *random process*, is a family of random variables $\{X(\omega; t)\}$, t lying in some index set T, where, for each fixed t, the random variable $X(\omega; t)$ is defined on a sample space S $(\omega \in S)$. If we consider a fixed choice of ω, then $X(\omega; t)$ defines a function on the set T, and each such function is termed a *realization* of the random function. If T is a subset of the real line $\Re$, then, for fixed ω, $X(\omega; t)$ is a signal and the random function $\{X(\omega; t)\}$ is called a *random signal, stochastic process*, or *random time function*. If T is a subset of the Euclidean plane $\Re^2$, then $\{X(\omega; t)\}$ is called a *random image* or *random field*, and, as a point in the plane, t may be written as $t = (u, v)$, so the process may be written as $\{X(\omega; u, v)\}$. Should a random process be defined only on the integers, it is sometimes called a *random time series*; if it is defined on the discrete Cartesian grid, it may be called a *random digital image*. More generally, t can be a point in n-dimensional Euclidean space $\Re^n$. In this case, each realization is a deterministic function of n variables.

To simplify notation we usually write $X(t)$ to denote a random function, keeping in mind the underlying probability mass governing process behavior. In particular, if we fix t and let ω vary, then $X(t)$ is a random variable on the sample space. A specific realization, which is a deterministic function on T, will often be denoted by a lower-case symbol, $x(t)$, or, if we wish to emphasize the choice of ω that leads to the particular realization, by $x_\omega(t)$. Unless otherwise specified, t can be either a scalar (a random signal) or a vector (a random image).

Since, for each t, a random function $X(t)$ is a random variable, for each t it possesses a probability distribution function

$$F(x; t) = P(X(t) \leq x) \tag{2.1}$$

called a *first-order distribution*. For the random functions that concern us, $X(t)$ will possess a *first-order density*

$$f(x;t) = \frac{d}{dx}F(x;t) \tag{2.2}$$

where the derivative might involve delta functions. Figure 2.1 shows four realizations of a random function. Corresponding to the four realizations, there are four outcomes of the random variable $X(t')$.

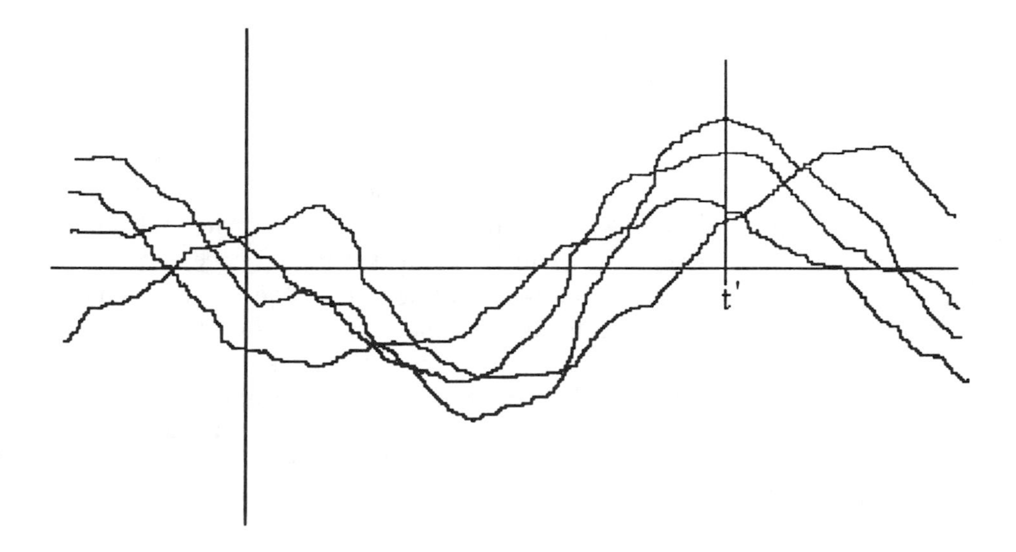

Figure 2.1 Four realizations of a random function showing four outcomes of $X(t')$.

***Example* 2.1.** A digital image is a function of two variables defined on some domain in the Cartesian grid, the domain often being a rectangle corresponding to the image frame. We denote a digital image by a matrix. To locate the position of the image frame in the grid, we use a bold character to represent the matrix entry situated at the grid origin. The remaining matrix positions correspond to the grid points that would be obtained were we to overlay the matrix on the grid with its bold entry over the origin. Consider the sample space $S = \{a, b, c, d, e\}$, with probabilities $P(a) = P(b) = 1/8$ and $P(c) = P(d) = P(e) = 1/4$. Then the digital images

$$x_a = \begin{pmatrix} 1 & -1 \\ \mathbf{0} & 1 \end{pmatrix} \quad x_b = \begin{pmatrix} 0 & 2 \\ \mathbf{2} & 0 \end{pmatrix} \quad x_c = \begin{pmatrix} 1 & 2 \\ \mathbf{0} & 1 \end{pmatrix} \quad x_d = \begin{pmatrix} 0 & -1 \\ \mathbf{0} & 1 \end{pmatrix} \quad x_e = \begin{pmatrix} 2 & 1 \\ \mathbf{1} & 2 \end{pmatrix}$$

are the realizations of a random image $X(\omega; t)$, with $\omega \in S$ and domain of definition

$T = \{(0, 1), (0, 0), (1, 1), (1, 0)\}$

The four first-order densities are

$$f(x; 0, 1) = 3/8\ \delta(x) + 3/8\ \delta(x-1) + 1/4\ \delta(x-2)$$

$$f(x; 0, 0) = 5/8\ \delta(x) + 1/4\ \delta(x-1) + 1/8\ \delta(x-2)$$

$$f(x; 1, 1) = 3/8\ \delta(x+1) + 1/4\ \delta(x-1) + 3/8\ \delta(x-2)$$

$$f(x; 1, 0) = 1/8\ \delta(x) + 5/8\ \delta(x-1) + 1/4\ \delta(x-2)$$

■

In practice it is common to index a random function by a random variable instead of elements in a sample space. Instead of considering the realizations to be dependent on observations coming from a sample space, it is more practical to suppose them to be chosen according to observations of a random variable. This convention is consistent with the customary desire to focus on random variables rather than on underlying probability spaces. Indeed, since each random variable Z defined on a probability space induces a probability measure on the Borel σ-algebra over the real line $\Re$, with the induced probability measure P_Z defined in terms of the original probability measure P by $P_Z(B) = P(Z \in B)$ for any event B, nothing is lost by indexing a random function by the values of a random variable.

Example 2.2. Let A and B be jointly distributed random variables with density $f_{A,B}(a, b)$ and consider the random function $X(t) = At + B$. Each realization of the random function is a straight line with slope and intercept determined by the random vector $(A, B)'$. The first-order probability distribution function is given by

$$F(x;t) = P(At + B \le x) = \int_{-\infty}^{\infty} \int_{-\infty}^{x-at} f_{A,B}(a,b)\, db\, da$$

■

For fixed t, the first-order distribution function completely describes the behavior of the random variable $X(t)$; however, in general we must consider joint distributions of the random variables determined by the random function. Hence, we require the *nth-order probability distributions*

$$F(x_1, x_2, \ldots, x_n; t_1, t_2, \ldots, t_n) = P(X(t_1) \le x_1, X(t_2) \le x_2, \ldots, X(t_n) \le x_n) \tag{2.3}$$

and the corresponding *nth-order densities* $f(x_1, x_2,..., x_n; t_1, t_2,..., t_n)$. In particular, there is a bivariate density $f(x_1, x_2; t_1, t_2)$ for each pair of points t_1 and t_2.

It is possible by integration to obtain the marginal densities from a given joint density. Hence, each nth-order density specifies the marginals for all subsets of $\{t_1, t_2,..., t_n\}$. Now, suppose we wish to give a complete characterization of the random function in terms of the various joint densities. If the point set is infinite, it is not generally possible to completely characterize the random function by knowing all finite joint densities; however, if the realizations are sufficiently well behaved, knowledge of the densities of all finite orders completely characterizes the random function — and we will always make this assumption. For practical manipulations, it is very useful to be able to characterize a random function by means of the joint densities of some finite order. In certain situations this is possible. For instance, such a characterization is certainly possible when the point set over which the process is defined is finite, as it is in the case of random vectors and random images with finite domains.

If for each point set $\{t_1, t_2,..., t_n\}$ the random variables $X(t_1), X(t_2),..., X(t_n)$ are independent, then the random function is characterized by its first-order densities since

$$f(x_1, x_2,..., x_n; t_1, t_2,..., t_n) = f(x_1; t_1) f(x_2; t_2) \cdots f(x_n; t_n) \tag{2.4}$$

An important class of random functions characterized by second-order distributions is the class of Gaussian random functions. A random function $X(t)$ is said to be *Gaussian*, or *normal*, if for any collection of n points $t_1, t_2,..., t_n$, the random variables $X(t_1), X(t_2),..., X(t_n)$ possess a multivariate normal distribution. A multivariate normal distribution is completely characterized by its mean vector and covariance matrix, and these are in turn determined from the first- and second-order densities of the variables.

2.2. Moments of a Random Function

A random function is a complicated mathematical entity and unless it can be fully characterized in terms of low-order joint distributions, we must rely on less complete descriptions, typically certain of its moments. A great deal of linear systems theory employs only second-order moment information. Mathematical tractability is gained by using only second-order information. The loss is that there is no distinction between random functions possessing identical second-order moments. Linear filters have a natural dependency on second-order information — and therefore a natural lack of discrimination relative to random processes differing only at higher orders. The moments of random functions are direct generalizations of moments defined for random variables.

2.2.1. Mean and Covariance Functions

The *expectation* (*mean function*) of a random function $X(t)$ is the deterministic function of t found by taking the expected value of the random variable $X(t)$ for each t:

$$\mu_X(t) = E[X(t)] = \int_{-\infty}^{\infty} x f(x;t)\,dx \tag{2.5}$$

The expectation depends only on the first-order density and contains no second-order information regarding the relationship between the random variables $X(t)$ and $X(t')$.

Another function depending on $X(t)$ in isolation is the *variance function*, defined by

$$\begin{aligned}\mathrm{Var}[X(t)] &= E[(X(t)-\mu_X(t))^2] \\ &= \int_{-\infty}^{\infty} (x-\mu_X(t))^2 f(x;t)\,dx\end{aligned} \tag{2.6}$$

For fixed t, $\mathrm{Var}[X(t)]$ is the variance of the random variable $X(t)$. As with a single random variable, the *standard deviation function* is defined by $\sigma_X(t) = \mathrm{Var}[X(t)]^{1/2}$.

In Fig. 2.2, several realizations of two distinct random functions are depicted. Though both processes have the same expectation, the one in part b has greater variance. In Fig. 2.3, both processes have the same expectation and variance functions, but there is greater variability between nearby pairs of the variable t in part a than in part b. Dependency between the random variables that comprise a random function cannot be revealed by measures that depend only on univariate distributions; higher-order moments are needed.

The *covariance function* of the random function $X(t)$ is defined by

$$\begin{aligned}K_X(t, t') &= E[(X(t)-\mu_X(t))(X(t')-\mu_X(t'))] \\ &= \int_{-\infty}^{\infty}\int_{-\infty}^{\infty} (x-\mu_X(t))(x'-\mu_X(t')) f(x,x';t,t')\,dx\,dx'\end{aligned} \tag{2.7}$$

Letting $t = t'$ in the covariance function yields the variance function:

$$K_X(t, t) = \mathrm{Var}[X(t)] \tag{2.8}$$

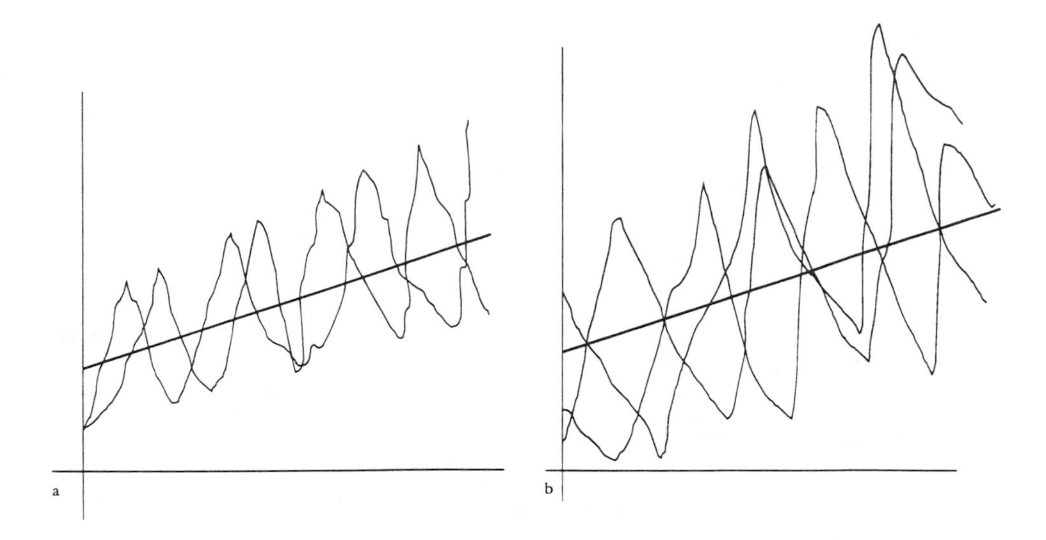

Figure 2.2 Realizations of two random functions: the processes have the same expectation, but the one in part b has greater variance.

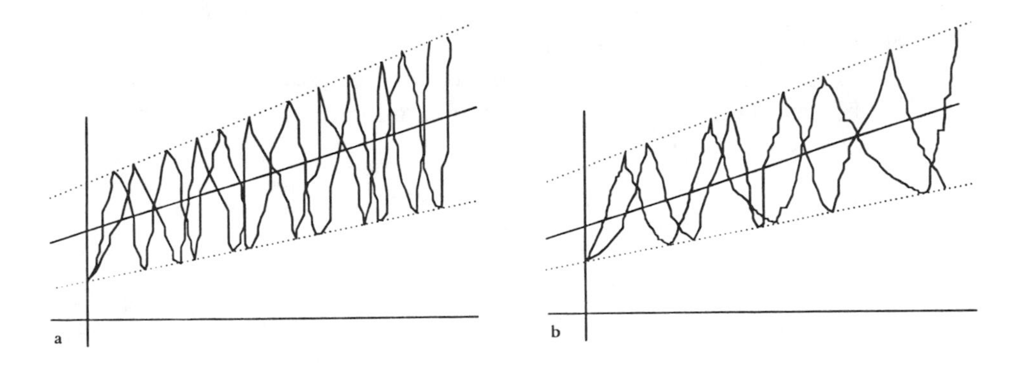

Figure 2.3 Realizations of two random functions: the processes have the same expectation and variance functions, but there is greater variability between nearby variables in part a.

It is seen directly from its definition that the covariance function is symmetric:

$$K_X(t, t') = K_X(t', t) \tag{2.9}$$

The *correlation-coefficient function* is defined by

$$\rho_X(t, t') = \frac{K_X(t, t')}{\sigma_X(t)\sigma_X(t')} \tag{2.10}$$

Since the covariance function is defined in accordance with the definition of covariance between two jointly distributed random variables, Theorem 1.14 implies that $|\rho_X(t, t')| \leq 1$ and $|\rho_X(t, t')| = 1$ if and only if there exist constants $a_{t,t'} \neq 0$ and $b_{t,t'}$ such that

$$P(X(t') = a_{t,t'}X(t) + b_{t,t'}) = 1 \tag{2.11}$$

The correlation coefficient is ±1 for a pair of points t and t' if and only if $X(t)$ and $X(t')$ are linearly related with probability 1.

The *autocorrelation function* is defined by

$$R_X(t, t') = E[X(t)X(t')] \tag{2.12}$$

From Eq. 1.177,

$$K_X(t, t') = R_X(t, t') - \mu_X(t)\mu_X(t') \tag{2.13}$$

If the mean of $X(t)$ is identically zero, then the covariance and autocorrelation functions are identical.

Given two random functions $X(t)$ and $Y(s)$, the *cross-covariance function* is defined as the covariance between pairs of random variables, one from each of the two processes:

$$K_{XY}(t, s) = E[(X(t) - \mu_X(t))(Y(s) - \mu_Y(s))]$$

$$= \int_{-\infty}^{\infty}\int_{-\infty}^{\infty}(x - \mu_X(t))(y - \mu_Y(s))f(x, y; t, s)\, dxdy \tag{2.14}$$

where $f(x, y; t, s)$ is the joint density for $X(t)$ and $Y(s)$. If the cross-covariance function is identically zero, then the random functions are said to be *uncorrelated*; otherwise, they are said to be *correlated*. As with the covariance function, there is symmetry:

$$K_{XY}(t, s) = K_{YX}(s, t) \tag{2.15}$$

The *cross-correlation coefficient* is defined by

$$\rho_{XY}(t, s) = \frac{K_{XY}(t,s)}{\sigma_X(t)\sigma_Y(s)} \tag{2.16}$$

and $|\rho_{XY}(t, s)| \leq 1$. The *cross-correlation function* is defined by

$$R_{XY}(t, s) = E[X(t)Y(s)] \tag{2.17}$$

and is related to the cross-covariance function by

$$K_{XY}(t, s) = R_{XY}(t, s) - E[X(t)]E[Y(s)] \tag{2.18}$$

General moments of a random function $X(t)$ are defined analogously to moments for random vectors. The *mixed moments* of order $p_1 + p_2 + \cdots + p_n$ are defined by

$$\mu'_{p_1, p_2, \ldots, p_n}(t_1, t_2, \ldots, t_n) = E[X(t_1)^{p_1} X(t_2)^{p_2} \cdots X(t_n)^{p_n}] \tag{2.19}$$

and the *mixed central moments* of the same order are defined by

$$\mu_{p_1, p_2, \ldots, p_n}(t_1, t_2, \ldots, t_n) =$$

$$E[(X(t_1) - \mu_X(t_1))^{p_1} (X(t_2) - \mu_X(t_2))^{p_2} \cdots (X(t_n) - \mu_X(t_n))^{p_n}] \tag{2.20}$$

In particular, the covariance results from $p_1 = p_2 = 1$.

***Example* 2.3.** Consider the random time function $X(Z; t) = I_{[Z,\infty)}(t)$, where Z is the standard normal variable and $I_{[Z,\infty)}(t)$, the indicator (characteristic) function for the random infinite interval $[Z, \infty)$, is defined by $I_{[Z,\infty)}(t) = 1$ if $t \in [Z, \infty)$ and $I_{[Z,\infty)}(t) = 0$ if $t \notin [Z, \infty)$. For each observation z of the random variable Z, $X(z; t)$ is a unit step function with a step at $t = z$. For fixed t, $X(t)$ is a binomial variable with $X(t) = 1$ if $t \geq Z$ and $X(t) = 0$ if $t < Z$. The density of $X(t)$ is characterized by the probabilities

$$P(X(t) = 1) = P(Z \leq t) = \frac{1}{2\pi} \int_{-\infty}^{t} e^{-z^2/2}\, dz = \Phi(t)$$

where Φ denotes the probability distribution function of Z, and

$$P(X(t) = 0) = P(Z > t) = 1 - \Phi(t)$$

The mean function for the process $X(t)$ is given by

$$\mu_X(t) = P(X(t) = 1) = \Phi(t)$$

To find the covariance function for $X(t)$, we first find the autocorrelation $R_X(t, t')$, recognizing that $X(t)X(t')$ is a binomial random variable. If $t < t'$, then $P(t' \geq Z \mid t \geq Z) = 1$, so that

$$\begin{aligned} P(X(t)X(t') = 1) &= P(t' \geq Z, t \geq Z) \\ &= P(t' \geq Z \mid t \geq Z)P(t \geq Z) \\ &= P(t \geq Z) \\ &= \Phi(t) \end{aligned}$$

A similar calculation shows that, for $t' \leq t$,

$$P(X(t)X(t') = 1) = \Phi(t')$$

Consequently,

$$P(X(t)X(t') = 1) = \Phi(\min(t, t'))$$

$$P(X(t)X(t') = 0) = 1 - \Phi(\min(\mathrm{t}, \mathrm{t}'))$$

Thus, the autocorrelation and covariance functions are given by

$$R_X(t, t') = E[X(t)X(t')] = \Phi(\min(t, t'))$$

$$\begin{aligned} K_X(t, t') &= E[X(t)X(t')] - \mu_X(t)_X\mu(t') \\ &= \Phi(\min(t, t')) - \Phi(t)\Phi(t') \end{aligned}$$

Setting $t = t'$ yields the variance:

$$\text{Var}[X(t)] = \Phi(t) - \Phi(t)^2$$

Factoring the variance yields

$$\begin{aligned}\text{Var}[X(t)] &= \Phi(t)(1 - \Phi(t)) \\ &= P(X(t) = 1)P(X(t) = 0)\end{aligned}$$

The variance can be interpreted intuitively. Since the probability mass of Z is centered near 0, the step in $I_{[Z,\infty)}(t)$ is likely to occur near 0. Hence, for t near 0, $X(t)$ should possess a relatively high variance. Note that $\text{Var}[X(t)] \to 0$ as $|t| \to \infty$. ■

***Example* 2.4.** Again consider a random unit step function, but now let $X(W; t) = I_{[W,\infty)}(t)$ where W is uniformly distributed over the closed interval $[0, 2]$. In this case, for fixed t,

$$\mu_X(t) = P(X(t) = 1) = P(W \le t) = \begin{cases} 0, & \text{for } t < 0 \\ t/2, & \text{for } 0 \le t \le 2 \\ 1, & \text{for } t > 2 \end{cases}$$

Proceeding as in Example 2.3, we obtain the autocorrelation

$$\begin{aligned}E[X(t)X(t')] &= P(X(t)X(t') = 1) \\ &= P(t' \ge W, t \ge W) \\ &= P(W \le \min(t, t'))\end{aligned}$$

The covariance is obtained by using this result in conjunction with the mean. ■

***Example* 2.5.** Again consider the random unit step function; however, this time let W be a binomial variable with equally likely outcomes 0 and 1. Thus, only two time functions are involved, $I_{[0,\infty)}(t)$ and $I_{[1,\infty)}(t)$. The mean function is

$$\mu_X(t) = \begin{cases} 0, & \text{for } t < 0 \\ 1/2, & \text{for } 0 \le t < 1 \\ 1, & \text{for } t \ge 1 \end{cases}$$

and the autocorrelation is

$$E[X(t)X(t')] = P(W \leq \min(t,t')) = \begin{cases} 0, & \text{if } \min(t,t') < 0 \\ 1/2, & \text{if } 0 \leq \min(t,t') < 1 \\ 1, & \text{if } 1 \leq \min(t,t') \end{cases}$$ ■

Example **2.6.** In Example 2.2 we considered the first-order distribution function for the random function $X(t) = At + B$. For this process,

$$\mu_X(t) = E[A]t + E[B]$$

$$E[X(t)X(t')] = E[A^2]tt' + E[AB](t + t') + E[B^2]$$

$$K_X(t, t') = (E[A^2] - E[A]^2)tt' + (E[AB] - E[A]E[B])(t + t') + E[B^2] - E[B]^2$$

$$= \text{Var}[A]tt' + \text{Cov}[A, B](t + t') + \text{Var}[B]$$

If A and B are uncorrelated, then

$$K_X(t, t') = \text{Var}[A]tt' + \text{Var}[B]$$ ■

Example **2.7.** This example concerns the *Poisson process*, which has applications in many areas, is especially important in imaging and queuing theory, and will be discussed more fully in Section 2.6. To appreciate the model, consider jobs arriving for service at a CPU and let $X(t)$ count the number of jobs arriving in the time interval $[0, t]$. If we assume that the numbers of arrivals in any finite set of nonoverlapping intervals are independent, that the probability of exactly one arrival in an interval of length t is $\lambda t + o(t)$ [λt plus a quantity very small in comparison to t], that the probability of two or more arrivals in an interval of length t is $o(t)$ [very small in comparison to t], and that λ is constant over all t intervals, then (as will be shown in Section 2.6),

$$P(X(t) = k) = e^{-\lambda t}\frac{(\lambda t)^k}{k!}$$

for $k = 0, 1, 2, \ldots$. [Mathematically, $o(t)$ represents any function $g(t)$ for which $\lim_{t\to 0} g(t)/t = 0$.] $X(t)$ is a Poisson-distributed random variable and both its mean and variance are equal to λt. Each realization of the Poisson process $X(t)$ results from a distribution of time points from a particular observation of the arrival process. Figure 2.4 shows realizations of two Poisson processes, λ being greater for the second process. ■

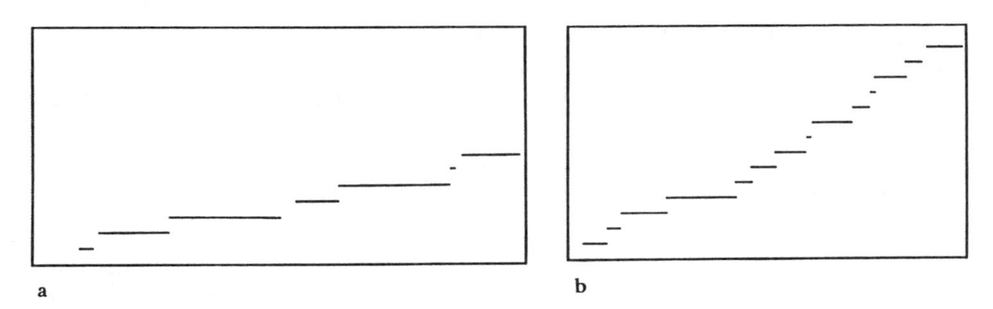

Figure 2.4 Realizations of two Poisson processes, λ being greater for the second process (b).

Example 2.8. Let the random Euclidean image $X(Z; u, v)$ be the indicator function of the right half-plane $\{(u, v): u \geq Z\}$, where Z is the standard normal random variable. $X(u, v)$ is a binary random image with mean, autocorrelation, and covariance

$$\mu_X(u, v) = P(X(u, v) = 1) = P(Z \leq u) = \Phi(u)$$

$$R_X((u, v), (u', v')) = P(X(u, v)X(u', v') = 1)$$

$$= P(Z \leq \min(u, u'))$$

$$= \Phi(\min(u, u'))$$

$$K_X((u, v), (u', v')) = \Phi(\min(u, u')) - \Phi(u)\Phi(u')$$

The similarity to Example 2.3 results from the fact that only the first variable is governed directly by the generating random variable Z. ■

Example 2.9. Let the Euclidean random image $X(R; u, v)$ be the indicator function of the random disk of radius R centered at the origin, where R possesses the exponential density $f(r) = be^{-br}I_{[0,\infty)}(r)$. Letting $s = (u^2 + v^2)^{1/2}$, the process mean, autocorrelation, covariance, and variance are given by

$$\mu_X(u, v) = P(R \geq s) = e^{-bs}$$

$$R_X((u, v), (u', v')) = P(R \geq s, R \geq s') = P(R \geq \max(s, s')) = e^{-b\max(s,s')}$$

$$K_X((u, v), (u', v')) = e^{-b\max(s,s')} - e^{-b(s+s')}$$

$$\mathrm{Var}[X(u, v)] = e^{-bs}(1 - e^{-bs})$$

The variance is a function of s, $\mathrm{Var}[X(0, 0)] = 0$, and $\mathrm{Var}[X(u, v)] \to 0$ as $s \to \infty$. The radius at which the variance is maximum can be found by differentiating the variance and setting the result equal to zero. Maximum variance occurs at $s = b^{-1}\log 2$. The correlation coefficient for $X(u, v)$ is given by

$$\rho_X((u, v), (u', v')) = \frac{e^{-b\max(s,s')} - e^{-b(s+s')}}{e^{-b(s+s')/2}\sqrt{(1-e^{-bs})(1-e^{-bs'})}}$$

If (u, v) and (u', v') lie on a common circle of radius $s = s'$, then their correlation is 1. This should be evident from the manner in which the process has been defined. ■

Example 2.10. Consider a discrete random image possessing four realizations:

$$x_a = \begin{pmatrix} 1 & 1 & 1 \\ 1 & 1 & 1 \\ \mathbf{0} & 1 & 1 \end{pmatrix} \quad x_b = \begin{pmatrix} 1 & 1 & 1 \\ 0 & 1 & 1 \\ \mathbf{0} & 0 & 1 \end{pmatrix} \quad x_c = \begin{pmatrix} 0 & 1 & 1 \\ 0 & 0 & 1 \\ \mathbf{0} & 0 & 0 \end{pmatrix} \quad x_d = \begin{pmatrix} 0 & 0 & 1 \\ 0 & 0 & 0 \\ \mathbf{0} & 0 & 0 \end{pmatrix}$$

The observations are made in accordance with the sample space $S = \{a, b, c, d\}$ and the probability assignments $P(a) = P(d) = 1/6$ and $P(b) = P(c) = 1/3$. The model describes partial diagonalized coverage of the image frame with the probability mass favoring coverage that is not very great or small. Straightforward calculations of the discrete expected values yield the mean image

$$\mu_X = \begin{pmatrix} 1/2 & 5/6 & 1 \\ 1/6 & 1/2 & 5/6 \\ \mathbf{0} & 1/6 & 1/2 \end{pmatrix}$$

We compute the covariance at (0, 1) and (1, 1). Since both pixels have the value 1 if and only if x_a is selected, $E[X(0, 1)X(1, 1)] = 1/6$. Thus, $K_X((0, 1), (1, 1)) = 1/12$. For pixels (2, 0) and (1, 1), $K_X((2, 0), (1, 1)) = 1/4$. Note that $\rho_X((2, 0), (1, 1)) = 1$, which makes sense since the values at (2, 0) and (1, 1) are identical for all realizations. ■

2.2.2. Mean and Covariance of a Sum

We often encounter the sum, $X(t) + Y(t)$, of two random functions. Linearity of the expected-value operator yields linearity of the expectation relative to random functions: for any real numbers a and b,

$$\mu_{aX+bY}(t) = a\mu_X(t) + b\mu_Y(t) \tag{2.21}$$

For the covariance of a sum,

$$\begin{aligned}K_{X+Y}(t, t') &= E[(X(t)+Y(t))(X(t')+Y(t'))]-(\mu_X(t)+\mu_Y(t))(\mu_X(t')+\mu_Y(t'))\\ &= R_X(t,t')+R_Y(t,t')+R_{XY}(t,t')+R_{YX}(t,t')\\ &\quad -\mu_X(t)\mu_X(t')-\mu_Y(t)\mu_Y(t')-\mu_X(t)\mu_Y(t')-\mu_Y(t)\mu_X(t')\\ &= K_X(t,t')+K_Y(t,t')+K_{XY}(t,t')+K_{YX}(t,t')\end{aligned} \tag{2.22}$$

where we have assumed that the relevant quantities exist. If $X(t)$ and $Y(t)$ happen to be uncorrelated, then the cross-covariance is identically zero and the preceding identity reduces to the covariance of the sum being equal to the sum of the covariances:

$$K_{X+Y}(t, t') = K_X(t, t') + K_Y(t, t') \tag{2.23}$$

The preceding relations generalize to sums of n random functions. Suppose

$$W(t) = \sum_{j=1}^{n} X_j(t) \tag{2.24}$$

Then, assuming the relevant quantities exist,

$$\mu_W(t) = \sum_{j=1}^{n} \mu_{X_j}(t) \tag{2.25}$$

$$K_W(t, t') = \sum_{i=1}^{n}\sum_{j=1}^{n} K_{X_i X_j}(t,t') \tag{2.26}$$

Should the X_j be mutually uncorrelated, then all terms in the preceding summation for which $i \neq j$ are identically zero. Thus, only the "diagonal" terms remain and the covariance reduces to

$$K_W(t, t') = \sum_{j=1}^{n} K_{X_j}(t,t') \tag{2.27}$$

2.3. Differentiation

Two major topics studied in the calculus are differentiation and integration. Given that a random function $X(t)$ possesses realizations that are deterministic functions, it is natural to extend the calculus to random processes. This section discusses differentiation and the next treats integration.

2.3.1. Differentiation of Random Functions

First consider ordinary differentiation (t a real variable). For the moment, assume that each realization of $X(t)$ is differentiable. Extension of deterministic differentiation to random functions (stochastic processes) is not completely straightforward because of the way in which the limit of the difference quotient defining the derivative interacts with randomness. One might propose to define the derivative of the stochastic process to be the process consisting of the derivatives of the realizations. This would lead to the process derivative being defined by

$$\frac{d}{dt}\{X(\omega;t)\} = \{\frac{d}{dt}X(\omega;t)\} \tag{2.28}$$

For the moment, let us proceed under this supposition. Let

$$\Delta_X(\omega; t, h) = \frac{X(\omega;t+h) - X(\omega;t)}{h} \tag{2.29}$$

be the difference quotient for the derivative (slope of the secant line from t to $t + h$). Then, for fixed ω (a fixed realization of the random process), the deterministic derivative is defined by

$$\frac{d}{dt}X(\omega;t) = \lim_{h\to 0}\Delta_X(\omega;t,h) \tag{2.30}$$

Taking the approach of realizationwise differentiation and letting $X'(t)$ denote the derivative process, we will attempt to find the mean of the derivative process.

Proceeding formally, without concern for mathematical rigor, and using the linearity of the expected value, we obtain

$$\mu_{X'}(t) = E[X'(t)]$$

$$= E\left[\lim_{h\to 0}\frac{X(\omega;t+h)-X(\omega;t)}{h}\right]$$

$$= \lim_{h\to 0} E\left[\frac{X(\omega;t+h)-X(\omega;t)}{h}\right]$$

$$= \lim_{h\to 0}\frac{E[X(\omega;t+h)]-E[X(\omega;t)]}{h}$$

$$= \frac{d}{dt}E[X(t)]$$

$$= \frac{d}{dt}\mu_X(t) \tag{2.31}$$

Although the preceding computation results in the mean of the derivative being the derivative of the mean (which is intuitively pleasing), the procedure is not rigorous. The difference quotient $\Delta_X(\omega;\, t,\, h)$ is a random variable and there is no justification for interchanging the expected value and limit operations. Some interpretation of the limit must be specified and interchange of the expected value and the limit must be justified in the context of that interpretation.

What is required is not abandonment of the intuitive nature of the foregoing procedure, but rather a more careful definition of the derivative of a random function. Although in practice it might sometimes be possible to obtain the derivative process by differentiating the realizations, a more subtle mathematical definition of limit is required for a rigorous development. We will shortly provide an appropriate limit concept.

First, however, we will proceed formally (nonrigorously) to find the covariance function of the derivative. In doing so we will employ the previously (formally) derived expression for the mean of the derivative process and will formally interchange the expected value with a partial derivative. Proceeding,

$$K_{X'}(u, v) = E[(X'(u)-\mu_{X'}(u))(X'(v)-\mu_{X'}(v))]$$

$$= E\left[\frac{\partial^2}{\partial u\partial v}(X(u)-\mu_X(u))(X(v)-\mu_X(v))\right]$$

$$= \frac{\partial^2}{\partial u\partial v}E[(X(u)-\mu_X(u))(X(v)-\mu_X(v))]$$

$$= \frac{\partial^2}{\partial u \partial v} K_X(u,v) \tag{2.32}$$

In words, the covariance of the derivative process results from taking a second-order partial derivative of the original covariance. Of course, the derivation of Eq. 2.32 lacks rigorous justification. Putting the entire discussion on sound mathematical footing requires the use of a suitable notion of convergence.

2.3.2. Mean-Square Differentiability

Random-process differentiation is made mathematically sound by defining the derivative of a random process via mean-square convergence. As originally defined (Eq. 1.206), mean-square convergence applies to discrete sequential convergence as $n \to \infty$; for differentiation, the random functions are functions of a continuous variable t and the limit is with respect to a continuous variable $h \to 0$. In this context, $X_h(t)$ converges to $X(t)$ in the mean-square if

$$\lim_{h \to 0} E[|X_h(t) - X(t)|^2] = 0 \tag{2.33}$$

For fixed h, $E[|X_h(t) - X(t)|^2]$ gives the *mean-square distance* between $X_h(t)$ and $X(t)$, so that mean-square convergence means that the distance between $X_h(t)$ and $X(t)$ is converging to 0. We will employ the common notation, $\text{l.i.m.}_{h \to \infty} X_h(t) = X(t)$ to denote mean-square convergence and refer to the convergence as *MS convergence*.

The random function $X(t)$ is said to be *mean-square* (*MS*) *differentiable* and $X'(t)$ is the *mean-square derivative* of $X(t)$ at point t if

$$X'(\omega; t) = \underset{h \to 0}{\text{l.i.m.}} \Delta_X(\omega; t, h)$$

$$= \underset{h \to 0}{\text{l.i.m.}} \frac{X(\omega; t+h) - X(\omega; t)}{h} \tag{2.34}$$

that is, if

$$\lim_{h \to 0} E\left[\left|\frac{X(\omega; t+h) - X(\omega; t)}{h} - X'(\omega; t)\right|^2\right] = 0 \tag{2.35}$$

For fixed t and h, both the difference quotient and $X'(\omega; t)$ are functions of ω (random variables). Because the limiting operation of Eq. 2.35 is with respect to an expectation involving these, it is well defined. The following theorem states that the formulas for the mean and covariance of the derivative that have been obtained nonrigorously in Eqs. 2.31 and 2.32 hold rigorously in terms of the MS derivative. It also states a converse.

Theorem 2.1. The random function $X(t)$ is MS differentiable on the interval T if and only if the mean is differentiable on T and the covariance possesses a second-order mixed partial derivative with respect to u and v on T. In the case of differentiability,

$$\text{(i)}\ \mu_{X'}(t) = \frac{d}{dt}\mu_X(t) \tag{2.36}$$

$$\text{(ii)}\ K_{X'}(u,v) = \frac{\partial^2}{\partial u \partial v} K_X(u,v) \quad \blacksquare \tag{2.37}$$

***Example* 2.11.** Let $X(Z; t) = (t - Z)^2$ be a parabola that is randomly translated by a random variable Z. Holding Z fixed and differentiating realizationwise yields the random signal $Y(Z; t) = 2(t - Z)$. We prove that $Y(Z; t)$ is actually the MS derivative. First,

$$E[|\Delta_X(Z; t, h) - Y(Z; t)|^2] = E\left[\left|\frac{(t+h-Z)^2 - (t-Z)^2}{h} - 2(t-Z)\right|^2\right] = E[h] = h$$

Letting $h \to 0$ shows that $X'(Z; t) = 2(t - Z)$. $\blacksquare$

***Example* 2.12.** Consider the random sine wave

$$X(t) = Z\sin(t - W)$$

where Z and W are independent random variables determining amplitude and phase. Differentiation of the realizations yields

$$Y(t) = Z\cos(t - W)$$

We demonstrate that this random function is, in fact, the MS derivative. Letting $\xi = \sin(h/2)/(h/2)$, trigonometric manipulation yields

$$\Delta_X(Z; t, h) - Y(Z; t) = Z\cos(t - W + h/2)\frac{\sin h/2}{h/2} - Z\cos(t - W)$$

$$= Z[\xi \cos(t + h/2) \cos W + \xi \sin(t + h/2) \sin W - \cos t \cos W - \sin t \sin W]$$

$$= Z[\cos W(\xi \cos(t + h/2) - \cos t) + \sin W(\xi \sin(t + h/2) - \sin t)]$$

Squaring and taking the expected value yields

$$E[|\Delta_X(Z; t, h) - Y(Z; t)|^2] = E[Z^2]\{E[\cos^2 W](\xi \cos(t + h/2) - \cos t)^2$$

$$+ E[\sin^2 W](\xi \sin(t + h/2) - \sin t)^2$$

$$+ 2E[\sin W \cos W](\xi \cos(t + h/2) - \cos t)(\xi \sin(t + h/2) - \sin t)\}$$

Since $\lim_{h \to 0} \xi = 1$ and since sine and cosine are continuous, the limit as $h \to 0$ of the preceding expression is 0 and therefore the MS derivative agrees with the process that results from differentiating the realizations. According to Theorem 2.1 (i), the mean of the derivative must equal the derivative of the mean. In fact,

$$\mu_X(t) = E[Z \sin(t - W)]$$

$$= E[Z]E[\sin t \cos W - \cos t \sin W]$$

$$= E[Z](E[\cos W] \sin t - E[\sin W] \cos t)$$

and

$$\mu_{X'}(t) = E[Z \cos(t - W)]$$

$$= E[Z](E[\cos W] \cos t + E[\sin W] \sin t)$$

so that part (i) of Theorem 2.1 is satisfied. We next compute the autocorrelation of $X(t)$:

$$E[X(u)X(v)] = E[Z^2 \sin(u - W) \sin(v - W)]$$

$$= E[Z^2 (\sin u \cos W - \cos u \sin W)(\sin v \cos W - \cos v \sin W)]$$

$$= E[Z^2](E[\cos^2 W] \sin u \sin v - E[\cos W \sin W] \sin u \cos v$$

$$- E[\sin W \cos W] \cos u \sin v + E[\sin^2 W] \cos u \cos v)$$

To obtain the covariance, subtract $E[X(u)]E[X(v)]$:

$$K_X(u, v) = (\sin u \sin v)(E[Z^2]E[\cos^2 W] - E[Z]^2 E[\cos W]^2)$$

$$+ (\cos u \cos v)(E[Z^2]E[\sin^2 W] - E[Z]^2 E[\sin W]^2)$$

$$- (\sin u \cos v)(E[Z^2]E[\sin W \cos W] - E[Z]^2 E[\cos W]E[\sin W])$$

$$- (\cos u \sin v)(E[Z^2]E[\sin W \cos W] - E[Z]^2 E[\cos W]E[\sin W])$$

Application of Theorem 2.1 (ii) gives

$$K_{X'}(u, v) = (\cos u \cos v)(E[Z^2]E[\cos^2 W] - E[Z]^2 E[\cos W]^2)$$

$$+ (\sin u \sin v)(E[Z^2]E[\sin^2 W] - E[Z]^2 E[\sin W]^2)$$

$$+ (\cos u \sin v)(E[Z^2]E[\sin W \cos W] - E[Z]^2 E[\cos W]E[\sin W])$$

$$+ (\sin u \cos v)(E[Z^2]E[\sin W \cos W] - E[Z]^2 E[\cos W]E[\sin W])$$

The same result is obtained by computing $K_{X'}(u, v)$ directly from $X'(t)$. ■

From the results of Examples 2.11 and 2.12, it should be clear why in practice we often assume that the MS derivative of a random function can be obtained by taking the derivatives of the realizations as the derivative process. Indeed, it was just such a methodology that caused us to discover the proper MS derivatives in the two examples, though, of course, we had to actually go through and check MS convergence. Such an approach is appropriate, though not foolproof, when the realizations are differentiable. The power of Theorem 2.1 is in the converse: MS differentiability is ensured if the mean of the original process is differentiable and the covariance of the original process possesses a second-order mixed partial derivative.

***Example* 2.13.** Consider the random unit step function $X(Z; t)$ of Example 2.3, Z being the standard normal variable. If we consider the derivatives of the realizations, then the resulting "derivative" process would possess realizations $y_z(t)$, each of which is zero except at $t = z$. For any particular z, if we were to remove the realization $x_z(t)$ from the process, the resulting process would be "differentiable" at $t = z$. Since $P(Z = z) = 0$, this new process is, from a probabilistic point of view, identical to the original. However, suppose we wish to consider differentiability over an interval T. Now the difficulty becomes evident: the probability that Z lies in the interval is not zero, and we cannot remove all realizations whose steps are in the interval. If we apply Theorem 2.1 to $X(t)$, the derivative of the mean is given by

$$\frac{d}{dt}\mu_X(t) = \frac{d}{dt}\Phi(t) = f_Z(t)$$

where $f_Z(t)$ is the density for the standard normal variable. To find the mixed partial derivative of the covariance, first assume $u < v$, fix v, and differentiate with respect to u:

$$\begin{aligned}\frac{\partial}{\partial u}K_X(u,v) &= \frac{\partial}{\partial u}[\Phi(\min(u,v)) - \Phi(u)\Phi(v)] \\ &= \frac{\partial}{\partial u}[\Phi(u) - \Phi(u)\Phi(v)] \\ &= f_Z(u) - f_Z(u)\Phi(v)\end{aligned}$$

On the other hand, if $v < u$, then

$$\begin{aligned}\frac{\partial}{\partial u}K_X(u,v) &= \frac{\partial}{\partial u}[\Phi(v) - \Phi(u)\Phi(v)] \\ &= -f_Z(u)\Phi(v)\end{aligned}$$

To apply Theorem 2.1 we need to next partially differentiate with respect to v. Thus, we consider $\partial K_X(u, v)/\partial u$ as a function of v, holding u fixed. Treated this way, $\partial K_X(u, v)/\partial u$ has a jump of $f_Z(u)$ at $v = u$, and thus, it is not differentiable with respect to v. According to Theorem 2.1 (ii), the random function $X(t)$ is not MS differentiable. Before we leave this example, we point out that the mixed partial derivative does exist in the generalized sense, i.e., as a delta function; indeed, since the delta function is only nonzero when $u = v$,

$$\frac{\partial^2}{\partial u \partial v}K_X(u,v) = f_Z(u)\delta(v-u) = f_Z(v)\delta(v-u) \qquad \blacksquare \tag{2.38}$$

Although the previous example demonstrates that the random unit step function is nondifferentiable in the MS sense, the fact that the covariance possesses a generalized derivative is of no small consequence. Such covariance functions will be discussed in detail in Section 2.6 and play an important role in both the theory and application of random functions.

The theory of MS differentiability has an immediate extension to random functions of two variables (random images): a partial derivative is an ordinary derivative applied to a function of two variables, with one of the variables being held fixed. The random image $Y(\omega; u, v)$ is the *MS partial derivative* of the random image $X(\omega; u, v)$ with respect to the variable u if

$$\lim_{h\to 0} E\left[\left|\frac{X(\omega;u+h,v)-X(\omega;u,v)}{h}-Y(\omega;u,v)\right|^2\right]=0 \tag{2.39}$$

so that

$$\underset{h\to 0}{\text{l.i.m.}}\frac{X(\omega;u+h,v)-X(\omega;u,v)}{h}=Y(\omega;u,v) \tag{2.40}$$

The partial derivative with respect to v is similarly defined. Moreover, a two-dimensional analogue of Theorem 2.1 holds.

2.4. Integration

As in the case of differentiation, extension of integration for deterministic functions to integration for random functions requires extension of the basic limiting definition in the framework of mean-square convergence.

In ordinary calculus the integral of the time function $x(t)$ is defined as a limit of Riemann sums: for any partition $a = t_0 < t_1 < t_2 < \ldots < t_n = b$ of the interval $[a, b]$,

$$\int_a^b x(t)\,dt = \lim_{\|\Delta t_k\|\to 0}\sum_{k=1}^{n} x(t_k')\Delta t_k \tag{2.41}$$

where $\Delta t_k = t_k - t_{k-1}$, $\|\Delta t_k\|$ is the maximum of the Δt_k over $k = 1, 2,\ldots, n$, t_k' is any point in the interval $[t_{k-1}, t_k]$, and the limit is taken to mean that the same value is obtained over all partitions, so long as $\|\Delta t_k\| \to 0$. The limit is, by definition, the value of the integral. In this section we use the method of Riemann sums to define the integral of a random function. The variable t is not restricted to a single dimension. Thus, the theory applies to random images as well as to random signals. To emphasize the multidimensional character of the integral, our development employs $t = (u, v)$ as a point in the Euclidean plane $\Re^2$ and a region of integration $T \subset \Re^2$.

Suppose a disjoint collection of rectangles R_k forms a partition $\Xi = \{R_k\}$ of T, meaning that $T = \cup_k R_k$. For each ω, we can form the Riemann sum corresponding to the realization of the random function $X(\omega; u, v)$ of two variables and the partition Ξ in a manner analogous to a deterministic function:

$$\Sigma_X(\omega; \Xi) = \sum_{k=1}^{n} X(\omega; u_k', v_k') A(R_k) \tag{2.42}$$

where $A(R_k)$ is the area of R_k and $(u'_k, v'_k) \in R_k$. Letting $\|R_k\|$ be the maximum of the rectangular dimensions, the limit can be taken over all partitions for which $\|R_k\| \to 0$ to give the integral:

$$\iint_T X(\omega; u, v)\, dudv = \lim_{\Xi, \|R_k\| \to 0} \Sigma_X(\omega; \Xi) \tag{2.43}$$

Relative to the random process $X(\omega; u, v)$, the limit in Eq. 2.43 is problematic, since the equation is stated for fixed ω and we really desire to take the limit with respect to the process. The difficulty can be grasped by recognizing that the sum on the right-hand side of Eq. 2.42 depends not only on the partition, but also on the realization (on ω). If we unfix ω, then, for each partition, the sum becomes a random variable $\Sigma_X(\omega; \Xi)$ dependent on ω (and possessing its own distribution). Hence, we must treat the limit of Eq. 2.43 as a limit of random variables.

As with differentiation, the integral is defined relative to mean-square convergence. The random function $X(\omega; u, v)$ is said to be *mean-square integrable* and possess the integral

$$I = \iint_T X(u, v)\, du\, dv \tag{2.44}$$

itself a random variable, if and only if

$$\lim_{\Xi, \|R_k\| \to 0} E[|I - \Sigma_X(\omega; \Xi)|^2] = 0 \tag{2.45}$$

where the limit is taken over all partitions $\Xi = \{R_k\}$ for which $\|R_k\| \to 0$. As with differentiation, the integral of a random function can often be obtained by integration of realizations; however, MS integrability depends on the limit of Eq. 2.45.

The basic MS integrability theorem is similar to Theorem 2.1. To avoid cumbersome notation, we state it for a single variable t; extension to random functions of two or more variables is straightforward. It is important to recognize that the theorem concerns integrands of the form $g(t, s)X(s)$, where $g(t, s)$ is a deterministic function of two variables. Hence, as stated in the theorem, the resulting integral is a random function of t, not simply a random variable. The intent is to consider $g(t, s)$ as the kernel of an integral transform with random function inputs.

Theorem 2.2. If the integral

$$Y(t) = \int_T g(t,s)X(s)\,ds \tag{2.46}$$

exists in the MS sense, then

$$\text{(i) } \mu_Y(t) = \int_T g(t,s)\mu_X(s)\,ds \tag{2.47}$$

$$\text{(ii) } K_Y(t, t') = \int_T\int_T g(t,s)g(t',s')K_X(s,s')\,ds\,ds' \tag{2.48}$$

Conversely, if the deterministic integrals in (i) and (ii) exist, then the integral defining $Y(t)$ exists in the MS sense. ■

Before proceeding to an example, we consider Theorem 2.2 from an intuitive perspective by formally computing the mean of $Y(t)$ given in (i) by passing to the limit in the defining Riemann sums without regard to rigorous mathematical justification. Then

$$\begin{aligned}\mu_Y(t) &= E\left[\lim_{\|\Delta s_k\|\to 0}\sum_{k=1}^{n} g(t,s_k')X(\omega;s_k')\Delta s_k\right] \\ &= \lim_{\|\Delta s_k\|\to 0}\sum_{k=1}^{n} g(t,s_k')E[X(\omega;s_k')]\Delta s_k \end{aligned} \tag{2.49}$$

where we have (without justification) exchanged the expected value operation with the limit and (justifiably) applied the linearity of the expectation E to the summation. Now, assuming the integrability of $g(t, s)\mu_X(s)$, passing to the limit yields

$$\mu_Y(t) = \int_T g(t,s)E[X(\omega;s)]\,ds$$

$$= \int_T g(t,s)\mu_X(s)\,ds \tag{2.50}$$

which is equation (i) of Theorem 2.2. Although we will not go through the details, equation (ii) of the theorem can likewise be formally "derived" by a similar (unjustified) interchange of the limit and the expected value. Indeed, the import of Theorem 2.2 is that such an interchange is rigorously justifiable if and only if there is MS integrability.

There is nothing about Theorem 2.2 that is dimensionally dependent. Hence, we can consider the variables t, t', s, and s' to be vector-valued, or, in the case of images, to be ordered pairs.

***Example* 2.14.** Consider the random sine wave $X(t) = Z\sin(t - W)$ of Example 2.12 and let $g(t, s) = 1$. Since both the mean and covariance are integrable, $X(t)$ is MS integrable, and since the kernel $g(t, s)$ is a constant, the integral is a random variable (not a random function of t). By Theorem 2.2 (i), if $T = [0, 2\pi]$, then

$$\mu_Y = \int_0^{2\pi} \mu_X(s)\,ds$$

$$= E[Z]\left(E[\cos W]\int_0^{2\pi}\sin t\,dt - E[\sin W]\int_0^{2\pi}\cos t\,dt\right) = 0$$

■

***Example* 2.15.** As noted prior to the theorem, the form of the theorem is appropriate to integral transforms. From the deterministic perspective, given a realization $x(t)$ that is square-integrable over $[0, 2\pi]$, $x(t)$ is represented by its trigonometric Fourier series,

$$x(t) = \frac{a_0}{2} + \sum_{k=1}^{\infty} a_k\cos kt + b_k\sin kt$$

where

$$a_k = \frac{1}{\pi}\int_0^{2\pi} x(t)\cos kt\,dt$$

$$b_k = \frac{1}{\pi} \int_0^{2\pi} x(t) \sin kt \, dt$$

and convergence is with respect to the L^2 norm, namely,

$$\lim_{n \to \infty} \int_0^{2\pi} \left| x(t) - \left(\frac{a_0}{2} + \sum_{k=1}^{n} a_k \cos kt + b_k \sin kt \right) \right|^2 dt = 0$$

The trigonometric Fourier coefficients represent a transform on $L^2[0, 2\pi]$, the space of square-integrable functions on $[0, 2\pi]$, and the representation of $x(t)$ as a trigonometic series provides the inverse transform. In the random function setting, we can still write down the finite trigonometric series, but the Fourier coefficients are random variables,

$$A_k = \frac{1}{\pi} \int_0^{2\pi} X(t) \cos kt \, dt$$

$$B_k = \frac{1}{\pi} \int_0^{2\pi} X(t) \sin kt \, dt$$

These integrals are interpreted in the mean-square sense and, according to the MS integrability theorem, their existence is determined by integrals (i) and (ii) of the theorem. In the case of the cosine coefficients, A_k exists as defined so long as the integrals

$$\mu_A(k) = \frac{1}{\pi} \int_0^{2\pi} \mu_X(s) \cos ks \, ds$$

$$K_A(k, k') = \frac{1}{\pi^2} \int_0^{2\pi} \int_0^{2\pi} K_X(s, s') \cos ks \cos k's' \, ds \, ds'$$

exist. If the integrals exist, then they give the mean and covariance functions of A_k. A similar comment applies to the sine coefficients. Existence of the mean and covariance integrals depends on the mean and covariance functions for $X(t)$. If both the mean and covariance functions are integrable, then both integrals exist. If all integrals exist, then the trigonometric Fourier integrals yield a transform $X(t) \to \{A_k, B_k\}$ taking a random function defined over continuous time to an infinite sequence of random variables. Insofar as inversion is concerned, one can write down the finite trigonometric series in terms of the A_k and B_k to produce a random function defined by the partial sum,

$$X_n(t) = \frac{A_0}{2} + \sum_{k=1}^{n} A_k \cos kt + B_k \sin kt$$

But what is the relation between the random functions $X_n(t)$ and $X(t)$? Certainly the relationship, if one exists, cannot be pointwise. It will be phrased in terms of mean-square convergence. We will return to this problem in the context of canonical representations of random functions, one of the most important topics pertaining to image and signal processing. ■

In Section 1.6, we showed that the covariance matrix of a random vector is nonnegative definite. Theorem 2.2 yields a corresponding property for covariance functions. If we let $g(t, s)$ simply be a function of t in Eq. 2.46 and integrate with respect to t, then the stochastic integral is just a random variable Y and Eq. 2.48 gives the variance of Y. Since the variance is nonnegative, the integral is nonnegative:

$$\int_T \int_T K_X(t,t')g(t)g(t')\,dt\,dt' \geq 0 \tag{2.51}$$

The inequality holds for any domain T and any function $g(t)$ for which the integral exists. A function of two variables for which this is true is said to be *nonnegative definite*. The requirement of nonnegative definiteness for a covariance function constraints the class of symmetric functions that can serve as covariance functions.

2.5. Mean Ergodicity

Like single random variables, random processes have parameters that need to be estimated. We would like to estimate parameters from a single observation of the random process, and this presents us with estimating from samples that are typically not random.

Key to many applications is estimation of the mean of a random process. Let us begin by considering a discrete random process, or time series, $X(k)$, $k = 1, 2,...,$ possessing constant mean $E[X(k)] = \mu$ and finite second moments. Our immediate goal is to determine under what conditions the mean μ can be estimated by averaging some finite collection of observations. Akin to a sample mean, for $n > 1$ we form the random variable

$$Y(n) = \frac{1}{n}\sum_{k=1}^{n} X(k) \tag{2.52}$$

and determine conditions for $Y(n)$ to converge in the mean-square to μ. Algebra yields

$$E[|Y(n) - \mu|^2] = \frac{1}{n^2} E\left[\left(\sum_{k=1}^{n} (X(k) - \mu)\right)^2\right]$$

$$= \frac{1}{n^2}\sum_{i=1}^{n}\sum_{j=1}^{n} E[(X(i) - \mu)(X(j) - \mu)]$$

$$= \frac{1}{n^2}\sum_{i=1}^{n}\sum_{j=1}^{n} K_X(i, j) \tag{2.53}$$

Thus, $Y(n)$ converges to μ in the mean-square if and only if

$$\lim_{n\to\infty} \frac{1}{n^2}\sum_{i=1}^{n}\sum_{j=1}^{n} K_X(i, j) = 0 \tag{2.54}$$

The salient point is that Eq. 2.54 gives necessary and sufficient conditions for MS convergence of the estimator in terms of a point limit of a deterministic function. Actually, MS convergence of the sample mean is a special case of Eq. 2.54 because in the case of the sample mean, the random variables $X(k)$ are independent and therefore uncorrelated, so that $K_X(i, j) = 0$ unless $i = j$, in which case $K_X(i, i) = \sigma_X^2$, the variance of the governing random variable of the random sample.

Since $E[X(k)] = \mu$ for $k = 1, 2, \ldots$, μ is the mean of $Y(n)$ and the covariance condition of Eq. 2.54 concerns convergence of $Y(n)$ to its mean in the mean-square. In fact, it is the same as Eq. 1.199 leading up to the weak law of large numbers. In the present context, Eq. 1.200, Chebyshev's inequality applied to $Y(n)$, continues to apply in the form

$$P(|Y(n) - \mu| \geq \varepsilon) \leq \frac{1}{\varepsilon^2 n^2}\sum_{i=1}^{n}\sum_{j=1}^{n} K_X(i, j) \tag{2.55}$$

for any $\varepsilon > 0$, which provides a bound on the precision of estimation. If Eq. 2.54 holds, then $Y(n)$ converges to μ in probability, since for any $\varepsilon > 0$,

$$\lim_{n\to\infty} P(|Y(n) - \mu| \geq \varepsilon) = 0 \tag{2.56}$$

which is just a restatement of Eq. 1.205 in the context of discrete time series.

Our goal is to generalize the condition of Eq. 2.54 to random functions of one or more continuous variables. To work in the context of random images, we consider a random process $X(\omega; u, v)$. For the moment, we make no assumptions regarding the mean of the process except that it exists. We further assume that the process is to be averaged over the square $T = [-r, r] \times [-r, r]$. The area of T is $4r^2$, so that averaging the process over T yields the integral

$$Y = \frac{1}{(2r)^2} \int_{-r}^{r}\int_{-r}^{r} X(u,v)\, dudv \tag{2.57}$$

Since Y is a random variable, not a random function, Theorem 2.2 (i) yields the constant

$$\mu_Y = \frac{1}{(2r)^2} \int_{-r}^{r}\int_{-r}^{r} \mu_X(u,v)\, dudv \tag{2.58}$$

Theorem 2.2 (ii) takes the form

$$\mathrm{Var}[Y] = \frac{1}{(2r)^4} \int_{-r}^{r}\int_{-r}^{r}\int_{-r}^{r}\int_{-r}^{r} K_X((u,v),(u',v'))\, dudvdu'dv' \tag{2.59}$$

Now let $X_0(u, v)$ be the *centered* random function corresponding to $X(u, v)$,

$$X_0(u, v) = X(u, v) - \mu_X(u, v) \tag{2.60}$$

Using Eqs. 2.57 and 2.58 for Y and μ_Y gives

$$\mathrm{Var}[Y] = E[|Y - \mu_Y|^2]$$

$$= E\left[\left|\frac{1}{(2r)^2} \int_{-r}^{r}\int_{-r}^{r} X_0(u,v)\, dudv\right|^2\right] \tag{2.61}$$

Setting Eqs. 2.59 and 2.61 equal yields

$$E\left[\left|\frac{1}{(2r)^2}\int_{-r}^{r}\int_{-r}^{r}X_0(u,v)\,dudv\right|^2\right] = \frac{1}{(2r)^4}\int_{-r}^{r}\int_{-r}^{r}\int_{-r}^{r}\int_{-r}^{r}K_X((u,v),(u',v'))\,dudvdu'dv' \tag{2.62}$$

Thus,

$$\underset{r\to\infty}{\text{l.i.m.}}\frac{1}{(2r)^2}\int_{-r}^{r}\int_{-r}^{r}X_0(u,v)\,dudv = 0 \tag{2.63}$$

if and only if the limit as $r \to \infty$ of the fourfold integral in Eq. 2.62 is equal to zero.

If the mean of $X(u, v)$ is constant, say μ_X, then Eq. 2.63 becomes

$$\underset{r\to\infty}{\text{l.i.m.}}\frac{1}{(2r)^2}\int_{-r}^{r}\int_{-r}^{r}X(u,v)\,dudv = \mu_X \tag{2.64}$$

The MS limit of the average of $X(u, v)$ over the square T is the mean of $X(u, v)$. Since $X(u, v)$ is a random function, this average is a random variable and use of an MS limit reflects this. This is no different than the need for MS convergence when considering the sample mean and is analogous to the MS convergence criterion for the discrete average of Eq. 2.52. Let $A[X; r]$ denote the average of $X(u, v)$ over the square T centered at the origin with sides of length $2r$ (Eq. 2.57). The foregoing conclusions are summarized in the *mean-ergodic theorem*. The theorem holds for random functions of any number of variables and we state it for a variable t of arbitrary dimension.

Theorem 2.3. If the random function $X(t)$ possesses a constant mean, then

$$\underset{r\to\infty}{\text{l.i.m.}}\,A[X;r] = \mu_X \tag{2.65}$$

if and only if

$$\lim_{r\to\infty}\frac{1}{(2r)^{2n}}\int_T\int_T K_X(u,v)\,dudv = 0 \tag{2.66}$$

where n is the number of variables in the argument of the random function and the integral is $2n$-fold. ■

Any random function satisfying Theorem 2.3 is said to be *mean-ergodic*. In analogy to the situation occurring for the mean of a single random variable, the import of the theorem can be stated as follows: just as the sample mean derived from a random sample provides an estimator of the random-variable mean that converges to the mean in the mean-square, for a mean-ergodic constant-mean random function $X(t)$, $A[X; r]$ provides an estimator of the mean μ_X of $X(t)$ that converges to μ_X in the mean-square. Practically, the mean of a mean-ergodic constant-mean random function can be estimated by averaging a single realization of the process over a sufficiently large finite region. Should this not be the case, then $\mu_X(t)$, the mean of $X(t)$, would have to be estimated by observing a number of independent realizations and averaging the observed values at t. In general, the necessary and sufficient condition of Theorem 2.3 is difficult to work with. Sufficient conditions are needed under which the limit holds so that we can conclude mean ergodicity. In Section 2.8 we discuss wide-sense stationary random functions and for these there are easy-to-apply sufficient conditions for mean ergodicity, but first we study some random functions that are important to image and signal processing.

2.6. Poisson Process

The Poisson process, which was briefly introduced in Example 2.7, has applications in many diverse areas and is especially important in imaging and the analysis of queues. It has a model-based formulation for both time and spatial processes, and it will used often in the remaining portions of the text.

2.6.1. One-dimensional Poisson Model

The one-dimensional Poisson model is mathematically described in terms of points arriving randomly in time and letting $X(t)$ count the number of points arriving in the interval $[0, t]$. Three assumptions are postulated:

(i) The numbers of arrivals in any finite set of nonoverlapping intervals are independent.

(ii) The probability of exactly one arrival in an interval of length t is $\lambda t + o(t)$.

(iii) The probability of two or more arrivals in an interval of length t is $o(t)$.

The *intensity* parameter λ is constant over all t intervals, and $o(t)$ represents any function $g(t)$ for which $\lim_{t\to 0} g(t)/t = 0$. Condition (ii) says that, for infinitesimal t, the probability of exactly one arrival in an interval of length t is λt plus a quantity very small in comparison to t and condition (iii) says that, for infinitesimal t, the probability of two or more arrivals in an interval of length t is very small in comparison to t. The random time points are called *Poisson points* and each realization of the *Poisson process* corresponds to a set of time points resulting from a particular observation of the arrival process.

Under the postulated conditions, $X(t)$ possesses a Poisson density with mean and variance equal to λt, namely,

$$P(X(t)=k) = e^{-\lambda t}\frac{(\lambda t)^k}{k!} \tag{2.67}$$

for $k = 0, 1, 2....$ To see this, break the interval $[0, t]$ into n equal subintervals and define the following events: A_n is the event that $X(t) = k$ and at least one subinterval contains more than one arrival; B_n is the event that k of the n subintervals contain exactly one arrival and the remaining $n - k$ subintervals contain no arrivals. By additivity,

$$P(X(t)=k) = P(A_n) + P(B_n) \tag{2.68}$$

Let C_i denote the event that the ith subinterval contains more than one arrival. Then

$$\begin{aligned} P(A_n) &\leq P\left(\bigcup_{i=1}^{n} C_i\right) \\ &\leq \sum_{i=1}^{n} P(C_i) \\ &= \sum_{i=1}^{n} o\left(\frac{t}{n}\right) \\ &= t\frac{o(t/n)}{t/n} \end{aligned} \tag{2.69}$$

so that $P(A_n) \to 0$ as $n \to \infty$. On the other hand, B_n corresponds to the binomial event of k successes out of n trials, with

$$P(B_n) = \binom{n}{k}\left[\frac{\lambda t}{n} + o\left(\frac{t}{n}\right)\right]^k \left[1 - \frac{\lambda t}{n} - o\left(\frac{t}{n}\right)\right]^{n-k} \tag{2.70}$$

Were there no $o(t/n)$ term, letting $n \to \infty$ would yield, according to the manner in which the binomial density converges to the Poisson density (Eq. 1.88),

$$\lim_{n\to\infty} P(B_n) = e^{-\lambda t}\frac{(\lambda t)^k}{k!} \tag{2.71}$$

In fact, it can be shown that, according to its definition, the presence of $o(t/n)$ does not affect this limit. Hence, letting $n \to \infty$ yields Eq. 2.67.

It follows from assumption (i) that the Poisson process has *independent increments*: if $t < t' < u < u'$, then $X(u') - X(u)$ and $X(t') - X(t)$ are independent. Using this independence, we find the covariance function. If $t < t'$, then the autocorrelation is

$$\begin{aligned} E[X(t)X(t')] &= E[X(t)^2 + X(t)(X(t') - X(t))] \\ &= E[X(t)^2] + E[X(t)]E[X(t') - X(t)] \\ &= E[X(t)^2] - E[X(t)]^2 + E[X(t)]E[X(t')] \\ &= \text{Var}[X(t)] + E[X(t)]E[X(t')] \\ &= \lambda t + \lambda^2 tt' \end{aligned} \tag{2.72}$$

Hence, for $t < t'$,

$$K_X(t, t') = \text{Var}[X(t)] = \lambda t \tag{2.73}$$

Interchanging the roles of t and t' yields

$$K_X(t, t') = \text{Var}[X(\min(t, t'))] = \lambda\min(t, t') \tag{2.74}$$

for all t and t'. The correlation coefficient function is given by

$$\rho_X(t, t') = \frac{\min(t, t')}{\sqrt{tt'}} \tag{2.75}$$

and $\rho_X(t, t') \to 0$ as $|t - t'| \to \infty$.

2.6.2. Derivative of the Poisson Process

Theorem 2.1 states that MS differentiability depends on differentiability of the covariance function, specifically, existence of the mixed partial derivative $\partial^2 K_X(u, v)/\partial u \partial v$. We have seen in Example 2.13 an instance of a random process that is not MS differentiable; nevertheless, in that example, the covariance possesses a generalized derivative in terms of a delta function (Eq. 2.38). The Poisson process provides a second instance of a random function that is not MS differentiable. Its covariance function possesses the generalized mixed partial derivative

$$\frac{\partial^2 K_X(t,t')}{\partial t \partial t'} = \lambda\delta(t-t') \tag{2.76}$$

Although we have defined differentiability in terms of MS convergence, it is possible to give meaning to a generalized differentiability, one for which differentiability of the process is related to the generalized mixed partial derivative of the covariance and the derivative of the mean. The point is that Theorem 2.1 can be applied in a generalized sense.

If a random process has covariance function $\lambda\delta(t - t')$ and constant mean λ (which is the derivative of the mean of the Poisson process), then we refer to the process as the *generalized derivative of the Poisson process*. Proceeding heuristically, the Poisson process $X(t)$ has step-function realizations, where steps occur at points in time randomly selected according to a Poisson density with parameter λ. For any realization, say $x(t)$, with steps at t_1, t_2,..., the usual generalized derivative is given by

$$x'(t) = \sum_{k=1}^{\infty} \delta(t-t_k) \tag{2.77}$$

If we assume that the derivative of the Poisson process consists of the process whose realizations agree with the derivatives of the realizations of the Poisson process itself, then the derivative process is given by

$$X'(t) = \sum_{k=1}^{\infty} \delta(t-Z_k) \tag{2.78}$$

where the Z_k form a sequence of random Poisson points. Since each realization of $X'(t)$ consists of a train of pulses, it is called the *Poisson impulse process*. If we now apply the generalized form of Theorem 2.1 that was alluded to previously, given that the mean and covariance of the Poisson process are λt and $\lambda\min(t, t')$, respectively, we conclude that the mean and covariance of the Poisson impulse process are given by

$$\mu_{X'}(t) = \lambda \tag{2.79}$$

$$K_{X'}(t, t') = \lambda\delta(t-t') \tag{2.80}$$

2.6.3. Properties of Poisson Points

Since the Poisson process $X(t)$ is generated by random Poisson points, a question arises: what is the time distribution between Poisson time points? More generally, what is the time distribution governing the kth Poisson point following a given Poisson point? To answer this question, let Y denote the random variable that, starting at time 0, gives the time at which the kth Poisson point occurs. Since $P(X(t) = x)$ gives the probability of x arrivals occurring in the interval $[0, t]$, the probability distribution function of Y is

$$\begin{aligned} F_Y(t) &= P(Y \le t) \\ &= P(X(t) \ge k) \\ &= \sum_{x=k}^{\infty} \frac{e^{-\lambda t}(\lambda t)^x}{x!} \end{aligned} \tag{2.81}$$

where the second equality follows from the fact that $Y \le t$ if and only if at least k arrivals occur in $[0, t]$. The density of Y is found by differentiating F_Y with respect to t:

$$\begin{aligned} f_Y(t) &= \frac{d}{dt} F_Y(t) \\ &= \sum_{x=k}^{\infty} \frac{e^{-\lambda t} x\lambda(\lambda t)^{x-1} - \lambda e^{-\lambda t}(\lambda t)^x}{x!} \\ &= \sum_{x=k}^{\infty} \frac{\lambda e^{-\lambda t}(\lambda t)^{x-1}}{(x-1)!} - \sum_{x=k}^{\infty} \frac{\lambda e^{-\lambda t}(\lambda t)^x}{x!} \\ &= \sum_{j=k-1}^{\infty} \frac{\lambda e^{-\lambda t}(\lambda t)^j}{j!} - \sum_{j=k}^{\infty} \frac{\lambda e^{-\lambda t}(\lambda t)^j}{j!} \end{aligned} \tag{2.82}$$

where the last equality results from letting $j = x - 1$ in the first sum and $j = x$ in the second. The first series is the same as the second except there is an extra term, $j = k - 1$, in the first series. The difference of the sums is simply that extra term. Consequently,

$$f_Y(t) = \frac{\lambda e^{-\lambda t}(\lambda t)^{k-1}}{(k-1)!} \tag{2.83}$$

which is a gamma distribution with $\alpha = k$ and $\beta = 1/\lambda$. We state the result as a theorem.

Theorem 2.4. The time distribution governing the arrival of the kth Poisson point following a given Poisson point is governed by a gamma distribution with $\alpha = k$ and $\beta = 1/\lambda$ (λt being the mean of the Poisson process). In particular, for $k = 1$, the interarrival time is governed by an exponential distribution with mean $1/\lambda$. ■

Poisson points are often said to model complete randomness, which means that intuitively they model a "uniform" distribution of points across the infinite interval $[0, \infty)$. It is often in this capacity that the Poisson process is used to model physical phenomena. This uniformity can be appreciated by considering Poisson points restricted to a finite interval. Consider the random set of Poisson points generated by a Poisson process of intensity λ and consider the subset of Poisson points lying in the nonempty interval $[a, b]$. Under the condition that there are n Poisson points in $[a, b]$ determining n random variables, these n random variables are independent and each is uniformly distributed over $[a, b]$. Note that we are not considering the points as being ordered in the interval, only that n random variables are defined by randomly assigning to each one of the points.

Suppose we now order the points within the interval, which, for simplicity, we take to be $[0, T]$. Let $\tau_1 < \tau_2 < \cdots < \tau_n$ be the n ordered Poisson points in $[0, T]$, where we still condition on there being n points in the interval. Then, as random variables, $\tau_1, \tau_2, \ldots, \tau_n$ can be shown to possess the same joint distribution as if they were the order statistics corresponding to a random sample of size n arising from the uniform distribution over $[0, T]$. The joint density of these order statistics is given by

$$f_{\tau_1, \tau_2, \ldots, \tau_n}(t_1, t_2, \ldots, t_n) = \begin{cases} n!/T^n, & \text{for } 0 \le t_1 \le t_2 \le \cdots \le t_n \le T \\ 0, & \text{otherwise} \end{cases} \tag{2.84}$$

We often encounter a function of Poisson points. Specifically, let $\{t_i\}$ be the set of Poisson points on the entire real line generated by a Poisson process with intensity λ and g be an integrable function. Consider the random variable defined as the summation of the functional values at the Poisson points. The expected value of the summation is

$$E\left[\sum_{i=1}^{\infty} g(t_i)\right] = \lambda \int_{-\infty}^{\infty} g(t)\, dt \tag{2.85}$$

To see this, first consider the Poisson points in a closed interval $[a, b]$ and suppose there are n of them that we label $s_1, s_2, \ldots, s_n$ without regard to order. Then

$$E\left[\sum_{i=1}^{n} g(s_i)\right] = \sum_{i=1}^{n} E[g(s_i)]$$

$$= \frac{n}{b-a}\int_a^b g(t)\,dt \tag{2.86}$$

where the second equality follows from Theorem 1.7 because the Poisson points are identically uniformly distributed on $[a, b]$. In fact, the number of points in $[a, b]$ is a random variable N. Letting

$$W = \sum_{a \le t_i \le b} g(t_i) \tag{2.87}$$

Eq. 2.86 gives the conditional expectation $E[W \mid N = n]$. According to Theorem 4.1 (Section 4.1), the expectation of $E[W \mid N = n]$ relative to the distribution of N equals $E[W]$. Hence,

$$E[W] = \sum_{n=0}^{\infty} E[W \mid N = n]P(N = n)$$

$$= \frac{1}{b-a}\int_a^b g(t)\,dt \sum_{n=0}^{\infty} nP(N = n)$$

$$= \lambda \int_a^b g(t)\,dt \tag{2.88}$$

where the last equality follows from $E[N] = \lambda(b - a)$. Letting $a \to -\infty$ and $b \to \infty$ yields Eq. 2.85.

Continuing to assume that $\{t_i\}$ is the set of Poisson points generated by a Poisson process with intensity λ and g is an integrable function, it can also be shown that

$$E\left[\prod_{i=1}^{\infty}(1 + g(t_i))\right] = \exp\left(\lambda \int_{-\infty}^{\infty} g(t)\,dt\right) \tag{2.89}$$

2.6.4. Axiomatic Formulation of the Poisson Process

So far we have based our discussion of the Poisson process on the conditions defining a Poisson arrival process, a key consequence being that the process possesses independent increments. In general, a random process $X(t)$, $t \geq 0$, is said to have *independent increments* if $X(0) = 0$ and, for any $t_1 < t_2 < \cdots < t_n$, the random variables $X(t_2) - X(t_1)$, $X(t_3) - X(t_2)$,..., $X(t_n) - X(t_{n-1})$ are independent. It has *stationary independent increments* if $X(t + r) - X(t' + r)$ is identically distributed to $X(t) - X(t')$ for any t, t', and r. When the increments are stationary, the increment distribution depends only on the length of time, $t - t'$, over which the increments are taken, not the specific points in time, $t + r$ and $t' + r$, As defined via the arrival model, the Poisson process has stationary independent increments.

Proceeding axiomatically, we can define a process $X(t)$ to be a *Poisson process* with *mean rate* λ if

P1. $X(t)$ has values in $\{0, 1, 2,...\}$.

P2. $X(t)$ has stationary independent increments.

P3. For $s > t$, $X(t) - X(s)$ has a Poisson distribution with mean $\lambda(t - s)$.

The axiomatic formulation P1 through P3 captures completely the arrival model.

By generalizing the axioms P1 through P3, we can arrive at a definition of Poisson points in space, a concept important to modeling grain and texture processes. Consider points randomly distributed in Euclidean space $\Re^n$ and let $N(D)$ denote the number of points in a domain D. The points are said to be *distributed in accordance with a spatial Poisson process with mean rate* λ if

PS1. For any disjoint domains D_1, D_2,..., D_r, the counts $N(D_1)$, $N(D_2)$,..., $N(D_r)$ are independent random variables.

PS2. For any domain D of finite volume, $N(D)$ possesses a Poisson distribution with mean $\lambda\nu(D)$, where $\nu(D)$ denotes the volume (measure) of D.

In analogy to the one-dimensional model, the spatial Poisson process can be modeled as a spatial point process. Under the independence assumption, if the probability of exactly one point in a domain of volume ν is $\lambda\nu + o(\nu)$ and the probability of more than a single point in a domain of volume ν is $o(\nu)$, then the points are distributed in accordance with a Poisson process, with $N(D)$ possessing a Poisson distribution with mean $\lambda\nu(D)$.

If $\{t_i\}$ is a Poisson point process in $\Re^n$ with intensity λ and s is a fixed n-vector, then it follows at once from PS1 and PS2 that $\{t_i + s\}$ is also a Poisson point process in $\Re^n$ with intensity λ. The next theorem states a more general version of this property.

Theorem 2.5. Suppose $\{t_i\}$ is a Poisson point process in $\Re^n$ with intensity λ and $\{s_i\}$ is a sequence of independent, identically distributed n-vectors that are also independent of the point process $\{t_i\}$. Then the point process $\{t_i + s_i\}$ is a Poisson point process with intensity λ. ■

2.7. Wiener Process and White Noise

This section treats two random processes that are important to both image and signal processing. The first, white noise, plays key roles in canonical representation, noise modeling, and design of optimal filters; the second, the Wiener process, is used to model random incremental behavior, in particular, Brownian motion. The processes are related because white noise is the generalized derivative of the Wiener process.

2.7.1. White Noise

A zero-mean random function $X(k)$ defined on a discrete domain (which for convenience we take to be the positive integers) is called *discrete white noise* if $X(k)$ and $X(j)$ are uncorrelated for $k \neq j$. If $X(k)$ is discrete white noise, then its covariance is given by

$$K_X(k,j) = E[X(k)X(j)] = \begin{cases} \mathrm{Var}[X(k)], & \text{if } k = j \\ 0, & \text{if } k \neq j \end{cases} \tag{2.90}$$

For any function g defined on the integers and for all k,

$$\sum_{i=1}^{\infty} K_X(k,i)g(i) = \mathrm{Var}[X(k)]g(k) \tag{2.91}$$

Suppose there were a similar process in the continuous setting. If $X(t)$ were such a random function defined over the real line, then the preceding equation would take the form

$$\int_{-\infty}^{\infty} K_x(t,t')g(t')dt' = I(t)g(t) \tag{2.92}$$

where $I(t)$ is a function of t that plays the role played by $\mathrm{Var}[X(k)]$ in Eq. 2.91. If we set

$$K_X(t, t') = I(t)\delta(t - t') \tag{2.93}$$

then we obtain Eq. 2.92. Hence, any zero-mean random function having a covariance of the form $I(t)\delta(t - t')$ is called *continuous white noise*. From a standard mathematical perspective, continuous white noise does not exist.

Although there exist mathematical means (generalized functions) to precisely characterize continuous white noise, the matter is beyond the scope of this book and, besides, as with delta functions, it is the manner in which the white noise model approximates certain physical processes that is important for application. It follows from the definition of white noise in terms of its covariance function $I(t)\delta(t - t')$ that continuous white noise processes have infinite variance (set $t = t'$) and uncorrelated variables. $I(t)$ is called the *intensity* of the white noise process.

For an approximation of continuous white noise in one dimension, it can be shown that there exists a normal, zero-mean, stochastic process $X(t)$ having covariance function

$$K_X(t, t') = be^{-b|t-t'|} \tag{2.94}$$

for $b > 0$. For very large values of b, this covariance function behaves approximately like the covariance of white noise, and therefore $X(t)$ is used to approximate white noise.

2.7.2. Random Walk

Brownian motion is modeled via the continuous-time Wiener process to be introduced shortly. Before discussing it, we consider a discrete-time process, the random walk, that is useful in its own right in the digital environment and represents a simplified version of the continuous-time process.

Suppose a particle, starting at the origin (in $\mathfrak{R}$) moves a unit length to the right or left. Movements are taken independently and for each movement the probabilities of going right or left are p and $q = p - 1$, respectively. Let $X(n)$ be the number of units the particle has moved to the right after n movements. $X(n)$ is called the one-dimensional *random walk*. The range of $X(n)$ is the set of integers $\{-n, -n + 2,..., n - 2, n\}$. To find the density for $X(n)$, let Y be the binomial random variable for n trials with probability p of success. For $X(n) = x$, there must be $(n + x)/2$ movements to the right and $(n - x)/2$ movements to the left. Hence, $X(n) = x$ if and only if $Y = (n + x)/2$. Consequently,

$$f_{X(n)}(x) = P\left(Y = \frac{n+x}{2}\right) = \binom{n}{\frac{n+x}{2}} p^{(n+x)/2} q^{(n-x)/2} \tag{2.95}$$

Owing to the relationship between $X(n)$ and Y, the moments of $X(n)$ can be evaluated in terms of moments of $Y = (n + X(n))/2$. Specifically, $X(n) = 2Y - n$ and

$$E[X(n)^m] = \sum_{y=0}^{n} (2y - n)^m P(Y = y) \tag{2.96}$$

Since Y is binomial, $E[Y] = np$ and $E[Y^2] = npq + n^2p^2$. Letting $m = 1$ and $m = 2$ yields

$$E[X(n)] = 2np - n \tag{2.97}$$

$$E[X(n)^2] = 4(npq + n^2p^2) + (1 - 4p)n^2 \tag{2.98}$$

The most important case is when movements to the right and left are equiprobable, so that $p = q = 1/2$. Then $E[X(n)] = 0$ and $\text{Var}[X(n)] = E[X(n)^2] = n$. It is clear from the model that the process has stationary independent increments. Therefore the covariance argument applied to the Poisson process in Eqs. 2.72 and 2.73 also applies here. Thus,

$$K_X(n, n') = \text{Var}[X(\min(n, n'))] = \min(n, n') \tag{2.99}$$

As for the correlation coefficient,

$$\rho_X(n, n') = \frac{\min(n, n')}{\sqrt{nn'}} \tag{2.100}$$

and $\rho_X(n, n') \to 0$ as $|n - n'| \to \infty$.

2.7.3. Wiener Process

In the random-walk analysis we have chosen the time interval between movements to be 1; however, the analysis can be adopted to any finite time interval — the smaller the interval, the more spasmodic the motion. Moreover, instead of restricting the motion to a single dimension, it could be analyzed in two dimensions, where the random walker can choose between four directions: left, right, up, and down. Taking a limiting situation, one can imagine an infinitesimal particle being continually acted upon by forces from its environment. The motion of such a particle can appear spasmodic and under suitable conditions such motion is referred to as *Brownian motion*.

Suppose a particle is experiencing Brownian motion and $X(t)$ is its displacement in a single dimension from its original initial position. Assuming the particle's motion results from a multitude of molecular impacts lacking any discernible regularity, it is reasonable to suppose that $X(t)$ possesses independent increments. Moreover, if it is assumed that the nature of the particle, the medium, and the relationship between the particle and its medium remain stable, and that the displacement over any interval of time depends only on the elapsed time, not on the moment the time period commences, then it is reasonable to postulate stationarity of the increments. Next, we assume that for any fixed t, the random variable $X(t)$ is normally distributed with mean zero, an assumption supported by empirical data. Note that if the forces act on the particle without directional bias, it is certainly plausible to expect the mean of the displacement process to be identically zero.

In accordance with the foregoing considerations, we define a *Wiener process* $X(t)$, $t \geq 0$, to be a random function satisfying the following conditions:

W1. $X(0) = 0$.

W2. $X(t)$ has stationary independent increments.

W3. $E[X(t)] = 0$.

W4. For any fixed t, $X(t)$ is normally distributed.

Based on W1 through W4, one can verify that, for $0 \leq t' < t$, the increment random variable $X(t) - X(t')$ has mean zero and variance $\sigma^2|t - t'|$, where σ^2 is a parameter to be empirically determined. In particular, $\text{Var}[X(t)] = \sigma^2 t$ for $t \geq 0$. As for the covariance, based on independent increments the same argument used for the Poisson and random walk processes can be applied to obtain

$$K_X(t, t') = \text{Var}[X(\min(t, t'))] = \sigma^2 \min(t, t') \tag{2.101}$$

Up to a multiplicative constant, the Poisson and Wiener processes have the same covariance. If we take the generalized mixed partial derivative of the Wiener process covariance, we obtain, as in the Poisson case, a constant times a delta function:

$$\frac{\partial^2 K_X(t-t')}{\partial t \partial t'} = \sigma^2 \delta(t - t') \tag{2.102}$$

Moreover, since the mean of the Wiener process is zero, so is the derivative of its mean. Consequently, assuming the generalized version of Theorem 2.1, we conclude that the derivative of the Wiener process is white noise.

2.8. Stationarity

In general, the nth order probability distributions of a random function at two different sets of time points need not have any particular relation to each other. This section discusses two situations in which they do. First, the covariance function of a random process $X(t)$ is generally a function of two variables; however, in some cases it is a function of the difference between the variables. For such processes, covariance and autocorrelation relations are simplified and a number of properties important to applications are satisfied, including the existence of simple conditions for ergodicity and simplified design of optimal linear filters. A stronger relation occurs when the nth order probability distribution itself is invariant under a translation of the time-point set.

2.8.1. Wide-Sense Stationarity

If the covariance function of the random function $X(t)$ can be written as

$$K_X(t, t') = k_X(\tau) \tag{2.103}$$

where $\tau = t - t'$ and, in addition, if $X(t)$ has a constant mean, then $X(t)$ is said to be *wide-sense stationary*, or simply *WS stationary*. The point t can be a scalar or a vector. In two dimensions (random images), the process takes the form $X(u, v)$ and Eq. 2.103 becomes

$$K_X((u, v), (u', v')) = k_X(\tau_1, \tau_2) \tag{2.104}$$

where $\tau_1 = u - u'$ and $\tau_2 = v - v'$.

If $X(t)$ is WS stationary, then its variance function is constant,

$$\mathrm{Var}[X(t)] = K_X(t, t) = k_X(t - t) = k_X(0) \tag{2.105}$$

Owing to the symmetry of $K_X(t, t')$, the covariance function of $X(t)$ is an even function,

$$k_X(-\tau) = k_X(t' - t) = k_X(t - t') = k_X(\tau) \tag{2.106}$$

Hence, $k_X(\tau) = k_X(|\tau|)$. The correlation-coefficient function reduces to a function of τ:

$$\rho_X(t - t') = \rho_X(\tau) = \frac{k_X(\tau)}{k_X(0)} \tag{2.107}$$

Since $|\rho_X(\tau)| \leq 1$,

$$|k_X(\tau)| \leq k_X(0) \tag{2.108}$$

The autocorrelation is also a function of τ: letting μ_X denote the constant mean of $X(t)$,

$$\begin{aligned} R_X(t, t') &= K_X(t,t') + \mu_X(t)\mu_X(t') \\ &= k_X(\tau) + \mu_X^2 \end{aligned}$$

$$= r_X(\tau) \tag{2.109}$$

A random function $X(t)$ is WS stationary if and only if its covariance function is *translation invariant*, which means that, for any increment h,

$$K_X(t + h, t' + h) = K_X(t, t') \tag{2.110}$$

To see this, suppose $X(t)$ is WS stationary. Then

$$K_X(t + h, t' + h) = k_X(t + h - (t' + h)) = k_X(t - t') = K_X(t, t') \tag{2.111}$$

Conversely, if the covariance function is translation invariant, then

$$K_X(t, t') = K_X(t - t', t' - t') = K_X(t - t', 0) \tag{2.112}$$

which is a function of $t - t'$.

***Example* 2.16.** Consider the square wave defined on [0, 2) by

$$g(t) = \begin{cases} 1, & \text{if } 0 \le t < 1 \\ 0, & \text{if } 1 \le t < 2 \end{cases}$$

and defined on $\Re$ by periodic extension. Let Z be a uniformly distributed random variable on the interval [0, 2] and define the random function $X(t) = g(t - Z)$. Each realization of the process is a translated copy of the square wave. Owing to the uniformity of Z, for each t, $X(t)$ is an equiprobable binary variable and $\mu_X = 1/2$. As for the covariance, first suppose $|t - t'| \le 1$. Then, since $X(t)X(t')$ is either equal to 1 or to 0, the autocorrelation function is given by

$$R_X(t, t') = P(X(t) = X(t') = 1)$$

To have $X(t) = X(t') = 1$, both t and t' must be located beneath a block of the wave (Fig. 2.5). Since Z is uniformly distributed over [0, 2], the probability of this happening is

$$P(X(t) = X(t') = 1) = \frac{1 - |t - t'|}{2}$$

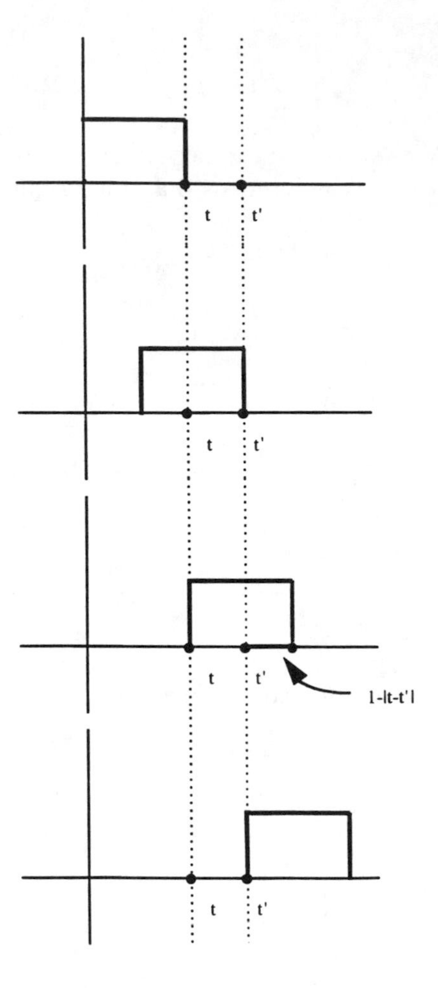

Figure 2.5 Square-wave realizations.

Next, for $1 < |t - t'| \leq 2$,

$$P(X(t) = X(t') = 1) = \frac{|t - t'| - 1}{2}$$

Hence, for $|t - t'| \leq 2$,

$$R_X(t, t') = \frac{|1 - |t - t'||}{2}$$

Subtraction of $\mu_X^2 = 1/4$ yields the covariance function

$$k_X(\tau) = \frac{2|1 - |\tau|| - 1}{4}$$

for $|\tau| \le 2$. It is defined over all τ by periodic extension. Owing to WS stationarity, $\text{Var}[X(t)] = k_X(0) = 1/4$. ■

***Example* 2.17.** Let $Y(t)$ be the Poisson process with mean λt and r be a positive constant. The Poisson increment process is defined by

$$X(t) = Y(t + r) - Y(t)$$

According to the Poisson model, $X(t)$ counts the number of points in $(t, t + r]$ and

$$\mu_X(t) = E[Y(t + r)] - E[Y(t)] = \lambda r$$

For the covariance $K_X(t, t')$, there are two cases: $|t - t'| \ge r$ and $|t - t'| < r$. If $|t - t'| \ge r$, then the intervals determined by t and t' are nonoverlapping and, owing to independent increments, $X(t)$ and $X(t')$ are independent and their covariance is 0. Suppose $|t - t'| < r$. First consider the case where $t < t'$. Then $t < t' < t + r < t' + r$ and we can apply the result of Eq. 2.72 together with the observation that, because the process counts the number of points in $(t, t + r]$, its mean is λr. We obtain

$$\begin{aligned}
K_X(t, t') &= E[(Y(t'+r) - Y(t'))(Y(t+r) - Y(t))] - E[X(t')]E[X(t)] \\
&= E[Y(t'+r)Y(t+r)] - E[Y(t'+r)Y(t)] - E[Y(t+r)Y(t')] + E[Y(t')Y(t)] - \lambda^2 r^2 \\
&= \lambda(t+r) + \lambda^2(t'+r)(t+r) - \lambda t - \lambda^2 t(t'+r) - \lambda t' - \lambda^2 t'(t+r) + \lambda t + \lambda^2 tt' - \lambda^2 r^2 \\
&= \lambda(r - (t' - t))
\end{aligned}$$

Owing to symmetry, interchanging the roles of t and t' $(t' < t)$ yields

$$K_X(t, t') = \lambda(r - |t - t'|)$$

when $|t - t'| < r$. Hence, $X(t)$ is WS stationary with

$$k_X(\tau) = \begin{cases} \lambda(r - |\tau|), & \text{if } |\tau| \le r \\ 0, & \text{if } |\tau| > r \end{cases}$$ ■

***Example* 2.18.** Again consider a stochastic process determined in part by distributions of Poisson points on the nonnegative axis. $X(t)$ is two-valued, -1 and 1, changes its value at every Poisson point, and remains constant between these changes. For now, assume for the process an initial value $X(0) = 1$. If k is the number of Poisson points in the interval $(0, t]$, then

$$X(t) = \begin{cases} 1, & \text{if } k \text{ is even} \\ -1, & \text{if } k \text{ is odd} \end{cases}$$

Since the probability that there are k points in the interval $(0, t]$ is $e^{-\lambda t}(\lambda t)^k/k!$,

$$P(X(t) = 1) = P(k \text{ even})$$

$$= P(k = 0) + P(k = 2) + P(k = 4) + \cdots$$

$$= \sum_{k=0}^{\infty} e^{-\lambda t} \frac{(\lambda t)^{2k}}{(2k)!}$$

$$= e^{-\lambda t} \cosh \lambda t$$

$$P(X(t) = -1) = P(k \text{ odd})$$

$$= \sum_{k=0}^{\infty} e^{-\lambda t} \frac{(\lambda t)^{2k+1}}{(2k+1)!}$$

$$= e^{-\lambda t} \sinh \lambda t$$

Hence,

$$\mu_X(t) = e^{-\lambda t} \cosh \lambda t - e^{-\lambda t} \sinh \lambda t = e^{-2\lambda t}$$

As for the autocorrelation, suppose $t < t'$ and let F_e and F_o be the events that there is an even number and an odd number of points in $(t, t']$, respectively. Then

$$R_X(t, t') = E[X(t)X(t')]$$

$$= P(X(t)X(t') = 1) - P(X(t)X(t') = -1)$$

$$= P(X(t) = X(t') = 1) + P(X(t) = X(t') = -1)$$

$$- P(X(t) = 1, X(t') = -1) - P(X(t) = -1, X(t') = 1)$$

$$= P(X(t)=1)P(X(t')=1|X(t)=1)$$

$$+ P(X(t)=-1)P(X(t')=-1|X(t)=-1)$$

$$- P(X(t)=1)P(X(t')=-1|X(t)=1)$$

$$- P(X(t)=-1)P(X(t')=1|X(t)=-1)$$

$$= e^{-\lambda t}\cosh\lambda t\, P(F_e) + e^{-\lambda t}\sinh\lambda t\, P(F_e)$$

$$- e^{-\lambda t}\cosh\lambda t\, P(F_o) - e^{-\lambda t}\sinh\lambda t\, P(F_o)$$

$$= e^{-\lambda t}\cosh\lambda t\, e^{-\lambda(t'-t)}\cosh\lambda(t'-t) + e^{-\lambda t}\sinh\lambda t\, e^{-\lambda(t'-t)}\cosh\lambda(t'-t)$$

$$- e^{-\lambda t}\cosh\lambda t\, e^{-\lambda(t'-t)}\sinh\lambda(t'-t) - e^{-\lambda t}\sinh\lambda t\, e^{-\lambda(t'-t)}\sinh\lambda(t'-t)$$

Interchanging the roles of t and t' yields

$$R_X(t, t') = e^{-2\lambda|t-t'|}$$

Now suppose $Y(0)$ is a binomial random variable with values -1 and 1, each having probability $1/2$, $Y(0)$ is independent of the Poisson process determining the changes of the random signal $X(t)$, and, for $t > 0$,

$$Y(t) = Y(0)X(t) = \begin{cases} X(t), & \text{if } Y(0) = 1 \\ -X(t), & \text{if } Y(0) = -1 \end{cases}$$

$Y(t)$ is called the *random telegraph signal* and it behaves similarly to $X(t)$, except that its initial value need not be 1. Since $E[Y(0)] = 0$, independence of $Y(0)$ and $X(t)$ implies

$$\mu_Y(t) = E[Y(0)]E[X(t)] = 0$$

Since $E[Y(0)^2] = 1$, $Y(t)$ has covariance

$$K_Y(t, t') = E[Y(t)Y(t')]$$

$$= E[Y(0)^2]E[X(t)X(t')]$$

$$= e^{-2\lambda|t-t'|}$$

Thus, $Y(t)$ is WS stationary. ■

***Example* 2.19.** Consider the random signal

$$X(t) = Z \cos bt + W \sin bt$$

where b is a fixed constant and where Z and W are uncorrelated, zero-mean random variables, each having variance σ^2. Then

$$\mu_X(t) = E[Z] \cos bt + E[W] \sin bt = 0$$

and $X(t)$ is WS stationary because

$$\begin{aligned} K_X(t, t') &= E[X(t)X(t')] \\ &= E[Z^2]\cos bt \cos bt' + E[W^2]\sin bt \sin bt' \\ &\quad + E[ZW](\cos bt \sin bt' + \sin bt \cos bt') \\ &= \sigma^2 \cos b(t - t') \end{aligned}$$

■

2.8.2. Mean-Ergodicity for WS Stationary Processes

Theorem 2.3 gives a necessary and sufficient condition in terms of the covariance function for a constant-mean random function to be mean-ergodic. In the single-variable case, if the random function happens to be WS stationary, then the necessary and sufficient condition given by the theorem and involving the double integral of the covariance takes the form

$$\frac{1}{(2r)^2} \int_{-r}^{r} \int_{-r}^{r} k_X(u - v)\, du dv = \frac{1}{(2r)^2} \int_{-r}^{r} \int_{-r-v}^{r-v} k_X(\tau)\, d\tau dv \tag{2.113}$$

where we have employed the change of variable $\tau = u - v$ and the integral on the right-hand side of the equation involves integration of $k_X(\tau)$ over the shaded region in Fig. 2.6. To change the order of integration, it is convenient to define the function $G(\tau, v)$ on the rectangle in the figure by $G(\tau, v) = k_X(\tau)$ on the shaded portion and $G(\tau, v) = 0$ elsewhere in the rectangle. Then the right-hand-side integral in Eq. 2.113 becomes

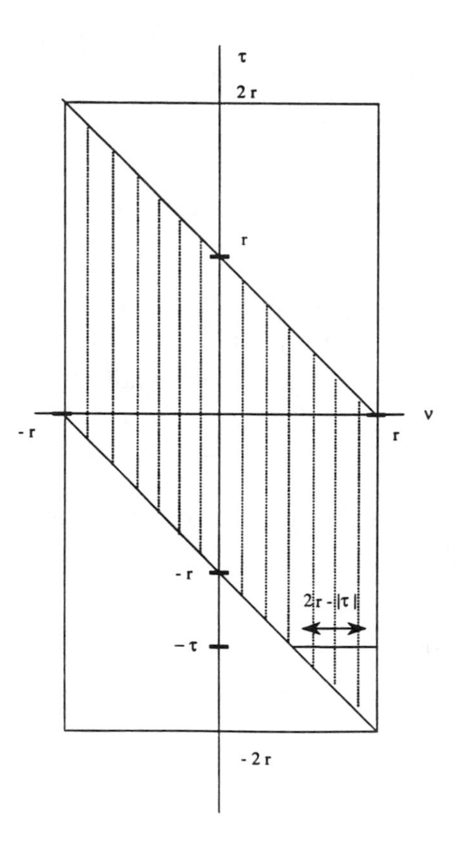

Figure 2.6 Region of integration.

$$\frac{1}{(2r)^2}\int_{-r}^{r}\int_{-2r}^{2r} G(\tau, v)\, d\tau dv = \frac{1}{(2r)^2}\int_{-2r}^{2r}\int_{-r}^{r} G(\tau, v)\, dv d\tau$$

$$= \frac{1}{(2r)^2}\int_{-2r}^{2r}(2r - |\tau|)k_X(\tau)\, d\tau \tag{2.114}$$

Hence, for a WS stationary stochastic process, Theorem 2.3 takes the following form.

Theorem 2.6. A WS stationary stochastic process is mean-ergodic if and only if

$$\lim_{r\to\infty}\frac{1}{2r}\int_{-2r}^{2r}\left(1-\frac{|\tau|}{2r}\right)k_X(\tau)\, d\tau = 0 \qquad \blacksquare \tag{2.115}$$

Since Theorem 2.3 is dimensionally independent, the preceding analysis applies to random images: a WS stationary random image $X(u, v)$ is mean-ergodic if and only if

$$\lim_{r\to\infty}\frac{1}{(2r)^2}\int_{-2r}^{2r}\int_{-2r}^{2r}\left(1-\frac{|\tau_1|}{2r}\right)\left(1-\frac{|\tau_2|}{2r}\right)k_X(\tau_1,\tau_2)\,d\tau_1 d\tau_2=0 \tag{2.116}$$

Extension to n variables is immediate.

Application of Theorem 2.6 is difficult if the mean-ergodic condition is left in the given form. While the limiting condition is most powerful because it is necessary and sufficient, what we really need are easy-to-apply sufficient conditions. We now provide these under the assumption of WS stationarity.

If the covariance of a WS stationary random function is integrable, then the limiting condition on the integral of the covariance in Theorem 2.6 is satisfied. Specifically,

$$\left|\left(1-\frac{|\tau|}{2r}\right)k_X(\tau)\right| \le |k_X(\tau)| \tag{2.117}$$

and

$$\left|\frac{1}{2r}\int_{-2r}^{2r}\left(1-\frac{|\tau|}{2r}\right)k_X(\tau)\,d\tau\right|\le\frac{1}{2r}\int_{-\infty}^{\infty}|k_X(\tau)|\,d\tau \tag{2.118}$$

Integrability of the covariance means that the integral on the right-hand side of the inequality is finite, so that division by $2r$ forces the expression to 0 as $r \to \infty$. An identical argument applies for random functions of n variables.

Another sufficient condition for a WS stationary random function (having finite variance) to be mean-ergodic is that

$$\lim_{|\tau|\to\infty} k_X(\tau)=0 \tag{2.119}$$

which clearly holds if $k_X(\tau) = 0$ for sufficiently large $|\tau|$. To verify the condition, note that, since $k_X(\tau) \to 0$ as $|\tau| \to \infty$, given any $c > 0$, there exists $M > 0$ such that $|k_X(\tau)| < c$ for $|\tau| > M$. Hence, since $|k_X(\tau)| < k_X(0)$, breaking up the integral of Theorem 2.6 yields

$$\left|\frac{1}{2r}\int_{-2r}^{2r}\left(1-\frac{|\tau|}{2r}\right)k_X(\tau)\,d\tau\right|\le\frac{1}{2r}\left(\int_{-M}^{M}|k_X(\tau)|\,d\tau+\int_{M<|\tau|<2r}|k_X(\tau)|\,d\tau\right)$$

$$\leq \frac{2Mk_X(0) + 2(2r - M)c}{2r} \tag{2.120}$$

The limit of the last expression as $r \to \infty$ is $2c$, so the limit of the integral of Theorem 2.6 must be bounded by $2c$. Since c is arbitrary, the limit of the integral must be less than any predetermined positive value and therefore must be 0. Because the argument is independent of dimension, it applies to random functions of any number of variables.

The limiting condition of Eq. 2.119 applies to the random telegraph signal of Example 2.18. Its covariance function is $e^{-2\lambda|\tau|}$, which tends to 0 as $|\tau| \to \infty$. Hence, the process is mean-ergodic.

2.8.3. Covariance-Ergodicity for WS Stationary Processes

Theorem 2.6 gives a necessary and sufficient condition for a WS stationary random function to be mean-ergodic. A constant-mean random function is mean-ergodic if its mean can be found as an MS limit of the average value of the random function. We now discuss conditions that ensure covariance-ergodicity of a Gaussian WS stationary random function. Since for any covariance function there exists a normally distributed random function possessing the given covariance, for the purposes of application the normality restriction is not as severe as it might first appear.

A WS stationary stochastic process $X(t)$ is said to be *covariance-ergodic* if

$$k_X(\tau) = \underset{r\to\infty}{\text{l.i.m.}} \frac{1}{2r} \int_{-r}^{r} X_0(t+\tau) X_0(t)\, dt \tag{2.121}$$

where $X_0(t) = X(t) - \mu_X$ is the centralization of $X(t)$. According to the definition of the MS limit, Eq. 2.121 is satisfied if and only if

$$\lim_{r\to\infty} E\left[\left| k_X(\tau) - \frac{1}{2r} \int_{-r}^{r} X_0(t+\tau) X_0(t)\, dt \right|^2\right] = 0 \tag{2.122}$$

If $X(t)$ is mean-ergodic, its mean can be estimated by averaging a single realization of $X(t)$ over a sufficiently large interval because the estimator formed by the random integral of $X(t)$ is MS convergent to the mean; if $X(t)$ is covariance-ergodic, its covariance can be estimated by averaging a single realization of $X_0(t + \tau)X_0(t)$ over a sufficiently large interval because the estimator formed by the random integral of $X_0(t + \tau)X_0(t)$ is MS convergent to the covariance. Generalization of the definition to n-dimensional random functions is immediate: let t and τ be vectors, the integral be n-fold, and the averaging factor be $(2r)^{-n}$.

To arrive at an analogue of Theorem 2.6 for covariance-ergodicity, consider the random function

$$Y(t) = X_0(t + \tau)X_0(t) \tag{2.123}$$

where t and τ can be variables of any dimension. We will demonstrate that, under the restriction of normality, Theorem 2.6 applies to $Y(t)$. To begin with, the mean of $Y(t)$ is

$$\begin{aligned} \mu_Y(t) &= E[X_0(t+\tau)X_0(t)] \\ &= K_X(t+\tau, t) \\ &= k_X(\tau) \end{aligned} \tag{2.124}$$

the last equality following from the WS stationarity of $X(t)$. The autocorrelation of $Y(t)$ is

$$R_Y(t, t') = E[X_0(t+\tau)X_0(t)X_0(t'+\tau)X_0(t')] \tag{2.125}$$

which is a mixed fourth moment of the random variables $X_0(t + \tau)$, $X_0(t)$, $X_0(t' + \tau)$, and $X_0(t')$. Assuming $X(t)$ is Gaussian, these random variables possess a multivariate normal distribution. It is known that the mixed fourth moment of four jointly normally distributed zero-mean random variables Z_1, Z_2, Z_3, and Z_4 is given by

$$\begin{aligned} E[Z_1Z_2Z_3Z_4] = {} & \mathrm{Cov}[Z_1, Z_2]\mathrm{Cov}[Z_3, Z_4] + \mathrm{Cov}[Z_1, Z_3]\mathrm{Cov}[Z_2, Z_4] \\ & + \mathrm{Cov}[Z_1, Z_4]\mathrm{Cov}[Z_2, Z_3] \end{aligned} \tag{2.126}$$

Hence,

$$\begin{aligned} R_Y(t, t') &= K_X(t+\tau, t)K_X(t'+\tau, t') + K_X(t+\tau, t'+\tau)K_X(t, t') + K_X(t+\tau, t')K_X(t, t'+\tau) \\ &= k_X^2(\tau) + k_X^2(t-t') + k_X(t-t'+\tau)k_X(t-t'-\tau) \end{aligned} \tag{2.127}$$

Combining the results for the mean and autocorrelation yields

$$K_Y(t, t') = R_Y(t, t') - \mu_Y(t)\mu_Y(t')$$

$$= k_X^2(t-t')+k_X(t-t'+\tau)k_X(t-t'-\tau) \tag{2.128}$$

Since $K_Y(t, t')$ is a function of $t - t'$, $Y(t)$ is WS stationary and Theorem 2.6 applies to $Y(t)$. Letting t denote the difference $t - t'$ in the covariance for $Y(t)$ and applying Theorem 2.6 yields the next theorem.

Theorem 2.7. A WS stationary Gaussian stochastic process $X(t)$ is covariance-ergodic if and only if

$$\lim_{r\to\infty}\frac{1}{2r}\int_{-2r}^{2r}\left(1-\frac{|t|}{2r}\right)[k_X^2(t)+k_X(t+\tau)k_X(t-\tau)]\,dt=0 \quad \blacksquare \tag{2.129}$$

To obtain a statement of Theorem 2.7 that applies to random images, apply the image version of Theorem 2.6 (Eq. 2.116) to the random function $Y(t) = X_0(t + \tau)X_0(t)$, where t is treated as an ordered pair. Extension to n variables is immediate. As in the case of Theorem 2.6, the given necessary and sufficient condition is difficult to employ and, as in the case of mean ergodicity, the covariance limiting condition of Eq. 2.119 is sufficient.

2.8.4. Strict-Sense Stationarity

A stronger form of stationarity concerns higher-order probabilistic information. The random function $X(t)$ is said to be *strict-sense stationary* (*SS stationary*) if, for any points $t_1, t_2,\ldots, t_n$, and for any increment h, its nth-order distribution function satisfies the relation

$$F(x_1, x_2,\ldots, x_n; t_1+h, t_2+h,\ldots, t_n+h) = F(x_1, x_2,\ldots, x_n; t_1, t_2,\ldots, t_n) \tag{2.130}$$

In terms of the nth order density,

$$f(x_1, x_2,\ldots, x_n; t_1+h, t_2+h,\ldots, t_n+h) = f(x_1, x_2,\ldots, x_n; t_1, t_2,\ldots, t_n) \tag{2.131}$$

Given any finite set of random variables from the random function, a spatial translation of each by a constant h results in a collection of random variables whose multivariate distribution is identical to that of the original collection. From a probabilistic perspective, the new collection is indistinguishable from the first. If we define the random vector

$$\mathbf{X}(t_1, t_2,\ldots, t_n) = \begin{pmatrix} X(t_1) \\ X(t_2) \\ \vdots \\ X(t_n) \end{pmatrix} \tag{2.132}$$

then $\mathbf{X}(t_1 + h, t_2 + h, \ldots, t_n + h)$ is identically distributed to $\mathbf{X}(t_1, t_2, \ldots, t_n)$.

If we specialize Eq. 2.130 to a single point, then it becomes

$$F(x; t+h) = F(x; t) \tag{2.133}$$

$X(t + h)$ is identically distributed to $X(t)$ and therefore the mean at $t + h$ must equal the mean at t for all h, which implies that the mean must be constant. If we next specialize Eq. 2.130 to two points, then it becomes

$$F(x_1, x_2; t_1 + h, t_2 + h) = F(x_1, x_2; t_1, t_2) \tag{2.134}$$

$\mathbf{X}(t_1 + h, t_2 + h)$ is identically distributed to $\mathbf{X}(t_1, t_2)$. Hence,

$$\mathrm{Cov}[X(t_1 + h), X(t_2 + h)] = \mathrm{Cov}[X(t_1), X(t_2)] \tag{2.135}$$

that is,

$$K_X(t_1 + h, t_2 + h) = K_X(t_1, t_2) \tag{2.136}$$

Therefore, the covariance function is translation invariant and $X(t)$ is WS stationary. In summary, SS stationarity implies WS stationarity.

Geometrically, if observers A and B observe a SS stationary process at points $t_1, t_2, \ldots, t_n$ and $t_1 + h, t_2 + h, \ldots, t_n + h$, respectively, then they observe "the same process" (more accurately, identically distributed random vectors). WS stationarity only requires that observers A and B see the same covariance at t_1, t_2 and $t_1 + h$, $t_2 + h$, respectively. If a property of the process depends only on the covariance function, then that property is indistinguishable to the two observers of the WS stationary process.

For a random image, Eq. 2.130 applies with $t_j = (u_j, v_j)$, $j = 1, 2, \ldots, n$, and $h = (h_1, h_2)$. Observation of a SS stationary image at a set of n spatial points is statistically unchanged if those points are all moved along a common vector. This is translation invariance in two dimensions. For WS stationarity of an image process, the covariance at any two spatial points is unchanged if the two points are translated along a common vector.

For a normally distributed random function, SS and WS stationarity are equivalent: since a normal process is completely described by its first- and second-order moments and these are translation invariant for a WS stationary process, the higher-order probability distribution functions must also be translation invariant.

The next example concerns a binary process generated by a *random set*. The concept of random set will be rigorously discussed later in the text, but for now let us simply assume that a closed set S_ω is generated based upon the outcomes ω in some sample space or on the outcome of some random variable or other random process. The collection of all possible sets together with their probabilistic description constitutes a random set. Corresponding to the set process is a binary random process defined by

$$X(\omega; t) = \begin{cases} 1, & \text{if } t \in S_\omega \\ 0, & \text{if } t \notin S_\omega \end{cases} \tag{2.137}$$

The process is a binary signal process if t is a time variable and a binary image process if t is a spatial variable. The distribution of $X(t)$ depends on the random set process.

***Example* 2.20.** Consider a random Poisson point process in the Euclidean plane $\Re^2$. If $N(D)$ denotes the number of points in a domain D, then

$$P(N(D) = k) = e^{-\lambda v(D)} \frac{(\lambda v(D))^k}{k!}$$

where $v(D)$ is the area of D and $\lambda v(D)$ is the expected number of points falling in D. At each point marked by the point process, there is placed a closed disk of radius r with the center at the point. Let the set $S \subset \Re^2$ be the union of all the disks and define the binary image process $X(t)$ according to Eq. 2.137. We can view S as a random set or $X(t)$ as a binary process. Geometrically, each realization of S appears as a partial coverage of the plane by disks of constant radius r. The greater the intensity λ, the greater the coverage (on average). To show that $X(t)$ is SS stationary, we need to show that every nth-order density is translation-invariant, namely, that it satisfies Eq. 2.131 for any points $t_1, t_2, \ldots, t_n$ and for any increment h. Since $X(t)$ is binary, we need to consider joint probabilities of the form

$$f(x_1, x_2, \ldots, x_n; t_1, t_2, \ldots, t_n) = P(X(t_1) = x_1, X(t_2) = x_2, \ldots, X(t_n) = x_n)$$

where $x_j = 0$ or $x_j = 1$ for $j = 1, 2, \ldots, n$. First consider the case of a single point t_1. $X(t_1) = 0$ if and only if $t_1 \notin S$, and $t_1 \notin S$ if and only if there exists no disk containing t_1. This happens if and only if no Poisson point falls within r of t_1; that is, if and only if no Poisson point falls in the disk $D(t_1)$ of radius r, centered at t_1. Hence,

$$\begin{aligned} P(X(t_1) = 0) &= P(N(D(t_1)) = 0) \\ &= \exp[-\lambda v(D(t_1))] \end{aligned}$$

$$P(X(t_1) = 1) = 1 - \exp[-\lambda v(D(t_1))]$$

where $\nu(D(t_1))$ denotes the area of $D(t_1)$. Next consider two points t_1 and t_2. $X(t_1) = 0$ and $X(t_2) = 0$ if and only if $t_1 \notin S$ and $t_2 \notin S$. If $D(t_j)$ is the disk of radius r centered at t_j, then $t_1 \notin S$ and $t_2 \notin S$ if and only if no Poisson points fall in the union $D(t_1) \cup D(t_2)$. Hence,

$$P(X(t_1) = 0, X(t_2) = 0) = P(N(D(t_1) \cup D(t_2)) = 0)$$

$$= \exp[-\lambda \nu(D(t_1) \cup D(t_2))]$$

Next, $X(t_1) = 0$ and $X(t_2) = 1$ if and only if $t_1 \notin S$ and $t_2 \in S$. But $t_1 \notin S$ and $t_2 \in S$ if and only if no Poisson points fall in $D(t_1)$ and at least one Poisson point falls in $D(t_2)$. Hence,

$$P(X(t_1) = 0, X(t_2) = 1) = P(N(D(t_1)) = 0, N(D(t_2)) \geq 1)$$

$$= P(N(D(t_2)) \geq 1 | N(D(t_1)) = 0)P(N(D(t_1)) = 0)$$

$$= P(N(D(t_2) - D(t_1)) \geq 1)P(N(D(t_1)) = 0)$$

$$= (1 - \exp[-\lambda \nu(D(t_2) - D(t_1))])\exp[-\lambda \nu(D(t_1)]$$

By symmetry,

$$P(X(t_1) = 1, X(t_2) = 0) = P(X(t_1) = 0, X(t_2) = 1)$$

Finally,

$$P(X(t_1) = 1, X(t_2) = 1) = 1 - P(X(t_1) = 0, X(t_2) = 0) - 2P(X(t_1) = 0, X(t_2) = 1)$$

Since $\nu(D(t_j + h)) = \nu(D(t_j))$ for any increment h, both the first-order and second-order densities are translation-invariant. Similar geometric arguments apply to any number of points $t_1, t_2, \ldots, t_n$. Hence, the binary process $X(t)$ is SS stationary. Note that the mean of the process is given in set terms by

$$\mu_X(t_1) = P(t_1 \in S)$$

and is given analytically by

$$\mu_X(t_1) = 1 - \exp[-\lambda \nu(D(t_1))]$$

$$= 1 - \exp[-\lambda \pi r^2]$$

The covariance function is given in set terms by

$$K_X(t_1, t_2) = P(t_1 \in S, t_2 \in S) - P(t_1 \in S)^2$$

and is given analytically by

$$K_X(t_1, t_2) = 1 - [P(t_1 \notin S, t_2 \notin S) + P(t_1 \notin S, t_2 \in S) + P(t_1 \in S, t_2 \notin S)] - P(t_1 \in S)^2$$

$$= 1 - P(t_1 \notin S, t_2 \notin S) - 2P(t_2 \in S | t_1 \notin S)P(t_1 \notin S) - P(t_1 \in S)^2$$

$$= 1 - \exp[-\lambda v(D(t_1) \cup D(t_2))]$$

$$- 2(1 - \exp[-\lambda v(D(t_2) - D(t_1))])\exp[-\lambda v(D(t_1)]$$

$$- (1 - \exp[-\lambda v(D(t_1))])^2$$

Since the process is SS stationary, it must be WS stationary. Indeed, the mean is constant and the covariance is a function of $t_1 - t_2$. In addition, if $|t_1 - t_2| \geq 2r$, then $K_X(t_1, t_2) = 0$, since for $|t_1 - t_2| \geq 2r$,

$$v(D(t_1) \cup D(t_2)) = v(D(t_1)) + v(D(t_2))$$

$$v(D(t_2) - D(t_1)) = v(D(t_2))$$

■

2.9. Estimation

If a WS stationary random function is both mean-ergodic and covariance-ergodic, then statistical estimation of its mean and covariance can be accomplished by observing a single trial, so long as the domain of observation is sufficiently large. This makes WS stationarity very appealing from a statistical point of view. Intuitively, if $X(t)$ is WS stationary, then its mean and variance are constant, its covariance function depends only on the difference $t - t'$, and its second-order probabilistic characteristics do not depend on the specific value of t. Rather than increase the precision of statistical estimation through observation of a large number of realizations, we can accomplish the same end by observing the process (in the case of a random signal) over an interval of sufficient duration. This will provide a large number of probabilistically equivalent observations. For the sake of expositional clarity, the estimation theory in the present section is stated in terms of random functions of a single variable; nonetheless, the operative propositions, Theorems 2.8 and 2.9, are dimensionally independent and the theory is applicable to higher dimensions by stating Theorems 2.8 and 2.9 in terms of random fields.

To estimate the mean and covariance of a WS stationary random function $X(t)$ by observing $X(t)$ from time $t = 0$ to time $t = r$, the mean and covariance estimators are

$$\hat{\mu}_X = \frac{1}{r}\int_0^r X(t)\,dt \tag{2.138}$$

$$\hat{k}_X(\tau) = \frac{1}{r-\tau}\int_0^{r-\tau} X_0(t+\tau)X_0(t)\,dt \tag{2.139}$$

where $X_0(t) = X(t) - \mu_X$. These estimators are continuous-time, random-function versions of the sample mean and sample covariance used in statistics for random samples. Indeed, if $X(k)$ is a discrete-time function and averaging is done over some finite number of observations, n for the mean and $n - \tau$ for the covariance, then the estimators reduce to

$$\hat{\mu}_X = \frac{1}{n}\sum_{k=1}^{n} X(k) \tag{2.140}$$

$$\hat{k}_X(\tau) = \frac{1}{n-\tau}\sum_{k=1}^{n-\tau} X_0(k+\tau)X_0(k) \tag{2.141}$$

Since $E[X(t)] = \mu_X$ and $E[X_0(t + \tau)X_0(t)] = k_X(\tau)$, bringing the expectation through the sums in Eqs. 2.140 and 2.141, or through the integrals in Eqs. 2.138 and 2.139, shows the estimators to be unbiased:

$$E[\hat{\mu}_X] = \mu_X \tag{2.142}$$

$$E[\hat{k}_X(\tau)] = k_X(\tau) \tag{2.143}$$

According to Eq. 1.249, the mean-square error of an unbiased estimator equals its variance and MS convergence is equivalent to convergence of the variance to 0. Since the mean and covariance estimators of Eqs. 2.138 and 2.139 are unbiased, they converge in the mean square if their variances converge to 0.

If the observation intervals are changed to $[0, r]$ in the mean- and covariance-ergodic theorems for WS stationary processes, then

$$\lim_{r\to\infty}\frac{1}{r}\int_0^r \left(1-\frac{\tau}{r}\right)k_X(\tau)\,d\tau = 0 \tag{2.144}$$

is a necessary and sufficient condition for mean ergodicity (as $r \to \infty$). If the random process is also normal, then

$$\lim_{r\to\infty}\frac{1}{r}\int_0^r\left(1-\frac{t}{r}\right)[k_X^2(t)+k_X(t+\tau)k_X(t-\tau)]\,dt=0 \tag{2.145}$$

is a necessary and sufficient condition for covariance ergodicity. These limiting conditions can be derived from the corresponding $[-r, r]$ conditions or be deduced from the next theorem. The theorem gives the variances of the mean and covariance estimators and, since these estimators are unbiased, the mean-square errors of the estimators.

Theorem 2.8. For a WS stationary random process, the estimator $\hat{\mu}_X$ of the mean is unbiased and has variance

$$\mathrm{Var}[\hat{\mu}_X]=\frac{2}{r}\int_0^r\left(1-\frac{\tau}{r}\right)k_X(\tau)\,d\tau \tag{2.146}$$

The estimator $\hat{k}_X(\tau)$ of the covariance is also unbiased and, if the random process is normal, then it has variance

$$\mathrm{Var}[\hat{k}_X(\tau)]=\frac{2}{r-\tau}\int_0^{r-\tau}\left(1-\frac{t}{r-\tau}\right)[k_X^2(t)+k_X(t+\tau)k_X(t-\tau)]\,dt \quad \blacksquare \tag{2.147}$$

Mean-ergodicity means

$$\lim_{r\to\infty}\mathrm{Var}[\hat{\mu}_X]=0 \tag{2.148}$$

Moreover, if $X(t)$ is Gaussian, then covariance-ergodicity means

$$\lim_{r\to\infty}\mathrm{Var}[\hat{k}_X(\tau)]=0 \tag{2.149}$$

For mean-ergodic processes, the mean estimator converges to the mean in the mean-square and for normal covariance-ergodic processes, the covariance estimator converges to the covariance in the mean-square. In both cases, estimation precision increases (in the MS sense) as the length of the observation interval tends to infinity.

***Example* 2.21.** The random telegraph signal $X(t)$ of Example 2.18 has covariance function $k_X(\tau) = e^{-2|\tau|}$. According to Eq. 2.146,

$$\text{Var}[\hat{\mu}_X] = \frac{2}{r}\int_0^r \left(1-\frac{\tau}{r}\right)e^{-2\tau}\,d\tau$$

Since $0 \le \tau/r \le 1$ for $0 \le \tau \le r$,

$$\text{Var}[\hat{\mu}_X] \le \frac{2}{r}\int_0^r e^{-2\tau}\,d\tau = \frac{1-e^{-2r}}{r}$$

$\text{Var}[\hat{\mu}_X]$ can be made arbitrarily small by choosing r sufficiently large. ■

The variance formulas for both the mean and covariance estimators contain the covariance. If the covariance is not known (which is likely the case), we can use a conservative estimate of the covariance in getting approximations of the variances in question. To be prudent, if one desires a degree of precision ε (the variance of the estimator less than ε), then, when using a (guessed at) approximation of the covariance, r might best be chosen to have the (now estimated) variance somewhat smaller than ε.

For digital computation, discrete estimators are employed in place of the integral estimators. A discrete mean estimator can be obtained by a standard integral approximation: partition $[0, r]$ into n subintervals of equal length r/n, let $t_1, t_2, \ldots, t_n$ be the midpoints of the n subintervals, and define the estimator

$$\hat{m}_X = \frac{1}{r}\sum_{i=1}^{n}\frac{r}{n}X(t_i) = \frac{1}{n}\sum_{i=1}^{n}X(t_i) \tag{2.150}$$

which is just the average of the random variables $X(t_1), X(t_2), \ldots, X(t_n)$.

For a discrete estimation rule for the covariance over the observation interval $[0, r]$, approximate the estimation rule for the continuous-time covariance estimator at the points $r/n, 2r/n, \ldots, mr/n, \ldots$ to obtain the estimator

$$\hat{k}_X\left(\frac{mr}{n}\right) = \frac{1}{n-m}\sum_{i=1}^{n-m}(X(t_{m+i})-\hat{m}_X)(X(t_i)-\hat{m}_X) \tag{2.151}$$

Theorem 2.9. The variances of the estimators $\hat{m}_X$ and $\hat{k}_X(mr/n)$ are

$$\text{Var}[\hat{m}_X] = \frac{1}{n}\left[k_X(0) + 2\sum_{i=1}^{n-1}\left(1-\frac{i}{n}\right)k_X\left(\frac{ir}{n}\right)\right] \tag{2.152}$$

$$\operatorname{Var}\left[\hat{k}_X\left(\frac{mr}{n}\right)\right]=\frac{1}{n-m}\left[k_X^2(0)+k_X^2\left(\frac{mr}{n}\right)\right.$$

$$\left.+2\sum_{i=1}^{n-m-1}\left(1-\frac{i}{n-m}\right)\left(k_X^2\left(\frac{ir}{n}\right)+k_X\left(\frac{i+m}{n}r\right)k_X\left(\frac{i-m}{n}r\right)\right)\right] \quad \blacksquare \qquad (2.153)$$

As with Theorem 2.8, if the covariance is unknown, an estimate of it can be used to approximate the covariances in Theorem 2.9. Also in analogy to Theorem 2.8, if $X(t)$ is mean-ergodic, then the variance of the discrete mean estimator tends to 0 as the observation interval increases in length, and, if $X(t)$ is Gaussian and covariance-ergodic, then the same can be said for the variance of the discrete covariance estimator.

Although the integral estimators for the mean and covariance of a WS stationary process are more appealing from a mathematical perspective, discrete estimators are often more practical. For digital processing, only discrete forms are relevant. And although we have presented Theorems 2.6 and 2.7 for continuous random functions, both mean- and covariance-ergodicity have immediate discrete analogues. Indeed, for mean-ergodicity this is precisely the meaning of Eq. 2.53. For estimation, the estimators of Eqs. 2.150 and 2.151 are applicable to discrete sampling.

2.10. Linear Systems

There are two basic system problems concerning random functions. First, given a system taking random function inputs, what can be said concerning output distributions in terms of input distributions? If Ψ is a system operating on random functions, our knowledge of Ψ concerns expression of the nth-order distributions of the output random function $\Psi(X)$ in terms of the distributions of the input random function X. For instance, what is the relationship between the input covariance function $K_X(t, t')$ and the output covariance function $K_{\Psi(X)}(s, s')$? Because we need to analyze the system as it applies to random functions, the problem is one of *analysis*.

The second basic system problem concerns *synthesis*: given our desire to transform random functions, where the transformation is perhaps constrained, can we find some *optimal* system? For instance, we may desire a system that filters an observation process to produce an output process that is a best estimate of some unobserved process and the system is constrained to linear. Linear systems are the most easily characterized.

2.10.1. Commutation of a Linear Operator with Expectation

If Ψ is a linear operator on a class of random functions, then, by superposition,

$$\Psi(a_1X_1 + a_2X_2) = a_1\Psi(X_1) + a_2\Psi(X_2) \tag{2.154}$$

For $Y = \Psi(X)$, we desire $\mu_Y(s)$ and $K_Y(s, s')$ in terms of $\mu_X(t)$ and $K_X(t, t')$, respectively. Schematically, we would like to find operations that would complete (on the bottom horizontal arrows) the following two commutative diagrams involving the expectation and covariance:

$$\begin{array}{ccc} X(t) & \xrightarrow{\Psi} & Y(s) \\ E\downarrow & & \downarrow E \\ \mu_X(t) & \longrightarrow & \mu_Y(s) \end{array} \tag{2.155}$$

$$\begin{array}{ccc} X(t) & \xrightarrow{\Psi} & Y(s) \\ \text{Cov}\downarrow & & \downarrow \text{Cov} \\ K_X(t,t') & \longrightarrow & K_Y(s,s') \end{array} \tag{2.156}$$

In Sections 2.3 and 2.4 we considered the cases where the operators were differentiation and integration, respectively, the major propositions being Theorems 2.1 and 2.2. We also considered a finite sum in Section 2.2. In all cases, the key to the development was interchange of the linear operator and the expected value operation,

$$E[\Psi(X)] = \Psi[E(X)] \tag{2.157}$$

In terms of the relation $Y(s) = \Psi(X)(s)$, the interchange can be written $\mu_Y(s) = \Psi(\mu_X)(s)$, and thus commutativity is achieved in the diagram of Eq. 2.155 with Ψ on the bottom arrow. Although interchange of expectation and a linear operator is not always valid, it is in all practical situations and henceforth we assume conditions to be such that interchange is justified.

2.10.2. Representation of Linear Operators

For the moment, let us focus on linear systems operating on deterministic functions. Of particular importance are integral operators defined in terms of a *kernel* (or *weighting function*) $g(s, t)$, these being of the form

$$y(s) = \int_T g(s,t)x(t)\,dt \tag{2.158}$$

where $x(t)$ belongs to some appropriate linear space of functions and the variables t and s can be scalars or vectors, the former for signals and the latter for images. Whereas $x(t)$ is defined over T, the output function $y(s)$ is defined over some set of values $s \in S$, where S need not equal T. An example of such a weighting function is the one defining the Laplace transform, e^{-st}. More generally, we can consider the integro-differential operator

$$y(s) = \sum_{k=0}^{\infty} \int_T g_k(s,t) x^{(k)}(t)\, dt \tag{2.159}$$

where $x^{(k)}(t)$ is the kth derivative of $x(t)$.

If we interpret a linear operator with weighting function $g(s, t)$ in the discrete sense, then it takes the form of a (possibly finite) sum,

$$y(n) = \sum_{k=-\infty}^{\infty} g(n,k) x(k) \tag{2.160}$$

Using delta functions to represent discrete impulses, we can write

$$y(n) = \int_{-\infty}^{\infty} g(n,t) x(t)\, dt \tag{2.161}$$

where

$$x(t) = \sum_{k=-\infty}^{\infty} x(k) \delta(t-k) \tag{2.162}$$

According to the principle of superposition, if Ψ is a linear operator on a linear function space $\mathcal{L}$, the functions $x_1(t), x_2(t),..., x_n(t)$ lie in $\mathcal{L}$,

$$x(t) = \sum_{k=1}^{n} a_k x_k(t) \tag{2.163}$$

and $y(s) = \Psi(x)(s)$, then

$$y(s) = \sum_{k=1}^{n} a_k y_k(s) \tag{2.164}$$

where $y_k(s) = \Psi(x_k)(s)$ for $k = 1, 2,..., n$. Superposition applies to finite sums of input functions; should a sum be infinite, and even converge, interchanging summation with the operator may not be valid, or, to achieve validity, the procedure might have to be interpreted in some specialized sense. For our purposes here we simply note that when the functions involved are "well-behaved," such interchange can often take place.

More generally, if a function $x(t)$ is represented as an integral,

$$x(t) = \int_U a(u)Q(t,u)\,du \tag{2.165}$$

and Ψ is a linear operator such that for each fixed u, $Q(t, u)$ is in the domain of Ψ, the question arises as to whether we can apply the principle of superposition to the integral, in the sense that the integral represents a generalized summation. Specifically, can we interchange the order of integration and application of Ψ and write

$$y(s) = \Psi(x)(s) = \int_U a(u)[\Psi_t(Q(t,u))](s)\,du \tag{2.166}$$

where the subscript t of Ψ_t denotes that Ψ is applied relative to the variable t (for fixed u). As with infinite sums, conditions under which such interchange is valid are outside the current discussion and validity depends on the function class involved. Nonetheless, three points can be made: (1) conditions can be imposed on the function class to make interchange valid; (2) for practical image and signal applications, especially in the digital case where all sums are finite, the required conditions impose no significant constraints on the linear model; and (3) interchange facilitates the use of weighting functions to represent linear system laws and the suggestiveness of such representations makes interchange of laws and integrals, at least in a formal manner, invaluable. Consequently, we will apply superposition freely to functions defined in terms of weighting functions, recognizing that for finite sums (or for weighting functions that are finite sums of delta functions), the application is mathematically rigorous.

If Ψ and $x(t)$ are defined by Eqs. 2.158 and 2.165, respectively, then, by superposition,

$$\begin{aligned} y(s) &= \int_T \int_U g(s,t)a(u)Q(t,u)\,du\,dt \\ &= \int_U a(u)\left(\int_T g(s,t)Q(t,u)\,dt\right)du \end{aligned} \tag{2.167}$$

and

$$\Psi_t Q(t,u)(s) = \int_T g(s,t)Q(t,u)\,dt \tag{2.168}$$

As an immediate application of the conventions regarding the interplay between the principle of superposition and functions defined in terms of integrals, consider representation of an arbitrary function in terms of an integral with a delta function kernel:

$$x(t) = \int_{-\infty}^{\infty} x(u)\delta(t-u)\,du \tag{2.169}$$

where, for notational convenience only, we have employed functions of a single variable. Application of the linear operator Ψ defined by Eq. 2.158 to $x(t)$ yields the output

$$y(s) = \int_{-\infty}^{\infty} x(u)\Psi_t\delta(t-u)(s)\,du \tag{2.170}$$

where $\Psi_t\delta(t - u)(s)$ is the response of the system at the point t to the unit impulse occurring at the point u; that is, $\Psi_t\delta(t - u)(s)$ is the impulse response function of the system. If $s = t$, then the impulse response can be written simply as $\Psi_t\delta(t - u)$.

For instance, for the differential operator $y(t) = x'(t)$, $s = t$ and, since

$$y(t) = -\int_{-\infty}^{\infty} \delta'(t-u)x(u)\,du \tag{2.171}$$

the impulse response function is $-\delta'(t - u)$. The nth order differential operator has impulse response $(-1)^n\delta^{(n)}(t - u)$, where $\delta^{(n)}$ is the nth derivative of δ.

2.10.3. Output Covariance

For random-function inputs, a key concern is completing the commuting diagram of Eq. 2.156. An operator defining the bottom arrow would provide a formulation of the output covariance of a linear system in terms of the input covariance. Regarding Eq. 2.155, suppose the linear operator Ψ is given by an integral representation via a weighting function $g(s, t)$,

$$\Psi(X)(s) = \int_T g(s,t)X(t)\,dt \tag{2.172}$$

Then

$$E\left[\int_T g(s,t)X(t)\,dt\right] = \int_T g(s,t)E[X(t)]\,dt \tag{2.173}$$

which is a formal statement of Theorem 2.2.

To avoid cumbersome notation, two conventions will be adopted. First, equations may be shortened by not including the variable s subsequent to the operation. Although this practice will result in equations with the variable s on the left and no explicitly stated variable s on the right, no confusion should result if one keeps the meaning of the operations in mind. A second convention will be omission of parentheses when the meaning of the operations is obvious. For instance, we may write ΨX instead of $\Psi(X)$.

For the centered random functions X_0 and Y_0, the identity $E\Psi = \Psi E$ yields

$$\begin{aligned} Y_0(s) &= Y(s) - \mu_Y(s) \\ &= \Psi[X(t) - \mu_X(t)](s) \\ &= \Psi[X_0(t)](s) \end{aligned} \tag{2.174}$$

Consequently,

$$\begin{aligned} K_Y(s, s') &= E[Y_0(s)Y_0(s')] \\ &= E[\Psi_t X_0(t)\Psi_{t'} X_0(t')] \\ &= E[\Psi_t \Psi_{t'} X_0(t) X_0(t')] \\ &= \Psi_t \Psi_{t'} E[X_0(t) X_0(t')] \\ &= \Psi_t \Psi_{t'} K_X(t,t') \end{aligned} \tag{2.175}$$

where Ψ_t denotes that Ψ is being applied relative to the variable t and $\Psi_t\Psi_{t'}$ indicates operator composition by Ψ with respect to t' and then with respect to t. Since the roles of Ψ_t and $\Psi_{t'}$ can be interchanged in the preceding computation, we obtain the next theorem, which states that the output covariance is obtained by a double application of Ψ to the input covariance, once with respect to t and once with respect to t'. Thus, the commuting diagram of Eq. 2.156 is now complete, thereby providing a characterization of the output covariance for a linear system. The same technique applies to the autocorrelation.

Theorem 2.10. If $X(t)$ is a random function for which $\Psi EX = E\Psi X$, then

(i) $\mu_{\Psi X}(s) = \Psi(\mu_X(t))$ (2.176)

(ii) $K_{\Psi X}(s, s') = \Psi_t \Psi_{t'} K_X(t, t') = \Psi_{t'} \Psi_t K_X(t, t')$ ■ (2.177)

If $g(t, u)$ is the impulse response function for Ψ and we let $Y = \Psi X$, then

$$Y(t) = \int_{-\infty}^{\infty} g(t,u)X(u)\,du \tag{2.178}$$

and the conclusions of Theorem 2.10 can be rewritten as

(i') $$\mu_Y(t) = \int_{-\infty}^{\infty} g(t,u)\mu_X(u)\,du \tag{2.179}$$

(ii') $$K_Y(t, t') = \int_{-\infty}^{\infty}\int_{-\infty}^{\infty} g(t,u)g(t',u')K_X(u,u')\,du\,du' \tag{2.180}$$

Letting $t = t'$ yields the output variance

$$\mathrm{Var}[Y(t)] = \int_{-\infty}^{\infty}\int_{-\infty}^{\infty} g(t,u)g(t,u')K_X(u,u')\,du\,du' \tag{2.181}$$

If the input random function is digital and consists of a (possibly infinite) collection of impulses defined on the integers, then with respect to $g(n, k)$, the impulse response function of the system, Theorem 2.10 takes the form

(i'') $$\mu_Y(n) = \sum_{k=-\infty}^{\infty} g(n,k)\mu_X(k) \tag{2.182}$$

(ii'') $$K_Y(n, m) = \sum_{k=-\infty}^{\infty}\sum_{l=-\infty}^{\infty} g(n,k)g(m,l)K_X(k,l) \tag{2.183}$$

For digital images, these relations become

$$\text{(i'')}\quad \mu_Y(i,j) = \sum_{r=-\infty}^{\infty}\sum_{s=-\infty}^{\infty} g(i,j,r,s)\mu_X(r,s) \tag{2.184}$$

$$\text{(ii'')}\quad K_Y(n,m,i,j) = \sum_{r=-\infty}^{\infty}\sum_{s=-\infty}^{\infty}\sum_{p=-\infty}^{\infty}\sum_{q=-\infty}^{\infty} g(n,m,r,s)g(i,j,p,q)K_X(r,s,p,q) \tag{2.185}$$

***Example* 2.22.** Let Ψ denote differentiation, so that its impulse response function is given by $g(t, u) = -\delta'(t - u)$. By Eq. 2.180,

$$K_Y(t,t') = \int_{-\infty}^{\infty}\int_{-\infty}^{\infty} \delta'(t-u)\delta'(t'-u')K_X(u,u')\,du\,du'$$

$$= \int_{-\infty}^{\infty} \delta'(t'-u')\left(\int_{-\infty}^{\infty} \delta'(t-u)K_X(u,u')\,du\right)du'$$

$$= -\int_{-\infty}^{\infty} \delta'(t'-u')\frac{\partial}{\partial t}K_X(t,u')\,du'$$

$$= \frac{\partial}{\partial t'}\frac{\partial}{\partial t}K_X(t,t')$$

which is in accordance with Theorem 2.1. Note two points. First, we could have applied Theorem 2.10 directly without recourse to the impulse response function: the relation $K_Y(t, t') = \partial/\partial t'\partial/\partial t K_X(t, t')$ follows at once from Theorem 2.10 (ii) by recognizing that Ψ_t and $\Psi_{t'}$ are partial derivatives with respect to t and t' for the differential operator. Second, whereas Theorem 2.1 holds for derivatives in the MS sense, the present result extends the theorem to generalized derivatives involving delta functions. ■

***Example* 2.23.** Consider the moving-average signal filter having mask $M = (\boldsymbol{a_1}, a_2, a_3)$, with bold indicating the origin position. The linear operator generated by M is defined by

$$Y(n) = a_1X(n) + a_2X(n+1) + a_3X(n+2)$$

Let $X(n)$ be white noise with unit variance. We find the output covariance first by direct computation and second by appeal to Theorem 2.10 (ii). The input covariance is

$$K_X(n,m) = \begin{cases} 1, & \text{if } n = m \\ 0, & \text{if } n \neq m \end{cases}$$

First apply the filter to $K_X(n, m)$ with respect to n. If (n, m) lies on the diagonal line marked D_1 in Fig. 2.7, then $K_X(n + 2, m) = 1$ and $K_X(n + k, m) = 0$ for $k \neq 2$.

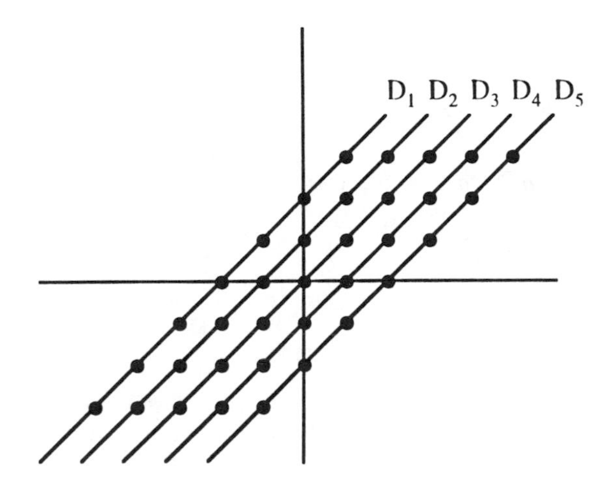

Figure 2.7 Covariance domain.

Thus,

$$\Psi_n K_X(n, m) = 0a_1 + 0a_2 + 1a_3 = a_3$$

For (n, m) on the diagonal D_2,

$$\Psi_n K_X(n, m) = 0a_1 + 1a_2 + 0a_3 = a_2$$

For (n, m) on the diagonal D_3,

$$\Psi_n K_X(n, m) = 1a_1 + 0a_2 + 0a_3 = a_1$$

For all other (n, m), $\Psi_n K_X(n, m) = 0$.

We now must apply the filter to $\Psi_n K_X(n, m)$ with respect to the second variable. For (n, m) on the diagonal D_5,

$$K_Y(n, m) = \Psi_m \Psi_n K_X(n, m) = 0a_1 + 0a_2 + a_1 a_3 = a_1 a_3$$

For (n, m) on the diagonal D_4,

$$K_Y(n, m) = 0a_1 + a_1 a_2 + a_2 a_3 = a_1 a_2 + a_2 a_3$$

For (n, m) on the diagonal D_3,

$$K_Y(n, m) = a_1^2 + a_2^2 + a_3^2$$

For (n, m) on the diagonal D_2,

$$K_Y(n, m) = a_2 a_1 + a_3 a_2 + 0a_3 = a_1 a_2 + a_2 a_3$$

For (n, m) on the diagonal D_1,

$$K_Y(n, m) = a_3 a_1 + 0a_2 + 0a_3 = a_1 a_3$$

For all other points (n, m), $K_Y(n, m) = 0$. From the geometry of the diagonals, $K_Y(n, m)$ depends only on $n - m$, and hence the output $Y(n)$ is WS stationary. More generally, a moving-average filter applied to white noise yields a WS stationary output process.

Let us reapproach the problem, at first not assuming X is white noise. In summation form the filter appears as

$$Y(n) = \sum_{k=-\infty}^{\infty} g(n, k) X(k)$$

where the impulse response function $g(n, k)$ is given by

$$g(n, k) = \begin{cases} a_{k-n+1}, & \text{for } k = n, n+1, n+2 \\ 0, & \text{otherwise} \end{cases}$$

Thus, according to Eq. 2.182,

$$\mu_Y(n) = a_1 \mu_X(n) + a_2 \mu_X(n+1) + a_3 \mu_X(n+2)$$

According to Eq. 2.183,

$$K_Y(n, m) = \sum_{k=n}^{n+2}\sum_{l=m}^{m+2} a_{k-n+1}a_{l-m+1}K_X(k,l)$$

$$= a_1^2 K_X(n,m) + a_1a_2K_X(n,m+1) + a_1a_3K_X(n,m+2)$$

$$+ a_2a_1K_X(n+1,m) + a_2^2K_X(n+1,m+1) + a_2a_3K_X(n+1,m+2)$$

$$+ a_3a_1K_X(n+2,m) + a_3a_2K_X(n+2,m+1) + a_3^2K_X(n+2,m+2)$$

Now assuming $X(n)$ is white noise, from the summation representation of $K_Y(n, m)$ we again observe WS stationarity. ■

Exercises for Chapter 2

1. Let Z_1 and Z_2 be independent standard normal random variables. Define

$$X(t) = \begin{cases} 0, & \text{if } t < Z_1, t < Z_2 \\ 1, & \text{if } t < Z_1, t \geq Z_2 \text{ or } t \geq Z_1, t < Z_2 \\ 2, & \text{if } t \geq Z_1, t \geq Z_2 \end{cases}$$

   Draw a typical realization of the random process $X(t)$. Find the mean, autocorrelation, covariance, variance, and correlation coefficient for $X(t)$.

2. Let Z_1 and Z_2 be independent exponential random variables with mean μ. Define

$$X(t) = \begin{cases} 0, & \text{if } t < Z_1, t < Z_2 \\ 1, & \text{if } t < Z_1, t \geq Z_2 \text{ or } t \geq Z_1, t < Z_2 \\ 0, & \text{if } t \geq Z_1, t \geq Z_2 \end{cases}$$

   Find the mean, autocorrelation, covariance, variance, and correlation coefficient for $X(t)$.

3. Let A, B, and C be three random variables and consider the stochastic process

$$X(t) = At^2 + Bt + C$$

   Find the mean, autocorrelation, covariance, variance, and correlation coefficient for $X(t)$. Find the covariance function when the coefficient random variables are independent.

4. Let A and B be jointly normally distributed random variables with means μ and 0, and variances σ^2 and ξ^2, respectively, and correlation coefficient ρ. Let $X(t) = A \cos t + B$. Find the mean, autocorrelation, covariance, variance, and correlation coefficient for the process. Find the covariance function when A and B are independent.

5. Let W be a binary random variable with $P(W = 1) = p$. Define the two-dimensional random process $X(u, v)$ in the following manner: (1) if $W = 0$, then $X(u, v)$ is identically 0; (2) if $W = 1$ and (u, v) lies in the closed disk of radius 1 centered at the origin, then $X(u, v) = 1$; (3) if $W = 1$ and (u, v) does not lie in the closed disk of radius 1 centered at the origin, then $X(u, v) = 0$. Find the mean, autocorrelation, covariance, variance, and correlation coefficient for the process.

6. Repeat Exercise 2.5 under the assumption that the disk is of random radius R and R is exponentially distributed with mean μ.

7. Redo Example 2.9 under the assumption that R is uniformly distributed over [0, 1].
8. Suppose W has a Laplace distribution with mean a and variance $2/b^2$. Define the two-dimensional random process $X(u, v)$ as follows: if $W \leq u < W + 1$, then $X(u, v) = 1$; otherwise, $X(u, v) = 0$. Find the mean, autocorrelation, and covariance for the process.
9. Consider a discrete random image having the following five equiprobable realizations:

$$x_a = \begin{pmatrix} 0 & 0 & 0 \\ 0 & 0 & 0 \\ \mathbf{1} & 0 & 0 \end{pmatrix} \quad x_b = \begin{pmatrix} 0 & 0 & 0 \\ 1 & 0 & 0 \\ \mathbf{0} & 1 & 0 \end{pmatrix} \quad x_c = \begin{pmatrix} 1 & 0 & 0 \\ 0 & 1 & 0 \\ \mathbf{0} & 0 & 1 \end{pmatrix}$$

$$x_d = \begin{pmatrix} 0 & 1 & 0 \\ 0 & 0 & 1 \\ \mathbf{0} & 0 & 0 \end{pmatrix} \quad x_e = \begin{pmatrix} 0 & 0 & 1 \\ 0 & 0 & 0 \\ \mathbf{0} & 0 & 0 \end{pmatrix}$$

Find the mean, variance, and autocorrelation functions.

10. Apply Eqs. 2.26 and 2.27 to $X_1(t) = At$, $X_2(t) = B \sin t$, and $X_3(t) = C \sin t$, where A, B, and C are arbitrary random variables.
11. Apply Theorem 2.1 to find the mean and covariance functions for the derivative of the random function $X(t)$ of Example 2.6 assuming the coefficient variables are correlated.
12. Apply Theorem 2.1 to find the mean and covariance functions for the derivative of $X(t) = A \sin t + B \cos t$ under the assumption that A and B are correlated.
13. Consider the random polynomial expansion

$$X(t) = \sum_{k=0}^{n} A_k t^n$$

Assuming the coefficient variables $A_0, A_1, \ldots, A_n$ are independent, apply Theorem 2.1 to find the mean and covariance functions for the derivative process.

14. Let $Y(s)$ be the Laplace transform of the random function $X(t)$ of Exercise 13,

$$Y(s) = \int_0^\infty X(t) e^{-st}\, dt$$

Apply Theorem 2.2 to find the mean and covariance of $Y(s)$.

15. Let $X(t) = At + B$, where A and B are uncorrelated, and define

$$Y(s) = \int_0^{2\pi} X(t) \cos st \, dt$$

Apply Theorem 2.2 to find the mean and covariance of $Y(s)$.
16. Reconsider the one-dimensional random walk of Section 2.7.2 by setting up a *barrier* at the origin: if the particle is at the origin and the outcome of the random variable indicates that the particle should move left, it simply stays at the origin. Find the probability mass function for $n = 6$.
17. Reconsider the random telegraph signal of Example 2.18 by letting $X(t)$ take on the values 1 and 0 instead of 1 and -1. Find the mean and covariance functions.
18. The two-dimensional binary random process of Example 2.20 can be simplified by considering it over a single dimension. In one dimension, a closed interval of length $2r$ is centered at each selected Poisson point. Derive the mean and covariance functions for this process $X(t)$ and, in doing so, show that $X(t_1)$ and $X(t_2)$ are uncorrelated if $|t_1 - t_2| \geq 2r$.
19. Referring to Exercise 2.14, if $Y(s)$ is the Laplace transform of the random telegraph signal of Example 2.18, derive the mean and covariance functions for $Y(s)$.
20. Let $X(n)$ and $Y(n)$ be independent random walk variables determined by the parameters p_1 and p_2, respectively. Treated as coordinates in the grid, they compose a random walk in two dimensions. Find their joint probability distribution function.
21. Suppose $X(n)$ is a discrete independent random function defined over the integers, $X(n)$ is binary with $P(X(n) = 1) = p$, and the operator Ψ is defined by

$$Y(n) = \Psi(X)(n) = \max\{Y(n-1), Y(n), Y(n+1)\}$$

Find the mean and covariance function for $Y(n)$.
22. Suppose $X(t)$ is a Poisson process with mean λt, each Poisson point has probability p of being selected when the Poisson points are recorded, and the recording of any point is independent of the recording of all other points as well as being independent of the Poisson process itself. Show that, if $Y(t)$ is the number of Poisson points recorded in the interval $[0, t]$, then $Y(t)$ is a Poisson process with mean $p\lambda t$. Hint: Owing to the independence assumptions and independent increments of the original Poisson process, the recording process also has independent increments. Hence, it needs to be shown that

$$P(Y(t) - Y(t') = k) = \frac{[p\lambda(t - t')]^k}{k!e^{p\lambda(t-t')}}$$

for $t > t' \geq 0$ and any integer k. To demonstrate this equality, note that, if n points have occurred in the interval $[t', t]$, then the number of recorded points satisfies a binomial distribution with parameters n and p; namely,

$$P(Y(t) - Y(t') = k | X(t) - X(t') = n) = \binom{n}{k} p^k (1-p)^{n-k}$$

23. Let $\lambda(t)$ be a nonnegative integrable function over $\Re$. The *nonuniform Poisson process* $X(t)$ with *intensity function* $\lambda(t)$ is characterized by three conditions: (1) $X(0) = 0$; (2) $X(t)$ has independent increments; and (3) for $t \geq s$ and $n \geq 0$,

$$P(X(t) - X(s) = n) = \frac{1}{n!}\left(\int_s^t \lambda(u)\,du\right)^n \exp\left[-\int_s^t \lambda(u)\,du\right]$$

Find the mean and covariance function for $X(t)$.

24. The moment-generating function is a two-sided Laplace transform of the density of a random variable; we can also take the Fourier transform of a density. The *characteristic function* of a random variable X is defined by

$$\varphi_X(u) = E[e^{juX}] = \int_{-\infty}^{\infty} e^{jux} f_X(x)\,dx$$

where $j = \sqrt{-1}$. If φ_X is absolutely integrable, meaning the integral of $|\varphi_X|$ over $\Re$ is finite, then the inverse Fourier transform can be taken, so that

$$f_X(x) = \frac{1}{2\pi}\int_{-\infty}^{\infty} e^{-jux} \varphi_X(u)\,du$$

For a Poisson random variable X with mean λ, show that $\varphi_X(u) = \exp[\lambda(e^{ju} - 1)]$.

25. For X exponentially distributed with mean $1/b$, show that $\varphi_X(u) = (1 - jbu)^{-1}$.

26. For n random variables $X_1, X_2, \ldots, X_n$, their *joint characteristic function* is defined by

$$\varphi_{X_1,X_2,\ldots,X_n}(u_1,u_2,\ldots,u_n) = E[\exp[j(u_1X_1 + u_2X_2 + \cdots + u_nX_n)]$$

$$= \int_{-\infty}^{\infty} \cdots \int_{-\infty}^{\infty} \exp[j(u_1x_1 + u_2x_2 + \cdots + u_nx_n)] f_{X_1,X_2,\ldots,X_n}(x_1,x_2,\ldots,x_n)\, dx_1 dx_2 \cdots dx_n$$

Any two sets of random variables possessing the same joint characteristic function are identically jointly distributed; that is, if $X_1, X_2,\ldots, X_n$ and $Y_1, Y_2,\ldots, Y_n$ have equal joint characteristic functions, then their multivariate probability distribution functions are equal. Show that if $X_1, X_2,\ldots, X_n$ are independent, then

$$\varphi_{X_1,X_2,\ldots,X_n}(u_1,u_2,\ldots,u_n) = \varphi_{X_1}(u_1)\, \varphi_{X_2}(u_2) \ldots \varphi_{X_n}(u_n)$$

27. If $X(t)$ is a random function determined by its finite-order joint distributions, then $X(t)$ must be determined by its finite-order joint characteristic functions. Hence, the equality given in the previous exercise can be used to describe independent random functions. Suppose $X(k)$, $k = 1, 2,\ldots$, is a discrete independent random function and

$$Y(k) = \sum_{i=1}^{k} X(i)$$

Show that

$$\varphi_{Y(1),Y(2),\ldots,Y(n)}(u_1,u_2,\ldots,u_n) =$$

$$\varphi_{X(1)}(u_1 + u_2 + \cdots + u_n)\, \varphi_{X(2)}(u_2 + \cdots + u_n) \ldots \varphi_{X(n)}(u_n)$$

thereby showing that $Y(k)$ can be fully described via the characteristic functions of $X(1), X(2),\ldots$. Hint: Letting $Y(0) = 0$, show that

$$\sum_{k=1}^{n} u_k Y(k) = \sum_{k=1}^{n} \left(\sum_{i=k}^{n} u_i \right) (Y(k) - Y(k-1))$$

Because independent displacements generate $Y(k)$, it is called a *random walk*.

28. If X_1, X_2,..., X_n are jointly normally distributed with means μ_1, μ_2,..., μ_n, respectively, and covariance matrix **K** given by Eq. 1.186, then it can be shown that their joint characteristic function is given by

$$\varphi_{X_1,X_2,\ldots,X_n}(u_1,u_2,\ldots,u_n) = \exp\left[j\sum_{k=1}^{n} u_k\mu_k - \frac{1}{2}\sum_{i=1}^{n}\sum_{k=1}^{n} u_i\sigma_{ik}^2 u_k\right]$$

Restate this property in terms of time points t_1, t_2,..., t_n as a necessary and sufficient condition for a random function to be normal. Based upon this condition, what would have to be proven to show that the Wiener process is normal?

29. Use Exercise 2.28 to prove the following theorem: If X_1, X_2,..., X_n are jointly normally distributed random variables with means μ_1, μ_2,..., μ_n, respectively, and covariance matrix **K** given by Eq. 1.186, then the random variables

$$\begin{aligned} Y_1 &= a_{11}X_1 + a_{12}X_2 + \cdots a_{1n}X_n \\ Y_2 &= a_{21}X_1 + a_{22}X_2 + \cdots a_{2n}X_n \\ &\vdots \\ Y_m &= a_{m1}X_1 + a_{m2}X_2 + \cdots a_{mn}X_n \end{aligned}$$

are jointly normally distributed with characteristic function

$$\varphi_{Y_1,Y_2,\ldots,Y_m}(u_1,u_2,\ldots,u_m) = \exp\left[j\sum_{k=1}^{n} u_k\xi_k - \frac{1}{2}\sum_{i=1}^{n}\sum_{k=1}^{n} u_i\eta_{ik}^2 u_k\right]$$

where

$$\xi_k = E[Y_k] = \sum_{l=1}^{n} a_{kl}\mu_l$$

$$\eta_{ik} = \mathrm{Cov}[Y_i, Y_k] = \sum_{r=1}^{n}\sum_{s=1}^{n} a_{ir}a_{ks}\sigma_{rs}^2$$

30. Use the previous exercise to prove that the Wiener process $X(t)$ is normal. Hint: Use the fact that the following random variables are normal for any set of time points $t_1 < t_2 < \cdots < t_n$:

$$\begin{aligned} Z_1 &= X(t_1) \\ Z_2 &= X(t_2) - X(t_1) \\ &\vdots \\ Z_n &= X(t_n) - X(t_{n-1}) \end{aligned}$$

31. In this exercise, we generalize the results of Example 2.23. Let $Z(n)$, $n = 0, 1, 2,\ldots$, be discrete white noise with constant variance σ^2 and, for $n \geq k$, define

$$X(n) = \sum_{j=0}^{k} a_j Z(n-j)$$

where $a_0 + a_1 + \ldots + a_k = 1$. The random function $X(n)$ is the output of the *moving-average filter* determined by the coefficients (weights) a_0, $a_1,\ldots, a_k$. Find the variance and covariance of $X(n)$, in the process showing that $X(n)$ is WS stationary.

32. Let $Z(n)$, $n = \ldots, -2, -1, 0, 1, 2,\ldots$, be discrete white noise with constant variance σ^2 and let a be a nonzero constant for which $|a| < 1$. Recursively define the *first-order autoregressive model* by

$$X(n) = aX(n-1) + Z(n)$$

Show that

$$X(n) = a^N X(n-N) + \sum_{k=0}^{N-1} a^k Z(n-k)$$

for $N = 1, 2,\ldots$. Letting $N \to \infty$ yields

$$X(n) = \sum_{k=0}^{\infty} a^k Z(n-k)$$

Find the mean of $X(n)$ and show that the covariance function for $X(n)$ is given by

$$K_X(n,n') = a^{|n-n'|} \frac{\sigma^2}{1-a^2}$$

thereby showing that $X(n)$ is WS stationary. Find $\mathrm{Var}[X(n)]$ and the correlation function.

33. In Exercise 2.32, the autoregressive model was defined with the input white-noise process being defined over all discrete time. Now suppose $Z(n)$, $n = 0, 1, 2,\ldots$, is discrete white noise with constant variance σ^2, $X(n)$ is defined as in Exercise 2.32 for $n = 1, 2,\ldots$, and $X(0) = 0$. Show that

$$X(n) = \sum_{k=0}^{n-1} a^k Z(n-k)$$

Find the mean of $X(n)$ and show that the covariance function for $X(n)$ is given by

$$K_X(n,n') = \left(\frac{1-a^{2\min(n,n')}}{1-a^2} \right) a^{|n-n'|} \sigma^2$$

thereby showing that $X(n)$ is not WS stationary. Find Var[$X(n)$] and the correlation function.

34. To simulate a Poisson variable with parameter λ, we can use the fact that the interarrival time for a Poisson process with mean λ possesses an exponential distribution with parameter λ. From this, deduce that a value x of the Poisson random variable can be generated by summing interarrival times T_1, T_2,... and letting x be the value for which $T_1 + T_2 \cdots + T_x \leq 1$ and $T_1 + T_2 \cdots + T_{x+1} > 1$, where each interarrival time is of the form $T_k = -\lambda^{-1} \log U_k$ and U_1, U_2,... are independently distributed uniform variables on [0, 1].
35. Run 100 simulations of the Poisson process with mean $\lambda = 1$. Average these simulations pointwise to estimate the mean and variance functions. The estimate of the mean should be close to the theoretical value of $\lambda = 1$.
36. Simulate the random telegraph signal of Example 2.18 with $\lambda = 1$. Since the process is both mean- and covariance-ergodic, its mean and covariance functions can be estimated from a single simulated realization. Perform the estimations and compare them to the theoretical functions.
37. Simulate the random-set process of Example 2.20 with $r = 1$. Find λ so that the mean of the process is the constant 1/2. Since the process is both mean- and covariance-ergodic, its mean and covariance functions can be estimated from a single simulated realization. Perform the estimations and compare them to the theoretical functions.

Chapter 3

Canonical Representation

3.1 Canonical Expansions

A random function is a complicated mathematical entity. For some applications, one need only consider second-order characteristics and not be concerned with the most delicate mathematical aspects; for others, it is beneficial to find convenient representations to facilitate the use of random functions. Specifically, given a random function $X(t)$, where the variable t can either be vector or scalar, we desire a representation of the form

$$X(t) = \mu_X(t) + \sum_{k=1}^{\infty} Z_k x_k(t) \tag{3.1}$$

where $x_1(t)$, $x_2(t)$,... are deterministic functions, Z_1, Z_2,... are uncorrelated zero-mean random variables, the sum may be finite or infinite, and some convergence criterion (meaning of the equality) is given. Equation 3.1 is said to provide a *canonical expansion* (*representation*) for $X(t)$. The Z_k, $x_k(t)$, and $Z_k x_k(t)$ are called *coefficients*, *coordinate functions*, and *elementary functions*, respectively. $\{Z_k\}$ is a discrete white-noise process, so that the sum in Eq. 3.1 is an expansion of the centered process $X(t) - \mu_X(t)$ in terms of white noise. Consequently, it is called a *discrete white-noise representation*.

If an appropriate canonical representation can be found, then dealing with a family of random variables defined over the domain of t is reduced to considering a discrete family of random variables. Equally as important is that, whereas there may be a high degree of correlation among the random variables composing the random function, the random variables in a canonical expansion are uncorrelated. The uncorrelatedness of the Z_k is useful in eliminating redundant information and designing optimal linear filters.

3.1.1. Fourier Representation and Projections

The development of random-function canonical expansions is closely akin to finding Fourier-series representations of vectors in an inner product space, in particular, the expansion of deterministic signals in terms of functions composing an orthonormal system. Such expansions represent signals in manners that facilitate certain types of analyses, such as the expression of a signal in terms of its frequency characteristics. To achieve Fourier representations in the random-process framework, we need to place the problem in the context of an inner product space.

For a vector space $\mathcal{V}$ to be an inner product space, there needs to be defined on pairs of vectors in $\mathcal{V}$ an *inner product* (*dot product*). Letting $\langle x, y\rangle$ denote the inner product in $\mathcal{V}$, $\langle x, y\rangle$ must satisfy five properties:

I1. $\langle x, x\rangle \geq 0$.

I2. $\langle x, x\rangle = 0$ if and only if x is the zero vector.

I3. $\langle x, y\rangle = \overline{\langle y, x\rangle}$

I4. $\langle x, y + z\rangle = \langle x, y\rangle + \langle x, z\rangle$.

I5. $\langle cx, y\rangle = c\langle x, y\rangle$ for any complex scalar c.

where the overbar denotes complex conjugation. For a real vector space, the conjugation in I3 can be dropped.

Thus far we have only considered real random variables. Our focus continues to be on real random functions; however, we do not want to restrict ourselves to canonical expansions in which the coefficients and coordinate functions are real-valued. Therefore, we must consider complex random variables. A complex random variable is of the form $X = X_1 + jX_2$, where $j = \sqrt{-1}$, and X_1 and X_2 are real random variables constituting the real and imaginary parts of X. The modulus (absolute value) $|X|$ of X is the real random variable defined by $|X|^2 = X\overline{X}$. X has mean $\mu_X = E[X] = E[X_1] + jE[X_2]$. It is immediate that $\overline{E[X]} = E[\overline{X}]$. The mixed second moment of two complex random variables, X and Y, is $E[X\overline{Y}]$. Their covariance is $E[(X-\mu_X)\overline{(Y-\mu_Y)}]$. It satisfies the relation

$$\begin{aligned}\operatorname{Cov}[Y, X] &= E[(Y-\mu_Y)\overline{(X-\mu_X)}] \\ &= \overline{E[(X-\mu_X)\overline{(Y-\mu_Y)}]} \qquad (3.2) \\ &= \overline{\operatorname{Cov}[X, Y]}\end{aligned}$$

X has second moment $E[|X|^2]$ and variance $E[|X - \mu_X|^2]$. X and Y are uncorrelated if $E[X\overline{Y}] = E[X]E[\overline{Y}]$. The basic properties of the mean and covariance hold for complex random variables (see Exercises 3.1 through 3.5).

The complex random variables defined on a probability space are functions on the space and therefore form a vector space relative to the addition of random variables and scalar multiplication of a random variable by a complex number. Restricting our attention to random variables having finite second moments, $E[|X|^2] < \infty$, these form a vector space that is a subspace of the original vector space. For any two random variables X and Y in the subspace, we define their inner product by $\langle X, Y\rangle = E[X\overline{Y}]$. Properties I1, I3, I4, I5 are clearly satisfied. Moreover, if we identify any two identically distributed random variables (as is typically done), then property I2 is satisfied and the family of finite-second-moment random variables on a probability space is an inner product space with inner product $E[X\overline{Y}]$. The (*mean-square*) norm of X is $\|X\| = E[|X|^2]^{1/2}$. A sequence of random variables X_n converges to X if and only if $E[|X_n - X|^2] \to 0$ as $n \to \infty$; that is, if and only if X_n converges to X in the mean-square.

Recalling the theory of inner products, some theorems hold only when the inner product space is actually a Hilbert space, the latter meaning that the inner product space is complete, in the sense that, if a sequence of vectors in the inner product space is Cauchy convergent, then it must be convergent. The mean-square inner product space of random variables on a probability space is a Hilbert space, so that all of the standard Fourier convergence theorems apply. From the perspective of approximation, which plays the central role in applications of Fourier methods to signal processing, the norm measures the distance between two vectors in an inner product space. Thus, for random variables, the distance from X to Y is $E[|X - Y|^2]^{1/2}$.

Since real random functions are our concern, we could proceed without considering complex random functions; however, such an approach would lead to a lack of symmetry in many expressions when working with canonical expansions, since there would be conjugates occurring in the representations. Therefore, although our examples will involve real random functions, the theory of canonical expansions will be developed in the context of complex random functions. Since a random function $X(t)$ is a collection of random variables, the definition of a complex random function in terms of real and imaginary parts, each being a real random function, follows at once from the definition a complex random variable. Specifically, $X(t) = X_1(t) + jX_2(t)$. Because the second factors in both the mixed second moment and covariance are conjugated, the definitions of the autocorrelation and covariance functions in Eqs. 2.12 and 2.7 are correspondingly adjusted with their second factors being conjugated. According to Eq. 3.2, the covariance function of a complex random function satisfies the relation $K_X(t', t) = \overline{K_X(t,t')}$, rather than Eq. 2.9. Similarly, $R_X(t', t) = \overline{R_X(t,t')}$. Analogous comments apply to the cross-correlation and cross-covariance functions.

If Var[$X(t)$] is finite, then the mean-square theory applies to $X(t)$ for each t, the inner product between $X(t)$ and $X(t')$ is given by the autocorrelation $R_X(t, t')$, and $X(t)$ has norm

$$\|X(t)\| = R_X(t, t)^{1/2} = (\mathrm{Var}[X(t)] + \mu_X(t)^2)^{1/2} \tag{3.3}$$

If $X(t)$ is centered, then $\|X(t)\|^2 = \mathrm{Var}[X(t)]$. This relation motivates us to employ zero-mean random functions when discussing canonical representations. No generality is lost since for an arbitrary random function we can always consider its centered version and, once we have a canonical representation for the centered random function, transposition of the mean yields a canonical representation for the original random function.

Now suppose $\{Z_1, Z_2,...\}$ is a complete orthonormal system of complex (or real) zero-mean random variables. In the present context orthogonality means uncorrelatedness and an orthonormal system is a collection of uncorrelated random variables each having variance 1. Then for any finite-second-moment random variable Y,

$$Y = \sum_{k=1}^{\infty} E[Y\overline{Z}_k]Z_k \tag{3.4}$$

where $E[Y\overline{Z}_k]$ is the Fourier coefficient of Y with respect to Z_k and where convergence of the partial sums of the series to Y as $n \to \infty$ is with respect to the mean-square norm.

If $X(t)$ is a zero-mean random function having finite variance and $\{Z_k\}$ is an orthonormal system of random variables, then, for fixed t, $X(t)$ possesses a Fourier representation in terms of $\{Z_k\}$. The Fourier coefficients of $X(t)$ are $\hat{x}_k(t) = E[X(t)\overline{Z}_k]$. The projection of $X(t)$ into the subspace spanned by Z_1, $Z_2,..., Z_m$ is given by

$$X_m(t) = \sum_{k=1}^{m} Z_k \hat{x}_k(t) \tag{3.5}$$

As t varies, each Fourier coefficient is a deterministic function of t and the projection $X_m(t)$ is a random function. Since the square of the norm is the variance, the following theorem follows at once from projection theory on inner product spaces.

Theorem 3.1. Suppose $\{Z_1, Z_2,...\}$ is a collection of uncorrelated zero-mean, unit-variance random variables (so that $\{Z_k\}$ is an orthonormal system) and suppose $X(t)$ is a zero-mean, finite-variance random function. Then

(i) for any coordinate functions $y_1(t), y_2(t),..., y_m(t)$,

$$\text{Var}[X(t) - X_m(t)] \le \text{Var}\left[X(t) - \sum_{k=1}^{m} Z_k y_k(t)\right] \tag{3.6}$$

(ii) $$\text{Var}[X(t) - X_m(t)] = \text{Var}[X(t)] - \sum_{k=1}^{m} |E[X(t)\overline{Z}_k]|^2 \tag{3.7}$$

(iii) $$\sum_{k=1}^{\infty} |E[X(t)\bar{Z}_k]|^2 \leq \text{Var}[X(t)] \quad \textit{(Bessel inequality)} \quad \blacksquare \tag{3.8}$$

An immediate consequence of Theorem 3.1(i) is that to minimize the mean-square error (variance) of the difference between $X(t)$ and finite approximations in terms of the Z_k, the coordinate functions should be given by the Fourier coefficients with respect to the Z_k, which is exactly the situation encountered in ordinary inner-product-space theory. The projection of $X(t)$ into the subspace spanned by $Z_1, Z_2, \ldots, Z_m$ gives the best mean-square approximation to $X(t)$ lying in the subspace. Hence, if we are going to find a canonical representation in the sense of Eq. 3.1 for which convergence is in the mean square, then the coordinate functions should be the Fourier coefficients $\hat{x}_k(t) = E[X(t)\bar{Z}_k]$.

Equality holds in the Bessel inequality [Theorem 3.1(iii)] for all $t \in T$ if and only if, for any $t \in T$, $X(t)$ is equal in the mean-square norm to its Fourier series expansion. If $X(t)$ does equal its Fourier series in the mean-square, i.e., if

$$\lim_{m \to \infty} \text{Var}[X(t) - X_m(t)] = 0 \tag{3.9}$$

for any $t \in T$, then we write

$$X(t) = \sum_{k=1}^{\infty} Z_k \hat{x}_k(t) \tag{3.10}$$

and we have a canonical (white-noise) representation of $X(t)$ in the sense of Eq. 3.1. When $X(t)$ equals its Fourier series in the mean-square, the Bessel inequality, now equality, is known as the *Parseval identity*,

$$\text{Var}[X(t)] = \sum_{k=1}^{\infty} |E[X(t)\bar{Z}_k]|^2 \tag{3.11}$$

The problem is to find an orthonormal system for which the representation of Eq. 3.10 holds. In the general theory of Hilbert spaces, if $\mathcal{H}$ is a Hilbert space and $\{u_k\}$ is an orthonormal system such that each $x \in \mathcal{H}$ is equal to is Fourier series (relative to the norm of $\mathcal{H}$), then the orthonormal system is called *complete*, completeness being equivalent to the Parseval identity holding for all $x \in \mathcal{H}$. But a system need not be complete to be useful. For instance, in the deterministic theory for $L^2(T)$, $T = [-\pi, \pi]$, the trigonometric system forms a complete orthonormal system but the cosine subsystem does not; nevertheless, even functions in the space are represented by their cosine series and for even functions the cosine orthonormal system is sufficient. While it is advantageous to have a complete system, it is often sufficient to find an orthonormal representation of a given function. For random functions, finding orthonormal representations is more difficult than in the deterministic L^2 theory. Given a random function $X(t)$, we will be satisfied to find a canonical representation by way of a decorrelated, zero-mean system for which Eq. 3.1 holds. Of course, some representations are more useful than others.

Thus far we have assumed that $\mathrm{Var}[Z_k] = 1$ to make the system orthonormal. In fact, it is sufficient to find a collection of mutually uncorrelated zero-mean random variables Z_k (not necessarily having unit variance). The system can be orthonormalized by dividing each Z_k by its norm. With respect to the orthonormalized random variables $Z_k/\mathrm{Var}[Z_k]^{1/2}$, and still assuming $X(t)$ to have zero mean, the orthonormal expansion of Eq. 3.1 becomes

$$X(t) = \sum_{k=1}^{\infty} \frac{Z_k}{\mathrm{Var}[Z_k]^{1/2}} \hat{x}_k(t) \tag{3.12}$$

where $\hat{x}_k(t)$ is the Fourier coefficient with respect to the normalized random variable,

$$\hat{x}_k(t) = E\left[X(t)\frac{\overline{Z}_k}{\mathrm{Var}[Z_k]^{1/2}}\right] \tag{3.13}$$

Factoring $\mathrm{Var}[Z_k]^{1/2}$ out of the expectation recovers Eq. 3.1; however, rather than being the Fourier coefficients, the coordinate functions $x_k(t)$ are now given by

$$x_k(t) = \frac{E[X(t)\overline{Z}_k]}{\mathrm{Var}[Z_k]} \tag{3.14}$$

Because requiring the coefficients in a canonical expansion to have unit variance would complicate some of the ensuing algorithms, we will only require orthogonality of the family $\{Z_k\}$, which is precisely the condition set forth at the outset. With a few minor changes, Theorem 3.1 still applies. In particular, choosing the $x_k(t)$ according to Eq. 3.14 provides the best subspace approximations and results in a Fourier series with nonnormalized coefficients. The partial sums still give the appropriate projections. If there is convergence in the mean-square to $X(t)$, then the series provides the desired representation in terms of white noise (albeit, white noise not possessing unit variance).

3.1.2. Expansion of the Covariance Function

If $X(t)$ is given by the canonical representation of Eq. 3.1, then

$$K_X(t,t') = E[(X(t)-\mu_X(t))\overline{(X(t')-\mu_X(t'))}]$$

$$= \left(\sum_{k=1}^{\infty} Z_k x_k(t)\right)\overline{\left(\sum_{i=1}^{\infty} Z_i x_i(t')\right)}$$

$$= E\left[\sum_{k=1}^{\infty}\sum_{i=1}^{\infty} Z_k \overline{Z}_i x_k(t)\overline{x_i(t')}\right] \tag{3.15}$$

Because the zero-mean random variables Z_k are uncorrelated, applying the expected value under the assumption that it can be interchanged with the infinite sum yields

$$K_X(t,t') = \sum_{k=1}^{\infty}\sum_{i=1}^{\infty} E[Z_k \overline{Z_i}] x_k(t)\overline{x_i(t')}$$

$$= \sum_{k=1}^{\infty} \mathrm{Var}[Z_k] x_k(t)\overline{x_k(t')} \tag{3.16}$$

This equation is called a *canonical expansion of the covariance function* in terms of the coordinate functions x_k. Given a canonical expansion of the covariance function in terms of linearly independent coordinate functions, there exists a canonical expansion of the random function itself in terms of the same coordinate functions (see Section 3.8). Setting $t = t'$ in the covariance expansion yields a variance expansion:

$$\mathrm{Var}[X(t)] = \sum_{k=1}^{\infty} \mathrm{Var}[Z_k] |x_k(t)|^2 \tag{3.17}$$

Continuing to assume that $X(t)$ is given by the canonical expansion of Eq. 3.1, suppose a linear operator Ψ is applied to $X(t)$ to obtain the output process $Y(s)$. The principle of superposition yields

$$Y(s) = \mu_Y(s) + \sum_{k=1}^{\infty} Z_k y_k(s) \tag{3.18}$$

where $\mu_Y = \Psi(\mu_X)$ and $y_k = \Psi(x_k)$. Owing to the canonical form of this expansion,

$$K_Y(s,s') = \sum_{k=1}^{\infty} \text{Var}[Z_k] y_k(s)\overline{y_k(s')} \tag{3.19}$$

$$\text{Var}[Y(s)] = \sum_{k=1}^{\infty} \text{Var}[Z_k] |y_k(s)|^2 \tag{3.20}$$

3.2. Karhunen-Loeve Expansion

In this section we discuss a theorem that provides a valid canonical representation throughout the domain over which the random function is defined. The theorem will play a central role in the chapter. It serves as a benchmark for data compression and provides an important special case of a general methodology for finding canonical expansions.

3.2.1. The Karhunen-Loeve Theorem

Finding a canonical representation for a random function $X(t)$ entails finding formulas to generate the coefficients and coordinate functions. As we will see in Section 3.7, there exists a general theory concerning how to obtain the coefficients and coordinate functions from linear functionals applied to $X(t)$ and $K_X(t, t')$, respectively. A particularly elegant instance of this method involves the eigenfunctions of the covariance function.

Theorem 3.2. (*Karhunen-Loeve*) Suppose $X(t)$ is a zero-mean random function on T and $w(t) \geq 0$ is a weight function such that

$$\int_T \int_T |K_X(t,s)|^2 \, w(t)w(s)\, dt ds < \infty \tag{3.21}$$

Then

(i) For the integral equation

$$\int_T K_X(t,s)u(s)w(s)\, ds = \lambda u(t) \tag{3.22}$$

there exists a discrete (finite or infinite) set of *eigenvalues* $\lambda_1 \geq \lambda_2 \geq \lambda_3 \geq \ldots 0$ with the corresponding *eigenfunctions* $u_1(t)$, $u_2(t)$, $u_3(t)$,... such that

$$\int_T K_X(t,s)u_k(s)w(s)\, ds = \lambda_k u_k(t) \tag{3.23}$$

for all k, and such that $\{u_k(t)\}$ is a deterministic orthonormal system on T relative to the weight function $w(t)$, meaning that

$$\int_T u_k(t)\overline{u_j(t)}w(t)\,dt = \begin{cases} 1, & \text{if } k = j \\ 0, & \text{if } k \neq j \end{cases} \tag{3.24}$$

(ii) For $k = 1, 2, \ldots$, the *generalized Fourier coefficients* of $X(t)$ relative to the set of eigenfunctions,

$$Z_k = \int_T X(t)\overline{u_k(t)}w(t)\,dt \tag{3.25}$$

are uncorrelated and $\mathrm{Var}[Z_k] = \lambda_k$.
(iii) $X(t)$ is represented by its canonical expansion in terms of the Z_k and the $u_k(t)$:

$$X(t) = \sum_{k=1}^{\infty} Z_k u_k(t) \tag{3.26}$$

in the mean-square sense. ■

There is a key difference between the general form of a canonical expansion given in Eq. 3.1 and the Karhunen-Loeve expansion of Eq. 3.26: the Karhunen-Loeve coordinate functions form an orthonormal system in $L^2[w(t)dt]$, the inner product being

$$\langle x, y \rangle = \int_T x(t)\overline{y(t)}w(t)\,dt \tag{3.27}$$

No such claim is made for general canonical expansions. In addition, the nonrandom orthonormal system $\{u_k(t)\}$ is determined by the covariance function of the random function and the uncorrelated random variables of the Karhunen-Loeve expansion are found by placing the random function into the inner product of Eq. 3.27 with the $u_k(t)$ to form the generalized Fourier coefficients of $X(t)$ given in Eq. 3.25. If a square-integrable realization $x(t)$ of $X(t)$ were placed into Eq. 3.25, then the resulting coefficients z_k would be the Fourier coefficients of $x(t)$ with respect to the orthonormal system $\{u_k(t)\}$, so that Eq. 3.26 would hold with $x(t)$ in place of $X(t)$ and z_k in place of Z_k, namely,

$$x(t) = \sum_{k=1}^{\infty} z_k u_k(t) \tag{3.28}$$

Whereas convergence in the Karhunen-Loeve expansion is relative to the mean-square norm, convergence in Eq. 3.28 is with respect to the norm arising from Eq. 3.27, that is, the norm for $L^2[w(t)dt]$. If each MS integral Z_k could be obtained by integrating each of the random function's realizations, then it would appear that Eq. 3.26 could be "derived from" Eq. 3.28. However, while all of this is intuitively appealing, the MS integral is not equivalent to a collection of deterministic integrals involving process realizations.

***Example* 3.1.** Consider a WS stationary random signal $X(t)$ with covariance function

$$K_X(t, t') = be^{-b|t-t'|}$$

on the interval $[0, a]$. Then Eq. 3.22 with a constant unit weight function takes the form

$$\int_0^a be^{-b|t-s|} u(s)\, ds = \lambda u(t)$$

It can be shown that this integral equation has the orthonormal solutions (eigenfunctions)

$$u_k(t) = \sqrt{\frac{2}{a + \lambda_k}} \sin\left[c_k\left(t - \frac{a}{2}\right) + \frac{k\pi}{2}\right]$$

for $k = 1, 2, \ldots$, with corresponding eigenvalues

$$\lambda_k = \frac{2b}{b^2 + c_k^2}$$

where $c_1, c_2, \ldots$ are the positive roots, arranged in ascending order, of the equation

$$\tan ac = -\frac{2bc}{b^2 - c^2}$$

According to Theorem 3.1,

$$X(t) = \mu_X(t) + \sum_{k=1}^{\infty} Z_k u_k(t)$$

where

$$Z_k = \int_0^a X(t)\sqrt{\frac{2}{a+\lambda_k}}\sin\left[c_k\left(t-\frac{a}{2}\right)+\frac{k\pi}{2}\right]dt$$

The mean $\mu_X(t)$ appears in the canonical expansion because $X(t)$ need not have zero mean. ■

***Example* 3.2.** The covariance function for the Wiener process is given in Eq. 2.101. For simplicity let $w(s) \equiv 1$ and $\sigma^2 = 1$. Then the integral equation of Eq. 3.22 over the interval $(0, T)$ is

$$\int_0^T \min(t,s)u(s)\,ds = \lambda u(t)$$

for $0 < t < T$. According to the definition of minimum, the integral equation becomes

$$\left(\int_0^t su(s)\,ds + t\int_t^T u(s)\,ds\right) = \lambda u(t)$$

with boundary condition $u(0) = 0$. Differentiation with respect to t yields

$$\int_t^T u(s)\,ds = \lambda u'(t)$$

with boundary condition $u'(T) = 0$. A second differentiation yields

$$\lambda u''(t) + u(t) = 0$$

Along with its boundary values, this second-order differential equation has solutions for

$$\lambda_k = \left(\frac{2T}{(2k-1)\pi}\right)^2$$

for $k = 1, 2, \ldots$ These form distinct eigenvalues for the integral equation and the corresponding basis functions are

$$u_k(t) = \sqrt{\frac{2}{T}} \sin\left[\frac{(k-1/2)\pi t}{T}\right]$$

The uncorrelated generalized Fourier coefficients are

$$Z_k = \int_0^T X(t)\sqrt{\frac{2}{T}} \sin\left[\frac{(2k-1)\pi t}{2T}\right] dt$$

and the Karhunen-Loeve expansion of the Wiener process is given by Eq. 3.26. ■

3.2.2. Discrete Karhunen-Loeve Expansion

In the discrete setting over the time points $t_1, t_2, \ldots,$ the integral equation of Eq. 3.22 becomes the system of summation equations

$$\sum_{j=1}^{\infty} K_X(t_i, t_j) u(t_j) w(t_j) = \lambda u(t_i) \tag{3.29}$$

for $i = 1, 2, \ldots$ If the weight function is identically 1 and the number of points t_i is finite, say n of them, then the sum has n terms and the system reduces to the finite system

$$\begin{aligned}
k_{11}u(t_1) + k_{12}u(t_2) + \cdots + k_{1n}u(t_n) &= \lambda u(t_1) \\
k_{21}u(t_1) + k_{22}u(t_2) + \cdots + k_{2n}u(t_n) &= \lambda u(t_2) \\
\vdots \qquad\qquad \vdots \qquad\qquad \vdots \qquad & \quad\ \vdots \\
k_{n1}u(t_1) + k_{n2}u(t_2) + \cdots + k_{nn}u(t_n) &= \lambda u(t_n)
\end{aligned} \tag{3.30}$$

where $k_{ij} = K_X(t_i, t_j)$. In matrix form,

$$\mathbf{Ku} = \lambda \mathbf{u} \tag{3.31}$$

where $\mathbf{K} = (k_{ij})$ is the covariance matrix (autocorrelation matrix, since $\mu_X = 0$) and

$$\mathbf{u} = \begin{pmatrix} u(t_1) \\ u(t_2) \\ \vdots \\ u(t_n) \end{pmatrix} \tag{3.32}$$

Consequently, the required eigenvalues and eigenfunctions are the eigenvalues and eigenvectors of the covariance matrix. Since **K** is symmetric, if λ_1 and λ_2 are distinct eigenvalues, then their respective eigenvectors will be orthogonal and the desired orthonormal eigenvectors can be found by dividing each by its own magnitude. On the other hand, if an eigenvalue has repeated eigenvectors, then these will be linearly independent and an algebraically equivalent set can be found by the Gram-Schmidt orthogonalization procedure. (Although we will not go into detail, analogous remarks can be applied to the continuous case, where Theorem 3.2(i) involves an integral equation.)

3.2.3. Canonical Expansions with Orthonormal Coordinate Functions

The Karhunen-Loeve expansion provides a canonical representation in terms of orthonormal coordinate functions derived from the covariance function of the random function to be expanded. We will now consider the problem from the other direction by beginning with an orthonormal function system and trying to find a canonical expansion for a zero-mean random function $X(t)$. In the process we will prove the Karhunen-Loeve theorem and demonstrate its centrality to canonical expansions having orthonormal coordinate functions.

If $\{u_k(t)\}$ is an orthonormal system, then we wish to find an expression of the form given in Eq. 3.26, only now $u_1(t)$, $u_2(t)$,... are given and the problem is to find the coefficients Z_1, Z_2,.... To discover the form of the coefficients, multiply both sides of the proposed expansion (Eq.3.26) by the conjugate of $u_k(t)$ and integrate over T, assuming that the order of integration and summation can be interchanged. Owing to the orthonormality of $\{u_k(t)\}$,

$$\begin{aligned} \int_T X(t)\overline{u_k(t)}\,dt &= \int_T \left(\sum_{i=1}^{\infty} Z_i u_i(t) \right) \overline{u_k(t)}\,dt \\ &= \sum_{i=1}^{\infty} \int_T Z_i u_i(t)\overline{u_k(t)}\,dt \\ &= Z_k \end{aligned} \tag{3.33}$$

for $k = 1, 2,\ldots$ Z_k is given by the generalized Fourier coefficient of $X(t)$ relative to $u_k(t)$, just as in the Karhunen-Loeve expansion. Equation 3.33 gives the form of $Z_1, Z_2,\ldots$ if there is MS convergence. Note that Eq. 3.33 does not presuppose that $Z_1, Z_2,\ldots$ are uncorrelated. Given an orthonormal function system $\{u_k(t)\}$, a series of the form given in Eq. 3.26 with $Z_1, Z_2,\ldots$ given as the generalized Fourier coefficients may or may not converge, and $Z_1, Z_2,\ldots$ may or may not be uncorrelated.

Regarding correlation of the coefficients, the inner product between Z_k and Z_i is

$$
\begin{aligned}
E[Z_k \bar{Z}_i] &= E\left[\int_T X(t)\overline{u_k(t)}\,dt\,\overline{\int_T X(s)\overline{u_i(s)}\,ds}\right] \\
&= E\left[\int_T\int_T X(t)\overline{X(s)u_k(t)}u_i(s)\,dt\,ds\right] \\
&= \int_T\int_T R_X(t,s)\overline{u_k(t)}u_i(s)\,dtds
\end{aligned}
\tag{3.34}
$$

For $k \neq i$, Z_k and Z_i are uncorrelated when this double integral vanishes. Moreover, the variance of Z_k is given by

$$
E[|Z_k|^2] = \int_T\int_T R_X(t,s)u_k(s)\overline{u_k(t)}\,dtds \tag{3.35}
$$

Assuming $Z_1, Z_2,\ldots$ are uncorrelated, to analyze convergence we apply Theorem 3.1 to $Z_1, Z_2,\ldots$, keeping in mind that $Z_1, Z_2,\ldots$ need not be orthonormal. Letting $V_k = \mathrm{Var}[Z_k]$, Eq. 3.14 gives the nonnormalized Fourier coefficients as

$$
\begin{aligned}
\frac{E[X(t)\bar{Z}_k]}{V_k} &= \frac{1}{V_k}E\left[X(t)\overline{\int_T X(s)\overline{u_k(s)}\,ds}\right] \\
&= \frac{1}{V_k}\int_T E[X(t)\overline{X(s)}]u_k(s)\,ds \\
&= \frac{1}{V_k}\int_T R_X(t,s)u_k(s)\,ds
\end{aligned}
\tag{3.36}
$$

(where the autocorrelation and covariance functions are equal). $X(t)$ has the Fourier series

$$X(t) \sim \sum_{k=1}^{\infty} \left(\frac{1}{V_k} \int_T R_X(t,s) u_k(s)\, ds \right) Z_k \tag{3.37}$$

relative to the system $\{Z_k\}$. According to Theorem 3.1(ii), $X(t)$ equals its Fourier series in the mean square if and only if

$$\sum_{k=1}^{\infty} \frac{1}{V_k} \left| \int_T R_X(t,s) u_k(s)\, ds \right|^2 = R_X(t,t) \tag{3.38}$$

where we recognize that $\mathrm{Var}[X(t)] = R_X(t, t)$ because $X(t)$ has zero mean. Equation 3.38 states that there is equality in the Bessel inequality. Recalling Eq. 3.17, we see that Eq. 3.38 is the canonical expansion for the variance corresponding to the canonical expansion defined by the Fourier series for $X(t)$. Hence, we can state the following theorem.

Theorem 3.3. Suppose $\{u_k(t)\}$ is an orthonormal function system over T and the generalized Fourier coefficients $Z_1, Z_2,\ldots$ of the zero-mean random function $X(t)$ with respect to $u_1(t), u_2(t),\ldots$ are uncorrelated. Then $X(t)$ possesses the canonical representation

$$X(t) = \sum_{k=1}^{\infty} \left(\frac{1}{V_k} \int_T R_X(t,s) u_k(s)\, ds \right) Z_k \tag{3.39}$$

if and only if the series of Eq. 3.38 provides a canonical expansion of $\mathrm{Var}[X(t)]$. The expansion is the Fourier series of $X(t)$ relative to $Z_1, Z_2,\ldots$. ■

Returning to the Karhunen-Loeve expansion, where for simplicity we let $w(s) \equiv 1$, Z_k is the generalized Fourier coefficient for the kth eigenfunction. Applying Eq. 3.23 to Eq. 3.34, which holds for any orthonormal system, shows that $Z_1, Z_2,\ldots$ are uncorrelated:

$$E[Z_k \overline{Z}_i] = \lambda_i \int_T u_i(t) \overline{u_k(t)}\, dt = 0 \tag{3.40}$$

for $k \neq i$. Equation 3.35 shows that $V_k = \lambda_k$. According to Eq. 3.23,

$$u_k(t) = \frac{1}{V_k} \int_T R_X(t,s) u_k(s)\, ds \tag{3.41}$$

so that the Fourier series of Eq. 3.39 takes the form of Eq. 3.26. We can now appeal to Theorem 3.3 to demonstrate the validity of the Karhunen-Loeve expansion.

From the theory of integral equations, *Mercer's theorem* states that, if $R_X(t, s)$ is continuous over $T \times T$, then

$$R_X(t, s) = \sum_{k=1}^{\infty} \lambda_k u_k(t) \overline{u_k(s)} \tag{3.42}$$

where convergence is uniform and absolute. Letting $s = t$ in Mercer's theorem yields

$$R_X(t, t) = \sum_{k=1}^{\infty} \lambda_k |u_k(t)|^2 \tag{3.43}$$

which is a canonical expansion of the covariance function (Eq. 3.38).

The ease with which MS convergence has been demonstrated shows that the depth of the Karhunen-Loeve expansion lies in the integral-equation theory pertaining to Eq. 3.22. We have shown convergence under continuity of the covariance function; the theorem holds for more general covariance functions in the context of generalized function theory.

A converse proposition holds. If $\{u_k(t)\}$ is an orthonormal system and $X(t)$ has a canonical representation of the form given in Eq. 3.26, then (as we have shown) $Z_1, Z_2,\ldots$ must be the generalized Fourier coefficients of $X(t)$ with respect to $u_1(t), u_2(t),\ldots$ and, because the Fourier representation of Eq. 3.39 is unique, $u_k(t)$ must be given by Eq. 3.41. This means that $u_1(t), u_2(t),\ldots$ must be eigenfunctions for Eq. 3.22 and $V_k = \lambda_k$. For the converse to hold, it must be that $u_1(t), u_2(t),\ldots$ constitute the eigenfunctions for all nonzero eigenvalues. This is assured because a canonical expansion of the variance must result from the function expansion and the variance expansion is given by Eq. 3.43, which according to Mercer's theorem involves all eigenfunctions.

Since it is always possible to find a weight function resulting in the square-integrability condition of Eq. 3.21, in theory the Karhunen-Loeve theorem is definitive. Unfortunately, we must solve an integral equation that may be very difficult to solve. Consequently, it might be worthwhile resorting to the more general methodology of Section 3.7, or to an algorithmic approach such as the one discussed in Section 3.12, even though the latter will provide a valid representation only on a set of sample data points.

In digital processing, the Karhunen-Loeve expansion is often applied over a finite system $X(1)$, $X(2)$,..., $X(n)$, so that the integral equation reduces to a finite matrix system $\mathbf{Ku} = \lambda\mathbf{u}$. Even here there is a practical difficulty, since $\mathbf{K}$ often must be estimated from observations, and variance-covariance estimation tends to require a large number of observations to obtain good precision. Moreover, the eigenvalues of $\mathbf{K}$ must be found and numerical computation of eigenvalues can be problematic.

3.2.4. Relation to Data Compression

To see why the Karhunen-Loeve expansion is important for compression, consider a discrete zero-mean random process $X(n)$, $n = 1, 2,...,$ and its Karhunen-Loeve expansion

$$X(n) = \sum_{k=1}^{\infty} Z_k u_k(n) \tag{3.44}$$

Let

$$X_m(n) = \sum_{k=1}^{m} Z_k u_k(n) \tag{3.45}$$

be the mth partial sum of the series. According to the Karhunen-Loeve theorem, $X_m(n)$ converges to $X(n)$ in the mean square as $m \to \infty$. Owing to the uncorrelatedness of the generalized Fourier coefficients, the mean-square error is given by

$$\begin{aligned} E[|X(n) - X_m(n)|^2] &= E[(X(n) - X_m(n))\overline{(X(n) - X_m(n))}] \\ &= E\left[\left(\sum_{k=m+1}^{\infty} Z_k u_k(n)\right)\overline{\left(\sum_{i=m+1}^{\infty} Z_i u_i(n)\right)}\right] \\ &= E\left[\sum_{k=m+1}^{\infty}\sum_{i=m+1}^{\infty} Z_k \overline{Z}_i u_k(n)\overline{u_i(n)}\right] \\ &= \sum_{k=m+1}^{\infty} \mathrm{Var}[Z_k]|u_k(n)|^2 \end{aligned} \tag{3.46}$$

Suppose the eigenfunctions are uniformly bounded, so that there exists a constant c such that $|u_k(n)| \leq c$ for all k. Then, since $\mathrm{Var}[Z_k] = \lambda_k$, the mean-square error is bounded in terms of the eigenvalues of the covariance matrix:

$$E[|X(n) - X_m(n)|^2] \leq c^2 \sum_{k=m+1}^{\infty} \lambda_k \tag{3.47}$$

Since the eigenvalues are decreasing, the best m terms to select to approximate $X(n)$ relative to the bound of Eq. 3.47 are the first m terms. If the tail-sum of the eigenvalues falls off rapidly, then the approximation gets good rapidly.

Intuitively, since $E[Z_k u_k(n)] = 0$ for all k, terms in the expansion of Eq. 3.44, those for which $\mathrm{Var}[Z_k] \approx 0$ contribute very little. Should we continue to assume $|u_k(n)| \leq c$ for all k, then

$$\mathrm{Var}[Z_k u_k(n)] \leq c^2 \lambda_k \tag{3.48}$$

and, since $\lambda_1 \geq \lambda_2 \geq \lambda_3 \geq \ldots$, the contribution of $Z_k u_k(n)$ decreases as k increases.

A better picture of the matter can be obtained if we define the *mean-square error* between the discrete random function $X(n)$ and the approximating random function $X_m(n)$ over all n rather than for a particular n. We define this error to be

$$\mathrm{E}[X, X_m] = \sum_{n=1}^{\infty} E[|X(n) - X_m(n)|^2] \tag{3.49}$$

which is the sum of the MS errors over all n. Applying this error to Eq. 3.46 yields

$$\begin{aligned}
\mathrm{E}[X, X_m] &= \sum_{n=1}^{\infty} \sum_{k=m+1}^{\infty} \mathrm{Var}[Z_k] |u_k(n)|^2 \\
&= \sum_{k=m+1}^{\infty} \mathrm{Var}[Z_k] \sum_{n=1}^{\infty} |u_k(n)|^2 \\
&= \sum_{k=m+1}^{\infty} \mathrm{Var}[Z_k]
\end{aligned}$$

$$= \sum_{k=m+1}^{\infty} \lambda_k \tag{3.50}$$

where the third equality follows from the fact that $\{u_k(n)\}_{k=1,2,...}$ is an orthonormal system of functions defined over the positive integers. The import of Eq. 3.50 is that the global mean-square error defined by $E[X, X_m]$ is equal to the tail-sum of the eigenvalues. $X_m(n)$ represents a compressed approximation of $X(n)$ that can be used to save computation and storage when used in place of $X(n)$ and Eq. 3.50 provides a measure of the error.

***Example* 3.3.** Let $\mathbf{X} = (X_1, X_2)'$ be a random vector for which X_1 and X_2 are jointly normal zero-mean random variables having common variance σ^2 and correlation coefficient $\rho > 0$. Then

$$\mathbf{K} = \begin{pmatrix} \sigma^2 & \rho\sigma^2 \\ \rho\sigma^2 & \sigma^2 \end{pmatrix}$$

The eigenvalues of $\mathbf{K}$ are found by solving $\det[\mathbf{K} - \lambda\mathbf{I}] = 0$. The two distinct solutions are

$$\lambda_1 = (1 + \rho)\sigma^2$$

$$\lambda_2 = (1 - \rho)\sigma^2$$

The corresponding orthonormal eigenvectors are

$$\mathbf{u}_1 = \begin{pmatrix} u_{11} \\ u_{12} \end{pmatrix} = \frac{1}{\sqrt{2}}\begin{pmatrix} 1 \\ 1 \end{pmatrix}$$

$$\mathbf{u}_2 = \begin{pmatrix} u_{21} \\ u_{22} \end{pmatrix} = \frac{1}{\sqrt{2}}\begin{pmatrix} 1 \\ -1 \end{pmatrix}$$

The generalized Fourier coefficients are

$$Z_1 = \mathbf{X}'\mathbf{u}_1 = \frac{1}{\sqrt{2}}(X_1 + X_2)$$

$$Z_2 = \mathbf{X}'\mathbf{u}_2 = \frac{1}{\sqrt{2}}(X_1 - X_2)$$

As it must,

$$\mathbf{X} = Z_1\mathbf{u}_1 + Z_2\mathbf{u}_2$$

According to the Karhunen-Loeve theorem,

$$\text{Var}[Z_1] = \lambda_1 = (1 + \rho)\sigma^2$$

$$\text{Var}[Z_2] = \lambda_2 = (1 - \rho)\sigma^2$$

Hence, if the variables are highly correlated with $\rho \approx 1$, then $\text{Var}[Z_2] \approx 0$ and $Z_2\mathbf{u}_2$ can be dropped from the expression for $\mathbf{X}$ with little effect on $\mathbf{X}$. ■

3.3. Noncanonical Representation

The Karhunen-Loeve expansion provides a canonical representation in terms of orthonormal coordinate functions. Representation in terms of orthonormal coordinate functions can still be useful when the generalized Fourier coefficients are correlated, in which case the representation is noncanonical.

3.3.1. Generalized Bessel Inequality

Given an orthonormal function system $\{u_k(t)\}$ and a zero-mean random function $X(t)$, if we desire a representation for $X(t)$ of the form given in Eq. 3.26, then Eq. 3.33 gives the necessary form of the coefficients $Z_1, Z_2,\ldots$ to be the generalized Fourier coefficients relative to the system $\{u_k(t)\}$. The theory of Section 3.2.3 then proceeds to obtain necessary and sufficient conditions for convergence under the assumption that $Z_1, Z_2,\ldots$ are uncorrelated. Specifically, $X(t)$ equals its Fourier series relative to the orthogonal system $Z_1, Z_2,\ldots$ (Eq. 3.37) if and only if there is equality in the Bessel inequality (Eq. 3.38) relative to $Z_1, Z_2,\ldots$. Suppose, however, that $Z_1, Z_2,\ldots$ are not uncorrelated. What then happens when we form the expansion of Eq. 3.26?

The expansion is not canonical because the coefficients are correlated. But it might be MS convergent to $X(t)$. Moreover, it may be useful even if it does not converge to $X(t)$. In fact, if $\{u_k(t)\}$ is a complete orthonormal system relative to $L^2(T)$, then each realization of $X(t)$ is equal in $L^2(T)$ to its deterministic Fourier series according to Eq. 3.28, where the coefficients in that expansion are given by a deterministic Fourier integral of the same form as the integral giving the generalized Fourier coefficients. It is not unreasonable to apply the expansion of Eq. 3.26 in an approximate sense if it does not converge to $X(t)$ or if convergence is uncertain. As with any approximate method, one must be prudent.

In the case where $Z_1, Z_2,\ldots$ are uncorrelated, the ordinary Bessel inequality plays a key role in determining MS convergence. If $Z_1, Z_2,\ldots$ are not necessarily uncorrelated, then a generalized Bessel inequality can be useful. For MS convergence, $Z_1, Z_2,\ldots$ must satisfy

$$\lim_{n\to\infty} E\left[\left|X(t)-\sum_{k=1}^{n} Z_k u_k(t)\right|^2\right]=0 \tag{3.51}$$

Temporarily ignoring the limit, integration of the mean-square difference over T (assuming the legitimacy of interchanging the order of expectation and integration) yields

$$\int_T E\left[\left|X(t)-\sum_{k=1}^{n} Z_k u_k(t)\right|^2\right]dt$$

$$=\int_T E\left[\left(X(t)-\sum_{k=1}^{n} Z_k u_k(t)\right)\overline{\left(X(t)-\sum_{i=1}^{n} Z_i u_i(t)\right)}\right]$$

$$=\int_T E\left[|X(t)|^2-\sum_{k=1}^{n} X(t)\overline{Z_k u_k(t)}-\sum_{k=1}^{n}\overline{X(t)}Z_k u_k(t)+\sum_{k=1}^{n}\sum_{i=1}^{n} Z_k\overline{Z}_i u_k(t)\overline{u_i(t)}\right]dt$$

$$=\int_T E[|X(t)|^2]\,dt-\sum_{k=1}^{n} E\left[\overline{Z}_k\int_T X(t)\overline{u_k(t)}\,dt\right]-\sum_{k=1}^{n} E\left[Z_k\overline{\int_T X(t)\overline{u_k(t)}\,dt}\right]$$

$$+\sum_{k=1}^{n}\sum_{i=1}^{n} E[Z_k\overline{Z}_i]\int_T u_k(t)\overline{u_i(t)}\,dt$$

$$=\int_T E[|X(t)|^2]\,dt-2\sum_{k=1}^{n} E[|Z_k|^2]+\sum_{k=1}^{n} E[|Z_k|^2]\int_T |u_k(t)|^2\,dt$$

$$=\int_T E[|X(t)|^2]\,dt-\sum_{k=1}^{n} E[|Z_k|^2] \tag{3.52}$$

where the last two equalities follow from the orthonormality of $\{u_k(t)\}$. Assume the second moment of $X(t)$ is integrable over T:

$$\int_T E[|X(t)|^2]\,dt<\infty \tag{3.53}$$

Since the left-hand side of Eq. 3.52 is nonnegative,

$$\sum_{k=1}^{\infty} E[|Z_k|^2] \leq \int_T E[|X(t)|^2]\,dt \tag{3.54}$$

which is the *generalized Bessel inequality*. There is equality,

$$\sum_{k=1}^{\infty} E[|Z_k|^2] = \int_T E[|X(t)|^2]\,dt \tag{3.55}$$

if and only if

$$\lim_{n\to\infty} \int_T E\left[\left|X(t) - \sum_{k=1}^{n} Z_k u_k(t)\right|^2\right] dt = 0 \tag{3.56}$$

If we let $\varepsilon_n(t)$ denote the expectation (MSE) in the preceding equation, then the equation states that the integral of $\varepsilon_n(t)$ converges to 0 as $n \to \infty$. Mean-square convergence of the series to $X(t)$ means that $\varepsilon_n(t) \to 0$ as $n \to \infty$. Convergence of the integral does not necessarily imply this. However, it does in the discrete case, where the integral over T becomes a summation. Then equality in the generalized Bessel inequality (Eq. 3.55) implies that $\Sigma_i \varepsilon_n(i) \to 0$ as $n \to \infty$, which in turn implies that

$$\lim_{n\to\infty} \sum_{k=1}^{n} Z_k u_k(i) = X(i) \tag{3.57}$$

in the mean square. This limit yields a canonical expansion if $Z_1, Z_2, \ldots$ are uncorrelated.

Referring to Eq. 3.35, the generalized Bessel inequality can be expressed as

$$\sum_{k=1}^{\infty} \int_T \int_T R_X(t,s) u_k(s) \overline{u_k(t)}\, dt\, ds \leq \int_T R_X(t,t)\, dt \tag{3.58}$$

3.3.2. Decorrelation

When $X(t)$ has a representation according to Eq. 3.26 and $Z_1, Z_2, \ldots$ are correlated, a decorrelation procedure can be used to arrive at an uncorrelated representation. We now give a decorrelation algorithm that replaces the expansion by one having uncorrelated coefficients. The decorrelation procedure does not depend on the orthogonality of the coordinate functions. Indeed, so long as a random function is given (in the mean-square sense) by a series of elementary functions, the methodology yields a new series representation having uncorrelated coefficients.

If $X(t)$ is equal (in the MS sense) to the expansion given in Eq. 3.26, then the following procedure yields a canonical representation:

D1. Set

$$V_1 = \mathrm{Var}[Z_1]$$

$$a_{21} = \mathrm{Cov}[Z_2 Z_1] V_1^{-1} \tag{3.59}$$

$$V_2 = \mathrm{Var}[Z_2] - |a_{21}|^2$$

D2. Proceeding recursively, for $k > 2$, set

$$a_{k1} = \mathrm{Cov}[Z_k Z_1] V_1^{-1} \tag{3.60}$$

$$a_{kj} = V_j^{-1}\left(\mathrm{Cov}[Z_k Z_j] - \sum_{i=1}^{j-1} a_{ki} a_{ji} V_i\right) \qquad (j = 2, 3, \ldots, k-1) \tag{3.61}$$

$$V_k = \mathrm{Var}[Z_k] - \sum_{i=1}^{k-1} |a_{ki}|^2 V_i \tag{3.62}$$

D3. Letting $\mu_k = E[Z_k]$, the desired uncorrelated random variables W_k are found by recursively solving the system

$$\begin{aligned} Z_1 - \mu_1 &= W_1 \\ Z_2 - \mu_2 &= a_{21} W_1 + W_2 \\ Z_3 - \mu_3 &= a_{31} W_1 + a_{32} W_2 + W_3 \\ &\vdots \\ Z_k - \mu_k &= a_{k1} W_1 + a_{k2} W_2 + \cdots + a_{k,k-1} W_{k-1} + W_k \\ &\vdots \end{aligned} \tag{3.63}$$

D4. For $k = 1, 2, \ldots$, $\mathrm{Var}[W_k] = V_k$.

D5. For $k = 1, 2, \ldots$, set

$$x_k(t) = u_k(t) + \sum_{i=k+1}^{\infty} a_{ik} u_i(t) \tag{3.64}$$

D6. The desired decorrelated canonical representation of $X(t)$ is given by

$$X(t) = \mu_X(t) + \sum_{k=1}^{\infty} W_k x_k(t) \tag{3.65}$$

***Example* 3.4.** Suppose

$$X(t) = Z_1 u_1(t) + Z_2(t) u_2(t)$$

where Z_1 and Z_2 are jointly normal zero-mean random variables having covariance matrix **K** of Example 3.3. Applying the decorrelation procedure yields $V_1 = \sigma^2$, $a_{21} = \rho\sigma^{-2}$, $V_2 = \sigma^2 - \rho^2\sigma^{-2}$, $W_1 = Z_1$, $W_2 = Z_2 - \rho\sigma^{-2}Z_1$, $x_1(t) = u_1(t) + \rho\sigma^{-2}u_2(t)$, and $x_2(t) = u_2(t)$. The desired representation is

$$X(t) = W_1[u_1(t) + \rho\sigma^{-2}u_2(t)] + W_2 u_2(t)$$

The representation is decorrelated because

$$E[W_1 W_2] = E[Z_1(Z_2 - \rho\sigma^{-2}Z_1)] = 0$$ ■

3.4. Trigonometric Representation

Trigonometric Fourier series are important because of the manner in which they characterize functions via frequency content. In this section we discuss trigonometric representation of random functions, both canonical and noncanonical. Trigonometric representation is most useful for WS stationary random functions.

3.4.1. Trigonometric Fourier Series

We return to the question raised in Example 2.15 concerning convergence of a trigonometric Fourier series in the random-function setting. We will assume the series is written in such a way that the orthonormal system is explicit, which means that the orthonormal system over $[0, 2\pi]$ consists of the deterministic functions $1/\sqrt{2\pi}$, $\cos kt/\sqrt{\pi}$, and $\sin kt/\sqrt{\pi}$. The generalized Fourier coefficients for $X(t)$ are

$$A_0 = \frac{1}{\sqrt{2\pi}} \int_0^{2\pi} X(t)\,dt$$

$$A_k = \frac{1}{\sqrt{\pi}} \int_0^{2\pi} X(t)\cos kt\,dt \tag{3.66}$$

$$B_k = \frac{1}{\sqrt{\pi}} \int_0^{2\pi} X(t)\sin kt\,dt$$

Representation depends on whether

$$X(t) = \frac{A_0}{\sqrt{2\pi}} + \sum_{k=1}^{\infty} A_k \frac{\cos kt}{\sqrt{\pi}} + B_k \frac{\sin kt}{\sqrt{\pi}} \tag{3.67}$$

in the mean-square.

To further consider the matter and to simplify some of the expressions, let us restrict our attention to the cosine expansion,

$$X(t) \sim \frac{A_0}{\sqrt{2\pi}} + \sum_{k=1}^{\infty} A_k \frac{\cos kt}{\sqrt{\pi}} \tag{3.68}$$

(noting that, in the deterministic setting, the cosine orthonormal system is complete relative to the subspace of even functions). According to Theorem 2.2, the generalized cosine coefficients exist as MS integrals and the moments

$$E[A_0] = \frac{1}{\sqrt{2\pi}} \int_0^{2\pi} E[X(t)]\,dt \tag{3.69}$$

$$E[A_k] = \frac{1}{\sqrt{\pi}} \int_0^{2\pi} E[X(t)]\cos kt\,dt \tag{3.70}$$

$$E[A_kA_j] = E\left[\frac{1}{\pi} \int_0^{2\pi}\int_0^{2\pi} X(t)X(s)\cos kt\cos js\,dt\,ds\right]$$

$$= \frac{1}{\pi} \int_0^{2\pi}\int_0^{2\pi} R_X(t,s)\cos kt\cos js\,dt\,ds \tag{3.71}$$

exist so long as these deterministic integrals exist. Assuming they do, for k = 0, 1, 2,...,

$$E[A_k] = E[X(t)]_k \tag{3.72}$$

where $E[X(t)]_k$ is the kth deterministic Fourier coefficient for the mean of $X(t)$. The second moments of the generalized Fourier coefficients are

$$E[|A_0|^2] = \frac{1}{2\pi}\int_0^{2\pi}\int_0^{2\pi} R_X(t,s)\,dt\,ds \tag{3.73}$$

$$E[|A_k|^2] = \frac{1}{\pi}\int_0^{2\pi}\int_0^{2\pi} R_X(t,s)\cos kt\cos ks\,dt\,ds \tag{3.74}$$

The variances of the generalized Fourier coefficients are obtained from the second moments and the means.

Equation 3.72 is interesting in regard to estimation of the (deterministic) Fourier coefficients of the mean of a random function. Suppose N random functions identically distributed to $X(t)$ are randomly observed and the kth Fourier coefficient of each is found, thereby giving a random sample $A_{k,1}$, $A_{k,2}$,..., $A_{k,N}$ for the random coefficient A_k. The sample mean $\hat{A}_k$ of $A_{k,1}$, $A_{k,2}$,..., $A_{k,N}$ provides an unbiased estimator for the mean of A_k. Because $E[A_k] = E[X(t)]_k$, $\hat{A}_k$ is also an unbiased estimator of the kth Fourier coefficient of $\mu_X(t)$. The variance of $\hat{A}_k$ is given by Var[$\hat{A}_k$] = Var[A_k]/N and from Eq. 3.74 it is seen that the variance depends on the autocorrelation function for $X(t)$.

3.4.2. Generalized Fourier Coefficients for WS Stationary Processes

For an orthonormal system $\{u_k(t)\}$ in $L^2(T)$ and a WS stationary random function $X(t)$, the inner product, squared norm, and generalized Bessel inequality in Eqs. 3.34, 3.35, and 3.58 become

$$E[Z_k\overline{Z}_i] = \int_T\int_T R_X(t-s)\overline{u_k(t)}u_i(s)\,dtds \tag{3.75}$$

$$E[|Z_k|^2] = \int_T\int_T R_X(t-s)\overline{u_k(t)}u_k(s)\,dtds \tag{3.76}$$

$$\sum_{k=1}^{\infty} \int_T \int_T R_X(t-s)\overline{u_k(t)}u_k(s)\,dtds \le l(T)r_X(0) \tag{3.77}$$

where $l(T)$ denotes the length of T. If $T = [t_1, t_2]$, then the substitution $\tau = t - s$ in Eqs. 3.75 through 3.77 yields

$$E[Z_k\overline{Z}_i] = \int_{t_1}^{t_2} \int_{t_1-s}^{t_2-s} r_X(\tau)\overline{u_k(\tau+s)}u_i(s)\,d\tau\,ds \tag{3.78}$$

$$E[|Z_k|^2] = \int_{t_1}^{t_2} \int_{t_1-s}^{t_2-s} r_X(\tau)\overline{u_k(\tau+s)}u_k(s)\,d\tau\,ds \tag{3.79}$$

$$\sum_{k=1}^{\infty} \int_{t_1}^{t_2} \int_{t_1-s}^{t_2-s} r_X(\tau)\overline{u_k(\tau+s)}u_k(s)\,d\tau\,ds \le (t_2 - t_1)r_X(0) \tag{3.80}$$

We consider a complex trigonometric Fourier series over the interval $[-T/2, T/2]$. The complex trigonometric orthonormal system consists of the functions

$$u_k(t) = \frac{\exp[j\omega_0 kt]}{\sqrt{T}} \tag{3.81}$$

where $\omega_0 = 2\pi/T$ and $k = \ldots, -2, -1, 0, 1, 2, \ldots$. The Fourier series of a random function $X(t)$ takes the form

$$X(t) \sim \sum_{k=-\infty}^{\infty} Z_k \frac{\exp[j\omega_0 kt]}{\sqrt{T}} \tag{3.82}$$

where

$$Z_k = \frac{1}{\sqrt{T}} \int_{-T/2}^{T/2} X(t)\overline{\exp[j\omega_0 kt]}\,dt$$

$$= \frac{1}{\sqrt{T}} \int_{-T/2}^{T/2} X(t)\exp[-j\omega_0 kt]\,dt \tag{3.83}$$

3.4.3. Mean-Square Periodic WS Stationary Processes

Periodicity plays an important role in the next theorem. A random function $X(t)$ is said to be *mean-square periodic* with period θ if

$$E[|X(t+\theta) - X(t)|^2] = 0 \tag{3.84}$$

for all t. As we now demonstrate, $X(t)$ is MS periodic if and only if its autocorrelation function is *doubly periodic*, namely, if and only if

$$R_X(t + m\theta, t' + n\theta) = R_X(t, t') \tag{3.85}$$

for all integers $m, n \geq 0$. First suppose $X(t)$ is MS periodic with period θ. The Cauchy-Schwarz inequality, together with MS periodicity, yields

$$|E[X(t)(\overline{X(t'+\theta)} - \overline{X(t')})]|^2 \leq E[|X(t)|^2]\, E[|(\overline{X(t'+\theta)} - \overline{X(t')}|^2\,]$$

$$= E[|X(t)|^2]E[|X(t'+\theta) - X(t')|^2] = 0 \tag{3.86}$$

Hence,

$$E[X(t)\overline{X(t'+\theta)}] = E[X(t)\overline{X(t')}] \tag{3.87}$$

Equivalently,

$$R_X(t, t'+\theta) = R_X(t, t') \tag{3.88}$$

Applying the same argument m times to the first variable and n times to the second variable yields Eq. 3.85. Conversely, if Eq. 3.85 holds, then

$$R_X(t+\theta, t+\theta) = R_X(t+\theta, t) = R_X(t, t+\theta) = R_X(t, t) \tag{3.89}$$

Hence,

$$E[|X(t+\theta) - X(t)|^2] = E[(X(t+\theta) - X(t))\overline{(X(t+\theta) - X(t))}] \tag{3.90}$$

$$= R_X(t+\theta, t+\theta) - R_X(t+\theta, t) - R_X(t, t+\theta) + R_X(t, t) = 0$$

and $X(t)$ is MS periodic. Note that if $X(t)$ is WS stationary, then Eq. 3.85 reduces to θ-periodicity of the autocorrelation function.

Theorem 3.4. If $X(t)$ is an MS periodic WS stationary random function with a period of length T, then it possesses the trigonometric canonical representation

$$X(t) = \sum_{k=-\infty}^{\infty} Z_k \frac{\exp[j\omega_0 kt]}{\sqrt{T}} \qquad \blacksquare \tag{3.91}$$

To prove the theorem, let

$$c_k = \frac{1}{\sqrt{T}} \int_{-T/2}^{T/2} r_X(\tau)\exp[-j\omega_0 k\tau]\,d\tau \tag{3.92}$$

be the kth Fourier coefficient for $r_X(\tau)$. Then

$$r_X(\tau) = \sum_{k=-\infty}^{\infty} c_k \frac{\exp[j\omega_0 k\tau]}{\sqrt{T}} \tag{3.93}$$

Uncorrelatedness of the coefficients follows by applying Eq. 3.78: for $k \neq n$,

$$E[Z_k \overline{Z_n}] = \frac{1}{T} \int_{-T/2}^{T/2} \int_{-T/2}^{T/2} r_X(\tau)\overline{\exp[j\omega_0 k(\tau + s)]}\exp[j\omega_0 ns]\,d\tau\,ds$$

$$= c_k \frac{1}{\sqrt{T}} \int_{-T/2}^{T/2} \exp[-j\omega_0 ks]\exp[j\omega_0 ns]\,ds = 0 \tag{3.94}$$

From Eq. 3.79,

$$E[|Z_k|^2] = c_k\sqrt{T} \tag{3.95}$$

To apply Theorem 3.3, we need to demonstrate that Eq. 3.38 holds in the present setting. Owing to periodicity,

$$\int_{-T/2}^{T/2} R_X(t-s)\frac{\exp[j\omega_0 ks]}{\sqrt{T}}\,ds = \frac{\exp[j\omega_0 kt]}{\sqrt{T}} \int_{-T/2-t}^{T/2-t} r_X(\tau)\exp[j\omega_0 k\tau]\,d\tau$$

$$= c_k\sqrt{T}\frac{\exp[j\omega_0 kt]}{\sqrt{T}} \tag{3.96}$$

Therefore,

$$\sum_{k=-\infty}^{\infty}\frac{1}{V_k}\left|\int_{-T/2}^{T/2}R_X(t-s)\frac{\exp[j\omega_0 ks]}{\sqrt{T}}ds\right|^2=\sum_{k=-\infty}^{\infty}\frac{1}{c_k\sqrt{T}}\left|c_k\sqrt{T}\frac{\exp[j\omega_0 ks]}{\sqrt{T}}\right|^2$$

$$=\sum_{k=-\infty}^{\infty}\frac{c_k}{\sqrt{T}} \tag{3.97}$$

$$= r_X(0)$$

Thus, Eq. 3.38 provides a canonical expansion of the variance and the representation follows by Theorem 3.3. From Eq. 3.96, $T^{-1/2}\exp[j\omega_0 kt]$ is an eigenfunction for the eigenvalue $c_k T^{1/2}$. Since the expansion provides a canonical representation of $X(t)$, these constitute the full set of eigenfunctions and Eq. 3.91 is the Karhunen-Loeve expansion of $X(t)$.

If $X(t)$ is WS stationary but not MS periodic, then MS convergence to $X(t)$ in Eq. 3.91 is only assured for $|t| < T/2$ and the Z_k are correlated. Here we apply the generalized Bessel inequality. For fixed s, let

$$c_k(s)=\frac{1}{\sqrt{T}}\int_{-T/2}^{T/2}R_X(t-s)\exp[-j\omega_0 kt]\,dt \tag{3.98}$$

be the kth Fourier coefficient for $R_X(t-s)$. Then, for $|t| < T/2$ and for fixed s,

$$R_X(t-s)=\sum_{k=-\infty}^{\infty}c_k(s)\frac{\exp[j\omega_0 kt]}{\sqrt{T}} \tag{3.99}$$

Expanding the left-hand side of Eq. 3.77, interchanging the order of integration and summation, and applying the preceding Fourier expansion at $t = s$ yields equality in the generalized Bessel inequality, namely,

$$\sum_{k=-\infty}^{\infty}\int_{-T/2}^{T/2}\int_{-T/2}^{T/2}R_X(t-s)\overline{u_k(t)}u_k(s)\,dt\,ds$$

$$=\sum_{k=-\infty}^{\infty}\frac{1}{T}\int_{-T/2}^{T/2}\int_{-T/2}^{T/2}R_X(t-s)\exp[-j\omega_0 kt]\exp[j\omega_0 ks]\,dt\,ds$$

$$= \sum_{k=-\infty}^{\infty} \frac{1}{\sqrt{T}} \int_{-T/2}^{T/2} c_k(s) \exp[j\omega_0 ks]\, ds \tag{3.100}$$

$$= \int_{-T/2}^{T/2} \left(\sum_{k=-\infty}^{\infty} c_k(s) \frac{\exp[j\omega_0 ks]}{\sqrt{T}} \right) ds$$

$$= Tr_X(0)$$

Equality in the generalized Bessel inequality implies MS convergence in the discrete case. As stated, the proposition also holds in the continuous case. Moreover, it can be shown that the correlation coefficients of the coefficients tend to 0 as $T \to \infty$. The coefficients can be decorrelated by the procedure given in the previous section.

3.5. Expansions as Transforms

Trigonometric Fourier coefficients serve as a transform for the space $L^2[0, 2\pi]$ of square-integrable deterministic functions on $[0, 2\pi]$. Specifically, they define a mapping

$$\Phi(x) = \{a_0, a_k, b_k\}_{k=1,2,...} \tag{3.101}$$

for any $x(t) \in L^2[0, 2\pi]$. Because the trigonometric system is complete, the Fourier series of $x(t)$ converges to $x(t)$ in $L^2[0, 2\pi]$ and the series acts as the inverse transform,

$$\Phi^{-1}(\{a_0, a_k, b_k\}_{k=1,2,...}) = x \tag{3.102}$$

Other examples of complete systems are orthogonal polynomial systems and the Walsh-function system. Each of these defines a transform pair (Φ, Φ^{-1}), where Φ^{-1} is defined by a series expansion. This section interprets expansions as transforms acting on classes of random functions. These play key roles in compression and classification.

3.5.1. Orthonormal Transforms of Random Functions

Suppose C is a class of zero-mean random functions and $\{u_k(t)\}$ a fixed deterministic orthonormal system. For any $X(t) \in C$, define the mapping $\Omega(X) = \{Z_1, Z_2,...\}$, where Z_k is the kth generalized Fourier coefficient of $X(t)$ with respect to $\{u_k(t)\}$ and it is assumed that each generalized Fourier integral exists in the mean-square sense. If $X(t)$ equals its expansion relative to $\{u_k(t)\}$ for all $X(t) \in C$, meaning Eq. 3.26 holds in the MS sense for $X(t) \in C$, then the expansion defines the inverse transform Ω^{-1} of the transform pair

(Ω, Ω^{-1}). Even if Ω is not invertible, it still represents a transform on C whose outputs are sequences of random variables.

A compressed transform Ω_m is defined on C by keeping only the first m generalized Fourier coefficients, namely,

$$\Omega_m(X) = \{Z_1, Z_2,..., Z_m\} \tag{3.103}$$

For $m = 1, 2,...$, let

$$X_m(t) = \sum_{k=1}^{m} Z_k u_k(t) \tag{3.104}$$

be the corresponding truncation of the full expansion. If the full expansion equals $X(t)$ in the mean square, then $X_m(t)$ is an approximation of $X(t)$ defined via the compressed sequence $\{Z_1, Z_2,..., Z_m\}$; if not, Ω_m is a still a transform on C and can be useful.

***Example* 3.5.** Consider the class of even functions in $L^2[0, 2\pi)$ and the cosine basis. If all realizations of the random function $X(t)$ are even and if $x(t)$ is a realization of $X(t)$, then $x(t)$ equals its Fourier representation (in the L^2 norm). Now suppose $x(t)$ is even, continuous, 2π-periodic on $\Re$, and has a piecewise continuous derivative. Then its cosine series converges uniformly to it on any closed interval, namely,

$$x(t) = \frac{a_0}{\sqrt{2\pi}} + \sum_{k=1}^{\infty} a_k \frac{\cos kt}{\sqrt{\pi}} \quad \text{(uniformly)}$$

Hence, for any $\varepsilon > 0$, there exists m such that, for all t,

$$\left| x(t) - \left(\frac{a_0}{\sqrt{2\pi}} + \sum_{k=1}^{m} a_k \frac{\cos kt}{\sqrt{\pi}} \right) \right| < \varepsilon$$

Compression is achieved by using only the coefficients $a_0, a_1,..., a_m$ instead of the full set: if one knows that $x(t)$ is to be transmitted, then one need only transmit $a_0, a_1,..., a_m$ for the receiver to obtain an ε approximation of $x(t)$ via the truncated Fourier series. The problem is that this ε approximation based on m terms depends on the specific realization. Relative to the random function $X(t)$, the error of the preceding equation is a random function of t depending on $X(t)$, with the generalized cosine coefficient A_k replacing a_k. The rate of convergence for any particular realization does not characterize the rate of convergence for $X(t)$. If the cosine series converges to $X(t)$ in the mean square, then

$$\lim_{m\to\infty} E\left[\left|X(t)-\left(\frac{A_0}{\sqrt{2\pi}}+\sum_{k=1}^{m} A_k \frac{\cos kt}{\sqrt{\pi}}\right)\right|^2\right]=0$$

However, here the error term is a function of t. The problem can be addressed by defining the process error for any m to be the integral of the expected square error,

$$\mathrm{E}[X, X_m] = \int_0^{2\pi} E\left[\left|X(t)-\left(\frac{A_0}{\sqrt{2\pi}}+\sum_{k=1}^{m} A_k \frac{\cos kt}{\sqrt{\pi}}\right)\right|^2\right] dt$$

If there is MS convergence of $X_m(t)$ to $X(t)$, then the expectation inside the preceding integral converges to 0. In general, convergence of a nonnegative integrand to 0 does not imply convergence of the integral, but it does if the integrand is bounded by a function that is integrable over the domain of integration. In any event, let us assume that the preceding integral converges to 0. Then there is equality in the generalized Bessel inequality (Eq. 3.55) and, according to Eqs. 3.52 and 3.74,

$$\begin{aligned}\mathrm{E}[X, X_m] &= \sum_{k=1}^{\infty} E[|A_k|^2] - \sum_{k=1}^{m} E[|A_k|^2] \\ &= \sum_{k=m+1}^{\infty} E[|A_k|^2] \\ &= \sum_{k=m+1}^{\infty} \frac{1}{\pi} \int_0^{2\pi}\int_0^{2\pi} R_X(t,s) \cos kt \cos ks \, dt \, ds\end{aligned}$$

This expression is a continuous analogue of Eq. 3.50; however, there the matter was simplified because the deterministic orthonormal system consisted of the eigenfunctions of the covariance integral equation of Eq. 3.22 (in which $K_X = R_X$ since $\mu_X = 0$). For compression via the transform $X(t) \to \{A_k\}_{k=0,1,2,\ldots}$, the key point is that $\mathrm{E}\,[X, X_m] \to 0$ as $m \to \infty$. However, we lack the conclusion obtained in Eq. 3.50, where the error is the tail-sum of the eigenvalues and the eigenvalues are decreasing. ■

The preceding considerations can be applied to the Karhunen-Loeve expansion. If C_K is the class of random functions having covariance function K, then each random function in C_K is equal in the mean-square sense to its Karhunen-Loeve expansion relative to the eigenfunctions $u_k(t)$ of the integral equation in Eq. 3.22. For $X(t) \in C_K$, a transform is defined by $\Xi_K(X) = \{Z_1, Z_2, ...\}$, Z_k being the kth generalized Fourier coefficient of $X(t)$ with respect to the eigenfunctions. Ξ_K^{-1} is defined by the Karhunen-Loeve expansion. Compression is achieved by

$$\Xi_{K,m}(X) = \{Z_1, Z_2, ..., Z_m\} \tag{3.105}$$

and Eq. 3.104 provides an approximation to $X(t)$. Indeed, this is simply a restatement of the kind of Karhunen-Loeve compression we have previously discussed for discrete random functions.

As we have so far defined the Karhunen-Loeve transform, it is restricted to the class C_K. If the covariance function for the random function $Y(t)$ is not K, then $Y(t) \notin C_K$. $\Xi_K(Y)$ is the sequence of generalized Fourier coefficients W_k for $Y(t)$ computed from the eigenfunctions $u_k(t)$ relative to K, not the eigenfunctions relative to the covariance function of $Y(t)$. Hence, the expansion whose elementary functions are $W_k u_k(t)$ is not the Karhunen-Loeve expansion for $Y(t)$. There is nothing strange here: with regard to $Y(t)$, the eigenfunctions relative to K simply form a given orthonormal system, not the Karhunen-Loeve system for $Y(t)$. For certain applications, it may be useful to employ the generalized Fourier coefficients for $Y(t)$ relative to these eigenfunctions; on the other hand, if we are not going to use the eigenfunctions corresponding to the covariance function for $Y(t)$, it might be better to employ the generalized Fourier coefficients relative to the trigonometric system, Walsh system, or some other commonly employed system of orthonormal functions. When using the terminology "Karhunen-Loeve transform," one must keep in mind that this transform is defined relative to a specific covariance function. There is no single Karhunen-Loeve transform and when a Karhunen-Loeve transform is used, the special properties of the expansion apply only to those random functions whose covariance functions are the same as the one from which the particular form of the transform has been derived.

3.5.2. Fourier Descriptors

Even if one cannot prove that an expansion over a given orthonormal system converges to the random function, the coefficient transform can still be used to generate image features known as *Fourier descriptors*. These can be used to discriminate among image processes so long as the multivariate distributions of the coefficients are sufficiently distinct. The manner in which Fourier descriptors are generated depends on the functional representation of the images under consideration and whether or not the transforms are continuous or discrete. We present the method generically by using the trigonometric system to represent a curve's projections onto both the x and y axes.

A planar arc γ is expressed parametrically as $\gamma = \gamma(t) = (x(t), y(t))$, where $x(t)$ and $y(t)$ are the coordinate functions of $\gamma(t)$ as the arc length t runs from 0 to T, the total arc length. As deterministic functions, $x(t)$ and $y(t)$ possess Fourier expansions. If we assume closed curves and continuously differentiable coordinate functions, then the trigonometric Fourier series of $x(t)$ and $y(t)$ are uniformly convergent to $x(t)$ and $y(t)$, respectively, and

$$x(t) = a_0 + \sum_{n=1}^{\infty} a_n \cos\frac{2n\pi t}{T} + b_n \sin\frac{2n\pi t}{T}$$

$$y(t) = c_0 + \sum_{n=1}^{\infty} c_n \cos\frac{2n\pi t}{T} + d_n \sin\frac{2n\pi t}{T} \tag{3.106}$$

If $x_m(t)$ and $y_m(t)$ denote the truncations of the respective Fourier series at m terms, then for any $\varepsilon > 0$ there exists M such that, for $m \geq M$, $|x_m(t) - x(t)| < \varepsilon$ and $|y_m(t) - y(t)| < \varepsilon$ for $t \in T$. For increasing m, the approximating curve $(x_m(t), y_m(t))$ gets uniformly closer to $\gamma(t)$.

While deterministic theory may provide motivation, classification of closed contours cannot be characterized by Fourier series of deterministic coordinate functions. The latter are merely realizations of random coordinate functions. The classification problem involves a set of random closed curves $\Gamma_1, \Gamma_2, \ldots, \Gamma_N$. Randomness enters in many ways, including variability in scanning, variability in the class of objects producing each shape, and random perturbations in object boundaries. Because curves are random, their coordinate functions are random functions. Thus, parameterization of a curve takes the form $\Gamma(t) = (X(t), Y(t))$ and the coordinate functions possess the random Fourier series

$$X(t) \sim A_0 + \sum_{n=1}^{\infty} A_n \cos\frac{2n\pi t}{T} + B_n \sin\frac{2n\pi t}{T}$$

$$Y(t) \sim C_0 + \sum_{n=1}^{\infty} C_n \cos\frac{2n\pi t}{T} + D_n \sin\frac{2n\pi t}{T} \tag{3.107}$$

where the random generalized Fourier coefficients for $X(t)$ are given by

$$A_0 = \frac{1}{T}\int_0^T X(t)\, dt$$

$$A_n = \frac{2}{T}\int_0^T X(t)\cos\frac{2n\pi t}{T}\,dt \tag{3.108}$$

$$B_n = \frac{2}{T}\int_0^T X(t)\sin\frac{2n\pi t}{T}\,dt$$

and the coefficients C_n and D_n are defined similarly. For $m = 1, 2,\ldots,$ let $X_m(t)$ and $Y_m(t)$ be the truncations at m terms of the series in Eq. 3.107. Whether or not these converge to the appropriate random coordinate functions, they nonetheless provide finite random vectors

$$\mathbf{A} = \begin{pmatrix} A_0 \\ A_1 \\ \vdots \\ A_m \end{pmatrix} \quad \mathbf{B} = \begin{pmatrix} B_1 \\ B_2 \\ \vdots \\ B_m \end{pmatrix} \quad \mathbf{C} = \begin{pmatrix} C_0 \\ C_1 \\ \vdots \\ C_m \end{pmatrix} \quad \mathbf{D} = \begin{pmatrix} D_1 \\ D_2 \\ \vdots \\ D_m \end{pmatrix} \tag{3.109}$$

that can be used for classification.

Each curve Γ_i has a *feature vector*

$$\mathbf{F}_i = \begin{pmatrix} F_{i1} \\ F_{i2} \\ \vdots \\ F_{in} \end{pmatrix} \tag{3.110}$$

associated with it, where $\mathbf{F}_i$ is formed from the four vectors of Eq. 3.109 resulting from Γ_i. The random vectors $\mathbf{F}_1, \mathbf{F}_2,\ldots, \mathbf{F}_N$ serve as descriptors of the images, so that F_{ij} is a *feature* of Γ_i. The association between a random curve and its feature vector defines a mapping by $\Gamma_i \rightarrow \mathbf{F}_i$. Feature-based classification can be characterized in the following manner: given an image realization, apply the feature mapping to the realization to compute a feature-vector realization $\mathbf{f}$ and, by comparing $\mathbf{f}$ with $\mathbf{F}_1, \mathbf{F}_2,\ldots, \mathbf{F}_N$, decide to which of the given N random image processes the realization belongs. One way to proceed is to define a distance function $D(\mathbf{f}, \mathbf{F}_i)$ and declare $\mathbf{f}$ to be a realization of $\mathbf{F}_k$ if $D(\mathbf{f}, \mathbf{F}_i)$ is minimized for $k = i$. We will address this kind of decision procedure in Section 4.11. For now one need only recognize that $D(\mathbf{f}, \mathbf{F}_i)$ depends on the separation between feature-vector distributions and that the greater the extent to which the multivariate distributions overlap, as measured by probability mass, the greater the probability of classification error. For Fourier descriptors, classification depends on the multivariate distributions of the generalized Fourier coefficients.

3.6. Transform Coding

This section treats coding of signals and images via matrix transforms. These transforms exploit interpixel correlation for the purpose of compression. The theory is developed for digital signals and is then adapted to images. Besides compression by means of a matrix transform, an overall transform-coding methodology has various stages on the source side: an image is blocked, a transform is applied, and then the transformed blocks must be compressed and quantized. The final source symbols are themselves coded via a method such as Huffman coding (Section 1.11). These operations are reversed on the receiver side to produce a reconstructed image. Application of the matrix transform achieves coding in the sense that the amount of data is compressed at the transform stage. The coding scheme is said to be *lossy* (as opposed to *lossless*), since after compression the transform cannot be inverted exactly; indeed, the purpose is to give up a small amount of information in return for a reduction in the number of bits that must be stored or transmitted.

A digital signal is compressed via a matrix transform by blocking it into strings of some fixed length n. Each piece of the signal is an n-vector and can be transformed by multiplying it by an $n \times n$ matrix. Consequently, digital signal compression involves matrix transformations on n-vectors. We assume that vectors are real.

3.6.1. Karhunen-Loeve Compression

Let $\mathbf{X} = (X_1, X_2,..., X_n)'$ be a zero-mean real random vector with covariance (autocorrelation) matrix $\mathbf{K_X}$. A zero mean is assumed for convenience and, should a vector not be zero-mean, it can be centralized by subtracting its mean vector. According to the Karhunen-Loeve theorem, if the vectors $\mathbf{u}_1, \mathbf{u}_2,..., \mathbf{u}_n$ are the orthonormalized eigenvectors of $\mathbf{K_X}$ corresponding to the eigenvalues $\lambda_1 \geq \lambda_2 \geq ... \geq \lambda_n$, then

$$\mathbf{X} = \sum_{i=1}^{n} Z_i \mathbf{u}_i \tag{3.111}$$

provides a canonical representation of $\mathbf{X}$, where $Z_1, Z_2,..., Z_n$ are uncorrelated and are the generalized Fourier coefficients of $\mathbf{X}$ with respect to $\mathbf{u}_1, \mathbf{u}_2,..., \mathbf{u}_n$,

$$Z_i = \mathbf{X}'\overline{\mathbf{u}}_i = \sum_{j=1}^{n} X_j \overline{u}_{ij} \tag{3.112}$$

where $\mathbf{u}_i = (u_{i1}, u_{i2},..., u_{in})'$. The data compression analysis of Section 3.2.4 applies: for $m < n$, data compression is achieved by approximating $\mathbf{X}$ by

$$\mathbf{X}_m = \sum_{i=1}^{m} Z_i \mathbf{u}_i \tag{3.113}$$

In the present matrix-vector setting, the mean-square error of Eq. 3.49 takes the form

$$\mathrm{E}[\mathbf{X}, \mathbf{X}_m] = \sum_{k=1}^{n} E[|X_k - X_{m,k}|^2] \tag{3.114}$$

where the components of $\mathbf{X}_m$ are $X_{m,1}, X_{m,2},..., X_{m,n}$. Application of Eq. 3.50 yields

$$\mathrm{E}[\mathbf{X}, \mathbf{X}_m] = \sum_{k=m+1}^{n} \lambda_k \tag{3.115}$$

As in the case of Eq. 3.50, since the eigenvalues are decreasing with increasing k, the error is minimized when keeping only m terms by keeping the first m terms.

For a matrix-transform perspective, let $\mathbf{Z}$ be the random vector whose components are $Z_1, Z_2,..., Z_n$. In terms of the matrix

$$\mathbf{U} = \begin{pmatrix} \mathbf{u}_1' \\ \mathbf{u}_2' \\ \vdots \\ \mathbf{u}_n' \end{pmatrix} = \begin{pmatrix} u_{11} & u_{12} & \cdots & u_{1n} \\ u_{21} & u_{22} & \vdots & u_{2n} \\ \vdots & \vdots & \ddots & \vdots \\ u_{n1} & u_{n2} & \cdots & u_{nn} \end{pmatrix} \tag{3.116}$$

whose rows are the orthonormal covariance-matrix eigenvectors, $\mathbf{Z} = \overline{\mathbf{U}}\mathbf{X}$. Since our intent in this section is to study coding by means of real matrix transforms, we assume that $\mathbf{U}$ is real; that is, we restrict ourselves to covariance matrices having real eigenvectors. This will avoid conjugation in the Karhunen-Loeve transform and give its equations a common form to the matrix transforms to which it will be compared. Under the assumption of real eigenvectors, Eq. 3.112 holds without conjugation and we have the Karhunen-Loeve transform (KLT) pair

$$\begin{aligned} \mathbf{Z} &= \mathbf{U}\mathbf{X} \\ \mathbf{X} &= \mathbf{U}'\mathbf{Z} \end{aligned} \tag{3.117}$$

Since the rows of **U** form an orthonormal system, **U** is a unitary matrix, meaning that its inverse is equal to its transpose, $\mathbf{U}^{-1} = \mathbf{U}'$. Since **U** is unitary, $|\det[\mathbf{U}]| = 1$. The covariance matrix of **Z** is diagonal, with the eigenvalues of $\mathbf{K_X}$ running down the diagonal,

$$\mathbf{K_Z} = \begin{pmatrix} \lambda_1 & 0 & \cdots & 0 \\ 0 & \lambda_2 & \cdots & 0 \\ 0 & 0 & \ddots & \vdots \\ 0 & 0 & \cdots & \lambda_n \end{pmatrix} \tag{3.118}$$

This output covariance matrix shows the decorrelation effect of the transform as well as the manner in which variances of decreasing size are strung down the diagonal.

To apply the transform for the purposes of compression, some number $r = n - m$ is chosen and only the first m terms of **Z** are retained; the rest are set to zero. In terms of the transform pair, instead of exactly reconstructing **X** by $\mathbf{X} = \mathbf{U}'\mathbf{Z}$, the inverse transform is applied to the truncated vector

$$\mathbf{Z}_m = (Z_1, Z_2, \ldots, Z_m, 0, \ldots, 0)' \tag{3.119}$$

Applying the inverse transform $\mathbf{U}'$ to $\mathbf{Z}_m$ yields the approximation of Eq. 3.113:

$$\mathbf{X}_m = \mathbf{U}'\mathbf{Z}_m \tag{3.120}$$

For each realization **x** of **X** we have the transform pair

$$\begin{aligned} \mathbf{z} &= \mathbf{U}\mathbf{x} \\ \mathbf{x} &= \mathbf{U}'\mathbf{z} \end{aligned} \tag{3.121}$$

If a realization **x** of **X** is observed and $\mathbf{z} = \mathbf{U}\mathbf{x}$ is computed, then **x** can be recovered by $\mathbf{x} = \mathbf{U}'\mathbf{z}$. For compression, **z** is truncated to $\mathbf{z}_m$, and, when desired, the compressed vector is decompressed to produce the reconstructed approximation of **x**,

$$\mathbf{x}_m = \mathbf{U}'\mathbf{z}_m \tag{3.122}$$

3.6.2. Transform Compression Using Arbitrary Orthonormal Systems

Application of the Karhunen-Loeve expansion has several drawbacks: the need for the covariance matrix or an estimate of it, the need to find the eigenvalues, and the need to construct the transform matrix **U** for the individual case at hand, rather than apply it in general across a wide class of images. If an orthonormal system can be found that works sufficiently well on the various random vectors of interest, then it is pragmatic to apply the matrix transform corresponding to this system.

Suppose $\mathbf{u}_1$, $\mathbf{u}_2$,..., $\mathbf{u}_n$ form an orthonormal system for the space of real deterministic n-vectors and matrix **U** is defined via Eq. 3.116. As with the Karhunen-Loeve transform, **U** is a unitary matrix. For any vector **x**, we have the transform pair of Eq. 3.121; however, to distinguish it from the Karhunen-Loeve transform, we let $\mathbf{y} = \mathbf{Ux}$, so that the inverse transform is given by $\mathbf{x} = \mathbf{U'y}$. Compression can be achieved via the compressed vector $\mathbf{y}_m$ and the decompression $\mathbf{x}_m = \mathbf{U'y}_m$. The worth of the compression scheme will depend on the orthonormal system chosen and the input random process, not some particular realization of the input process. For infinite orthonormal systems, the random process may or may not be represented by the infinite orthonormal expansion involving the generalized Fourier coefficients. Representation is determined by whether or not the infinite series converges in the mean-square to the random process. Here, however, we are concerned with finite linear combinations, so that convergence is not an issue. *Ipso facto*, we have the transform pair $\mathbf{Y} = \mathbf{UX}$ and $\mathbf{X} = \mathbf{U'Y}$.

To evaluate transform performance, let **X** be a zero-mean random vector with covariance matrix $\mathbf{K_X}$ and let $\mathbf{Y} = \mathbf{UX}$. **Y** has covariance matrix

$$\begin{aligned}\mathbf{K_Y} &= E[\mathbf{YY'}]\\ &= E[\mathbf{UX(UX)'}]\\ &= E[\mathbf{UXX'U'}]\\ &= \mathbf{U}E[\mathbf{XX'}]\mathbf{U'}\\ &= \mathbf{UK_XU'}\end{aligned} \tag{3.123}$$

Unlike the components of the vector **Z** from the Karhunen-Loeve transform, the components of **Y** are not likely to be uncorrelated.

Although $\mathbf{K_Y}$ is not diagonal, orthonormal transforms tend to decorrelate the input components so that terms off the diagonal, which represent covariances, tend to be small. Decorrelation diminishes redundancy between vector components. Moreover, if we define the *energy* of a vector to be the sum of the variances of its components, orthonormal transforms leave the energy invariant and tend to *pack* the energy of the coefficients into a small number of terms. We say "tend to" pack because the exact nature of the energy (variance) concentration depends on the random vector **X** and the matrix **U**. Together, decorrelation and energy packing tend to produce output matrices akin to the output matrix resulting from the Karhunen-Loeve expansion: small terms off the diagonal and most of the variance packed into the upper diagonal elements. Hence, we can expect the compression scheme $\mathbf{x} \to \mathbf{y} \to \mathbf{y}_m \to \mathbf{x}_m$ to work fairly well.

To provide measures of decorrelation and energy packing, let the covariance matrices of **X** and **Y** be given by

$$\mathbf{K}_{\mathbf{X}} = \begin{pmatrix} k_{11} & k_{12} & \cdots & k_{1n} \\ k_{21} & k_{22} & \cdots & k_{2n} \\ \vdots & \vdots & \ddots & \vdots \\ k_{n1} & k_{n2} & \cdots & k_{nn} \end{pmatrix} \tag{3.124}$$

$$\mathbf{K}_{\mathbf{Y}} = \begin{pmatrix} w_{11} & w_{12} & \cdots & w_{1n} \\ w_{21} & w_{22} & \cdots & w_{2n} \\ \vdots & \vdots & \ddots & \vdots \\ w_{n1} & w_{n2} & \cdots & w_{nn} \end{pmatrix} \tag{3.125}$$

Energy is conserved because

$$\sum_{i=1}^{n} k_{ii} = \sum_{i=1}^{n} w_{ii} \tag{3.126}$$

To quantify the degree to which the transform packs the energy into the lower-order coefficients, for $m = 1, 2, \ldots, n$, define the *energy-packing coefficient*

$$\varepsilon(m) = \frac{\sum_{i=1}^{m} w_{ii}}{\sum_{i=1}^{n} w_{ii}} \tag{3.127}$$

Note that $\varepsilon(m)$ depends on **U** and $\mathbf{K}_{\mathbf{X}}$. To quantify the degree to which **U** decorrelates the representation, define the *decorrelation coefficient*

$$\delta = 1 - \frac{\sum_{i=1}^{n} \sum_{j=1, j \neq i}^{n} |w_{ij}|}{\sum_{i=1}^{n} \sum_{j=1, i \neq j}^{n} |k_{ij}|} \tag{3.128}$$

δ also depends on $\mathbf{U}$ and $\mathbf{K_X}$. For the Karhunen-Loeve transform (for $\mathbf{X}$), δ = 1 and

$$\varepsilon(m) = \frac{\sum_{i=1}^{m} \lambda_i}{\sum_{i=1}^{n} \lambda_i} \tag{3.129}$$

The energy-packing and decorrelation coefficients are evaluated relative to a transform and an input covariance matrix. For comparing transforms, it is useful to compute the coefficients relative to some common model (or models) of the input covariance matrix. If we assume a WS stationary input random vector $\mathbf{X} = (X_1, X_2, \ldots, X_n)'$, then the input covariance matrix can be expressed in terms of the common variance σ^2 and the covariances σ_k^2 = $\mathrm{Cov}[X_i, X_{i+k}]$; specifically,

$$\mathbf{K_X} = \begin{pmatrix} \sigma^2 & \sigma_1^2 & \sigma_2^2 & \cdots & \sigma_{n-1}^2 \\ \sigma_1^2 & \sigma^2 & \sigma_1^2 & \cdots & \sigma_{n-2}^2 \\ \sigma_2^2 & \sigma_1^2 & \sigma^2 & \cdots & \sigma_{n-3}^2 \\ \vdots & \vdots & \vdots & \ddots & \vdots \\ \sigma_{n-1}^2 & \sigma_{n-2}^2 & \sigma_{n-3}^2 & \cdots & \sigma^2 \end{pmatrix} \tag{3.130}$$

If we assume a unit variance, then the components of the covariance matrix are the correlation coefficients,

$$\mathbf{K_X} = \begin{pmatrix} 1 & \rho_1 & \rho_2 & \cdots & \rho_{n-1} \\ \rho_1 & 1 & \rho_1 & \cdots & \rho_{n-2} \\ \rho_2 & \rho_1 & 1 & \cdots & \rho_{n-3} \\ \vdots & \vdots & \vdots & \ddots & \vdots \\ \rho_{n-1} & \rho_{n-2} & \rho_{n-3} & \cdots & 1 \end{pmatrix} \tag{3.131}$$

For slowly varying signals, a common model employed to evaluate transform coding efficiency is based on a Markov model (Section 5.1). For such a model, the distribution of any random variable conditioned on observations of preceding variables in a sequence is only dependent on the previous observation. In this case, the k-step correlation coefficient ρ_k can be expressed in terms of the single-step correlation coefficient $\rho_1 = \rho$. Specifically, $\rho_k = \rho^{|k|}$. For this model, the covariance matrix becomes

$$\mathbf{K}_\mathbf{X} = \begin{pmatrix} 1 & \rho & \rho^2 & \cdots & \rho^{n-1} \\ \rho & 1 & \rho & \cdots & \rho^{n-2} \\ \rho^2 & \rho & 1 & \cdots & \rho^{n-3} \\ \vdots & \vdots & \vdots & \ddots & \vdots \\ \rho^{n-1} & \rho^{n-2} & \rho^{n-3} & \cdots & 1 \end{pmatrix} \tag{3.132}$$

Note that correlation drops off exponentially relative to distance between points. We will employ this model to evaluate transform performance.

***Example* 3.6.** The eigenvalues of the covariance matrix $\mathbf{K}_\mathbf{X}$ of Eq. 3.132 are

$$\lambda_k = \frac{1-\rho^2}{1+\rho^2 - 2\rho\cos\omega_k}$$

$k = 1, 2, \ldots, n$, where $\omega_1, \omega_2, \ldots, \omega_n$ are the n solutions of the equation

$$\tan n\omega = \frac{-(1-\rho^2)\sin\omega}{(1+\rho^2)\cos\omega - 2\rho}$$

The resulting basis vectors $\mathbf{u}_1, \mathbf{u}_2, \ldots, \mathbf{u}_n$ are given by

$$u_{ki} = \sqrt{\frac{2}{n+\lambda_k}}\sin\left[\omega_k\left(i - \frac{n-1}{2}\right) + \frac{(k+1)\pi}{2}\right]$$

for $k, i = 1, 2, \ldots, n$. The following table gives the eigenvalues for $n = 8$ and $\rho = 0.91$:

k	1	2	3	4	5	6	7	8
λ_k	6.358	0.931	0.298	0.148	0.093	0.068	0.055	0.049

The eigenvalues appear on the diagonal of the output covariance matrix; all off-diagonal elements are 0. The decorrelation coefficient is $\delta = 1$ and the energy-packing coefficients for $m = 1, 2, \ldots, 8$ are given in the following table:

m	1	2	3	4	5	6	7	8
$\varepsilon(m)$	0.795	0.911	0.948	0.967	0.979	0.987	0.994	1.000

The decorrelation and energy-packing coefficients can be compared to the same coefficients for fixed-matrix transforms. ■

3.6.3. Walsh-Hadamard Transform

For $n = 2^k$, $k = 1, 2,\ldots$, the *Hadamard matrix* of size N can be defined recursively by

$$\mathbf{H}(n) = \frac{1}{\sqrt{2}}\begin{pmatrix} \mathbf{H}(n/2) & \mathbf{H}(n/2) \\ \mathbf{H}(n/2) & -\mathbf{H}(n/2) \end{pmatrix} \tag{3.133}$$

with recursion initiated by $\mathbf{H}(1) = (1)$. Entries in $\mathbf{H}(n)$ are either 1 or -1. For coding purposes, the rows are reordered so there is increasing oscillation between 1 and -1 down the rows. For $n = 1$, 2, and 3, $\mathbf{H}(2)$, $\mathbf{H}(4)$, and $\mathbf{H}(8)$ are taken to be

$$\mathbf{H}(2) = \frac{1}{\sqrt{2}}\begin{pmatrix} 1 & 1 \\ 1 & -1 \end{pmatrix} \tag{3.134}$$

$$\mathbf{H}(4) = \frac{1}{2}\begin{pmatrix} 1 & 1 & 1 & 1 \\ 1 & 1 & -1 & -1 \\ 1 & -1 & -1 & 1 \\ 1 & -1 & 1 & -1 \end{pmatrix} \tag{3.135}$$

$$\mathbf{H}(8) = \frac{1}{\sqrt{8}}\begin{pmatrix} 1 & 1 & 1 & 1 & 1 & 1 & 1 & 1 \\ 1 & 1 & 1 & 1 & -1 & -1 & -1 & -1 \\ 1 & 1 & -1 & -1 & -1 & -1 & 1 & 1 \\ 1 & 1 & -1 & -1 & 1 & 1 & -1 & -1 \\ 1 & -1 & -1 & 1 & 1 & -1 & -1 & 1 \\ 1 & -1 & -1 & 1 & -1 & 1 & 1 & -1 \\ 1 & -1 & 1 & -1 & -1 & 1 & -1 & 1 \\ 1 & -1 & 1 & -1 & 1 & -1 & 1 & -1 \end{pmatrix} \tag{3.136}$$

For a random vector $\mathbf{X}$ of dimension $n = 2^k$, the *Walsh-Hadamard transform* (*WHT*) is defined by $\mathbf{Y} = \mathbf{H}(n)\mathbf{X}$. The rows of $\mathbf{H}(n)$ are the basis signals for the WHT. For instance, if $\mathbf{y} = (y_1, y_2, y_3, y_4)' = \mathbf{H}(4)\mathbf{x}$, then

$$\mathbf{x} = \frac{1}{2}\left[y_1\begin{pmatrix}1\\1\\1\\1\end{pmatrix} + y_2\begin{pmatrix}1\\1\\-1\\-1\end{pmatrix} + y_3\begin{pmatrix}1\\-1\\-1\\1\end{pmatrix} + y_4\begin{pmatrix}1\\-1\\1\\-1\end{pmatrix}\right] \tag{3.137}$$

Since the Hadamard matrix **H** is symmetric, the transform pair is **Y** = **HX** and **X** = **HY**.

Example 3.7. For a 4-point domain, let $\mathbf{x}_1 = (8, 7, 9, 6)'$ and $\mathbf{x}_2 = (12, 10, 6, 4)'$. Then

$$\mathbf{y}_1 = \mathbf{H}\mathbf{x}_1 = \frac{1}{2}\begin{pmatrix}1&1&1&1\\1&1&-1&-1\\1&-1&-1&1\\1&-1&1&-1\end{pmatrix}\begin{pmatrix}8\\7\\9\\6\end{pmatrix} = \begin{pmatrix}15\\0\\-1\\2\end{pmatrix}$$

$$\mathbf{y}_2 = \mathbf{H}\mathbf{x}_2 = \frac{1}{2}\begin{pmatrix}1&1&1&1\\1&1&-1&-1\\1&-1&-1&1\\1&-1&1&-1\end{pmatrix}\begin{pmatrix}12\\10\\6\\4\end{pmatrix} = \begin{pmatrix}16\\6\\0\\2\end{pmatrix}$$

Compression is achieved by keeping only lower components of $\mathbf{y}_1$ and $\mathbf{y}_2$, that is, by using the vectors $\mathbf{y}_{1,m}$ and $\mathbf{y}_{2,m}$ obtained from $\mathbf{y}_1$ and $\mathbf{y}_2$, respectively, by setting to 0 all components above m. Decompression is achieved by taking the inverse transform. If we keep only a single component, then the decompressed vectors are

$$\mathbf{H}\mathbf{y}_{1,1} = \mathbf{H}\begin{pmatrix}15\\0\\0\\0\end{pmatrix} = \begin{pmatrix}7.5\\7.5\\7.5\\7.5\end{pmatrix}$$

$$\mathbf{H}\mathbf{y}_{2,1} = \mathbf{H}\begin{pmatrix}16\\0\\0\\0\end{pmatrix} = \begin{pmatrix}8\\8\\8\\8\end{pmatrix}$$

For $\mathbf{x}_1$, compression followed by decompression has yielded a constant vector whose entries are the average of the $\mathbf{x}_1$ entries and the approximation is fairly good. For $\mathbf{x}_2$, the decompression is not very good. It represents an averaging; however, the original vector $\mathbf{x}_2$ is decreasing fairly rapidly. If instead of keeping only the first component of $\mathbf{y}_2$, we keep the first two to obtain the compressed vector $\mathbf{y}_{2,2}$, then decompression yields

$$\mathbf{H}\mathbf{y}_{2,2} = \mathbf{H}\begin{pmatrix} 16 \\ 6 \\ 0 \\ 0 \end{pmatrix} = \begin{pmatrix} 11 \\ 11 \\ 5 \\ 5 \end{pmatrix}$$

If the gap between the second and third components of $\mathbf{x}_2$ represents a jump between fairly flat regions, then $\mathbf{H}\mathbf{y}_{2,2}$ provides a decent estimate of $\mathbf{y}_2$. ■

Transform performance is based on the overall random process to which it is applied. Even if overall performance is good, performance on any given signal can be poor. Strong energy packing in the lower components does not guarantee that a particular realization does not depend on a suppressed component. For the WHT in four dimensions, should a nonnull input vector $\mathbf{x}$ be a scalar multiple of the fourth row of $\mathbf{H}(4)$, then its transform vector $\mathbf{Hx}$ has only one nonzero component, the fourth, and therefore keeping only lower-order components results in the decompressed vector being null.

For the WHT, the first component of $\mathbf{Hx}$ contains constant signal content and succeeding coefficients contain increasing oscillatory content. This is analogous to frequency representation with trigonometric Fourier series. The WHT coefficients are called *sequency* coefficients. Eliminating higher-order coefficients of $\mathbf{HX}$ suppresses fine detail and preserves less varying signal content. Specific detail loss is a function of the WHT basis signals. When suppressing all components above a certain one, there is an implicit assumption that most (or most desirable) signal content is contained in the basis vectors of the lower components. To the extent that these represent low oscillation (frequency) content, truncation of the matrix-transformed vector acts as a low-pass filter.

***Example* 3.8.** If, for $n = 8$, we apply the WHT to a random vector possessing the covariance matrix of Eq. 3.132 with $\rho = 0.91$, we obtain the output covariance matrix

$$\mathbf{K}_Y = \begin{pmatrix} 6.344 & 0.000 & -0.261 & 0.000 & -0.066 & 0.000 & -0.131 & 0.000 \\ 0.000 & 0.796 & 0.000 & 0.261 & 0.000 & -0.091 & 0.000 & 0.131 \\ -0.261 & 0.000 & 0.275 & 0.000 & -0.001 & 0.000 & 0.091 & 0.000 \\ 0.000 & 0.261 & 0.000 & 0.226 & 0.000 & 0.001 & 0.000 & 0.066 \\ -0.066 & 0.000 & -0.001 & 0.000 & 0.094 & 0.000 & -0.001 & 0.000 \\ 0.000 & -0.091 & 0.000 & 0.001 & 0.000 & 0.094 & 0.000 & 0.001 \\ -0.131 & 0.000 & 0.091 & 0.000 & -0.001 & 0.000 & 0.091 & 0.000 \\ 0.000 & 0.131 & 0.000 & 0.066 & 0.000 & 0.001 & 0.000 & 0.080 \end{pmatrix}$$

The decorrelation coefficient is $\delta = 0.949$ and, depending on m, the energy-packing coefficient is given in the following table:

m	1	2	3	4	5	6	7	8
$\varepsilon(m)$	0.793	0.893	0.927	0.955	0.967	0.979	0.990	1.000

Performance is slightly worse than for the Karhunen-Loeve transform for the corresponding covariance matrix (Example 3.6). ■

3.6.4 Discrete Cosine Transform

The *discrete cosine transform* (*DCT*) is a discrete version of the cosine transform. The basis functions for the DCT are digital samplings of the cosine-transform basis functions. The DCT provides a frequency-based spectral representation of the signal. Compression suppresses high-frequency content, thereby producing a low-pass filter upon decoding.

The $n \times n$ DCT is defined by the matrix $\mathbf{C}$ with the components

$$c_{km} = c_0 \sqrt{\frac{2}{N}} \cos \frac{m(k+1/2)\pi}{N} \tag{3.138}$$

for k, $m = 0, 1, 2, \ldots, N-1$, where $c_0 = 2^{-1/2}$ if $m = 0$ and $c_0 = 1$ if $m \neq 0$. There are two forms of the DCT, *odd* and *even*, having to do with the way a signal block is periodically extended. Equation 3.138 gives the even form of the DCT, which for reasons having to do with block boundaries generally provides superior performance. The DCT basis signals for $n = 8$ are depicted in Fig. 3.1.

For slowly varying signals, the DCT tends to provide a reasonable approximation to the Karhunen-Loeve transform. Since the DCT is not model dependent, it does not require a good estimate of the covariance

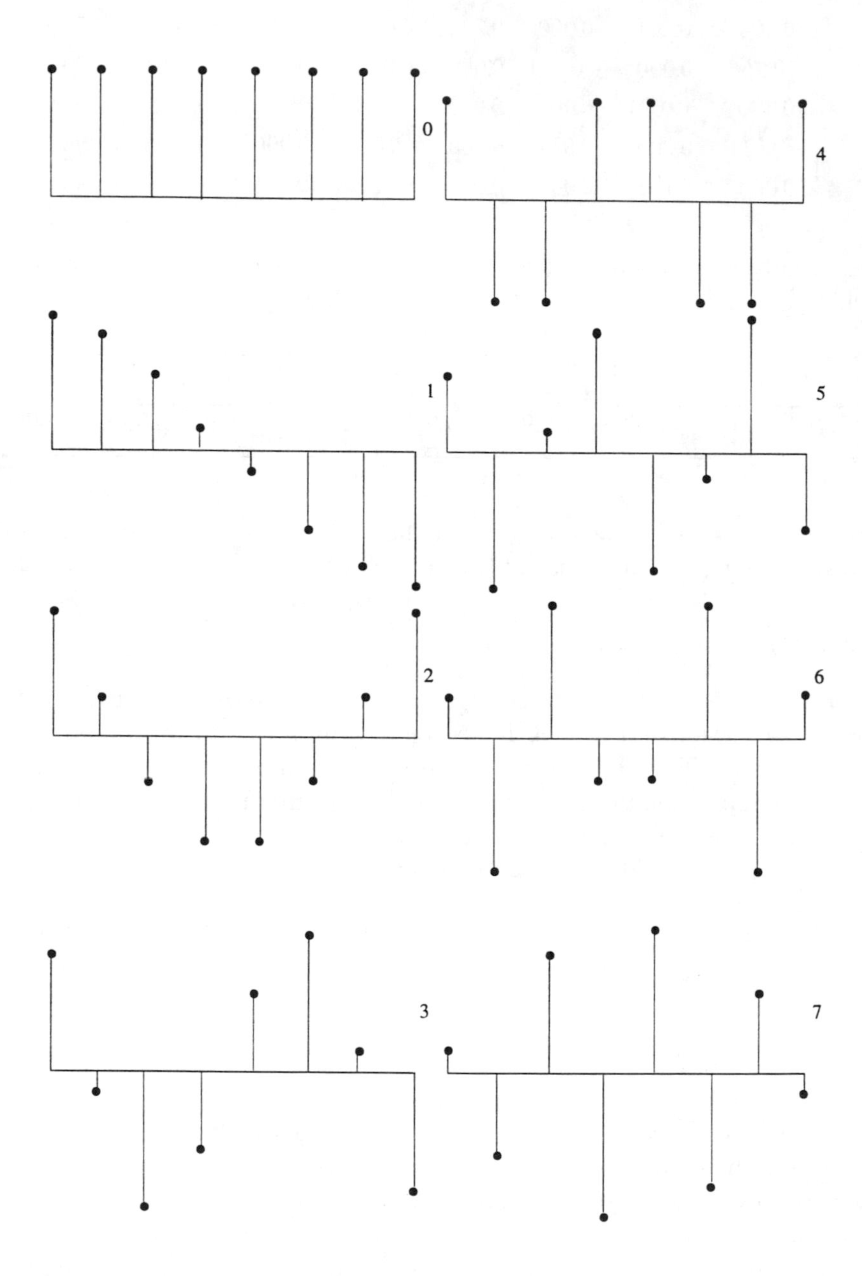

Figure 3.1 DCT basis signals for $n = 8$.

matrix or computation of covariance-matrix eigenvalues. Since DCT performance is close to the Karhunen-Loeve transform over a wide range of practical images, it is an excellent candidate for many applications.

***Example* 3.9.** If, for $n = 8$, we apply the DCT to a random vector possessing the covariance matrix of Eq. 3.132 with $\rho = 0.91$, we obtain the output covariance matrix

$$\mathbf{K_Y} = \begin{pmatrix} 6.344 & 0.000 & -0.291 & 0.000 & -0.066 & 0.000 & -0.021 & 0.000 \\ 0.000 & 0.930 & 0.000 & -0.027 & 0.000 & -0.008 & 0.000 & -0.002 \\ -0.291 & 0.000 & 0.312 & 0.000 & -0.001 & 0.000 & 0.000 & 0.000 \\ 0.000 & -0.027 & 0.000 & 0.149 & 0.000 & -0.001 & 0.000 & 0.000 \\ -0.066 & 0.000 & -0.001 & 0.000 & 0.094 & 0.000 & 0.000 & 0.000 \\ 0.000 & -0.008 & 0.000 & -0.001 & 0.000 & 0.068 & 0.000 & 0.000 \\ -0.021 & 0.000 & 0.000 & 0.000 & 0.000 & 0.000 & 0.055 & 0.000 \\ 0.000 & -0.002 & 0.000 & 0.000 & 0.000 & 0.000 & 0.000 & 0.049 \end{pmatrix}$$

The decorrelation coefficient is $\delta = 0.981$ and, depending on m, the energy-packing coefficient is given in the following table:

m	1	2	3	4	5	6	7	8
$\varepsilon(m)$	0.793	0.909	0.948	0.967	0.979	0.987	0.994	1.000

Performance is improved compared with the WHT for the same input covariance, and is very close to the KLT. This is not surprising if we compare the basis signals for the DCT with the basis signals for the KLT corresponding to the present input covariance. The DCT matrix and the KLT matrix for $n = 8$ are given respectively by

$$\mathbf{C}(8) = \begin{pmatrix} 0.354 & 0.354 & 0.354 & 0.354 & 0.354 & 0.354 & 0.354 & 0.354 \\ 0.490 & 0.416 & 0.278 & 0.098 & -0.098 & -0.278 & -0.416 & -0.490 \\ 0.462 & 0.191 & -0.191 & -0.462 & -0.462 & -0.191 & 0.191 & 0.462 \\ 0.416 & -0.098 & -0.490 & -0.278 & 0.278 & 0.490 & 0.098 & -0.416 \\ 0.354 & -0.354 & -0.354 & 0.354 & 0.354 & -0.354 & -0.354 & 0.354 \\ 0.278 & -0.490 & 0.098 & 0.416 & -0.416 & -0.098 & 0.490 & -0.278 \\ 0.191 & -0.462 & 0.462 & -0.191 & -0.191 & 0.462 & -0.462 & 0.191 \\ 0.098 & -0.278 & 0.416 & -0.490 & 0.490 & -0.416 & 0.278 & -0.098 \end{pmatrix}$$

$$
\mathbf{KLT} = \begin{pmatrix}
0.327 & 0.349 & 0.365 & 0.372 & 0.372 & 0.365 & 0.349 & 0.327 \\
0.473 & 0.424 & 0.293 & 0.104 & -0.104 & -0.293 & -0.424 & -0.473 \\
0.469 & 0.218 & -0.169 & -0.451 & -0.451 & -0.169 & 0.218 & 0.469 \\
0.428 & -0.076 & -0.483 & -0.279 & 0.279 & 0.483 & 0.076 & -0.428 \\
0.366 & -0.341 & -0.358 & 0.350 & 0.350 & -0.358 & -0.341 & 0.366 \\
0.288 & -0.486 & 0.091 & 0.415 & -0.415 & -0.091 & 0.486 & -0.288 \\
0.199 & -0.463 & 0.459 & -0.190 & -0.190 & 0.459 & -0.463 & 0.199 \\
0.101 & -0.279 & 0.415 & -0.489 & 0.489 & -0.415 & 0.279 & -0.101
\end{pmatrix} \blacksquare
$$

***Example* 3.10.** The random walk of Section 2.7.2 has covariance function $K_X(n, n') = \min(n, n')$. The covariance matrix for the random vector $\mathbf{X}$ over the first N time points is

$$
\mathbf{K} = \begin{pmatrix}
1 & 1 & 1 & \cdots & 1 \\
1 & 2 & 2 & \cdots & 2 \\
1 & 2 & 3 & \cdots & 3 \\
\vdots & \vdots & \vdots & \ddots & \vdots \\
1 & 2 & 3 & \cdots & N
\end{pmatrix}
$$

The eigenvalues of $\mathbf{K}$ have been found numerically for $N = 32$ and the KLT matrix determined. Using simulation, 512 realizations of the random walk have been generated and compressed via the KLT using values $m = 1, 2,\ldots, 32$. The same 512 realizations have been compressed using the WHT and DCT. Figures 3.2 through 3.4 show three realizations and the corresponding compressions for the three transforms for $m = 2$, 4, and 7, respectively. In the figures, parts a through c compare the WHT to the KLT and parts d through f compare the DCT to the KLT. Sum-of-squares errors (as functions of m) for the three realizations are shown in Fig. 3.5. To compare the overall performance of the three transforms, we employ the estimator

$$
\hat{\mathrm{E}}[\mathbf{X}, \mathbf{X}_m] = \frac{1}{512} \sum_{i=1}^{512} \sum_{k=1}^{32} | X_{i,k} - X_{i,m,k} |^2
$$

for the error of Eq. 3.114, where $X_{i,k}$ and $X_{i,m,k}$ are the kth components of the ith signal and of the partial inverse of the ith signal for m, respectively. Error curves are shown in Fig. 3.6, which includes the error curve when using an estimate of $\mathbf{K}$ based on 10 realizations (to show the problem with using covariance estimates without a sufficient sample size). Using 50 realizations to estimate $\mathbf{K}$, performance is close to that when using $\mathbf{K}$ itself. ■

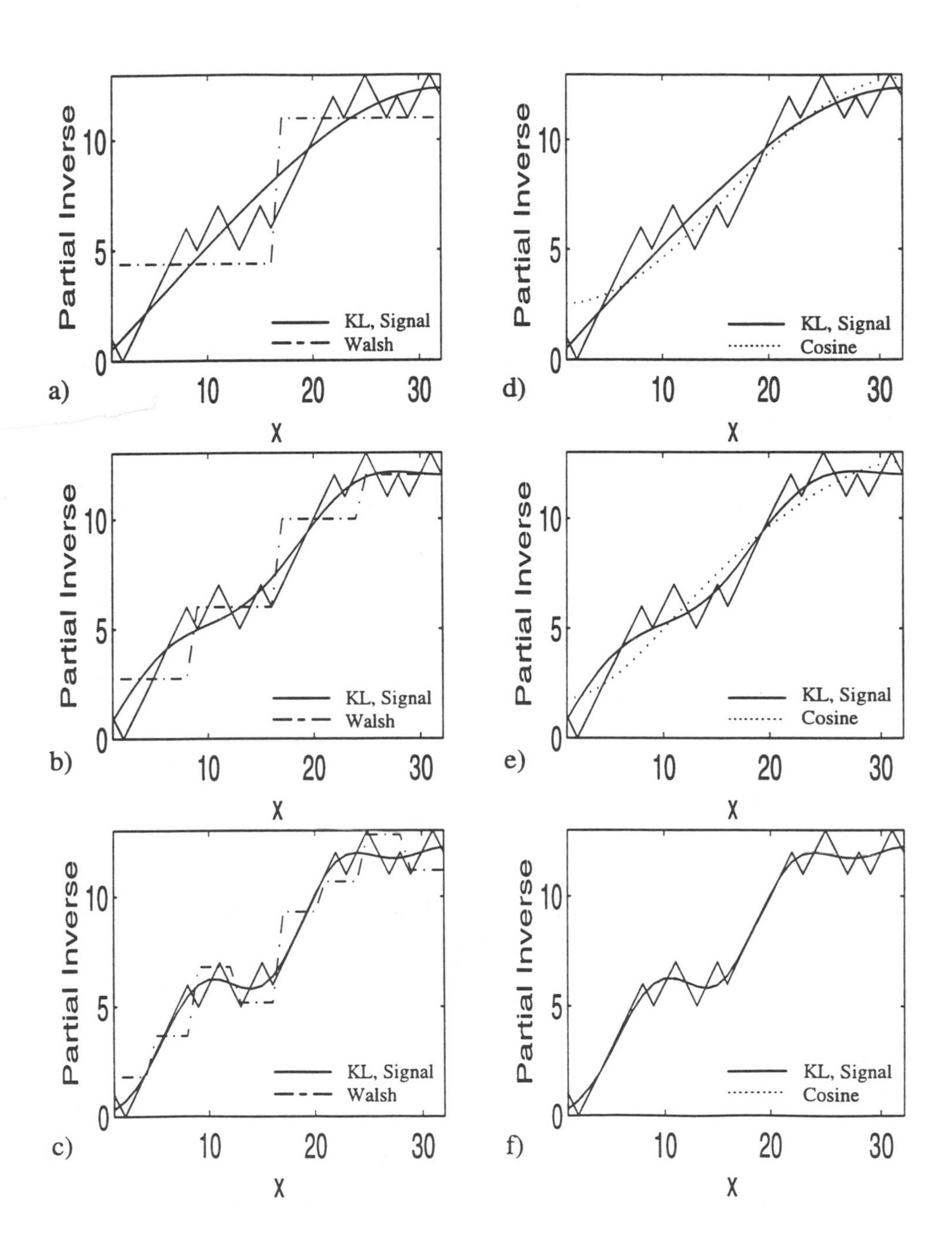

Figure 3.2 Realization of a random walk and compressed signals for m = 2, 4, 7. Parts a, b, c compare WHT to KLT; parts d, e, f compare DCT to KLT.

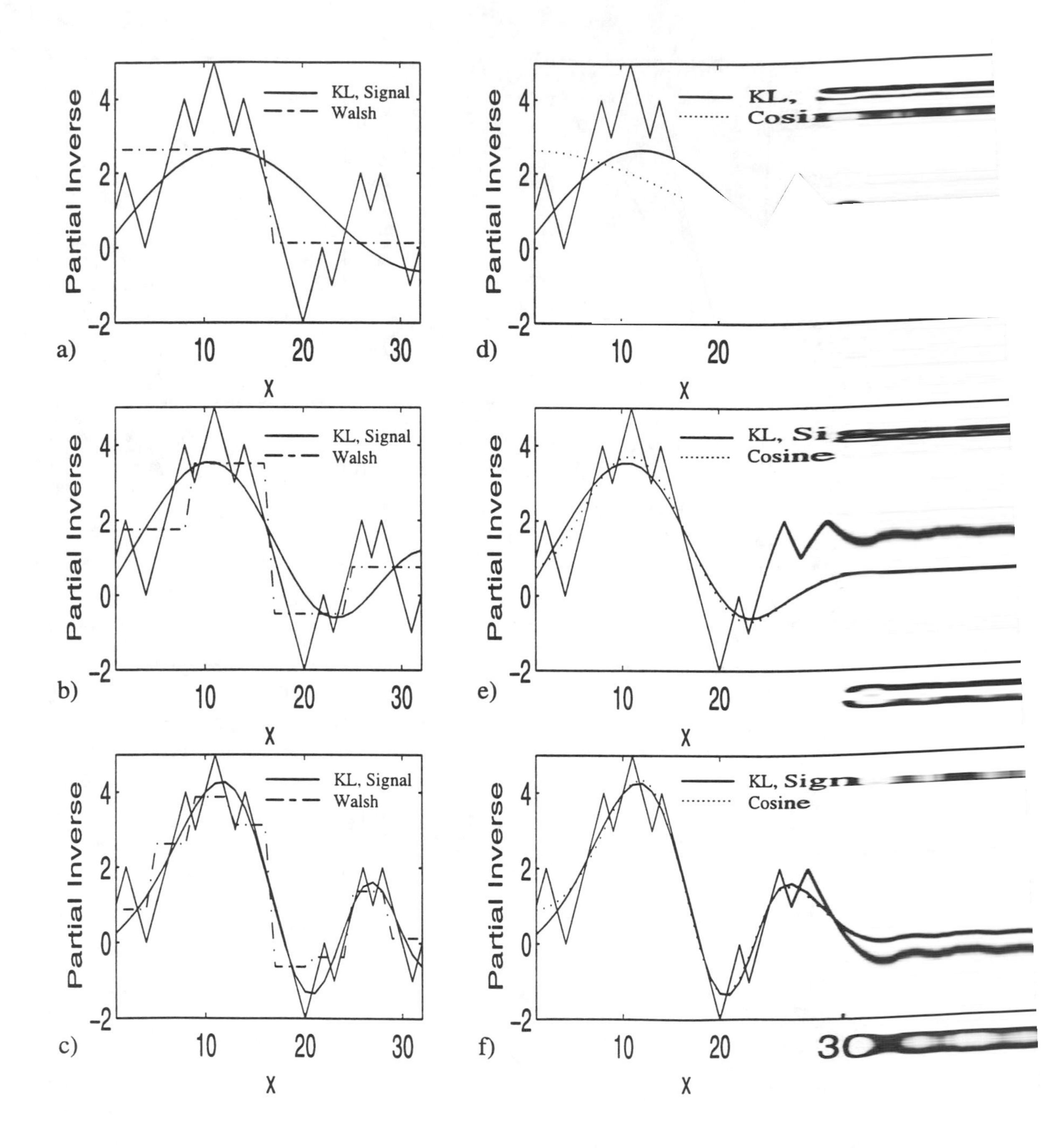

Figure 3.3 Realization of a random walk and compressed signals for $m = 2, 4, 7$. Parts a, b, c compare WHT to KLT; parts d, e, f compare DCT to KLT.

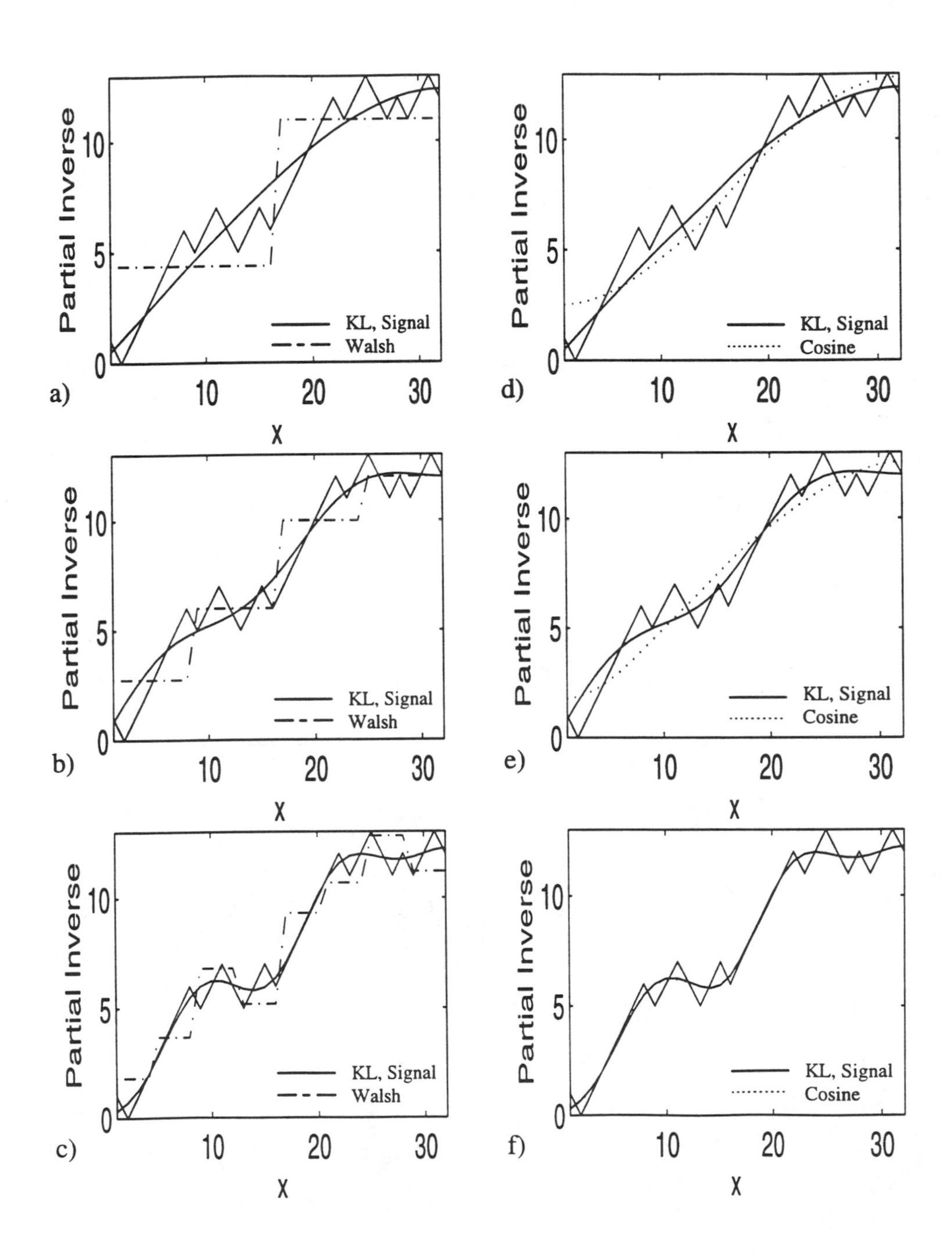

Figure 3.2 Realization of a random walk and compressed signals for m = 2, 4, 7. Parts a, b, c compare WHT to KLT; parts d, e, f compare DCT to KLT.

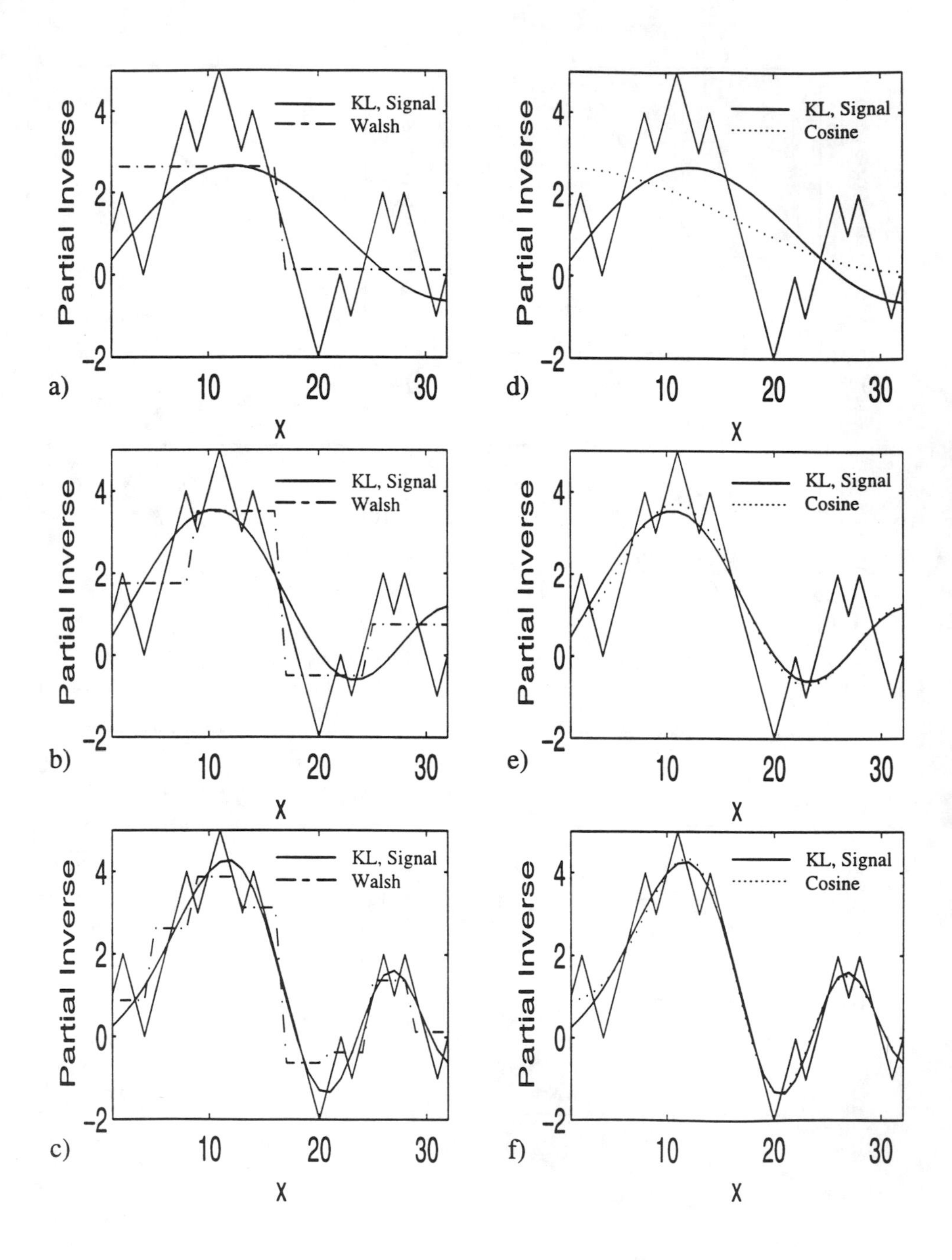

Figure 3.3 Realization of a random walk and compressed signals for $m = 2, 4, 7$. Parts a, b, c compare WHT to KLT; parts d, e, f compare DCT to KLT.

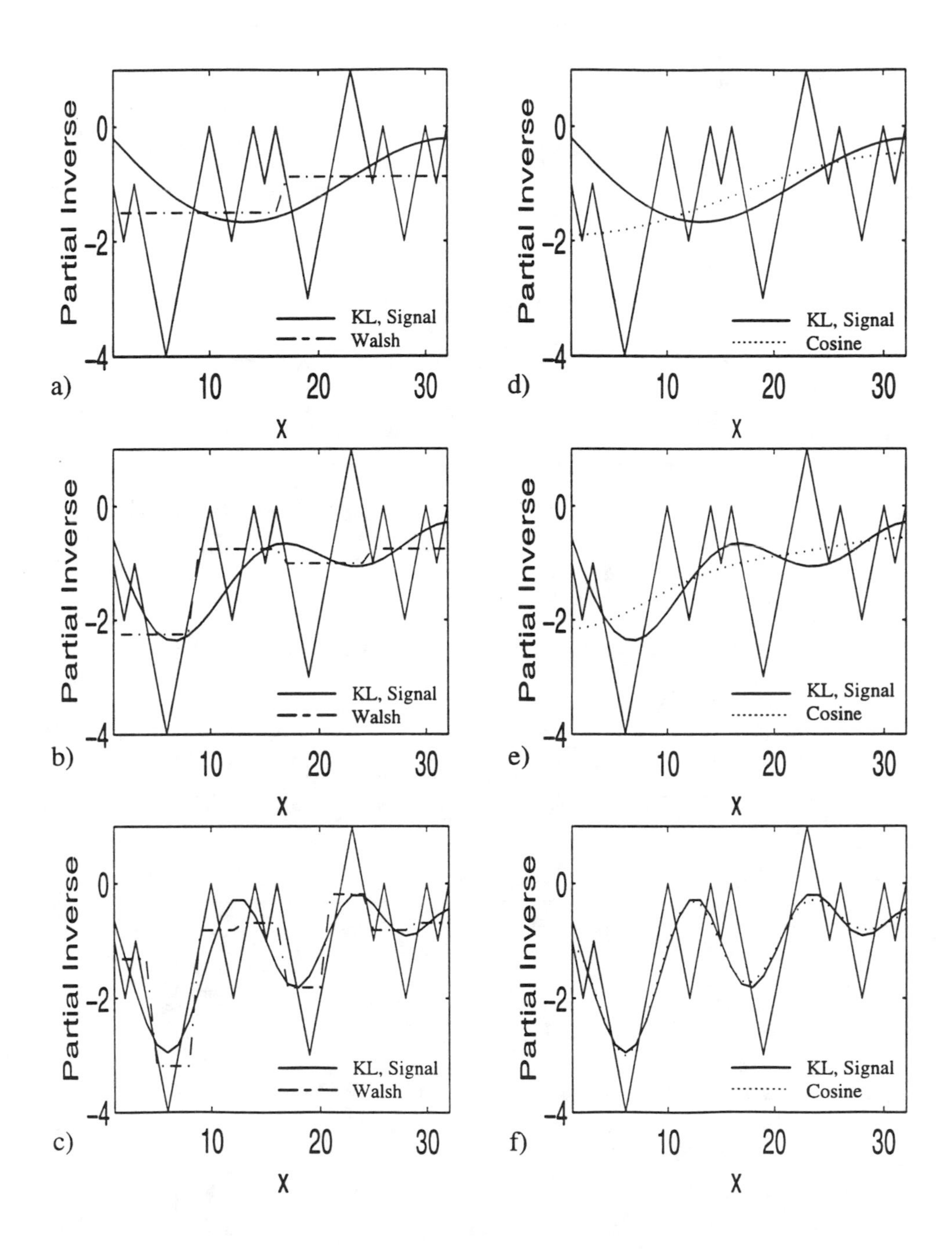

Figure 3.4 Realization of a random walk and compressed signals for m = 2, 4, 7. Parts a, b, c compare WHT to KLT; parts d, e, f compare DCT to KLT.

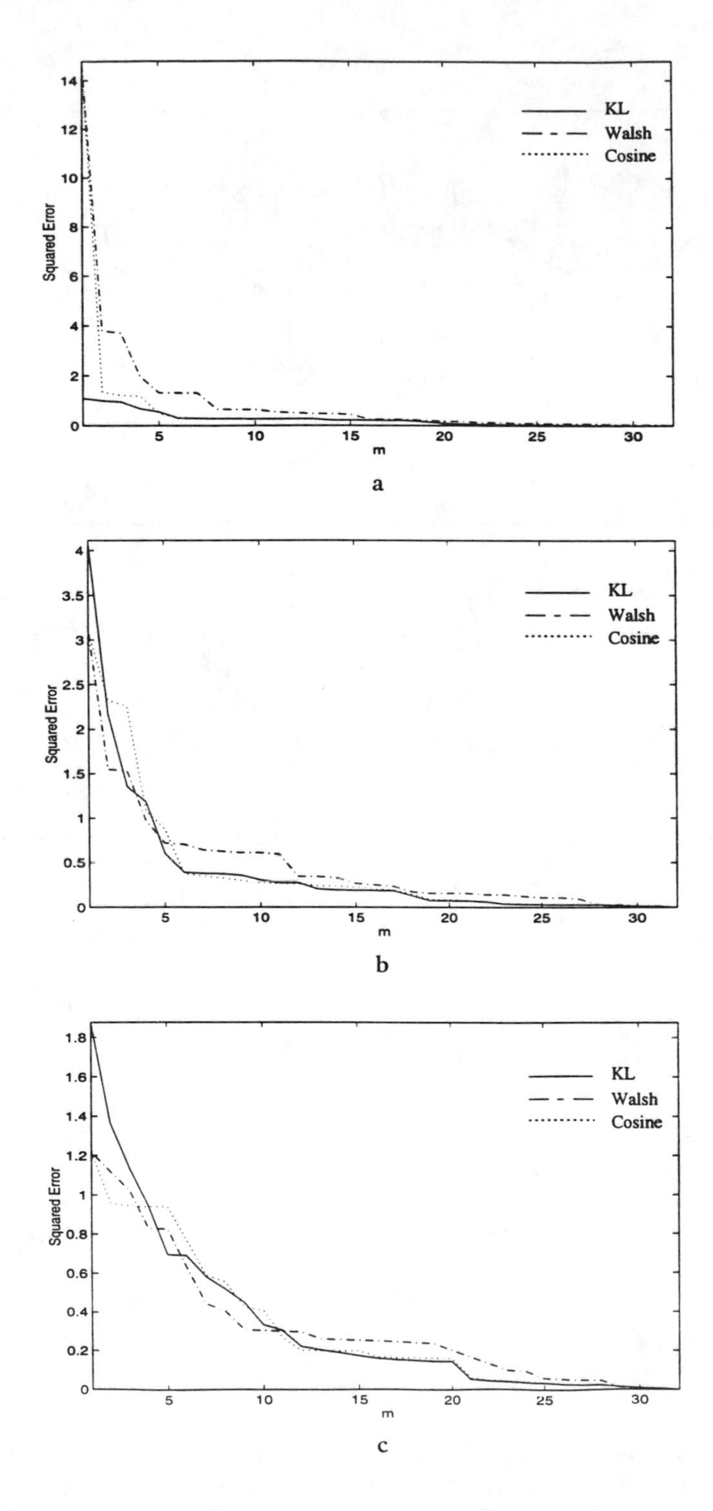

Figure 3.5 Errors for the compressions of Examples 3.2 through 3.4.

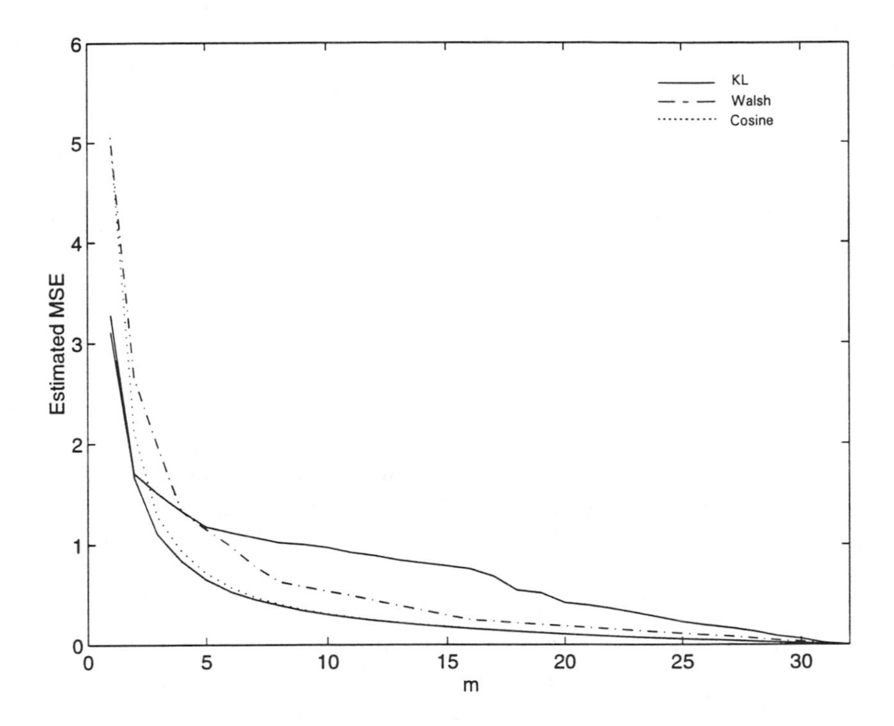

Figure 3.6 Estimated MSE (512 realizations) for compression of the random walk using WHT, DCT, KLT employing exact covariance matrix, and KLT employing covariance-matrix estimate based on 10 realizations.

As defined in Eq. 3.127, the energy-packing coefficient provides a measure of compression goodness based on keeping lower components, which is consistent with the KLT and often provides good performance for the WHT and DCT. If we wish to keep *m* components of a transformed vector based on minimizing the sum of the their variances, it may be best to keep some other set of *m* components. Moreover, the best components to keep may depend on some other criterion besides variance minimization.

3.6.5. Transform Coding for Digital Images

Matrix-transform coding is adapted to images by blocking an image into $n \times n$ blocks and applying the two-dimensional version of the transform to each block. To obtain the two-dimensional transform corresponding to an $n \times n$ unitary matrix **U** with orthonormal rows, let

$$\mathbf{B} = \begin{pmatrix} \mathbf{v}_1 & \mathbf{v}_2 & \cdots & \mathbf{v}_n \end{pmatrix} \tag{3.139}$$

be an $n \times n$ image block with columns $\mathbf{v}_1, \mathbf{v}_2, \ldots, \mathbf{v}_n$. Multiplication of $\mathbf{B}$ by $\mathbf{U}$ has the effect of compressing the data in each column vector into its lower components because

$$\mathbf{UB} = \begin{pmatrix} \mathbf{Uv}_1 & \mathbf{Uv}_2 & \cdots & \mathbf{Uv}_n \end{pmatrix} \tag{3.140}$$

Postmultiplication by $\mathbf{U}'$ then does the same on the rows of $\mathbf{UB}$. The transform pair

$$\begin{aligned} \mathbf{A} &= \mathbf{UBU}' \\ \mathbf{B} &= \mathbf{U}'\mathbf{AU} \end{aligned} \tag{3.141}$$

provides a compression/decompression scheme. Compression is achieved by keeping only the upper-left components of the transform block $\mathbf{A}$. If m rows and columns of $\mathbf{A}$ are kept and all other components are set to 0 to form a compressed transform block $\mathbf{A}_m$, then coding is accomplished by the sequence $\mathbf{B} \rightarrow \mathbf{A} \rightarrow \mathbf{A}_m \rightarrow \mathbf{B}_m$. The two-dimensional transform methodology gives a representation in terms of orthonormal basis images: there exists a set of basis images $\mathbf{I}_{ij}$, $i, j = 1, 2, \ldots, n$, such that the components of $\mathbf{A}$ provide the coefficients for the expression of $\mathbf{B}$ as a linear combination of the basis images.

For the WHT, $\mathbf{H} = \mathbf{H}'$, so the forward transform is $\mathbf{A} = \mathbf{HBH}$. Inversion is by $\mathbf{B} = \mathbf{HAH}$. The 4×4 WHT basis images are depicted in Fig. 3.7, where white and black represent $+1$ and -1, respectively. Applying the DCT is similar except that the transpose must be used. The 8×8 DCT basis images are shown in Fig. 3.8. For slowly varying images, energy tends to be packed into the upper left corner of $\mathbf{A}$, so that keeping only some set of upper-left components can provide satisfactory compression.

***Example* 3.11.** The compression achieved by transform coding depends on the number of components in a transform block turned to zero. So far we have considered *zonal coding*: a zone of the block is kept and the remainder suppressed (set to zero). *Threshold coding* uses an adaptive procedure for component selection. One way is to pass any component of any transform block exceeding a fixed global threshold. The heuristic is that large components represent significant information. Compression differs from block to block since blocks have different numbers of components exceeding the threshold. Rather than global thresholding, in componentwise thresholding each component in the transform block is compared to its own threshold. This can lead to better images upon decoding by allowing flexibility in the coding protocol. Compression varies from block to block and the decoder must be supplied the locations of the passed components.

Figure 3.7 4 × 4 WHT basis images.

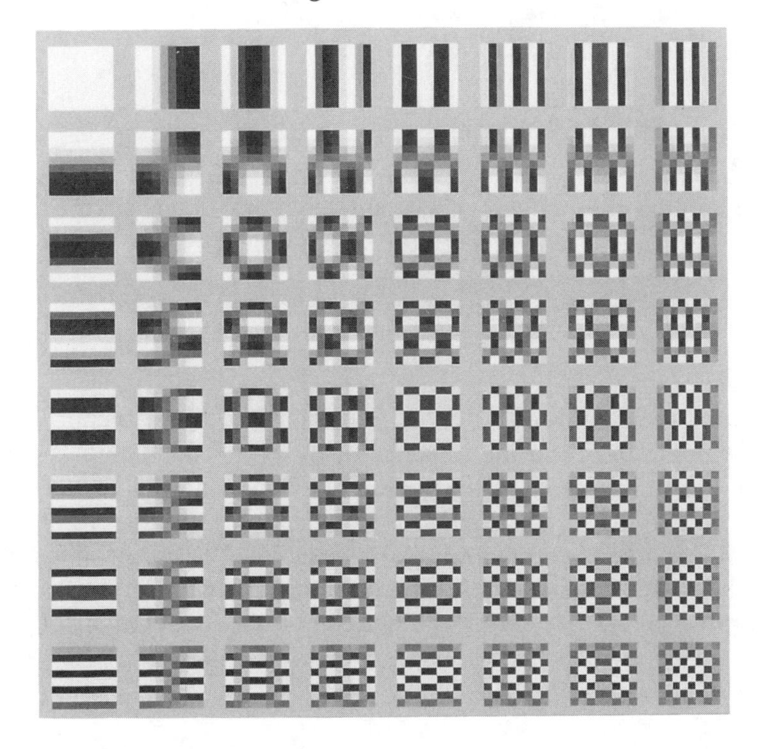

Figure 3.8 8 × 8 DCT basis images.

As of this writing, the dominant standard for threshold coding is the standard of the *Joint Photographic Experts Group (JPEG)*. JPEG has three coding algorithms. The one normally called "JPEG" (known as the *sequential baseline system*) involves 8×8 blocking of the image, DCT computation, and coding/quantization. Before application of the DCT to a block, its gray values are shifted down by subtracting $M/2$ from each pixel value (M being the number of gray levels). Subsequent to the DCT, compression of the transform block and quantization are accomplished by a threshold coding method that incorporates both. The resulting compressed block is then coded by a modified Huffman code.

For a block **B**, pixelwise threshold coding and quantization of a transform block

$$\underline{\mathbf{B}} = \begin{pmatrix} \omega_{1,1} & \omega_{1,2} & \cdots & \omega_{1,8} \\ \omega_{2,1} & \omega_{2,2} & \cdots & \omega_{2,8} \\ \vdots & \vdots & \ddots & \vdots \\ \omega_{8,1} & \omega_{8,2} & \cdots & \omega_{8,8} \end{pmatrix}$$

are accomplished jointly. A *quantization matrix*

$$\mathbf{Q} = \begin{pmatrix} q_{1,1} & q_{1,2} & \cdots & q_{1,8} \\ q_{2,1} & q_{2,2} & \cdots & q_{2,8} \\ \vdots & \vdots & \ddots & \vdots \\ q_{8,1} & q_{8,2} & \cdots & q_{8,8} \end{pmatrix}$$

is defined and the components of a quantized block $\underline{\mathbf{B}}_Q$ are defined by

$$\omega'_{i,j} = quan\left[\frac{\omega_{i,j}}{q_{i,j}}\right]$$

where *quan* is the round-off function. Threshold coding is achieved because, for small values of $\omega_{i,j}$ and large values of $q_{i,j}$, $\omega'_{i,j} = 0$. There is a standard JPEG quantization matrix stored in the JPEG file; however, a user has the option of specifying a quantization matrix more suitable to a particular application. Prior to applying the inverse DCT, $\underline{\mathbf{B}}_Q$ must be pixelwise multiplied by **Q**. Excluding Huffman coding following quantization, DCT coding to inverse DCT decoding is described by the following block sequence:

$$\mathbf{B} \rightarrow \underline{\mathbf{B}} \rightarrow \underline{\mathbf{B}}_Q \rightarrow \underline{\mathbf{B}}_{\bullet} \rightarrow \mathbf{B}_{\approx}$$

where $\underline{\mathbf{B}}$ is the DCT of **B**, $\underline{\mathbf{B}}_Q$ is the quantized compression of $\underline{\mathbf{B}}$, $\underline{\mathbf{B}}_\bullet$ is the pixelwise product of $\underline{\mathbf{B}}_Q$ and **Q**, and $\mathbf{B}_\approx$ is the decoded approximation of **B**. Figure 3.9 shows a 512 × 512, 8-bit/pixel image and a 0.5-bit/pixel JPEG-compressed image. Figure 3.10 shows blow-ups of the eye for the two images of Fig. 3.9. Notice the effects of compression along the edges, especially the lower edge where the original has intricate detail. ■

3.6.6. Optimality of the Karhunen-Loeve Transform

The decreasing order of the eigenvalues and full decorrelation of the coefficients make Karhunen-Loeve-based digital signal compression very appealing. However, might there be some other orthonormal set of vectors besides the eigenvectors of the covariance matrix (the Karhunen-Loeve basis set) that yields less error when employing an equivalent amount of compression by suppressing higher-order components? To wit, is the Karhunen-Loeve transform optimal when it is used to compress a random vector for which it has been derived? Indeed, as we will now discuss, it is optimal relative to the mean-square error expression of Eq. 3.114: if any other orthonormal matrix transform is applied to **X** and $\mathbf{X}_m$ is reconstructed by the inverse transform, then E[**X**, $\mathbf{X}_m$] is minimal for the Karhunen-Loeve transform.

The error representation of Eq. 3.50 was derived from the componentwise MS error expression of Eq. 3.46. In deriving the componentwise error form we made use of the uncorrelatedness of the Karhunen-Loeve coefficients. We need to proceed slightly differently for an arbitrary orthonormal matrix transform. So as not to needlessly restrict the demonstration, we do it for complex basis vectors. Letting the components of $\mathbf{X}_m$ be $X_{m,1}, X_{m,2},\ldots, X_{m,n}$, for the matrix transform $\mathbf{Y} = \overline{\mathbf{U}}\mathbf{X}$, we obtain

$$\mathrm{E}[\mathbf{X},\mathbf{X}_m] = \sum_{j=1}^{n} E[|X_j - X_{m,j}|^2]$$

$$= \sum_{j=1}^{n} E\left[\left|\sum_{k=m+1}^{n} Y_k u_{kj}\right|^2\right]$$

$$= \sum_{j=1}^{n} E\left[\left(\sum_{k=m+1}^{n} Y_k u_{kj}\right)\overline{\left(\sum_{l=m+1}^{n} Y_l u_{lj}\right)}\right]$$

$$= \sum_{j=1}^{n}\sum_{k=m+1}^{n}\sum_{l=m+1}^{n} E[Y_k \overline{Y_l}]u_{kj}\overline{u}_{lj}$$

Figure 3.9 Original and JPEG DCT compressed images.

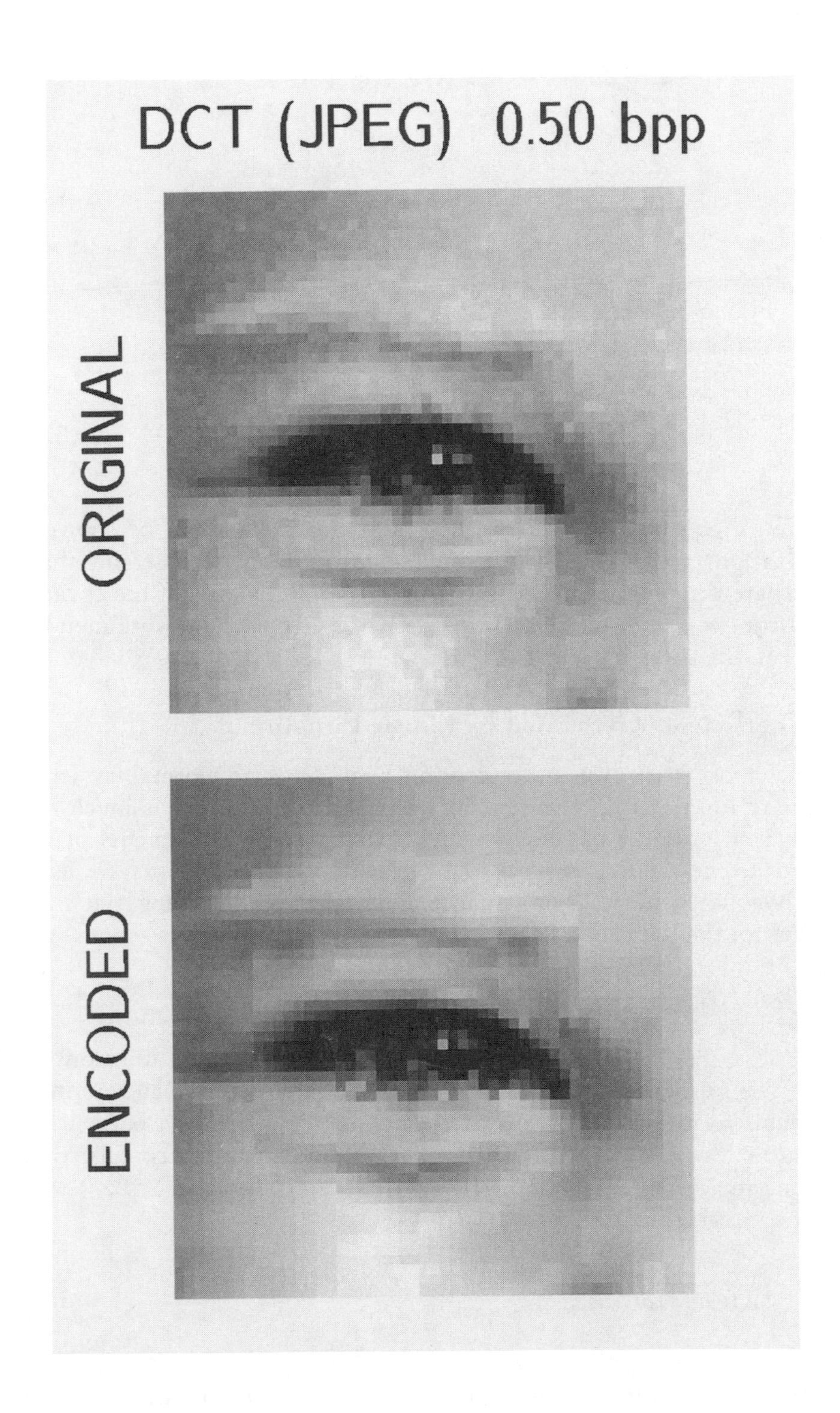

Figure 3.10 Enlarged portions of original and compressed images.

$$= \sum_{k=m+1}^{n} \sum_{l=m+1}^{n} E[Y_k \overline{Y}_l] \sum_{j=1}^{n} u_{kj} \overline{u}_{lj}$$

$$= \sum_{k=m+1}^{n} \sum_{l=m+1}^{n} E[Y_k \overline{Y}_l](\mathbf{u}_k{}' \overline{\mathbf{u}}_l)$$

$$= \sum_{k=m+1}^{n} E[Y_k \overline{Y}_k]$$

$$= \sum_{k=m+1}^{n} \text{Var}[Y_k] \tag{3.142}$$

the seventh equality following from orthonormality. It can be shown that minimization of the preceding variance sum under the constraint that $\mathbf{u}_1$, $\mathbf{u}_2,\ldots, \mathbf{u}_n$ are orthonormal yields the sum of the eigenvalues of the covariance matrix from $m + 1$ to n. Since this sum is the error for the Karhunen-Loeve transform, the latter is optimal.

3.7. Coefficients Generated by Linear Functionals

For the most part, we have concentrated on orthonormal systems of coordinate functions. In this section we see that the theory is much richer. Although an orthonormal coordinate-function system has benefits, it is not required for a canonical expansion. There can be benefits to using a nonorthogonal system. For instance, one can avoid solving the integral equation for the Karhunen-Loeve expansion.

3.7.1. Coefficients from Integral Functionals

According to Eq. 3.33, for an orthonormal set of coordinate functions $u_1(t)$, $u_2(t),\ldots,$ the expansion coefficients Z_1, $Z_2,\ldots$ are determined by integral functionals relative to the coordinate functions. Rather than begin with an orthonormal system, suppose $a_1(t)$, $a_2(t),\ldots$ form an arbitrary collection of functions in $L^2(T)$. Assuming $X(t)$ has mean zero, for $k = 1, 2,\ldots,$ we can form the integral functionals

$$Z_k = \int_T X(t)\overline{a_k(t)}\, dt \tag{3.143}$$

Theorem 2.2 holds for complex random functions, the only difference being that the factor $g(t', s')$ in Eq. 2.48 is conjugated. Hence, the integral in Eq.

3.143 holds in the MS sense if and only if the corresponding integrals in Eqs. 2.47 and 2.48 exist. Since $\mu_X(t) = 0$, $E[Z_k] = 0$. Thus, we only need be concerned with the integral of Eq. 2.48, which here is

$$E[Z_k \overline{Z}_j] = E\left[\int_T X(t)\overline{a_k(t)}\,dt \overline{\int_T X(s)\overline{a_j(s)}\,ds}\right]$$

$$= E\left[\int_T\int_T \overline{a_k(t)}a_j(s)X(t)\overline{X(s)}\,dtds\right]$$

$$= \int_T\int_T \overline{a_k(t)}a_j(s)K_X(t,s)\,dtds \tag{3.144}$$

We assume that $K_X(t, s) \in L^2(T \times T)$, meaning that Eq. 3.21 holds without a weight function. As in Theorem 3.2, one can employ a weight function to enlarge the class of square-integrable functions. By the Cauchy-Schwarz inequality,

$$\int_T\int_T |\overline{a_k(t)}a_j(s)K_X(t,s)|\,dtds \le \left(\int_T\int_T |\overline{a_k(t)}a_j(s)|^2\,dtds\right)^{1/2}\left(\int_T\int_T |K_X(t,s)|^2\,dtds\right)^{1/2} \tag{3.145}$$

$$= \left(\int_T |a_k(t)|^2\,dt\right)^{1/2}\left(\int_T |a_j(s)|^2\,ds\right)^{1/2}\left(\int_T\int_T |K_X(t,s)|^2\,dtds\right)^{1/2}$$

which is finite. Thus, $a_k(t)a_j(t)K_X(t, s)$ is absolutely integrable, which ensures the existence of the integral in Eq. 3.144 and MS integrability in Eq. 3.143.

$Z_1, Z_2, \ldots$ are uncorrelated if and only if

$$\int_T\int_T \overline{a_k(t)}a_j(s)K_X(t,s)\,dtds = 0 \tag{3.146}$$

for $j \neq k$, and the variance of Z_k is given by

$$V_k = \int_T\int_T \overline{a_k(t)}a_k(s)K_X(t,s)\,dtds \tag{3.147}$$

If $X(t)$ has the canonical expansion of Eq. 3.1, then according to Eq. 3.14 the coordinate functions are given by

$$x_k(t) = \frac{1}{V_k} E\left[X(t) \overline{\int_T X(s) \overline{a_k(s)}\, ds} \right]$$

$$= \frac{1}{V_k} E\left[\int_T X(t)\overline{X(s)} a_k(s)\, ds \right]$$

$$= \frac{1}{V_k} \int_T a_k(s) K_X(t,s)\, ds \tag{3.148}$$

Like $a_1(t)$, $a_2(t)$,..., the coordinate functions $x_1(t)$, $x_2(t)$,... lie in $L^2(T)$, since

$$\int_T |x_k(t)|^2\, dt = \frac{1}{V_k^2} \int_T \left| \int_T a_k(s) K_X(t,s)\, ds \right|^2 dt$$

$$\le \frac{1}{V_k^2} \int_T \left(\int_T |a_k(s) K_X(t,s)|\, ds \right)^2 dt$$

$$\le \frac{1}{V_k^2} \int_T \left(\int_T |a_k(s)|^2\, ds \right)\left(\int_T |K_X(t,s)|^2\, ds \right) dt$$

$$= \frac{1}{V_k^2} \int_T |a_k(s)|^2\, ds \int_T \int_T |K_X(t,s)|^2\, dsdt \tag{3.149}$$

which is finite because $a_k(s)$ and $K_X(t, s)$ are square-integrable. From Eq. 3.148,

$$\int_T \overline{a_j(t)} x_k(t)\, dt = \frac{1}{V_k} \int_T \int_T \overline{a_j(t)} a_k(s) K_X(t,s)\, dtds \tag{3.150}$$

Therefore, if we are given $a_1(t)$, $a_2(t)$,... and define $x_1(t)$, $x_2(t)$,... by Eq. 3.148, it follows from Eqs. 3.146, 3.147, and 3.150 that the coefficients are orthogonal if and only if

$$\int_T \overline{a_j(t)} x_k(t)\, dt = \delta_{jk} \tag{3.151}$$

where $\delta_{jk} = 1$ if $j = k$ and $\delta_{jk} = 0$ if $j \neq k$. We say that $a_j(t)$ and $x_k(t)$ are *bi-orthogonal*. For $j \neq k$, $a_j(t)$ and $x_k(t)$ are orthogonal in $L^2(T)$.

Looking at Eq. 3.150 a bit differently and using the conjugate symmetry of the covariance function,

$$\frac{1}{V_j} \int_T \int_T \overline{a_j(t)} a_k(s) K_X(t,s)\, dt ds = \frac{1}{V_j} \int_T a_k(s) \left(\int_T \overline{a_j(t) K_X(s,t)}\, dt \right) ds$$

$$= \int_T a_k(s) \overline{x_j(s)}\, ds \tag{3.152}$$

Putting Eqs. 3.150 and 3.152 together yields

$$V_k \int_T \overline{a_j(t)} x_k(t)\, dt = V_j \int_T a_k(t) \overline{x_j(t)}\, dt \tag{3.153}$$

Two questions need to be addressed. First, how do we find a sequence of functions $a_1(t)$, $a_2(t)$,... so that Eq. 3.151 is satisfied; second, if it is satisfied, what are conditions for the resulting expansion to converge to $X(t)$ in the mean square?

3.7.2. Generating Bi-Orthogonal Function Systems

We consider the problem of finding function sequences satisfying Eqs. 3.148 and 3.151. To simplify notation and clarify the linear structure of the overall procedure, we rewrite Eqs. 3.147, 3.148, 3.151, and 3.153 in terms of the inner product on $L^2(T)$:

$$V_k = \langle\langle K_X(t, s), a_k(t)\rangle, \overline{a_k(s)}\rangle \tag{3.154}$$

$$x_k(t) = V_k^{-1}\langle K_X(t, s), \overline{a_k(s)}\rangle \tag{3.155}$$

$$\langle x_k(t), a_j(t)\rangle = \delta_{jk} \tag{3.156}$$

$$V_k\langle x_k(t), a_j(t)\rangle = V_j\langle a_k(t), x_j(t)\rangle \tag{3.157}$$

where the variables are included because $K_X(t, s)$ is a function of two variables.

To generate sequences of functions in $L^2(T)$ satisfying Eq. 3.155 and 3.156, let $h_1(t)$, $h_2(t)$,... be an arbitrary function sequence in $L^2(T)$. Define $a_1(t) = h_1(t)$. Find V_1 and the coordinate function $x_1(t)$ from Eqs. 3.154 and

3.155, respectively. Then define $a_2(t) = c_{21}a_1(t) + h_2(t)$, where c_{21} is determined by the bi-orthogonality condition $\langle x_1(t), a_2(t)\rangle = 0$. Applying this condition yields

$$c_{21}\langle x_1(t), a_1(t)\rangle + \langle x_1(t), h_2(t)\rangle = 0 \tag{3.158}$$

Since $\langle x_1(t), a_1(t)\rangle = 1$, $c_{21} = -\langle x_1(t), h_2(t)\rangle$. V_2 and $x_2(t)$ are found from Eqs. 3.154 and 3.155. Proceeding recursively, suppose we have found $a_1(t)$, $a_2(t), \ldots, a_{n-1}(t), x_1(t), x_2(t), \ldots, x_{n-1}(t), V_1, V_2, \ldots, V_{n-1}$. Define

$$a_n(t) = \sum_{i=1}^{n-1} c_{ni}a_i(t) + h_n(t) \tag{3.159}$$

where $c_{n1}, c_{n2}, \ldots, c_{n,n-1}$ are determined by bi-orthogonality for $k = 1, 2, \ldots, n-1$. Applying bi-orthogonality to $a_n(t)$ yields the system

$$\sum_{i=1}^{n-1} c_{ni}\langle x_k(t), a_i(t)\rangle + \langle x_k(t), h_n(t)\rangle = 0 \tag{3.160}$$

for $k = 1, 2, \ldots, n-1$. Applying the bi-orthogonality conditions within the sums yields

$$c_{nk} = -\langle x_k(t), h_n(t)\rangle \tag{3.161}$$

for $k = 1, 2, \ldots, n-1$. V_n and $x_n(t)$ are found from Eqs. 3.154 and 3.155. By construction,

$$\langle x_k(t), a_n(t)\rangle = 0 \tag{3.162}$$

for $k < n$. From Eq. 3.157, $\langle x_n(t), a_k(t)\rangle = 0$ for $k < n$ and Eq. 3.156 is satisfied.

3.7.3. Complete Function Systems

Given a bi-orthogonal function system, we need to determine if the expansion with the coefficients of Eq. 3.143 is MS convergent to $X(t)$. Using inner product notation, that equation can be written as

$$Z_k = \langle X(t), a_k(t)\rangle \tag{3.163}$$

There exist necessary and sufficient conditions for convergence. We say that a sequence $a_1(t), a_2(t), \ldots$ of square-integrable functions is *complete relative to the random function* $X(t)$ if, whenever $b(t)$ is a square-integrable function such that

$$\langle\langle K_X(t, s), b(t)\rangle, \overline{a_k(s)}\rangle = 0 \tag{3.164}$$

for $k = 1, 2, \ldots$, it cannot be that

$$\langle\langle K_X(t, s), b(t)\rangle, \overline{b(s)}\rangle > 0 \tag{3.165}$$

Theorem 3.5. If $\{a_k(t)\}$ is a sequence of square-integrable functions, $Z_k = \langle X(t), a_k(t)\rangle$, $x_k(t) = V_k^{-1}\langle K_X(t, s), \overline{a_k(s)}\rangle$, and $\{x_k(t)\}$ and $\{a_k(t)\}$ are bi-orthogonal function sequences, then $X(t)$ is equal in the mean-square to its canonical expansion determined by Z_k and $x_k(t)$ if and only if $\{a_k(t)\}$ is complete relative to $X(t)$. ■

Let us interpret the theorem for the orthonormal eigenfunctions $u_1(t)$, $u_2(t), \ldots$ of the Karhunen-Loeve system. Suppose for these that $b(t)$ satisfies Eq. 3.164, namely,

$$\int_T \int_T \overline{b(t)} u_k(s) K_X(t, s)\, dt ds = 0 \tag{3.166}$$

for $k = 1, 2, \ldots$. If we assume that the covariance function is continuous, then Mercer's theorem (Eq. 3.42) applies and

$$\begin{aligned}
\int_T \int_T \overline{b(t)} b(s) K_X(t, s)\, dt ds &= \int_T \int_T \overline{b(t)} b(s) \left(\sum_{i=1}^{\infty} \lambda_i u_i(t) \overline{u_i(s)} \right) dt ds \\
&= \sum_{i=1}^{\infty} \int_T \lambda_i \overline{b(t)} u_i(t)\, dt \int_T b(s) \overline{u_i(s)}\, ds \qquad (3.167) \\
&= \sum_{i=1}^{\infty} \int_T \int_T \overline{b(t)} K_X(t, v) u_i(v)\, dv dt \int_T b(s) \overline{u_i(s)}\, ds
\end{aligned}$$

where the last equality holds because $u_i(t)$ is an eigenfunction for the eigenvalue λ_i. According to Eq. 3.166, each double integral in the series is zero, so that Eq. 3.167 reduces to zero and $u_1(t)$, $u_2(t), \ldots$ form a complete system relative to $X(t)$.

Theorem 3.5 is quite definitive. Since $L^2(T)$ is a Hilbert space, every continuous linear functional on $L^2(T)$ is defined by an inner product with some function in $L^2(T)$; that is, if ϕ is a continuous linear functional on $L^2(T)$, then there exists a function $a(t) \in L^2(T)$ such that $\phi(x) = \langle x, a\rangle$ for all $x \in L^2(T)$. Hence, the square-integrable functions $a_1(t)$, $a_2(t)$,... in the definition of a complete function system can be replaced by a sequence ϕ_1, ϕ_2,... of continuous linear functionals on $L^2(T)$. Theorem 3.5 then provides necessary and sufficient conditions for convergence in terms of the completeness of ϕ_1, ϕ_2,....

The theory we have provided concerns ordinary (nongeneralized) covariance functions. As has typically been the case, the theory extends to generalized covariance functions. The next example illustrates the matter.

***Example* 3.12.** Let $X(t)$ be white noise of constant intensity κ over the interval $[0, T]$. $X(t)$ has mean zero and covariance $K_X(t, t') = \kappa\delta(t - t')$. For $k = ..., -2, -1, 0, 1, 2,...$, let $a_k(t) = \exp[-j\omega_k t]$, where $\omega_k = 2k\pi/T$. From Eqs. 3.154 and 3.155,

$$V_k = \kappa \int_0^T \int_0^T \exp[j\omega_k (s - t)]\delta(t - s)\, dtds$$

$$= \kappa T$$

$$x_k(t) = \frac{1}{\kappa T} \int_0^T \exp[j\omega_k s]\kappa\delta(t - s)\, ds$$

$$= \frac{\exp[j\omega_k t]}{T}$$

The functions $x_k(t)$ and $a_i(t)$ are bi-orthogonal, since

$$\langle x_k(t), a_i(t)\rangle = \frac{1}{T} \int_0^T \exp[j(\omega_k - \omega_i)t]\, dt = \delta_{ki}$$

To show that $a_1(t)$, $a_2(t)$,... form a complete system, suppose $b(t)$ is square-integrable and

$$\int_0^T \int_0^T \overline{b(t)} a_k(s) K_X(t, s)\, dtds = 0$$

Then

$$0 = \int_0^T \int_0^T \overline{b(t)} \exp[-j\omega_k s] \kappa \delta(t-s)\, dt ds$$

$$= \kappa \int_0^T \overline{b(t)} \exp[-j\omega_k t]\, dt$$

for $k = ..., -2, -1, 0, 1, 2,....$ Hence, $b(t)$ is identically zero, the impossibility of the strict inequality in Eq. 3.165 is obvious, and the function system is complete relative to $X(t)$. Therefore $X(t)$ and $K_X(t, t')$ have the canonical expansions

$$X(t) = \frac{1}{T} \sum_{k=-\infty}^{\infty} Z_k \exp[j\omega_k t]\, dt$$

$$K_X(t, t') = \frac{\kappa}{T} \sum_{k=-\infty}^{\infty} \exp[j\omega_k (t - t')]$$

where

$$Z_k = \int_0^T X(t) \exp[-j\omega_k t]\, dt$$

■

3.8. Canonical Expansion of the Covariance Function

In Section 3.1 we saw that a canonical expansion of the covariance function can be obtained from a canonical expansion of the random function (Eqs. 3.15 and 3.16). We commented that, given a canonical expansion of the covariance function in terms of linearly independent coordinate functions, there exists a canonical expansion of the random function in terms of the same coordinate functions. In this section we demonstrate how to derive a canonical expansion of a random function from a canonical expansion of its covariance function.

3.8.1. Canonical Expansions from Covariance Expansions

For a zero-mean random function $X(t)$, assume that the covariance function possesses the canonical expansion

$$K_X(t, s) = \sum_{k=1}^{\infty} V_k x_k(t)\overline{x_k(s)} \tag{3.168}$$

having linearly independent coordinate functions. Then we can find uncorrelated zero-mean random variables $Z_1, Z_2,...$ such that

$$X(t) = \sum_{k=1}^{\infty} Z_k x_k(t) \tag{3.169}$$

with $V_k = \text{Var}[Z_k]$. We will demonstrate how this can be achieved by finding functions $a_1(t)$, $a_2(t)$,... that together with $x_1(t)$, $x_2(t)$,... satisfy the bi-orthogonality relation of Eq. 3.156 and the representation of Eq. 3.155 (equivalently, Eq. 3.148). Once $a_1(t)$, $a_2(t)$,... have been found, Eq. 3.169 provides a canonical expansion for $X(t)$ with $Z_k = \langle X(t), a_k(t)\rangle$ if the series is MS convergent to $X(t)$. Before showing how to find appropriate functions $a_1(t)$, $a_2(t)$,..., we first consider MS convergence given that $a_1(t)$, $a_2(t)$,... have been found.

Equations 3.148 and 3.36 both give Fourier coefficients of $X(t)$ relative to a set of uncorrelated random variables $Z_1, Z_2,...$. The formulations are the same except that $a_k(s)$ appears in Eq. 3.148 and $u_k(s)$ appears in Eq. 3.36. Since Theorem 3.3 is based solely on Theorem 3.1 and the formulation of the Fourier coefficients for $Z_1, Z_2,...$, we can apply it in the present case with $a_k(s)$ in place of $u_k(s)$, and Z_k given by Eq. 3.143 rather than Eq. 3.33. Here it states that $X(t)$ has the canonical expansion of Eq. 3.169 if and only if the variance of $X(t)$ has the expansion

$$\text{Var}[X(t)] = \sum_{k=1}^{\infty} V_k |x_k(t)|^2 \tag{3.170}$$

Setting $t = s$ in Eq. 3.168 yields this variance expansion as a corollary of the canonical expansion of the covariance function. In fact, since a canonical expansion of a random function yields a canonical expansion of the covariance, which in turn yields a canonical expansion of the variance, we have the following general theorem.

Theorem 3.6. If $\{a_k(t)\}$ is a sequence of square-integrable functions, $Z_k = \langle X(t), a_k(t)\rangle$, $x_k(t) = V_k^{-1}\langle K_X(t, s), \overline{a_k(s)}\rangle$, and $\{x_k(t)\}$ and $\{a_k(t)\}$ are bi-orthogonal function sequences, then the following conditions are equivalent: (i) $X(t)$ is equal in the mean-square to its canonical expansion determined by Z_k and $x_k(t)$; (ii) $K_X(t,s)$ has a canonical expansion in terms of V_k, $x_k(t)$, and $x_k(s)$; and (iii) $\text{Var}[X(t)]$ has a canonical expansion in terms of V_k and $x_k(t)$. ■

The fact that the three conditions of Theorem 3.6 are equivalent is not surprising. The first states that $X(t)$ is equal to its Fourier series relative to $Z_1, Z_2,\ldots$, the third states Parseval's identity for the norm $\|X(t)\|$, and the second states Parseval's identity for the inner product $\langle X(t), X(s)\rangle$, namely,

$$\begin{aligned}
\langle X(t), X(s)\rangle &= E[X(t)\overline{X(s)}] \\
&= R_X(t, s) \\
&= \sum_{k=1}^{\infty} V_k x_k(t)\overline{x_k(s)} \qquad (3.171) \\
&= \sum_{k=1}^{\infty} V_k E[X(t)\overline{V_k^{-1} Z_k}]\overline{E[X(s)\overline{V_k^{-1} Z_k}]} \\
&= \sum_{k=1}^{\infty} \langle X(t), V_k^{-1/2} Z_k\rangle \overline{\langle X(s), V_k^{-1/2} Z_k\rangle}
\end{aligned}$$

Two vectors in a Hilbert space equal their respective Fourier series if and only if both forms of Parseval's identity apply to them.

In light of Theorem 3.6, the role played by the integral equation in Eq. 3.21 becomes clear: its eigenfunctions supply both $\{a_k(t)\}$ and $\{x_k(t)\}$, which are *ipso facto* bi-orthogonal, and Mercer's Theorem gives a canonical expansion of the covariance.

We now return to the task of finding functions $a_1(t), a_2(t),\ldots$ that, together with the functions $x_1(t), x_2(t),\ldots$ in the covariance expansion of Eq. 3.168, satisfy the conditions of Theorem 3.6. To do so, begin with an arbitrary sequence of square-integrable functions $h_1(t), h_2(t),\ldots$ and define a new sequence of functions $g_1(t), g_2(t),\ldots$ in the following manner. Let $g_1(t) = h_1(t)$. Then define

$$g_2(t) = c_{21} g_1(t) + h_2(t) \qquad (3.172)$$

subject to the condition $\langle x_1, g_2\rangle = 0$ (where we have dropped the variable t in the inner product because only a single variable is currently involved). Applying g_2 to x_1 yields

$$c_{21} = -\frac{\langle x_1, h_2\rangle}{\langle x_1, g_1\rangle} \qquad (3.173)$$

To avoid a problem here owing to division by $\langle x_1, g_1\rangle$, h_1 must be chosen so that $\langle x_1, g_1\rangle \neq 0$. Proceeding recursively, given $g_1, g_2,..., g_{n-1}$, define

$$g_n(t) = \sum_{i=1}^{n-1} c_{ni} g_i(t) + h_n(t) \tag{3.174}$$

subject to the conditions $\langle x_i, g_n\rangle = 0$ for $i = 1, 2,..., n-1$. Also assume that $h_1, h_2,...$ are chosen so that $\langle x_n, g_n\rangle \neq 0$ for $n = 1, 2,....$ Applying g_n to x_k for $k = 1, 2,..., n-1$ yields

$$0 = \sum_{i=1}^{k} c_{ni} \langle x_k, g_i\rangle + \langle x_k, h_n\rangle \tag{3.175}$$

Solving this system for the coefficients c_{ni} yields

$$c_{n1} = -\frac{\langle x_1, h_n\rangle}{\langle x_1, g_1\rangle} \tag{3.176}$$

$$c_{nk} = -\frac{\langle x_k, h_n\rangle + \sum_{j=1}^{k-1} c_{nj} \langle x_k, g_j\rangle}{\langle x_k, g_k\rangle} \qquad (k = 2, 3,..., n-1) \tag{3.177}$$

We now define the desired functions $a_1(t), a_2(t),...$ by

$$a_n(t) = \sum_{i=n}^{\infty} d_{ni} g_i(t) \tag{3.178}$$

where the coefficients d_{ni} are found to satisfy the bi-orthogonality condition of Eq. 3.156. Applying a_n to x_k under these conditions yields

$$\sum_{i=n}^{k} d_{ni} \langle x_k, g_i\rangle = \delta_{nk} \tag{3.179}$$

for $n = 1, 2,...$ and $k = n, n+1,....$ For $n = 1, 2,...,$ we solve this system for the desired d_{nk}:

$$d_{nn} = \frac{1}{\langle x_n, g_n\rangle} \tag{3.180}$$

$$d_{nk} = -\frac{\sum_{i=n}^{k-1} d_{ni}\langle x_k, g_i\rangle}{\langle x_k, g_k\rangle} \qquad (k = n+1, n+2, \ldots) \tag{3.181}$$

3.8.2. Constructing Canonical Expansions for Covariance Functions

Given a canonical expansion for the covariance function of a random function, we can derive a canonical expansion of the random function itself. We now present a method for finding a canonical expansion of a covariance function via its trigonometric Fourier series. Other methods exist.

Let $X(t)$ be a random function of a single real variable t and the Fourier series of its covariance function $K_X(t, s)$ over the square $[-T, T] \times [-T, T]$ be given by

$$K_X(t, s) = \sum_{m-\infty}^{\infty}\sum_{n=-\infty}^{\infty} a_{mn} \exp[j(\omega_m t - \omega_n s)] \tag{3.182}$$

where $\omega_m = m\pi/T$ and $\omega_n = n\pi/T$. Applying the preceding expression with the variables reversed and conjugating yields

$$\overline{K_X(s,t)} = \sum_{m-\infty}^{\infty}\sum_{n=-\infty}^{\infty} \bar{a}_{mn} \exp[j(\omega_n t - \omega_m s)]$$

$$= \sum_{m-\infty}^{\infty}\sum_{n=-\infty}^{\infty} \bar{a}_{mn} \exp[j(\omega_m t - \omega_n s)] \tag{3.183}$$

where the second equality follows from the double sum being over all pairs (m, n). Comparing the preceding two equations shows that $a_{mn} = \bar{a}_{nm}$. This means that the coefficients a_{mn} can be expressed (in an infinite number of ways) by an expansion

$$a_{mn} = \sum_{k=0}^{\infty} V_k c_{mk} \bar{c}_{nk} \tag{3.184}$$

where V_0, V_1,... are real. Placing these representations into Eq. 3.182 yields

$$K_X(t, s) = \sum_{m-\infty}^{\infty}\sum_{n=-\infty}^{\infty}\left(\sum_{k=0}^{\infty} V_k c_{mk}\bar{c}_{nk}\right)\exp[j(\omega_m t - \omega_n s)]$$

$$= \sum_{k=0}^{\infty} V_k \left(\sum_{m=-\infty}^{\infty} c_{mk} \exp[j\omega_m t] \right) \left(\sum_{n=-\infty}^{\infty} \bar{c}_{nk} \exp[-j\omega_n s] \right) \tag{3.185}$$

A canonical expansion of the covariance function results from letting

$$x_k(t) = \sum_{m=-\infty}^{\infty} c_{mk} \exp[j\omega_m t] \tag{3.186}$$

Note that the variances V_k depend on the choice of decomposition in Eq. 3.184.

3.9. Integral Canonical Expansions

Rather than represent a random function as a sum involving discrete white noise and a countable collection of coordinate functions, we may be able to represent it as an integral involving continuous white noise:

$$X(t) = \mu_X(t) + \int_{\Xi} Z(\xi)x(t, \xi)\, d\xi \tag{3.187}$$

$Z(\xi)$ is white noise over the interval Ξ, which is the domain of the parameter ξ. The coordinate functions are replaced by the function $x(t, \xi)$, which is a function of t for each fixed value ξ of the parameter. If the integral is equal in the mean square to $X(t)$, then it is called an *integral canonical expansion (representation)* or a *continuous white-noise integral representation*. One can envision the uncorrelated elementary functions to be the infinitesimal random functions $Z(\xi)x(t, \xi)d\xi$; however, one must be careful to recognize the heuristic nature of such an interpretation.

The mathematical subtleties are greater in the continuous setting. With discrete canonical expansions, we can employ the L^2 theory, while with continuous canonical expansions we must use generalized functions. This is analogous to the classical Fourier theory where discrete Fourier representation takes place in the context of L^2 but integral Fourier representation requires generalized functions. This difficulty is evident at the outset because an integral canonical expansion involves white noise, which has a generalized covariance function. Owing to the use of generalized functions, we will proceed formally, ignoring mathematical questions, such as integrability.

3.9.1. Construction via Integral Functional Coefficients

Given the integral canonical representation of Eq. 3.187 for a zero-mean random function $X(t)$, an integral canonical expansion for the covariance function can be derived. From Eq. 2.93, the white noise $Z(\xi)$ has covariance function

$$K_Z(\xi, \xi') = I(\xi)\delta(\xi - \xi') \tag{3.188}$$

where $I(\xi)$ is the intensity of the white noise. Applying Theorem 2.2 in the generalized sense yields

$$\begin{aligned} K_X(t, t') &= E\left[\int_\Xi Z(\xi)x(t,\xi)\,d\xi \overline{\int_\Xi Z(\xi')x(t',\xi')\,d\xi'}\right] \\ &= \int_\Xi\int_\Xi K_Z(\xi,\xi')x(t,\xi)\overline{x(t',\xi')}\,d\xi d\xi' \\ &= \int_\Xi\int_\Xi x(t,\xi)\overline{x(t',\xi')}I(\xi)\delta(\xi-\xi')\,d\xi d\xi' \\ &= \int_\Xi I(\xi)x(t,\xi)\overline{x(t',\xi)}\,d\xi \end{aligned} \tag{3.189}$$

Letting $t' = t$ yields the variance expression

$$\text{Var}[X(t)] = \int_\Xi I(\xi)|x(t,\xi)|^2\,d\xi \tag{3.190}$$

In analogy with discrete representation, given an integral canonical expansion of the covariance function, there exists an integral canonical expansion for the random function.

To better appreciate the relationship between the random function, the white-noise coefficients, and the coordinate functions, consider the cross-covariance between $X(t)$ and $Z(\xi)$:

$$\begin{aligned} K_{XZ}(t, \xi) &= E[X(t)\overline{Z(\xi)}] \\ &= E\left[\overline{Z(\xi)}\int_\Xi Z(\xi')x(t,\xi')\,d\xi'\right] \end{aligned}$$

$$= \int_{\Xi} K_Z(\xi', \xi) x(t, \xi')\, d\xi'$$

$$= \int_{\Xi} I(\xi')\delta(\xi' - \xi) x(t, \xi')\, d\xi' \tag{3.191}$$

$$= I(\xi) x(t, \xi)$$

Dividing by the intensity of the white noise yields

$$x(t, \xi) = \frac{K_{XZ}(t, \xi)}{I(\xi)} \tag{3.192}$$

The expression corresponds to Eq. 3.14 for discrete canonical expansions. It (and its discrete version) says that the coordinate function for ξ is null at t if $X(t)$ is uncorrelated with $Z(\xi)$; in fact, if $X(t)$, as a random function, is uncorrelated with $Z(\xi)$, then $x(t, \xi) \equiv 0$.

As with discrete expansions, to construct an integral canonical expansion for a zero-mean random function $X(t)$, we can consider integral functionals formed from a class of functions $a(t, \xi)$, $\xi \in \Xi$, namely,

$$Z(\xi) = \int_T X(t)\overline{a(t, \xi)}\, dt \tag{3.193}$$

According to Eq. 3.192, if $Z(\xi)$ is formed in this manner, then

$$x(t, \xi) = \frac{1}{I(\xi)} E\left[X(t) \overline{\int_T X(s)\overline{a(s, \xi)}\, ds} \right]$$

$$= \frac{1}{I(\xi)} \int_T a(s, \xi) K_X(t, s)\, ds \tag{3.194}$$

Hence, according to Theorem 2.2,

$$K_Z(\xi, \xi') = \int_T \int_T K_X(t, t')\overline{a(t, \xi)} a(t', \xi')\, dt dt'$$

$$= I(\xi') \int_T \overline{a(t, \xi)} x(t, \xi')\, dt \tag{3.195}$$

Along with the covariance expression for white noise given in Eq. 3.188, this implies that

$$\int_T \overline{a(t,\xi)}x(t,\xi')\,dt = \delta(\xi-\xi') \tag{3.196}$$

which takes the place of the bi-orthogonality relation of Eq. 3.151. Equations 3.194 and 3.196 correspond to Eqs. 3.155 and 3.156, respectively.

Assuming $Z(\xi)$ is defined in this way, Eqs 3.194 and 3.196 follow from the integral canonical representation of $X(t)$. In fact, if we assume that these two equations hold, then it follows from Eq. 3.195, which only depends on the definition of $Z(\xi)$, that the random function $Z(\xi)$ is white noise. Consequently, Eqs. 3.194 and 3.196 constitute necessary and sufficient conditions for $Z(\xi)$ to be white noise. However, they are not sufficient to conclude that the random function is equal to the corresponding integral of Eq. 3.187. To clarify this point, Eqs. 3.194 and 3.196 have been derived from the integral form of Eq. 3.187 under the assumption that $Z(\xi)$ is white noise. The exact argument holds if we begin with the integral, define $Z(\xi)$ by Eq. 3.193, assume $Z(\xi)$ is white noise, and call the integral $X(t)$. What needs to be demonstrated is that, given a random function $X(t)$, it equals the integral of Eq. 3.187 under the assumptions that $Z(\xi)$ is defined by Eq. 3.193, $x(t, \xi)$ is defined by Eq. 3.194, and Eq. 3.196 holds [so that $Z(\xi)$ is white noise].

A third condition,

$$\int_\Xi x(t,\xi)\overline{a(t',\xi)}\,d\xi = \delta(t-t') \tag{3.197}$$

combines with Eqs. 3.194 and 3.196 to provide necessary and sufficient conditions for $X(t)$ to be given by the canonical expansion of Eq. 3.187. Indeed, with this latter assumption,

$$\begin{aligned}
\int_\Xi Z(\xi)x(t,\xi)\,d\xi &= \int_\Xi\int_T X(s)\overline{a(s,\xi)}x(t,\xi)\,d\xi ds \\
&= \int_\Xi X(s)\left(\int_T \overline{a(s,\xi)}x(t,\xi)\,d\xi\right) ds \\
&= \int_T X(s)\delta(t-s)\,ds \\
&= X(t)
\end{aligned} \tag{3.198}$$

One should recognize the lack of symmetry between Eqs. 3.196 and 3.197. The functions $a(t, \xi)$ and $x(t, \xi)$ are functions of t parameterized by ξ. The assumption in Eq. 3.196 is that, relative to the variable t, the integral exists in a generalized sense to yield the delta function $\delta(\xi - \xi')$ involving the expansion parameter; in Eq. 3.197, the integral exists in a generalized sense relative to the parameter ξ and is equal to $\delta(t - t')$.

To find an expression for the intensity of the white noise in the present setting, integrate Eq. 3.188 and then substitute the first integral in Eq. 3.195 to obtain

$$
\begin{aligned}
I(\xi) &= \int_{\Xi} K_Z(\xi, \xi')\, d\xi' \\
&= \int_{\Xi}\int_{T}\int_{T} K_X(t, t')\overline{a(t, \xi)}a(t', \xi')\, dt dt' d\xi'
\end{aligned}
\tag{3.199}
$$

3.9.2. Construction from a Covariance Expansion

Suppose we are given an integral canonical expansion for the covariance function according to Eq. 3.189 and wish to derive an integral canonical expansion of the random function itself. It can be shown that there exists white noise $Z(\xi)$ possessing the given intensity $I(\xi)$ in the equation and having cross-covariance given by Eq. 3.191. For this white noise process,

$$
\begin{aligned}
E\left[\left|X(t) - \int_{\Xi} Z(\xi)x(t, \xi)\, d\xi\right|^2\right]
&= E\left[\left(X(t) - \int_{\Xi} Z(\xi)x(t, \xi)\, d\xi\right)\overline{\left(X(t) - \int_{\Xi} Z(\xi)x(t, \xi)\, d\xi\right)}\right] \\
&= E[|X(t)|^2] - E\left[\int_{\Xi} X(t)\overline{Z(\xi)x(t, \xi)}\, d\xi\right] - E\left[\int_{\Xi} \overline{X(t)}Z(\xi)x(t, \xi)\, d\xi\right] \\
&\quad + E\left[\int_{\Xi}\int_{\Xi} Z(\xi)\overline{Z(\xi')}x(t, \xi)\overline{x(t, \xi')}\, d\xi d\xi'\right] \\
&= K_X(t, t) - \int_{\Xi} K_{XZ}(t, \xi)\overline{x(t, \xi)}\, d\xi - \int_{\Xi} \overline{K_{XZ}(t, \xi)}x(t, \xi)\, d\xi
\end{aligned}
$$

$$+ \int_{\Xi}\int_{\Xi} K_Z(\xi,\xi')x(t,\xi)\overline{x(t,\xi')}\, d\xi d\xi'$$

$$= K_X(t, t) - \int_{\Xi} I(\xi)x(t,\xi)\overline{x(t,\xi)}\, d\xi \; - \; \int_{\Xi} \overline{I(\xi)x(t,\xi)}x(t,\xi)\, d\xi$$

$$+ \int_{\Xi}\int_{\Xi} I(\xi)\delta(\xi-\xi')x(t,\xi)\overline{x(t,\xi')}\, d\xi d\xi'$$

$$= K_X(t, t) - \int_{\Xi} I(\xi)|x(t,\xi)|^2\, d\xi \tag{3.200}$$

the last equality following upon integration relative to $d\xi'$ in the double integral. Setting $t = t'$ in the integral canonical expansion for the covariance function shows the last expression to be 0.

3.10. Power Spectral Density

Wide-sense stationary random functions have Fourier-integral canonical representations. These expansions are related to integral canonical expansions of the covariance function by means of the power spectral density of the random function. The original spectral analysis of WS stationary random functions is due to Wiener and Khinchin.

3.10.1. The Power-Spectral-Density/Autocorrelation Transform Pair

Let $r_X(\tau)$ be the autocorrelation function of a WS stationary random function $X(t)$ of a single variable $t \in \mathfrak{R}$. Assume $r_X(\tau)$ is integrable. The *power spectral density* of $X(t)$ is the Fourier transform of $r_X(\tau)$:

$$S_X(\omega) = \int_{-\infty}^{\infty} r_X(\tau)e^{-j\omega\tau}\, d\tau \tag{3.201}$$

The integrability of $r_X(\tau)$ guarantees that its Fourier transform is defined for all $\omega \in \mathfrak{R}$. If $X(t)$ is centered, then $k_X(\tau) = r_X(\tau)$, and $S_X(\omega)$ can be defined via the covariance function. Since $r_X(\tau) = \overline{r}_x(-\tau)$, $S_X(\omega)$ is real valued. If $X(t)$ is real valued, then $S_X(\omega)$ is an even function, because it is the Fourier transform of an even, real-valued function.

If $S_X(\omega)$ is integrable, then the inversion formula,

$$r_X(\tau) = \frac{1}{2\pi} \int_{-\infty}^{\infty} S_X(\omega) e^{j\omega\tau} \, d\omega \tag{3.202}$$

holds almost everywhere (and holds for all $\tau \in \Re$ if $r_X(\tau)$ is continuous). We then have the transform pair, $r_X(\tau) \leftrightarrow S_X(\omega)$.

If $X(t)$ has zero mean, a supposition we make for the remainder of this subsection, and there is inversion for $\tau = 0$, then

$$\begin{aligned} \text{Var}[X(t)] &= k_X(0) \\ &= \frac{1}{2\pi} \int_{-\infty}^{\infty} S_X(\omega) \, d\omega \end{aligned} \tag{3.203}$$

There is a direct relationship between the power spectral density and the truncated Fourier integral of $X(t)$ defined by

$$\hat{X}_T(\omega) = \int_{-T}^{T} X(t) e^{-j\omega t} \, dt \tag{3.204}$$

$\hat{X}_T(\omega)$ is complex valued, has zero mean, and has variance

$$\begin{aligned} \text{Var}[\hat{X}_T(\omega)] &= E[|\hat{X}_T(\omega)|^2] \\ &= E\left[\int_{-T}^{T} X(t) e^{-j\omega t} \, dt \int_{-T}^{T} \overline{X(s)} e^{j\omega s} \, ds \right] \\ &= \int_{-T}^{T} \int_{-T}^{T} k_X(t-s) e^{-j\omega(t-s)} \, dt ds \end{aligned} \tag{3.205}$$

Applying the same transformation that yielded Eq. 2.114 from Eq. 2.113, yields

$$\text{Var}[\hat{X}_T(\omega)] = 2T \int_{-2T}^{2T} \left(1 - \frac{|\tau|}{2T}\right) k_X(\tau) e^{-j\omega\tau} \, d\tau \tag{3.206}$$

Assuming $\tau k_X(\tau)$ is integrable, dividing by $2T$ and letting $T \to \infty$ in the preceding equation yields the next theorem. The theorem has an important consequence: since Var[$\hat{X}_T(\omega)$] is nonnegative, $S_X(\omega)$ is also nonnegative.

Theorem 3.7. If $X(t)$ is a zero-mean WS stationary random function and $k_X(\tau)$ and $\tau k_X(\tau)$ are integrable, then

$$\lim_{T\to\infty} \frac{1}{2T} \operatorname{Var}[\hat{X}_T(\omega)] = S_X(\omega) \quad \blacksquare \tag{3.207}$$

It can be shown that

$$\frac{|\hat{X}_T(\omega)|^2}{2T} = \int_{-2T}^{2T} \left(1 - \frac{|\tau|}{2T}\right) A_T(\tau) e^{-j\omega\tau}\, d\tau \tag{3.208}$$

where

$$A_T(\tau) = \begin{cases} \dfrac{1}{2T+\tau} \displaystyle\int_{-T-\tau}^{T} X(t+\tau)\overline{X(t)}\, dt, & \text{for } \tau < 0 \\[2ex] \dfrac{1}{2T-\tau} \displaystyle\int_{-T}^{T-\tau} X(t+\tau)\overline{X(t)}\, dt, & \text{for } \tau > 0 \end{cases} \tag{3.209}$$

Moreover, if $X(t)$ is WS stationary and covariance ergodic, then the covariance function of $A_T(\tau)$ converges in the mean-square to $k_X(\tau)$ as $T \to \infty$. If one is not careful in taking the limit as $T \to \infty$ in Eq. 3.208, then one can erroneously conclude that $(2T)^{-1}|\ \hat{X}_T(\omega)\ |^2$ converges to $S_X(\omega)$ in the mean-square. For instance, if $X(t)$ is a WS stationary Gaussian random function, then the variance of $(2T)^{-1}|\ \hat{X}_T(\omega)\ |^2$ converges to $S_X{}^2(\omega)$ as $T \to \infty$. Hence, $(2T)^{-1}|\ \hat{X}_T(\omega)\ |^2$ cannot converge to the deterministic function $S_X(\omega)$.

What can be proven is that, for $\omega_0 < \omega_1$, integrating $(2T)^{-1}|\ \hat{X}_T(\omega)\ |^2$ from ω_0 to ω_1 and passing to the limit yields the mean-square limit

$$\underset{T\to\infty}{\text{l.i.m.}} \frac{1}{2T} \int_{\omega_0}^{\omega_1} |\hat{X}_T(\omega)|^2\, d\omega = \int_{\omega_0}^{\omega_1} S_X(\omega)\, d\omega \tag{3.210}$$

While it is true that

$$\lim_{\omega_1 \to \omega_0} \frac{1}{\omega_1 - \omega_0} \int_{\omega_0}^{\omega_1} S_X(\omega)\, d\omega = S_X(\omega_0) \tag{3.211}$$

and therefore

$$\lim_{\omega_1 \to \omega_0} \underset{T \to \infty}{\text{l.i.m.}} \frac{1}{2T(\omega_1 - \omega_0)} \int_{\omega_0}^{\omega_1} |\hat{X}_T(\omega)|^2\, d\omega = S_X(\omega_0) \tag{3.212}$$

the limits as $\omega_1 \to \omega_0$ and $T \to \infty$ cannot be interchanged to obtain $S_X(\omega_0)$ as the limit of $(2T)^{-1}|\hat{X}_T(\omega_0)|^2$ as $T \to \infty$.

Although we have only been considering the power spectral density for random functions of a single variable, the discussion extends to random functions of more than one variable by employing the Fourier transform in multiple dimensions.

Given a real-valued function, the question arises as to whether or not there is a random function for which it is the power spectral density.

Theorem 3.8. (i) If a real-valued function $S(\omega)$ is nonnegative and integrable, then it is the power spectral density of a WS stationary random function $X(t)$. $S(\omega)$ is even if and only if $X(t)$ is real valued. (ii) A function $r(\tau)$ is the autocorrelation function for a WS stationary random function if and only if it is *positive semidefinite*, meaning that, for any finite number of time points $t_1 < t_2 < \cdots < t_n$ and complex numbers $a_1, a_2, \ldots, a_n$,

$$\sum_{k=1}^{n}\sum_{i=1}^{n} a_k \bar{a}_i r(t_k - t_i) \geq 0 \tag{3.213}$$

Moreover, $r(\tau)$ is even if and only if the process is real valued. ■

3.10.2. Power Spectral Density and Linear Operators

For an integrable deterministic function $h(t)$, consider the linear operator defined on WS stationary inputs $X(t)$ by the MS convolution-type integral

$$Y(t) = \int_{-\infty}^{\infty} h(\tau) X(t - \tau)\, d\tau \tag{3.214}$$

(under the assumption that the integral exists in the MS sense). The mean of $Y(t)$ is

$$\mu_Y = \int_{-\infty}^{\infty} h(\tau) E[X(t-\tau)]\, d\tau$$

$$= \mu_X \int_{-\infty}^{\infty} h(\tau)\, d\tau \tag{3.215}$$

$$= \mu_X H(0)$$

where the system function $H(\omega)$ is the Fourier transform of $h(\tau)$. The autocorrelation function for $Y(t)$ can be expressed as a convolution product:

$$r_Y(\tau) = E\left[Y(t+\tau) \overline{\int_{-\infty}^{\infty} X(t-v) h(v)\, dv} \right]$$

$$= \int_{-\infty}^{\infty} E[Y(t+\tau)\overline{X(t-v)}]\overline{h(v)}\, dv$$

$$= \int_{-\infty}^{\infty} r_{YX}(\tau+v)\overline{h(v)}\, dv \tag{3.216}$$

$$= \int_{-\infty}^{\infty} r_{YX}(\tau-v)\overline{h(-v)}\, dv$$

$$= (r_{YX} * h_0)(\tau)$$

where h_0 is defined by $h_0(v) = \overline{h(-v)}$.

Letting S_{YX} denote the Fourier transform of the cross-correlation function r_{YX}, taking the Fourier transform in the preceding equation yields

$$S_Y(\omega) = S_{YX}(\omega) \int_{-\infty}^{\infty} \overline{h(-v)} e^{-j\omega v}\, dv$$

$$= S_{YX}(\omega) \overline{\int_{-\infty}^{\infty} h(-v) e^{j\omega v}\, dv}$$

$$= \overline{H(\omega)} S_{YX}(\omega) \tag{3.217}$$

The cross-correlation function is given by

$$r_{YX}(\tau) = E[Y(t)\overline{X(t-\tau)}]$$

$$= E\left[\overline{X(t-\tau)}\int_{-\infty}^{\infty} X(t-v)h(v)\,dv\right]$$

$$= \int_{-\infty}^{\infty} h(v)r_X(\tau - v)\,dv$$

$$= (h * r_X)(\tau) \tag{3.218}$$

Taking the Fourier transform yields

$$S_{YX}(\omega) = H(\omega)S_X(\omega) \tag{3.219}$$

(showing that S_{YX} need not be real valued). Inserting this equation into Eq. 3.217 yields

$$S_Y(\omega) = |H(\omega)|^2\, S_X(\omega) \tag{3.220}$$

The import of this relation is that, for WS stationary processes, the convolution-integral operator can be interpreted, relative to the power spectral densities of the input and output processes, as a system operator determined by multiplication by $|H(\omega)|^2$.

To appreciate the terminology "power spectral density," for $\omega_0 < \omega_1$, consider the transfer function defined by $H(\omega) = 1$ if $\omega \in (\omega_0, \omega_1]$ and $H(\omega) = 0$ otherwise. According to Eq. 3.220, $S_Y(\omega) = S_X(\omega)$ if $\omega \in (\omega_0, \omega_1]$ and $S_Y(\omega) = 0$ otherwise. The average of the output power in the random signal $Y(t)$ is given by $r_Y(0) = E[|Y(t)|^2]$ and (assuming integrability of the power spectral densities),

$$r_Y(0) = \frac{1}{2\pi}\int_{-\infty}^{\infty} S_Y(\omega)\,d\omega = \frac{1}{2\pi}\int_{\omega_0}^{\omega_1} S_X(\omega)\,d\omega \tag{3.221}$$

Integrating $S_X(\omega)$ across the frequency band $(\omega_0, \omega_1]$ gives the average output power.

3.10.3. Integral Representation of WS Stationary Random Functions

The power spectral density plays a central role in the integral canonical representation of WS stationary random functions. Assuming $X(t)$ has zero mean, if we let $\tau = t - t'$, then Eq. 3.202 becomes

$$k_X(t-t') = \frac{1}{2\pi}\int_{-\infty}^{\infty} S_X(\omega)e^{j\omega(t-t')}\,d\omega$$

$$= \frac{1}{2\pi}\int_{-\infty}^{\infty} S_X(\omega)e^{j\omega t}\,\overline{e^{j\omega t'}}\,d\omega \tag{3.222}$$

which is the WS stationary form of the integral canonical expansion of the covariance function given Eq. 3.189, with $x(t, \omega) = e^{j\omega t}/2\pi$. Accordingly, there exists a white noise process $Z(\omega)$ with intensity $2\pi S_X(\omega)$ such that $X(t)$ has the integral canonical expansion

$$X(t) = \frac{1}{2\pi}\int_{-\infty}^{\infty} Z(\omega)e^{j\omega t}\,d\omega \tag{3.223}$$

There is some subtlety here. The inversion integral of Eq. 3.202 depends on the integrability of the power spectral density, and according to Theorem 2.2, the MS integrability of the integral in Eq. 3.223 depends on the existence of the inversion integral. The problem exists even though the covariance function is assumed to be integrable. This, and the fact that many important random functions possess generalized covariance functions, compels us to abandon the assumption of covariance-function integrability and proceed in a generalized sense from the outset when considering integral canonical expansions of WS stationary random functions. Given the need to employ generalized functions when dealing with the Fourier transform, the necessity of this approach should not be surprising. Moreover, it is consistent with the preceding section.

***Example* 3.13.** In the generalized framework, white noise $Z(t)$ of unit intensity has covariance $\delta(\tau)$. Applying the Fourier transform yields its power spectral density,

$$S_Z(\omega) = \int_{-\infty}^{\infty} \delta(\tau)e^{-j\omega\tau}\,d\tau = 1$$

Applying the inverse transform yields

$$k_Z(\tau) = \frac{1}{2\pi}\int_{-\infty}^{\infty} e^{j\omega\tau}\, d\omega = \delta(\tau)$$

We recognize the transform pair $\delta(\tau) \leftrightarrow 1$.

Recall the Poisson increment process $X(t)$ of Example 2.17. $X(t)$ counts the number of Poisson points in the interval $[t, t + r)$. $X(t)$ has covariance function

$$k_X(\tau) = \max[\lambda(r - |\tau|), 0]$$

where λt is the mean of the driving Poisson process. Let $X_r(t)$ be the centered version of $X(t)/r$. Then

$$k_{X_r}(\tau) = \frac{\lambda}{r}\max\left[1 - \frac{|\tau|}{r}, 0\right]$$

which is a triangular function whose maximum value λ/r occurs at $\tau = 0$. As $r \to 0$, this covariance function tends to a delta function. The power spectral density of $X_r(t)$ is

$$S_{X_r}(\omega) = \lambda\left(\frac{\sin r\omega/2}{r\omega/2}\right)^2$$

As $r \to 0$, the fraction tends to 1. Hence, the limit of the power spectral density tends to the constant λ, which is the power spectral density of white noise. ■

The preceding section gives three conditions that together are sufficient for a canonical representation. Let

$$a(t, \omega) = \exp[j\omega t] \tag{3.224}$$

and define $Z(\omega)$ according to Eq. 3.193,

$$Z(\omega) = \int_{-\infty}^{\infty} X(t)e^{-j\omega t}\, dt \tag{3.225}$$

Then the three conditions of Eqs. 1.194, 1.196, and 1.197 become

$$x(t,\omega) = \frac{1}{I(\omega)} \int_{-\infty}^{\infty} k_X(t-s)\exp[j\omega s]\,ds$$

$$= \frac{1}{I(\omega)} \exp[j\omega t] \int_{-\infty}^{\infty} k_X(\tau)\exp[-j\omega\tau]\,d\tau \qquad (3.226)$$

$$= \frac{S_X(\omega)}{I(\omega)} \exp[j\omega t]$$

$$\int_{-\infty}^{\infty} \overline{a(t,\omega)}x(t,\omega')\,dt = \frac{S_X(\omega)}{I(\omega)} \int_{-\infty}^{\infty} \exp[-j\omega t]\exp[j\omega' t]\,dt$$

$$= \frac{S_X(\omega)}{I(\omega)} \int_{-\infty}^{\infty} \exp[-j(\omega-\omega')t]\,dt \qquad (3.227)$$

$$= \frac{2\pi S_X(\omega)}{I(\omega)} \delta(\omega-\omega')$$

$$\int_{-\infty}^{\infty} x(t,\omega)\overline{a(t',\omega)}\,d\omega = \frac{S_X(\omega)}{I(\omega)} \int_{-\infty}^{\infty} \exp[j\omega t]\exp[-jt'\omega]\,d\omega$$

$$= \frac{S_X(\omega)}{I(\omega)} \int_{-\infty}^{\infty} \exp[-j\omega(t'-t)\omega]\,d\omega \qquad (3.228)$$

$$= \frac{2\pi S_X(\omega)}{I(\omega)} \delta(t-t')$$

According to Eq. 3.199,

$$I(\omega) = \int_{\Xi}\int_{T}\int_{T} K_X(t,t')\overline{a(t,\omega)}a(t',\omega')\,dt\,dt'\,d\omega'$$

$$= \int_{-\infty}^{\infty}\int_{-\infty}^{\infty}\int_{-\infty}^{\infty} \exp[-j\omega t]\exp[j\omega' t']k_X(t-t')\,dt\,dt'\,d\omega'$$

$$= \int_{-\infty}^{\infty}\int_{-\infty}^{\infty}\int_{-\infty}^{\infty} \exp[-j\omega t]\exp[j\omega'(t-\tau)]k_X(\tau)\,dt d\tau d\omega'$$

$$= \int_{-\infty}^{\infty} k_X(\tau)\,d\tau \int_{-\infty}^{\infty} \exp[-j\omega'\tau]\,d\omega' \int_{-\infty}^{\infty} \exp[-j(\omega-\omega')t]\,dt \qquad (3.229)$$

$$= 2\pi \int_{-\infty}^{\infty} k_X(\tau)\,d\tau \int_{-\infty}^{\infty} \exp[-j\omega'\tau]\delta(\omega-\omega')\,d\omega'$$

$$= 2\pi \int_{-\infty}^{\infty} k_X(\tau)\exp[-j\omega\tau]\,d\tau$$

$$= 2\pi S_X(\omega)$$

Hence, Eqs. 3.227 and 3.228 reduce to the required expressions,

$$x(t, \omega) = \frac{\exp[j\omega t]}{2\pi} \qquad (3.230)$$

and Eq. 3.223 provides an integral canonical expansion for $X(t)$ in terms of white noise of intensity $2\pi S_X(\omega)$ (Eq. 3.225).

3.11. Canonical Representation of Vector Random Functions

In many situations we are concerned with vector-valued functions. In a stochastic framework these become vector-valued random functions. An important case will occur in the next chapter when we observe a collection of random functions over discrete time and wish to estimate a different collection of random functions at time instances based on the observed random processes.

3.11.1. Vector Random Functions

The general form of a vector random function is

$$\mathbf{X}(t) = (X_1(t), X_2(t),..., X_n(t))' \qquad (3.231)$$

where t can be of any dimension. If the domain of t is finite, say $T = \{t_1, t_2,..., t_m\}$, then a full probabilistic description of $\mathbf{X}(t)$ is given by the probability distribution function

$$F(x_{11}, x_{12},..., x_{1n}, x_{21}, x_{22},..., x_{nm}; t_1, t_2,..., t_m)$$

$$= P(X_i(t_j) \leq x_{ij} \text{ for } i = 1, 2,..., n \text{ and } j = 1, 2,..., m) \tag{3.232}$$

or by the corresponding densities. Marginal distributions can be obtained by integration. If T is infinite, then it is not generally possible to completely characterize $\mathbf{X}(t)$ by finite joint distributions; however, as in the case of scalar random functions, if the realizations are sufficiently well behaved, knowledge of all finite-order distributions completely characterizes $\mathbf{X}(t)$. We make this assumption. Analysis of vector random functions is facilitated by recognizing that we can treat $X_i(t)$ as a function of two variables, i and t, so that various definitions applying to random functions apply to the components of vector random functions. This observation is independent of the dimension of t.

Viewing $\mathbf{X}(t)$ as a random function, $X_i(t)$, of two variables, the mean function $E[X_i(t)]$ is a function of $i = 1, 2,..., n$ and $t \in T$. It is intuitive to equivalently treat it componentwise as

$$E[\mathbf{X}(t)] = (E[X_1(t)], E[X_2(t)],..., E[X_n(t)])' \tag{3.233}$$

Letting $\mu_{\mathbf{X},i}(t)$ denote the mean of $X_i(t)$, $E[\mathbf{X}(t)]$ takes the form

$$\mu_{\mathbf{X}}(t) = (\mu_{\mathbf{X},1}(t), \mu_{\mathbf{X},2}(t),..., \mu_{\mathbf{X},n}(t))' \tag{3.234}$$

The autocorrelation and covariance functions of $\mathbf{X}(t)$ are defined by

$$R_{i,j}^{\mathbf{X}}(t,t') = E[X_i(t)X_j(t')] \tag{3.235}$$

$$K_{i,j}^{\mathbf{X}}(t,t') = E[(X_i(t) - \mu_{\mathbf{X},i}(t))(X_j(t') - \mu_{\mathbf{X},j}(t'))]$$

$$= R_{i,j}^{\mathbf{X}}(t,t') - \mu_{\mathbf{X},i}(t)\mu_{\mathbf{X},j}(t') \tag{3.236}$$

respectively. The covariance function can be viewed as the matrix composed of covariance and cross-covariance functions of vector components, namely,

$$\mathbf{K}_{\mathbf{X}}(t, t') = \begin{pmatrix} K_{1,1}^{\mathbf{X}}(t,t') & K_{1,2}^{\mathbf{X}}(t,t') & \cdots & K_{1,n}^{\mathbf{X}}(t,t') \\ K_{2,1}^{\mathbf{X}}(t,t') & K_{2,2}^{\mathbf{X}}(t,t') & \cdots & K_{2,n}^{\mathbf{X}}(t,t') \\ \vdots & \vdots & \ddots & \vdots \\ K_{n,1}^{\mathbf{X}}(t,t') & K_{n,2}^{\mathbf{X}}(t,t') & \cdots & K_{n,n}^{\mathbf{X}}(t,t') \end{pmatrix} \tag{3.237}$$

One can analogously view the autocorrelation function as a matrix, $\mathbf{R}_{\mathbf{X}}(t, t')$. Properties of $\mathbf{K}_{\mathbf{X}}(t, t')$ are easily derived from corresponding properties of the individual covariance functions composing $\mathbf{K}_\mathbf{X}(t, t')$. Similar definitions and comments apply to the cross-covariance function of two vector random functions $\mathbf{X}(t)$ and $\mathbf{Y}(s)$. We define

$$K_{i,j}^{\mathbf{XY}}(t,s) = E[(X_i(t) - \mu_{\mathbf{X},i}(t))(Y_j(s) - \mu_{\mathbf{Y},j}(s))] \tag{3.238}$$

$\mathbf{K}_{\mathbf{XY}}(t, s)$ is defined as a matrix in like manner to $\mathbf{K}_\mathbf{X}(t, t')$.

3.11.2. Canonical Expansions for Vector Random Functions

A canonical expansion for a vector random function is defined in the same manner as a scalar random function of two variables, the difference being that the coordinate functions depend on the variables i and t. For $i = 1, 2,\ldots, n$,

$$X_i(t) = \mu_{\mathbf{X},i}(t) + \sum_{k=1}^{\infty} Z_k x_{ki}(t) \tag{3.239}$$

where $Z_1, Z_2,\ldots$ are uncorrelated zero-mean random variables and Eq. 3.14 becomes

$$x_{ki}(t) = \frac{E[X_i(t)\overline{Z_k}]}{V_k} \tag{3.240}$$

where $V_k = \mathrm{Var}[Z_k]$. Letting

$$\mathbf{x}_k(t) = (x_{k1}(t), x_{k2}(t),\ldots, x_{kn}(t))' \tag{3.241}$$

the canonical expansion takes the vector form

$$\mathbf{X}(t) = \mu_\mathbf{X}(t) + \sum_{k=1}^{\infty} Z_k \mathbf{x}_k(t) \tag{3.242}$$

A canonical expansion of the covariance function follows from Eq. 3.16:

$$K_{i,j}^{\mathbf{X}}(t,t') = \sum_{k=1}^{\infty} V_k x_{ki}(t)\overline{x_{kj}(t')} \tag{3.243}$$

In matrix form, a canonical expansion of the covariance function is a sum of matrices,

$$\mathbf{K}_{\mathbf{X}}(t, t') = \sum_{k=1}^{\infty} V_k \begin{pmatrix} x_{k1}(t)\overline{x_{k1}(t')} & x_{k1}(t)\overline{x_{k2}(t')} & \cdots & x_{k1}(t)\overline{x_{kn}(t')} \\ x_{k2}(t)\overline{x_{k1}(t')} & x_{k2}(t)\overline{x_{k2}(t')} & \cdots & x_{k2}(t)\overline{x_{kn}(t')} \\ \vdots & \vdots & \ddots & \vdots \\ x_{kn}(t)\overline{x_{k1}(t')} & x_{kn}(t)\overline{x_{k2}(t')} & \cdots & x_{kn}(t)\overline{x_{kn}(t')} \end{pmatrix} \tag{3.244}$$

Matrix representation provides compact notation and facilitates algebraic operations; nevertheless, it should be remembered that $\mathbf{K}_{\mathbf{X}}(t, t')$ is composed of scalar covariance functions whose properties depend on them being scalar covariance functions of i and t. A consequence of this relationship is that a canonical expansion of a vector random function can be obtained from a canonical expansion of its covariance function.

The formation of canonical expansions by means of integral functionals extends to the vector case by treating the vector random function as a function of two variables, i and t. Interpreting Eq. 3.143 in this manner, given a collection $\{a_{ki}(t)\}$, $i = 1, 2, \ldots, n$ and $k = 1, 2, \ldots$, of square-integrable functions, we define the coefficient random variable

$$Z_k = \sum_{i=1}^{n} \int_T X_i(t)\overline{a_{ki}(t)}\, dt \tag{3.245}$$

The coordinate functions (Eq. 3.148) are given by

$$x_{ki}(t) = \frac{1}{V_k} \sum_{j=1}^{n} \int_T a_{kj}(s) K_{i,j}^{\mathbf{X}}(t, s)\, ds \tag{3.246}$$

The bi-orthogonality relations (Eq. 3.151) become

$$\sum_{i=1}^{n} \int_T a_{ji}(t)\overline{x_{ki}(t)}\, dt = \delta_{jk} \tag{3.247}$$

Theorem 3.5 applies with the appropriate redefinition of completeness and the method for generating bi-orthogonal function systems extends to the vector setting. Theorem 3.6 also extends to vector random functions.

3.11.3. Finite Sets of Random Vectors

For digital processing, a vector random function needs a finite domain. It is composed of m random vectors, $\mathbf{X}_1, \mathbf{X}_2, \ldots, \mathbf{X}_m$. There is a total of mn random variables, which we assume possess zero means and finite second moments. For $j = 1, 2, \ldots, m$,

$$\mathbf{X}_j = (X_{j1}, X_{j2},..., X_{jn})' \tag{3.248}$$

The mn random variables span a subspace S of dimension $N \leq mn$. From among the mn random variables we can select a linearly independent subset that span S, and from these we can apply the Gram-Schmidt orthogonalization procedure to obtain N uncorrelated (orthogonal) random variables, $Z_1, Z_2,..., Z_N$, that span S. Since $X_{ji} \in S$ for $j = 1, 2,..., m$ and $i = 1, 2,..., n$, there exist scalars $x_{1ji}, x_{2ji},..., x_{Nji}$ such that

$$X_{ji} = \sum_{k=1}^{N} x_{kji} Z_k \tag{3.249}$$

where, for $k = 1, 2,..., N$, x_{kji} is the nonnormalized Fourier coefficient of X_{ji},

$$x_{kji} = \frac{E[X_{ji}\overline{Z_k}]}{E[|Z_k|^2]} \tag{3.250}$$

Letting

$$\mathbf{x}_k(j) = (x_{kj1}, x_{kj2},..., x_{kjN})' \tag{3.251}$$

yields a finite canonical expansion corresponding to Eq. 3.242, namely,

$$\mathbf{X}_j = \sum_{k=1}^{N} Z_k \mathbf{x}_k(j) \tag{3.252}$$

3.12. Canonical Representation over a Discrete Set

Throughout the chapter our main interest has been canonical representation of a random function $X(t)$ over its domain of definition. We now consider the situation in which $X(t)$ only needs to be represented over some (possibly infinite) discrete set of points $\{t_k\}$. Although $X(t)$ might be defined over a continuous set, its canonical representation need only converge to $X(t)$ on $\{t_k\}$. The following theorem provides a recursive procedure for finding an appropriate collection of uncorrelated zero-mean random variables and the corresponding coordinate functions. The expansion is actually finite at each t_k. Thus, if $\{t_k\}$ is finite, then so is the canonical representation at the sample points.

Theorem 3.9. If $X(t)$ is a zero-mean random function having finite variance and $\{t_k\}$ is a discrete point set within the domain of $X(t)$, then a canonical representation of $X(t)$ over $\{t_k\}$ is determined recursively by the following equations, in which V_k = Var[Z_k] and the recursion equations apply for k = 2, 3,...:

(i) $Z_1 = X(t_1)$

$$V_1 = K_X(t_1, t_1) \tag{3.253}$$

$$x_1(t) = V_1^{-1} K_X(t, t_1)$$

(ii)
$$Z_k = X(t_k) - \sum_{j=1}^{k-1} Z_j x_j(t_k) \tag{3.254}$$

(iii)
$$V_k = K_X(t_k, t_k) - \sum_{j=1}^{k-1} V_j |x_j(t_k)|^2 \tag{3.255}$$

(iv)
$$x_k(t) = V_k^{-1} \left[K_X(t, t_k) - \sum_{j=1}^{k-1} V_j x_j(t) \overline{x_j(t_k)} \right] \tag{3.256}$$

If the set $\{t_k\}$ is infinite, then the representation converges in the mean-square; if it is finite, then the expansion reduces to a finite sum. In either case, the representation is exact at each t_k. In particular, $x_k(t_k)$ = 1 for k = 1, 2,.... and $x_k(t_j)$ = 0 for $k > j$. ■

***Example* 3.14.** We apply Theorem 3.9 to a real random vector $(X(1), X(2))'$. Proceeding recursively, according to the theorem, with t_1 = 1 and t_2 = 2:

$$Z_1 = X(t_1) = X(1)$$

$$V_1 = K_X(t_1, t_1) = \text{Var}[X(1)]$$

$$x_1(i) = \text{Var}[X(1)]^{-1} K_X(i,1),\ i = 1, 2$$

$$Z_2 = X(t_2) - Z_1 x_1(t_2)$$

$$= X(2) - X(1)\text{Var}[X(1)]^{-1} K_X(2,1)$$

$$V_2 = K_X(t_2, t_2) - V_1 |x_1(t_2)|^2$$

$$= \text{Var}[X(2)] - \text{Var}[X(1)]^{-1} K_X(2,1)^2$$

$$x_2(i) = V_2^{-1}[K_X(i,2) - V_1 x_1(i) x_1(2)],\ i = 1, 2$$

All quantities needed for $x_2(i)$ are available. Let $X(1)$ and $X(2)$ have variance 1 and covariance c. Then $Z_1 = X(1)$, $V_1 = 1$, $x_1(i) = K_X(i, 1)$, $Z_2 = X(2) - cX(1)$, $V_2 = 1 - c^2$, and

$$x_2(i) = (1-c^2)^{-1}[K_X(i,2) - x_1(i)x_1(2)]$$

$$= (1-c^2)^{-1}[K_X(i,2) - cK_X(i,1)]$$

Note that Z_1 and Z_2 are orthogonal:

$$E[Z_1 Z_2] = E[X(1)(X(2) - cX(1))]$$

$$= E[X(1)X(2)] - cE[X(1)^2] = 0$$

Let us check to see if

$$X(i) = Z_1 x_1(i) + Z_2 x_2(i)$$

The expression for $x_2(i)$ gives $x_2(1) = 0$. Thus, with $i = 1$, the canonical expansion yields

$$Z_1 x_1(1) = X(1)K_X(1, 1) = X(1)$$

Finally, putting $i = 2$ into the canonical expansion yields

$$X(1)x_1(2) + (X(2) - cX(1))x_2(2) = X(1)c + (X(2) - cX(1))(1 - c^2)^{-1}(1 - c^2) = X(2) \quad \blacksquare$$

Example 3.15. Consider a zero-mean WS stationary random signal $X(t)$ with covariance function

$$K_X(t, t') = be^{-b|t - t'|}$$

We will find a canonical representation of $X(t)$ over the set of nonnegative integers $\{0, 1, 2,...\}$. Applying Theorem 3.9 with $t_k = k - 1$ gives

$$Z_1 = X(0)$$

$$V_1 = K_X(0, 0) = b$$

$$x_1(t) = b^{-1}K_X(t,t_1) = e^{-b|t|}$$

$$Z_2 = X(1) - Z_1x_1(1) = X(1) - X(0)e^{-b}$$

$$V_2 = K_X(1,1) - bx_1(1)^2 = b(1-e^{-2b})$$

$$x_2(t) = b^{-1}(1-e^{-2b})^{-1}(be^{-b|t-1|} - be^{-b|t|}e^{-b})$$

$$= (1-e^{-2b})^{-1}(e^{-b|t-1|} - e^{-b(1+|t|)})$$

$$Z_3 = X(2) - [Z_1x_1(2) + Z_2x_2(2)]$$

$$= X(2) - \left[X(0)e^{-2b} + (X(1) - X(0)e^{-b})\frac{e^{-b} - e^{-3b}}{1-e^{-2b}}\right]$$

$$= X(2) - X(0)e^{-2b} - (X(1) - X(0)e^{-b})e^{-b}$$

$$= X(2) - X(1)e^{-b}$$

$$V_3 = b - bx_1(2)^2 - b(1-e^{-2b})x_2(2)^2$$

$$= b[1-e^{-4b} - (1-e^{-2b})^{-1}(e^{-b} - e^{-3b})^2]$$

$$= b(1-e^{-2b})$$

$$x_3(t) = b^{-1}(1-e^{-2b})^{-1}\{be^{-b|t-2|} - be^{-b|t|}e^{-2b} - b(1-e^{-2b})$$

$$\times[(1-e^{-2b})^{-1}(e^{-b|t-1|} - e^{-b(1+|t|)})(1-e^{-2b})^{-1}(e^{-b} - e^{-3b})]\}$$

$$= (1-e^{-2b})^{-1}[e^{-b|t-2|} - e^{-b(1+|t-1|)}]$$

Although we will not go through the details, one could employ mathematical induction together with the recursion equations of Theorem 3.9 to show that, for $k = 2, 3, \ldots$,

$$Z_k = X(k-1) - e^{-b}X(k-2)$$

$$V_k = b(1-e^{-2b})$$

$$x_k(t) = (1-e^{-2b})^{-1}[e^{-b|t-k+1|} - e^{-b(1+|t-k+2|)}]$$

Note that, for $n = 0, 1, 2,...$ and $k = 1, 2,..., n + 1$,

$$x_k(n) = e^{-(n-k+1)b}$$

For $k > n + 1$, Theorem 3.9 yields $x_k(n) = 0$. Thus, at the nonnegative integers, the canonical expansion takes the form

$$X(0) = Z_1 x_1(0) = Z_1$$

$$X(1) = Z_1 x_1(1) + Z_2 x_2(1) = Z_1 e^{-b} + Z_2$$

$$X(2) = Z_1 x_1(2)+Z_2 x_2(2)+Z_3 x_3(2)=Z_1 e^{-2b}+Z_2 e^{-b}+Z_3$$

$$\vdots$$

$$X(n) = \sum_{k=1}^{n+1} Z_k e^{-(n-k+1)b}$$

Because the Z_k form an uncorrelated collection of zero-mean random variables, applying the expansion for X at n and m with $n > m$ yields the covariance

$$K_X(n,m) = \text{Cov}\left[\sum_{k=1}^{n+1} Z_k e^{-(n-k+1)b}, \sum_{k=1}^{m+1} Z_k e^{-(n-k+1)b}\right]$$

$$= E\left[\left(\sum_{k=1}^{n+1} Z_k e^{-(n-k+1)b}\right)\left(\sum_{k=1}^{m+1} Z_k e^{-(m-k+1)b}\right)\right]$$

$$- E\left[\sum_{k=1}^{n+1} Z_k e^{-(n-k+1)b}\right] E\left[\sum_{k=1}^{m+1} Z_k e^{-(m-k+1)b}\right]$$

$$= \sum_{k=1}^{n+1}\sum_{j=1}^{m+1} E[Z_k Z_j] e^{-(n-k+1)b-(m-j+1)b}$$

$$= \sum_{k=1}^{m+1} V_k e^{-b(n+m-2k+2)}$$

$$= e^{-b(n-m)} \sum_{k=1}^{m+1} V_k e^{-2b(m-k+1)}$$

$$= be^{-b(n-m)} \left[e^{-2bm} + \sum_{k=2}^{m+1} (1-e^{-2b}) e^{-2b(m-k+1)} \right]$$

$$= be^{-b(n-m)}$$

where we have applied the variance results $V_1 = b$ and $V_k = b(1 - e^{-2b})$ for $k > 1$, and where the last equality follows from evaluating the sum via its telescoping property. The result agrees (as it must) with the assumed covariance of the random function. ■

Exercises for Chapter 3

1. Taking into account that the mean of a sum of real random variables is the sum of the means, prove the same proposition for complex random variables from the definition of a complex random variable in terms of its real and imaginary parts.
2. If $X = X_1 + jX_2$ is a complex random variable, show that

$$\mathrm{Var}[X] = \mathrm{Var}[X_1] + \mathrm{Var}[X_2]$$

3. For a complex random variable X and a complex number a, show that

$$\mathrm{Var}[aX] = |a|^2\mathrm{Var}[X]$$

4. If $X = X_1 + jX_2$, then we can consider the joint density of X_1 and X_2, which we denote by $f_X(x_1, x_2)$. Defining expectation of X by

$$E[X] = \int_{-\infty}^{\infty}\int_{-\infty}^{\infty}(x_1 + jx_2)f_X(x_1, x_2)\,dx_1 dx_2$$

show that $E[X] = E[X_1] + j\,E[X_2]$.

5. Show that, if $X_1, X_2,\ldots, X_n$ are complex random variables and $a_1, a_2,\ldots, a_n$ are complex numbers, then

$$\mathrm{Var}\left[\sum_{k=1}^{n} a_k X_k\right] = \sum_{k=1}^{n}\sum_{i=1}^{n} a_k\bar{a}_i \mathrm{Cov}[X_k, X_i]$$

Supposing $X_1, X_2,\ldots, X_n$ are uncorrelated, deduce that

$$\mathrm{Var}\left[\sum_{k=1}^{n} a_k X_k\right] = \sum_{k=1}^{n} |a_k|^2\,\mathrm{Var}[X_k]$$

6. Demonstrate Eqs. 2.25, 2.26, and 2.27 for complex random functions.
7. For any elements x and y in an inner product space, and complex number c, show that $\langle x, cy\rangle = \bar{c}\langle x, y\rangle$.
8. The *Cauchy-Schwarz inequality* states that, in an inner product space,

$$|\langle x, y\rangle| \le \|x\|\,\|y\|$$

From the Schwarz inequality, derive the *triangle inequality*,

$$\|x+y\| \leq \|x\| + \|y\|$$

Express the Schwarz and triangle inequalities for the second-moment inner product.

9. Show that the elements x and y in an inner product space are orthogonal if and only if

$$\|x+y\|^2 = \|x\|^2 + \|y\|^2$$

10. For a domain $T \subset \Re^n$, the space $L^2(T)$ is composed of all square-integrable functions on T, namely, those functions $x(t)$ for which

$$\int_T |x(t)|^2 \, dt < \infty$$

Show that $L^2(T)$ is an inner product space under the inner product

$$\langle x, y\rangle = \int_T x(t)\overline{y(t)}\, dt$$

where the integral of a complex function is defined via its real and imaginary parts by

$$\int_T (x_1 + jx_2)(t)\, dt = \int_T x_1(t)\, dt + j\int_T x_2(t)\, dt$$

Axiom L2 is satisfied if we identify any two functions $x(t)$ and $y(t)$ for which

$$\int_T |x(t) - y(t)|\, dt = 0$$

11. State the Schwarz and triangle inequalities for $L^2(T)$.
12. Show that the real trigonometric system is orthogonal in $L^2(-\pi, \pi)$. The nonnormalized system is composed of the following functions: 1, $\cos t$, $\cos 2t$,..., $\sin t$, $\sin 2t$,....
13. Suppose x_1, x_2,... compose a linearly independent set in an inner product space. Let $u_1 = x_1$, $u_2 = x_2 - c_1x_1$,..., and

$$u_{n+1} = x_{n+1} - \sum_{k=1}^{n} c_k u_k$$

where

$$c_k = \frac{\langle x_{n+1}, u_k \rangle}{\langle u_k, u_k \rangle}$$

for $k = 1, 2, \ldots, n$. Show that $u_1, u_2, \ldots$ comprise an orthogonal system. This method of construction is known as the *Gram-Schmidt orthogonalization procedure*. It is a corollary of the Gram-Schmidt orthogonalization procedure that every finite-dimensional subspace of an inner product space has an orthonormal basis.

14. Apply the Gram-Schmidt process to the functions $1, t, t^2, \ldots$ in $L^2(-1, 1)$. Stop at $n = 3$.
15. Show that, if $\{u_1, u_2, \ldots, u_n\}$ is a finite orthonormal set in an inner product space and

$$x = \sum_{k=1}^{n} c_k u_k$$

then $c_k = \langle x, u_k \rangle$ for $k = 1, 2, \ldots, n$, and

$$\|x\| = \sum_{k=1}^{n} |c_k|$$

$\langle x, u_k \rangle$ is the *Fourier coefficient* of x relative to u_k.

16. Show that every finite orthonormal set is independent.
17. Suppose $\mathcal{W}$ is an n-dimensional subspace of the inner product space $\mathcal{V}$ and $x \in \mathcal{V}$. Show that there exists a unique decomposition $x = y + z$, where $y \in \mathcal{W}$ and z is orthogonal to $\mathcal{W}$, meaning that $\langle z, w \rangle = 0$ for every $w \in \mathcal{W}$. In particular, show that

$$y = \sum_{k=1}^{n} \langle x, u_k \rangle u_k$$

where $\{u_1, u_2, \ldots, u_n\}$ is an orthonormal basis for $\mathcal{W}$. Show that, if $v \in \mathcal{V}$, $v \neq z$, and $x - v \in \mathcal{W}$, then

$$\|v\|^2 = \|z\|^2 + \|v - z\|^2$$

Conclude that $\|v\| \geq \|z\|$. The norm of z is the distance from x to $\mathcal{W}$ and y is the orthogonal projection of x into $\mathcal{W}$.

18. Suppose $u_1, u_2, \ldots$ comprise an orthonormal system in an inner product space $\mathcal{V}$ and $x \in \mathcal{V}$. Deduce from the preceding exercise that, if $c_1, c_2, \ldots, c_n$ are scalars, then

$$\left\| x - \sum_{k=1}^{n} \langle x, u_k \rangle u_k \right\| \leq \left\| x - \sum_{k=1}^{n} c_k u_k \right\|$$

This is the general form of Eq. 3.6.

19. Suppose $u_1, u_2, \ldots$ comprise an orthonormal system in an inner product space $\mathcal{V}$ and $x \in \mathcal{V}$. Show that

$$\left\| x - \sum_{k=1}^{m} \langle x, u_k \rangle u_k \right\| = \|x\| - \sum_{k=1}^{m} \left| \langle x, u_k \rangle \right|^2$$

This is the general form of Eq. 3.7.

20. Show *Bessel's inequality*,

$$\sum_{k=1}^{\infty} \left| \langle x, u_k \rangle \right|^2 \leq \|x\|$$

Show that when there is equality in Bessel's inequality, meaning that

$$\sum_{k=1}^{\infty} \left| \langle x, u_k \rangle \right|^2 = \|x\|$$

(which is now called *Parseval's identity*), then x is equal to its Fourier series,

$$x = \sum_{k=1}^{\infty} \langle x, u_k \rangle u_k$$

21. An orthonormal system $\mathcal{U} = \{u_1, u_2, \ldots\}$ in an inner product space $\mathcal{V}$ is said to be *complete* if every vector is equal to its Fourier series relative to $\mathcal{U}$. Show that the following statements are equivalent: (i) $\mathcal{U}$ is complete; (ii) Parseval's identity holds for every $x \in \mathcal{V}$; and (iii) the following form of Parseval's identity holds for every pair of vectors $x, y \in \mathcal{V}$:

$$\langle x, y\rangle = \sum_{k=1}^{\infty} \langle x, u_k\rangle\overline{\langle y, u_k\rangle}$$

22. The *Grammian* of a finite set $S = \{x_1, x_2, \ldots, x_n\}$ of elements in an inner product space is defined by the matrix

$$\mathbf{G} = \begin{pmatrix} \langle x_1, x_1\rangle & \langle x_1, x_2\rangle & \cdots & \langle x_1, x_n\rangle \\ \langle x_2, x_1\rangle & \langle x_2, x_2\rangle & \cdots & \langle x_2, x_n\rangle \\ \vdots & \vdots & \ddots & \vdots \\ \langle x_n, x_1\rangle & \langle x_n, x_2\rangle & \cdots & \langle x_n, x_n\rangle \end{pmatrix}$$

Show that the elements of S are linearly independent if and only if the Gram determinant, det[$\mathbf{G}$], is nonzero.

23. A random process $X(t)$ is said to be *continuous* in the mean-square at the point t if

$$\lim_{h\to 0} E[|X(t+h) - X(t)|^2] = 0$$

Show that $X(t)$ is MS continuous at t if and only if $R_X(t_1, t_2)$ is continuous at the point $t_1 = t_2 = t$. Hint: Expand the mean-square expression in the limit.

24. Based on Exercise 3.23, show that the Wiener process is MS continuous.
25. Show that a WS stationary random process is continuous for all t if and only if its autocorrelation function $r_X(\tau)$ is continuous at $\tau = 0$.
26. Because $E[X\overline{Y}]$ defines an inner product for finite-second-moment random variables, various MS limit properties hold. Reinterpret the appropriate inner-product space proofs in the current context to prove that, if

$$\underset{n\to\infty}{\text{l.i.m.}}\, X_n = X \quad \text{and} \quad \underset{n\to\infty}{\text{l.i.m.}}\, Y_n = Y$$

then the following limit properties hold

(i) $\lim_{n\to\infty} E[X_n] = E[X]$

(ii) $\lim_{n\to\infty} E[|X_n|^2] = E[|X|^2]$

(iii) $\lim_{n\to\infty} E[X_n Y_n] = E[XY]$

(iv) $\underset{n\to\infty}{\text{l.i.m.}}\ aX_n + bY_n = aX + bY$ for any constants a, b.

27. Show that the MS derivative is a linear operator; namely, if $X(t)$ and $Y(t)$ are MS differentiable and a and b are constants, then

$$\frac{d}{dt}(aX(t)+bY(t)) = aX'(t)+bY'(t)$$

28. Suppose $\mathbf{X} = (X_1, X_2, X_3)'$ is a random vector for which X_1, X_2, and X_3 are jointly normal zero-mean random variables possessing covariance matrix

$$\mathbf{K} = \begin{pmatrix} \sigma^2 & \alpha & 0 \\ \alpha & \sigma^2 & \alpha \\ 0 & \alpha & \sigma^2 \end{pmatrix}$$

Find the Karhunen-Loeve expansion for $\mathbf{X}$.

29. Suppose $\mathbf{X} = (X_1, X_2, X_3)'$ is a random vector for which X_1, X_2, and X_3 are jointly normal zero-mean random variables possessing covariance matrix

$$\mathbf{K} = \begin{pmatrix} \sigma^2 & 0 & \alpha \\ 0 & \sigma^2 & 0 \\ \alpha & 0 & \sigma^2 \end{pmatrix}$$

Find the Karhunen-Loeve expansion for $\mathbf{X}$.

30. Consider a random function $X(t)$. Let $N(t)$ be white noise with covariance function $\sigma^2\delta(t - t')$, $X(t)$ and $N(t)$ be uncorrelated, and $Y(t) = X(t) + N(t)$. Show that the covariance functions of $X(t)$ and $Y(t)$ possess the same eigenfunctions, so that the Karhunen-Loeve expansions of $X(t)$ and $Y(t)$ possess the same coordinate functions, and that the eigenvalues are related by $\lambda_{X,n} + \sigma^2 = \lambda_{Y,n}$, where $\lambda_{X,n}$ and $\lambda_{Y,n}$ are the eigenvalues corresponding to $X(t)$ and $Y(t)$, respectively.

31. We have stated in the text that the following statement is not a valid proposition: if $g_n(t)$ is a nonnegative function on the interval T and

$$\lim_{n\to\infty} \int_T g_n(t)\,dt = 0$$

then $g_n(t) \to 0$ for (almost all) $t \in T$. Let $I_{[a,b]}(t)$ be the indicator function for the interval $[a, b]$ relative to the interval $[0, 1]$, which means that

$[a, b] \subset [0, 1]$, $I_{[a,b]}(t) = 1$ if $a \le t \le b$, and $I_{[a,b]}(t) = 0$ if $0 \le t \le 1$ but $t \notin [a, b]$. Show that the following sequence of functions provides a counterexample on $[0, 1]$ to the proposed proposition: $I_{[0,1/2]}(t)$, $I_{[1/2,1]}(t)$, $I_{[0,1/4]}(t)$, $I_{[1/4,1/2]}(t)$, $I_{[1/2,3/4]}(t)$, $I_{[3/4,1]}(t)$, $I_{[0,1/8]}(t)$, $I_{[1/8,1/4]}(t)$,....

32. We have stated in the text that the proposed proposition of Example 3.31 is valid for functions defined over discrete domains. Prove that, if $g_n(k)$ is a nonnegative function of $k = 1, 2,...$ and

$$\lim_{n \to \infty} \sum_{k=1}^{\infty} g_n(k) = 0$$

then $g_n(k) \to 0$ for all k as $n \to \infty$.

33. Show that the converse of the proposed proposition in Exercise 3.31 is also not valid by considering the sequence of functions defined by

$$g_n(t) = \begin{cases} n, & \text{if } 0 \le t \le 1/n \\ 0, & \text{otherwise} \end{cases}$$

In the text we stated that this converse is valid under the added assumption that there exists an integrable function bounding all functions $g_n(t)$. Prove the following simpler result: the converse is valid under the added assumptions that T is an interval of finite length and there exists a constant A such that $g_n(t) \le A$ for all n and t.

34. The first part of this exercise requires familiarity with sets of measure zero. Prove that, if the limit of the integral is 0 in Exercise 3.31 and it is known that the sequence $\{g_n(t)\}$ converges almost everywhere in T, then it must converge almost everywhere to 0. This proposition can be used for an alternate proof of the Karhunen-Loeve theorem using the generalized Bessel inequality, after having shown that Z_1, Z_2,... are orthogonal. First show that the series in Eq. 3.26 is MS convergent by showing it to be Cauchy convergent. To do this, show that, for $n \ge m$,

$$E\left[\left|\sum_{k=m}^{n} Z_k u_k(t)\right|^2\right] \le \sum_{k=m}^{\infty} E[|Z_k|^2]|u_k(t)|^2$$

and apply Mercer's theorem. Then show there is equality in the generalized Bessel inequality.

35. The *Rademacher functions* are square-type waveforms defined recursively on the interval $[0, 1)$ and then extended periodically to all $\Re$. On $[0, 1)$ they are defined by

$$r_0(t) = 1$$

$$r_1(t) = \begin{cases} 1, & \text{if } 0 \le t < 1/2 \\ -1, & \text{if } 1/2 \le t < 1 \end{cases}$$

$$\vdots$$

$$r_{n+1}(t) = r_1(2^n t) \text{ for } n > 1$$

Each Rademacher function alternates between 1 and −1 on subintervals of length 2^{-n} over [0, 1). The Rademacher functions form an orthonormal system in $L^2[0, 1)$, but the system is not complete. The *Walsh functions* are given by the set of all finite products of the Rademacher functions and they form a complete orthonormal system. The Walsh functions are defined by $w_0(t) = 1$ and, for $n \ge 1$,

$$w_n(t) = r_{n_1+1}(t) r_{n_2+1}(t) \cdots r_{n_p+1}(t)$$

where $n_1 < n_2 < \ldots < n_p$ and

$$n = \sum_{i=1}^{p} 2^{n_i}$$

The notation is cumbersome, its only purpose being to specify some ordering to the manner in which all products of the Rademacher functions are to be assembled. The first eight Walsh functions are given by $w_0 = r_0$, $w_1 = r_1$, $w_2 = r_2$, $w_3 = r_1 r_2$, $w_4 = r_3$, $w_5 = r_1 r_3$, $w_6 = r_2 r_3$, and $w_7 = r_1 r_2 r_3$. Write out the first four terms of the expansion of a random function $X(t)$ in terms of the Walsh orthonormal system. Write down the mean and correlation expressions for the first four random coefficients.

36. Apply the mean and correlation expressions found in Exercise 3.35 for the Walsh system to the random function of Example 2.5.
37. Apply the mean and correlation expressions found in Exercise 3.35 for the Walsh system to the random function of Example 2.6 under the assumptions that A and B are uncorrelated and $\text{Var}[A] = \text{Var}[B] = 1$.
38. We used Mercer's theorem (Eq. 3.42) to prove convergence of the Karhunen-Loeve expansion when the covariance function is continuous. If the covariance function is an ordinary (nongeneralized) function, then the series in Eq. 3.42 converges in the L^2 norm on $T \times T$,

$$\lim_{n\to\infty}\int_T\int_T\left|K_X(t,t')-\sum_{k=1}^{n}\lambda_k u_k(t)\overline{u_k(t')}\right|^2 dtdt' = 0$$

Use this to demonstrate convergence of the Karhunen-Loeve expansion to $X(t)$.

39. Suppose Z_1, Z_2, and Z_3 are jointly normal zero-mean random variables having the covariance matrix

$$\mathbf{K}=\begin{pmatrix} k_{11} & k_{12} & k_{13}\\ k_{21} & k_{22} & k_{23}\\ k_{31} & k_{32} & k_{33}\end{pmatrix}$$

and

$$X(t) = Z_1u_1(t) + Z_2u_2(t) + Z_3u_3(t)$$

Write out the decorrelation procedure in terms of **K**.

40. Find an expression for the variance of Z_k in Eq. 3.83.
41. Find the eigenvalues for the covariance matrix of Eq. 3.132 when ρ = 0.95, except let the matrix dimensions be 4×4. Find the corresponding Karhunen-Loeve transform matrix, the output covariance matrix, and the decorrelation and energy-packing coefficients. Suppose now that the Karhunen-Loeve transform derived for $\rho = 0.95$ is applied to a random vector possessing the same covariance matrix, except that $\rho = 0.91$. In this case, find the output covariance matrix and the decorrelation and energy-packing coefficients.
42. The *Haar transform* is based on sampling square-wave basis functions defined over the interval [0, 1). For $0 \le k < \log_2 n$ and $1 \le m \le 2^r$, these are defined by

$$h_{0,0}(t)=\frac{1}{\sqrt{n}}$$

$$h_{k,m}(t)=\begin{cases} n^{-1/2}2^{r/2}, & \text{for } (m-1)2^{-r}\le t<(m-1/2)2^{-r}\\ -n^{-1/2}2^{r/2}, & \text{for } (m-1/2)2^{-r}\le t<m2^{-r}\\ 0, & \text{elsewhere}\end{cases}$$

The Haar system is a complete orthonormal system. For n = 8, sampling the first eight Haar functions yields the matrix

$$\mathbf{T} = \begin{pmatrix} a & a & a & a & a & a & a & a \\ a & a & a & a & -a & -a & -a & -a \\ 1/2 & 1/2 & -1/2 & -1/2 & 0 & 0 & 0 & 0 \\ 0 & 0 & 0 & 0 & 1/2 & 1/2 & -1/2 & -1/2 \\ 2a & -2a & 0 & 0 & 0 & 0 & 0 & 0 \\ 0 & 0 & 2a & -2a & 0 & 0 & 0 & 0 \\ 0 & 0 & 0 & 0 & 2a & -2a & 0 & 0 \\ 0 & 0 & 0 & 0 & 0 & 0 & 2a & -2a \end{pmatrix}$$

where $a = 1/\sqrt{8}$. Find the output covariance matrix when the Haar transform is applied to a random vector possessing the covariance matrix of Eq. 3.132 with $\rho = 0.91$. Find the decorrelation and energy-packing coefficients.

43. The *discrete sine transform* (*DST*) is obtained by sampling sine waves of increasing frequency. The $n \times n$ DST is defined by the matrix **S** with components

$$c_{km} = \sqrt{\frac{2}{n+1}} \sin \frac{(k+1)(m+1)\pi}{n+1}$$

for $k, m = 0, 1, 2,\ldots, n-1$. Compute the DST matrix for $n = 8$ and find the output covariance matrix when the DST is applied to a random vector possessing the covariance matrix of Eq. 3.132 with $\rho = 0.91$. Find the decorrelation and energy-packing coefficients.

44. Redo Example 3.10 for the Hadamard transform with $N = 8$. Consider compression using 1, 2,..., 7 components.

45. Redo Example 3.10 for the Haar transform with $N = 8$. Consider compression using 1, 2,..., 7 components.

46. Consider the discrete-time process $X(n)$ defined by $X(n) = Y(1) + Y(2) + \cdots + Y(n)$ for $n = 1, 2,\ldots, 32$, where $Y(n)$ is a binary random variable with $P(Y(n) = 1) = 0.25$. Simulate the process $X(n)$ 100 times for $n = 1, 2,\ldots, 32$. Apply the WHT and DCT to each with the signal blocked into lengths of 8 points. Use the realizations to estimate the error of Eq. 3.114 for compression using 2 points out of 8 and using 4 points out of 8 for both the WHT and DCT.

47. Express all of the equations in Section 3.6 without the restriction to real matrices.

48. Reconsider Example 3.12 for the covariance function

$$K_X(t, t') = g(t)g(t')\delta(t - t')$$

letting $a_k(t) = \exp[j\omega_k t]/g(t)$.

49. Comparing Eqs. 3.182 and 3.183, we concluded that $a_{mn} = \bar{a}_{nm}$ and then claimed that this meant that the coefficients a_{mn} can be expressed by an expansion of the form given in Eq. 3.184. Prove this claim.
50. We only demonstrated the existence of a canonical expansion of the covariance function for functions of a single variable. Show that the method extends to functions of N variables. Begin with the Fourier series of the covariance function, which for N variables takes the form

$$K_X(t_1, t_2,..., t_N; s_1, s_2,..., s_N) =$$

$$\sum_{m_1=-\infty}^{\infty} \cdots \sum_{m_N=-\infty}^{\infty} \sum_{n_1=-\infty}^{\infty} \cdots \sum_{n_N=-\infty}^{\infty} a_{m_1,...,m_N,n_1,...,n_N} \exp\left[j\left(\sum_{p=1}^{N} \omega_{m_p} t_p - \sum_{p=1}^{N} \omega_{n_p} s_p \right) \right]$$

51. Show that, if $X(t)$ is WS stationary and $r_X(\tau) = e^{-a|\tau|}$, $a > 0$, then

$$S_X(\omega) = \frac{2a}{a^2 + \omega^2}$$

What can be concluded about the process regarding ergodicity?
52. Find the power spectral density for a WS stationary process $X(t)$ for which $r_X(\tau) = \exp[-a\tau^2]$, where $a > 0$.
53. Find the power spectral density for a WS stationary process $X(t)$ for which $r_X(\tau) = \exp[-a\tau^2] \cos \omega\tau$, where $a > 0$.
54. According to Example 2.17, the Poisson increment process $X(t)$ is WS stationary. Suppose the Poisson process has mean λt and the increment interval is of length $r = 1/\lambda$. Show that

$$S_X(\omega) = \frac{1}{\lambda}\left(\frac{\sin \omega/2\lambda}{\omega/2\lambda} \right)^2$$

Referring to Example 3.13, show that $X_r(t)$ has the claimed power spectral density.
55. Let $Y(t)$ be the binary random function that is randomly defined, with equal probability, to be 1 or -1 on each interval of length r along the entire real line, with one of the intervals being $[0, r)$. Values for different intervals are independent. The mean of $Y(t)$ is identically 0. If t and t' lie in the same interval, then $E[Y(t)Y(t')] = 1$; otherwise, $E[Y(t)Y(t')] = E[Y(t)]E[Y(t')] = 0$. Let A be a uniformly distributed random variable on $[0, r)$ and consider the randomly shifted random function $X(t) = Y(t - A)$. Show that $X(t)$ is WS stationary with autocorrelation function

$$r_X(\tau) = 1 - \frac{|\tau|}{a}$$

for $|\tau| < a$ and $r_X(\tau) = 0$ otherwise. Find the power spectral density.

56. Let $X(t)$ be WS stationary and

$$Y(n) = \frac{1}{n}\sum_{k=1}^{n} X(k)$$

Show that

$$E[Y(n)^2] = \frac{1}{2\pi n^2}\int_{-\infty}^{\infty} S_X(\omega)\frac{\sin^2 n\omega/2}{\sin^2 \omega/2}\,d\omega$$

57. Since $k_X(\tau)$ is an even function when $X(t)$ is real, the power spectral density can be expressed as a cosine transform. Express the spectral-density/covariance transform pair in terms of a cosine transform.

58. For discrete random functions, canonical expansions take the form

$$X(n) = \mu_X(n) + \sum_{k=1}^{\infty} Z_k x_k(n)$$

where $n = 1, 2, \ldots$. For instance, Eq. 3.44 gives the discrete Karhunen-Loeve expansion. For generating coefficients for discrete canonical expansions by linear functions, Eq. 3.143 is replaced by

$$Z_k = \sum_{n=1}^{\infty} X(n)\overline{a_k(n)}$$

where the sequence $\{a_k(n)\}$ is a sequence in l^2, meaning that

$$\sum_{n=1}^{\infty} |a_k(n)|^2 < \infty$$

For any two sequences $\{c(n)\}, \{d(n)\} \in l^2$, their inner product in l^2 is defined by

$$\langle c, d\rangle = \sum_{n=1}^{\infty} c(n)\overline{d(n)}$$

Show that, in the discrete setting, Eq. 3.144 becomes

$$E[Z_k \overline{Z_j}] = \sum_{n=1}^{\infty} \sum_{m=1}^{\infty} \overline{a_k(n) a_j(m) K_X(n,m)}$$

Find the discrete versions of Eqs. 3.146, 3.147, 3.148, 3.151, and 3.153.

59. Consider completeness for the discrete Karhunen-Loeve expansion.
60. Find the orthogonality condition for the coefficients of Eq. 3.245.
61. Find the variance of the coefficient defined by Eq. 3.245.
62. Fill in the steps to obtain Eq. 3.246.
63. For vector canonical expansions, find the formula corresponding to Eq. 3.153.
64. Extend the procedure of Section 3.7.2 to vector canonical expansions.
65. Define completeness for vector random functions and state the appropriate extension of Theorem 3.5.
66. State the extension of Theorem 3.6 to vector random functions. Explain it in terms of Fourier series in Hilbert spaces.
67. Apply Theorem 3.9 to the random vector of Example 3.3.
68. Apply Theorem 3.9 to a random vector $\mathbf{X} = (X_1, X_2, X_3)'$ possessing the covariance matrix of Exercise 3.39.
69. Apply Theorem 3.9 to a random vector $\mathbf{X} = (X_1, X_2, ..., X_n)'$ having covariance matrix

$$\mathbf{K} = \begin{pmatrix} \sigma_1^2 & 0 & \cdots & 0 \\ 0 & \sigma_2^2 & \cdots & 0 \\ \vdots & \vdots & \ddots & \vdots \\ 0 & 0 & \cdots & \sigma_n^2 \end{pmatrix}$$

What do you conclude?

70. Apply Theorem 3.9 to the Poisson process with mean λt over the integers 0, 1, 2,....

Chapter 4

Optimal Filtering

4.1. Optimal Mean-Square-Error Filters

A fundamental problem in engineering is *estimation* (*prediction*) of an outcome of an unobserved random variable based on outcomes of a set of observed random variables. In the context of random processes, we wish to estimate values of a random function $Y(s)$ based on observation of a random function $X(t)$. From a filtering perspective, we desire a system which, given an input $X(t)$, produces an output $\hat{Y}(s)$ that *best* estimates $Y(s)$, where the goodness of the estimator is measured by a probabilistic error measure between the estimator and the random variable it estimates.

For two random variables, X to be observed and Y to be estimated, we desire a function of X, say $\psi(X)$, such that $\psi(X)$ minimizes the *mean-square error* (*MSE*)

$$MSE\langle\psi\rangle = E[|Y - \psi(X)|^2] \tag{4.1}$$

If such an *estimation rule* ψ can be found, then $\psi(X)$ is called an *optimal mean-square-error estimator* of Y in terms of X. To make the estimation problem mathematically or computationally tractable, or to obtain an estimation rule with desirable properties, we often restrict the class of estimation rules over which the MSE minimum is to be achieved. The constraint trade-off is a higher MSE in return for a tractable design or desirable filter properties. In this section we examine the theoretically best solution over all functions of X. Subsequently, we focus on linear estimation in order to discover systems that provide optimal linear filtering. The theory, both linear and nonlinear, applies to complex random variables; however, our exposition assumes that all random variables are real valued.

4.1.1. Conditional Expectation

If two random variables X and Y possess joint density $f(x, y)$ and an observation of X, say $X = x$, is given, then the density $f(y|x)$ of the conditional random variable Y given $X = x$ is defined by Eq. 1.135. The conditional expectation (or mean) $E[Y|x]$ (or $\mu_{Y|x}$) is defined by Eq. 1.137. Letting x vary, the plot of $E[Y|x]$ is the *regression curve* of Y on X. For an observed value x of X, $E[Y|x]$ is a parameter; however, since X is a random variable, the observation of x is random and therefore the conditional expectation can be viewed as a random variable that is a function of X, in which case it is written $E[Y|X]$ (or $\mu_{Y|X}$).

Example **4.1.** Let the random variables X and Y be uniformly jointly distributed over the region bounded by the curves $y = x^2$, $y = 0$, and $x = 1$. The marginal densities are given by

$$f_X(x) = 3\int_0^{x^2} dy = 3x^2$$

for $0 \leq x \leq 1$, and 0 elsewhere, and

$$f_Y(y) = 3\int_{\sqrt{y}}^{1} dx = 3(1-\sqrt{y})$$

for $0 \leq y \leq 1$, and 0 elsewhere. For $0 < x \leq 1$, the conditional density $f(y|x)$ is defined by $f(y|x) = x^{-2}$ for $0 \leq y \leq x^2$ and $f(y|x) = 0$ for all other y. Uniformity of the joint density has resulted in uniformity of the conditional random variable: the conditioning $X = x$ has resulted in the uniform distribution of $Y|x$ over the interval $[0, x^2]$. Hence, the conditional mean must lie on the midpoint of the interval and the regression curve is given by $E[Y|x] = x^2/2$. Since this relation applies to every possible outcome of X, $\mu_{Y|X} = E[Y|X] = X^2/2$.

Since $E[Y|X]$ is a random variable, it has its own probability distribution function,

$$F_{E[Y|X]}(z) = P(E[Y|X] \leq z)$$

$$= P(X \leq \sqrt{2z})$$

$$= \int_0^{\sqrt{2z}} 3z^2\, dz = (2z)^{3/2}$$

for $0 \leq z \leq 1/2$. Differentiation yields the density of the conditional mean, $f_{E[Y|X]}(z) = 3\sqrt{2z}$. Thus, the expected value of the conditional mean is

$$E[E[Y|X]] = \int_0^{1/2} 3\sqrt{2}z^{3/2}\, dz = \frac{3}{10}$$

which is the mean of the unconditioned variable Y. Since $E[Y] = E[E[Y|X]]$, $E[Y|X]$ provides unbiased estimation of $E[Y]$. This is a general property of $E[Y|X]$. ■

Theorem 4.1. The expectation of the conditional expectation satisfies the property

$$E[E[Y|X]] = E[Y] \quad ■ \tag{4.2}$$

The continuous and discrete forms of Theorem 4.1 are

$$E[Y] = \int_{-\infty}^{\infty} E[Y|x] f_X(x)\, dx \tag{4.3}$$

$$E[Y] = \sum_x E[Y|x] P(X = x) \tag{4.4}$$

respectively. We demonstrate the discrete case:

$$\begin{aligned} E[E[Y|X]] &= \sum_x E[Y|x] P(X = x) \\ &= \sum_x \sum_y y P(Y = y | X = x) P(X = x) \\ &= \sum_y y \sum_x P(X = x, Y = y) \\ &= \sum_y y P(Y = y) \\ &= E[Y] \end{aligned} \tag{4.5}$$

If X and Y are independent, then Theorem 4.1 reduces to $E[Y] = E[Y]$.

***Example* 4.2.** Suppose a binary image is formed as a union of randomly sized and uniformly randomly rotated rectangles whose locations are random except for the constraint of disjointness. Let μ_N be the mean of the number N of rectangles in the image and the dimensions of the rectangles be given by the independent random variables H and W possessing gamma distributions with parameters α_1, β_1 and α_2, β_2, respectively. We compute the expectation of the image area A. Fixing $N = n > 0$,

$$E[A|n] = E\left[\sum_{k=1}^{n} H_k W_k\right]$$

$$= \sum_{k=1}^{n} E[H_k]E[W_k]$$

$$= n\alpha_1\alpha_2\beta_1\beta_2$$

Note that the relation also holds for $n = 0$. In terms of the random variable N, this identity takes the form $E[A|N] = N\alpha_1\alpha_2\beta_1\beta_2$, so that applying Theorem 4.1 yields

$$E[A] = E[E[A|N]]$$

$$= \sum_{n=0}^{\infty} n\alpha_1\alpha_2\beta_1\beta_2 P(N = n)$$

$$= \mu_N\alpha_1\alpha_2\beta_1\beta_2$$

Theorem 4.1 is interpreted to mean that averaging $E[A|n]$ over all n yields $E[A]$. ■

Theorem 4.1 provides the expectation of the conditional expectation; we next consider its variance. Given $X = x$, the conditional variance $\text{Var}[Y|x]$ is defined in the usual manner according to Eq. 1.138. Relative to the random variable X, the conditional variance is a random function of X and is written $\text{Var}[Y|X]$. Since the variance is the difference between the second moment and the squared mean,

$$\text{Var}[Y|X] = E[Y^2|X] - E[Y|X]^2 \tag{4.6}$$

Taking expectations and applying Theorem 4.1 yields

$$E[\text{Var}[Y|X]] = E[E[Y^2|X]] - E[E[Y|X]^2]$$

$$= E[Y^2] - E[E[Y|X]^2] \tag{4.7}$$

Moreover,

$$\text{Var}[E[Y|X]] = E[E[Y|X]^2] - E[E[Y|X]]^2$$

$$= E[E[Y|X]^2] - E[Y]^2 \tag{4.8}$$

Together, the two preceding equations yield the next theorem.

Theorem 4.2. The variance of the conditional expectation is given by

$$\text{Var}[E[Y|X]] = \text{Var}[Y] - E[\text{Var}[Y|X]]. \quad \blacksquare \tag{4.9}$$

It follows that $\text{Var}[E[Y|X]] \le \text{Var}[Y]$.

4.1.2. Optimal Nonlinear Filter

Prior to considering the general problem of finding a best MSE estimator of Y given X among all possible estimation rules based on X, we consider a special case: the best constant predictor of a random variable Y. Relative to a constant c, the MSE $E[|Y - c|^2]$ is minimized if $c = \mu_Y$, which means that

$$E[|Y - \mu_Y|^2] \le E[|Y - c|^2] \tag{4.10}$$

for all c. To see this, first observe that linearity of the expected-value operator yields

$$E[|Y - c|^2] = E[Y^2] - 2cE[Y] + c^2 \tag{4.11}$$

Differentiating with respect to c and setting the derivative equal to 0 yields $c = E[Y]$. Since the second derivative of $E[|Y - c|^2]$ with respect to c equals 2, $E[Y]$ is a minimum.

We now return to estimation (prediction) of Y by a function $\psi(X)$ of X, our goal being to obtain a best MSE estimator $\hat{Y} = \psi(X)$. When ψ is chosen from all among all possible functions, the resulting estimator is called the *optimal nonlinear mean-square-error filter* (*estimator*). As just demonstrated, in the presence of no observations of X it is best to estimate Y by its expected value $E[Y]$. Correspondingly, if we are given the observation $X = x$, then the best MSE estimate for the conditional random variable is its mean $E[Y|x]$. Hence, it would appear that the best MSE estimator for Y is $E[Y|X]$. The problem in proving the validity of this supposition is that $E[Y|X]$ is a random variable and therefore a rigorous demonstration must minimize the MSE relative to $E[Y|X]$ as a random variable.

Theorem 4.3. The conditional expectation $E[Y|X]$ is an optimal MSE estimator for the random variable Y based on the random variable X. Specifically, for any estimation rule ψ,

$$E[|Y - E[Y|X]|^2] \le E[|Y - \psi(X)|^2] \quad \blacksquare \tag{4.12}$$

To prove the theorem, consider an arbitrary estimation rule ψ. Then

$$E[|Y-\psi(X)|^2 \mid X] = E[|Y-E[Y|X]+E[Y|X]-\psi(X)|^2 \mid X]$$

$$= E[(Y-E[Y|X])^2 \mid X]$$

$$+\ 2E[(Y-E[Y|X])(E[Y|X]-\psi(X)) \mid X]$$

$$+\ E[(E[Y|X]-\psi(X))^2 \mid X]$$

$$= A + B + C \tag{4.13}$$

where the first equality results from subtracting and adding $E[Y|X]$, the second from grouping the terms, expanding the square and applying the linearity of E, and the third from simply naming the summands. We first consider B. Given X, $E[Y|X] - \psi(X)$, which is a function of X and which appears as a factor in the expression B, is a constant. Therefore it can be taken outside the expected value to yield

$$B = 2(E[Y|X]-\psi(X))E[Y-E[Y|X] \mid X]$$

$$= 2(E[Y|X]-\psi(X))(E[Y|X]-E[Y|X]) = 0 \tag{4.14}$$

where the second equality follows from the linearity of E. Now, since $C \geq 0$,

$$E[(Y-\psi(X))^2 \mid X] \geq A = E[(Y-E[Y|X])^2 \mid X] \tag{4.15}$$

Taking the expected value of both sides and applying Theorem 4.1 gives the result.

4.1.3. Optimal Filter for Jointly Normal Random Variables

An important case arises when X and Y are jointly normal with marginal means μ_X and μ_Y, marginal standard deviations σ_X and σ_Y, and correlation coefficient ρ. The marginals X and Y are normally distributed, and straightforward algebra yields the conditional density

$$f(y|x) = \frac{1}{\sqrt{2\pi\sigma_Y^2(1-\rho^2)}} \exp\left\{-\frac{1}{2}\left[\frac{y-\left(\mu_Y+\rho\frac{\sigma_Y}{\sigma_X}(x-\mu_X)\right)}{\sigma_Y\sqrt{1-\rho^2}}\right]^2\right\} \tag{4.16}$$

Since this density is Gaussian, for fixed $X = x$, the conditional random variable is normally distributed with conditional mean

$$\mu_{Y|x} = \mu_Y + \rho\frac{\sigma_Y}{\sigma_X}(x - \mu_Y) \tag{4.17}$$

which, as x varies, is a straight line. In this case the best MSE estimator is linear.

***Example* 4.3.** It is often prudent to filter a received signal to best estimate the actual transmitted signal. We consider a simplified setting in which a single numerical value is transmitted and a single value is received. The receiver knows from experience that the sent signal Y is normally distributed with mean μ_Y and standard deviation σ_Y, and that, if the transmitted signal is y, then the received signal $X|y$ is normally distributed with mean $\mu_{X|y} = y$ and fixed standard deviation (independent of y), which for simplicity we assume to be $\sigma_{X|y} = 1$. Thus, the received signal is treated as a random variable X and the conditional expectation $E[X|y]$ is equal to the actual transmitted signal. From a modeling perspective, the receiver receives a corrupted signal and bases the estimate of the transmitted signal on both the actual instrument reading and prior knowledge concerning the sender's tendencies.

Given that the receiver records the datum $X = x$, we desire the best MSE estimate of Y, the value transmitted. This estimate is given by the conditional expectation $E[Y|x]$. To find $E[Y|x]$, we need to first find the conditional density, which is given by

$$\begin{aligned}
f(y|x) &= \frac{f(x,y)}{f_X(x)} \\
&= \frac{f_Y(y)f(x|y)}{f_X(x)} \\
&= \frac{1}{f_X(x)}\left(\frac{1}{\sqrt{2\pi}\sigma_Y}\exp\left[-\frac{1}{2}\left(\frac{y-\mu_Y}{\sigma_Y}\right)^2\right]\frac{1}{\sqrt{2\pi}}\exp\left[-\frac{1}{2}(x-y)^2\right]\right) \\
&= \frac{1}{2\pi\sigma_Y f_X(x)}\exp\left[-\frac{1}{2}\left(\left(\frac{y-\mu_Y}{\sigma_Y}\right)^2 + (x-y)^2\right)\right]
\end{aligned}$$

Using some algebra, we change the form of the exponential to obtain

$$-\frac{1}{2\sigma_Y^2}\left[(y-\mu_Y)^2+\sigma_Y^2(x-y)^2\right]$$

$$=-\frac{1}{2\sigma_Y^2}(y^2-2y\mu_Y+y^2\sigma_Y^2-2xy\sigma_Y^2+\mu_Y^2+x^2\sigma_Y^2)$$

$$=-\frac{1+\sigma_Y^2}{2\sigma_Y^2}\left(y-\frac{\mu_Y+x\sigma_Y^2}{1+\sigma_Y^2}\right)^2-\frac{1}{2\sigma_Y^2}(\mu_Y^2+x^2\sigma_Y^2)+\frac{\mu_Y^2+2x\sigma_Y^2+x^2\sigma_Y^4}{2\sigma_Y^2(1+\sigma_Y^2)}$$

The last two summands in the preceding expression are functions of x and we denote their sum by $A(x)$. Substitution back into the expression for the conditional density yields

$$f(y|x)=\frac{\exp[A(x)]}{2\pi\sigma_Y f_X(x)}\exp\left[-\frac{1}{2}\frac{\left(y-\frac{\mu_Y+x\sigma_Y^2}{1+\sigma_Y^2}\right)^2}{\frac{\sigma_Y^2}{1-\sigma_Y^2}}\right]$$

where the leading factor does not involve y. Since $f(y|x)$ is a density in the variable y and has the form of a univariate normal density, the conditional random variable $Y|x$ is normal. From the form of $f(y|x)$, the conditional expectation and variance are given by

$$E[Y|x]=\frac{\mu_Y+x\sigma_Y^2}{1+\sigma_Y^2}$$

$$\mathrm{Var}[Y|x]=\frac{\sigma_Y^2}{1+\sigma_Y^2}$$

the former giving the best MSE estimate of the transmitted signal. Rewriting the conditional expectation as

$$E[Y|x]=\frac{1}{1+\sigma_Y^2}\mu_Y+\frac{\sigma_Y^2}{1+\sigma_Y^2}x$$

shows the best MSE estimate to be a weighted average of the sender's mean μ_Y and the received signal x. If the variance of Y is large, then the coefficient of μ_Y is small, the coefficient of x is near 1, and the estimate is essentially the received signal. This is reasonable since a wide dispersion of the sender's signals means the receiver should not heavily weigh the sender's tendencies. If the variance of Y is near 0, then the coefficient of μ_Y is near 1, the coefficient of x is near 0, and the estimate is essentially the mean of the sender's signals. This, too, is reasonable since the low dispersion of the sender's signals causes the receiver to doubt the validity of a received signal that varies too far from the mean of the sender's signals. ■

Although it is straightforward to obtain the conditional-mean estimator for jointly Gaussian random variables, in general the problem is more difficult and it may be necessary to reduce the class of estimators over which minimization of MSE takes place. If C is a class of estimation rules, then the optimal MSE estimator of the random variable Y in terms of X *relative to the class* C is the function $\psi(X)$ such that $\psi \in C$ and

$$E[|Y - \psi(X)|^2] \leq E[|Y - \xi(X)|^2] \tag{4.18}$$

for all $\xi \in C$. According to Theorem 4.3, if C consists of all functions, then the best MSE estimator is given by $E[Y|X]$. Looking at the mathematical description of the conditional mean, we see two salient points: (1) its direct computation requires knowledge of the joint density of X and Y, and (2) for joint densities that are complicated mathematically, evaluation of the conditional mean may be practically unfeasible. Consequently, we often constrain optimization to a convenient class of estimators, for instance, linear estimators.

4.1.4. Multiple Observation Variables

Rather than base prediction of a random variable Y on a single observed random variable, it can be estimated via an estimation rule defined on n variables $X_1, X_2,..., X_n$. Then Y is estimated by a function $\psi(X_1, X_2,..., X_n)$ and ψ is chosen to minimize

$$MSE\langle\psi\rangle = E[|Y - \psi(X_1, X_2,..., X_n)|^2]$$

$$= \int_{-\infty}^{\infty} \cdots \int_{-\infty}^{\infty} |y - \psi(x_1, x_2, \ldots, x_n)|^2 \, f(x_1, x_2, \ldots, x_n, y) \, dx_1 dx_2 \cdots dx_n dy \tag{4.19}$$

where the integral is $(n + 1)$-fold and $f(x_1, x_2,..., x_n, y)$ is the joint distribution of $X_1, X_2,..., X_n, Y$. Theorem 4.3 applies; however, here the conditional mean is

$$E[Y| x_1, x_2,..., x_n] = \int_{-\infty}^{\infty} y f(y|x_1, x_2, \ldots, x_n) \, dy \tag{4.20}$$

where the conditional density for Y given $x_1, x_2, \ldots, x_n$ is defined by

$$f(y|x_1, x_2, \ldots, x_n) = \frac{f(x_1, x_2, \ldots, x_n, y)}{f(x_1, x_2, \ldots, x_n)} \tag{4.21}$$

$f(x_1, x_2, \ldots, x_n)$ being the joint distribution of $X_1, X_2, \ldots, X_n$. The optimal MSE estimator for Y given $X_1, X_2, \ldots, X_n$ is the conditional expectation $E[Y|\ X_1, X_2, \ldots, X_n]$.

In the case of digital processing, all random variables possess finite ranges, so that $MSE\langle\psi\rangle$ in Eq. 4.19 reduces to an $(n + 1)$-fold sum over all $(x_1, x_2, \ldots, x_n, y)$ for which the joint probability mass function is positive. The conditional expectation is given by

$$E[Y|\, x_1, x_2, \ldots, x_n] = \sum_{k=0}^{m-1} kP(Y = k|\, X_1 = x_1, X_2 = x_2, \ldots, X_n = x_n) \tag{4.22}$$

where Y is assumed to take on the values $0, 1, \ldots, m - 1$. For binary images, $m = 2$ and the conditional expectation reduces to

$$E[Y|\, x_1, x_2, \ldots, x_n] = P(Y = 1|\, x_1, x_2, \ldots, x_n) \tag{4.23}$$

4.1.5. Bayesian Parametric Estimation

In discussing parametric estimation in Chapter 1 we have taken the classical approach that the parameter to be estimated is unknown and that all knowledge concerning the parameter must be extracted from observational data. A different approach is to suppose that the parameter is a random variable depending on the state of nature and that we are in possession of a *prior distribution* describing the behavior of the parameter. The problem is then to find a best estimator of the parameter, conditioned on the observations. Estimation involves optimally filtering the observations.

In *classical parametric estimation*, there is a density $f(x; \theta_1, \theta_2, \ldots, \theta_m)$ having unknown parameters $\theta_1, \theta_2, \ldots, \theta_m$, and the purpose of the estimation procedure is to find estimators $\hat{\theta}_1, \hat{\theta}_2, \ldots, \hat{\theta}_m$ for $\theta_1, \theta_2, \ldots, \theta_m$, respectively. Each estimator is a function of a random sample $X_1, X_2, \ldots, X_n$ of random variables governed by the density. In *Bayesian estimation*, the parameters are random variables, $\Theta_1, \Theta_2, \ldots, \Theta_m$, governed by a *prior density* $\pi(\theta_1, \theta_2, \ldots, \theta_m)$. The purpose of the estimation procedure is to form estimators of $\Theta_1, \Theta_2, \ldots, \Theta_m$ based on the prior density and the random sample. Heuristically, the distribution governing the random variables in the sample depends on the particular values of $\Theta_1, \Theta_2, \ldots, \Theta_m$ determining the density during the sampling procedure. Because the random variable is conditioned on the values of $\Theta_1, \Theta_2, \ldots, \Theta_m$, we write the density as a conditional density, $f(x \mid \theta_1, \theta_2, \ldots, \theta_m)$. In this section we will restrict our attention to a single parameter. Extension to several parameters will be evident once we have covered vector estimation.

For the density $f(x \mid \theta)$, with random parameter Θ possessing prior density $\pi(\theta)$, we define the *risk function* $E_{\hat{\theta}}[|\Theta - \hat{\theta}|^2]$, where the subscript indicates that the expectation is taken with respect to the distribution of $\hat{\theta}$. Because Θ is a random variable, so is the risk function. To compare two estimators, we take the *Bayes (mean) risk* of the estimator $\hat{\theta}$, which is defined by

$$B(\hat{\theta}) = E_{\Theta}[E_{\hat{\theta}}[|\Theta - \hat{\theta}|^2]] \tag{4.24}$$

where the outer and inner expectations are taken with respect to the distributions of Θ and $\hat{\theta}$, respectively. $B(\hat{\theta})$ expresses the expected risk and estimators are compared with respect to their Bayes risks. $\hat{\theta}_1$ is *better* than $\hat{\theta}_2$ if $B(\hat{\theta}_1) < B(\hat{\theta}_2)$. For a given prior distribution $\pi(\theta)$ and a class C of estimators, the *Bayes (best) estimator* is the one with minimum Bayes risk among all estimators in C.

Given a random sample $X_1, X_2,..., X_n$, the sample random variables can be considered to be jointly distributed with the parameter Θ. Thus, they are jointly governed by a multivariate probability density $f(x_1, x_2,..., x_n, \theta)$. Since our goal is to form an estimator of Θ based on $X_1, X_2,..., X_n$, we require the conditional density of Θ, given $X_1, X_2,..., X_n$. This density, known as the *posterior density*, is given by

$$f(\theta \mid x_1, x_2,..., x_n) = \frac{f(x_1, x_2, \ldots, x_n, \theta)}{f(x_1, x_2, \ldots, x_n)}$$

$$= \frac{\pi(\theta) f(x_1, x_2, \ldots, x_n | \theta)}{\int_{-\infty}^{\infty} f(x_1, x_2, \ldots, x_n, \theta)\, d\theta}$$

$$= \frac{\pi(\theta) \prod_{k=1}^{n} f(x_k | \theta)}{\int_{-\infty}^{\infty} \left(\pi(\theta) \prod_{k=1}^{n} f(x_k | \theta) \right) d\theta} \tag{4.25}$$

where the joint density factors into a product of the marginal densities because the random variables in a random sample are independent. The posterior density expresses our understanding regarding the random parameter Θ in light of the random sample, whereas the prior distribution expresses that understanding before sampling.

A Bayes estimator is determined by minimizing the Bayes risk relative to a prior distribution. The Bayes risk of an estimator $\hat{\theta}$ is given by

$$B(\hat{\theta}) = \int_{-\infty}^{\infty} E_{\hat{\theta}}[|\theta - \hat{\theta}|^2]\pi(\theta)\, d\theta$$

$$= \int_{-\infty}^{\infty}\left(\int_{-\infty}^{\infty}\cdots\int_{-\infty}^{\infty}|\theta - \hat{\theta}(x_1, x_2, \ldots, x_n)|^2 f(x_1, x_2, \ldots, x_n|\theta)\, dx_1 dx_2 \cdots dx_n\right)\pi(\theta)\, d\theta$$

$$= \int_{-\infty}^{\infty}\cdots\int_{-\infty}^{\infty}|\theta - \hat{\theta}(x_1, x_2, \ldots, x_n)|^2 f(x_1, x_2, \ldots, x_n, \theta)\, dx_1 dx_2 \cdots dx_n d\theta$$

$$= E[|\Theta - \hat{\theta}(X_1, X_2, \ldots, X_n)|^2] \qquad (4.26)$$

According to Theorem 4.3 (applied to n random variables $X_1, X_2, \ldots, X_n$), the Bayes risk is minimized by the conditional expectation. Hence, the Bayes estimator is

$$\hat{\theta} = E[\Theta \mid X_1, X_2, \ldots, X_n] \qquad (4.27)$$

***Example* 4.4.** It is common to model success and failure by a random variable X taking on the values $X = 1$ and $X = 0$, with probabilities p and $1 - p$, respectively. But these probabilities might fluctuate over time. In such a case, the success probability may be modeled by a beta distribution with parameters α and β. The success probability is a random variable P possessing prior beta density

$$\pi(p) = \frac{p^{\alpha-1}(1-p)^{\beta-1}}{\mathrm{B}(\alpha, \beta)}$$

for $0 \le p \le 1$ and $\pi(p) = 0$ otherwise. $\mathrm{B}(\alpha, \beta)$ is the beta function. X has density

$$f(x \mid p) = p^x(1-p)^{1-x}$$

where x equals 0 or 1. If $X_1, X_2, \ldots, X_n$ comprise a random sample, then

$$\pi(p)\prod_{k=1}^{n} f(x_k|p) = \frac{p^{n\bar{x}+\alpha-1}(1-p)^{n-n\bar{x}+\beta-1}}{B(\alpha,\beta)}$$

where $n\bar{x} = x_1 + x_2 + \cdots + x_n$. Therefore,

$$\int_{-\infty}^{\infty} \pi(p)\prod_{k=1}^{n} f(x_k|p)\,dp = \frac{1}{B(\alpha,\beta)}\int_0^1 p^{n\bar{x}+\alpha-1}(1-p)^{n-n\bar{x}+\beta-1}\,dp$$

$$= \frac{B(n\bar{x}+\alpha, n-n\bar{x}+\beta)}{B(\alpha,\beta)}$$

The posterior density is

$$f(p \mid x_1, x_2,\ldots, x_n) = \frac{p^{n\bar{x}+\alpha-1}(1-p)^{n-n\bar{x}+\beta-1}}{B(n\bar{x}+\alpha, n-n\bar{x}+\beta)}$$

for $0 \leq p \leq 1$, which is a beta density with parameters $n\bar{x}+\alpha$ and $n-n\bar{x}+\beta$. Hence, the Bayes estimator is the mean of this beta distribution,

$$\hat{p} = \frac{n\bar{X}+\alpha}{n+\alpha+\beta}$$

where $\bar{X}$ is the sample mean. As $n \to \infty$, $\hat{p} \to \bar{X}$. ∎

In the previous example, the Bayes estimator converges to the maximum-likelihood estimator as $n \to \infty$. This is typical of Bayes estimators. In fact, it can be shown that the difference between the two estimators is small when compared to $n^{-1/2}$. If n is not large and the sample values are compatible with the prior distribution, then the Bayes and maximum-likelihood estimators will be close; however, if n is small and the sample values differ markedly from those that would be expected given the prior distribution, then the two estimators can differ considerably.

***Example* 4.5.** If one lacks any credible prior information regarding the state of nature and still desires to take a Bayesian approach, then it is reasonable to employ a uniform prior distribution. If we again consider the success probability P, as in the previous example, then the uniform prior density is given by $\pi(p) = 1$ for $0 \leq p \leq 1$, and $\pi(p) = 0$ otherwise. The density $f(x \mid p)$ governing a single observation is the same. Hence, the posterior density is

$$f(p \mid x_1, x_2, \ldots, x_n) = \frac{p^{n\bar{x}}(1-p)^{n-n\bar{x}}}{\int_0^1 p^{n\bar{x}}(1-p)^{n-n\bar{x}}\, dp}$$

This is a beta density with parameters $n\bar{x}+1$ and $n-n\bar{x}+1$. The Bayes estimator is the mean of the posterior density, which in this case is

$$\hat{p} = \frac{n\overline{X}+1}{n+2}$$

Again, as $n \to \infty$, $\hat{p} \to \overline{X}$. ■

4.2. Optimal Finite-Observation Linear Filters

There are various reasons for constraining the class over which filter optimization takes place, such as design facilitation, image models, and filter properties. Linearity is a key property for both mathematical and engineering reasons. Linear-filter design is a parametric problem: among a (relatively small) class of parameters defining a filter class, find a mathematically tractable formulation (integral equation, matrix equation, etc.) for a parameter vector yielding a filter having minimum mean-square error. For the next eight sections we consider design of optimal linear filters. For these, optimization is embedded in the theory of projections in Hilbert spaces. The general theory of linear optimization via projections is due to Kolmogorov.

4.2.1. Linear Filters and the Orthogonality Principle

Given a finite number of observation random variables $X_1, X_2, \ldots, X_n$ and the random variable Y to be estimated, the problem of optimal linear filtering is to find a linear combination of the observation random variables that best estimates Y relative to mean-square error: find constants $a_1, a_2, \ldots, a_n$, and b that minimize the mean-square error

$$MSE\langle\psi_{\mathbf{A}}\rangle = E\left[\left|Y - \sum_{k=1}^{n} a_k X_k + b\right|^2\right] \tag{4.28}$$

where $\mathbf{A} = (a_1, a_2, \ldots, a_n, b)'$ and $\psi_{\mathbf{A}}$ is defined by

$$\psi_{\mathbf{A}}(X_1, X_2, \ldots, X_n) = \sum_{k=1}^{n} a_k X_k + b \tag{4.29}$$

If $b = 0$, then minimizing MSE yields the optimal *homogeneous* linear MSE filter; if b is arbitrary, then minimization gives the optimal *nonhomogeneous* linear MSE filter. Since the nonhomogeneous filter can be treated as a special case of the homogeneous filter by introducing the constant variable $X_0 = 1$, we consider in detail only the latter, in which case $b = 0$ in Eqs. 4.28 and 4.29.

The minimizing values $\hat{a}_1, \hat{a}_2, \ldots, \hat{a}_n$ give the *optimal linear MSE filter (estimator)*

$$\hat{Y} = \psi_{\hat{\mathbf{A}}}(X_1, X_2, \ldots, X_n) = \sum_{k=1}^{n} \hat{a}_k X_k \tag{4.30}$$

where $\hat{\mathbf{A}} = (\hat{a}_1, \hat{a}_2, \ldots, \hat{a}_n)'$. $\hat{Y}$ is the optimal linear filter if and only if

$$E[|Y - \hat{Y}|^2] \leq E\left[\left|Y - \sum_{k=1}^{n} a_k X_k\right|^2\right] \tag{4.31}$$

for all possible choices of $a_1, a_2, \ldots, a_n$.

For random variables having finite second moments (which we assume), design of optimal linear filters involves projections into Hilbert subspaces. Because we will be considering real random variables, the inner product between random variables U and V is $E[UV]$. Letting $\mathbf{X} = (X_1, X_2, \ldots, X_n)'$, we denote the span of $X_1, X_2, \ldots, X_n$ by $S_{\mathbf{X}}$. Since the sum in Eq. 4.31 is a linear combination of $X_1, X_2, \ldots, X_n$, the arbitrariness of the coefficients means that $\hat{Y}$ is the optimal linear filter if the norm $\|Y - U\|$ is minimized over all $U \in S_{\mathbf{X}}$ by letting $U = \hat{Y}$. The distance between Y and $S_{\mathbf{X}}$ is minimized by $\hat{Y}$. Since $S_{\mathbf{X}}$ is finite-dimensional, there exists a unique estimator $\hat{Y}$ minimizing this distance. If $X_1, X_2, \ldots, X_n$ are linearly independent, then we could proceed by applying Gram-Schmidt orthogonalization to $X_1, X_2, \ldots, X_n$ to produce an orthonormal set $\{Z_1, Z_2, \ldots, Z_n\}$ spanning $S_{\mathbf{X}}$ and then express the projection $\hat{Y}$ of Y into $S_{\mathbf{X}}$ as the Fourier expansion

$$\hat{Y} = \sum_{k=1}^{n} E[YZ_k]Z_k \tag{4.32}$$

If $X_1, X_2, \ldots, X_n$ are not linearly independent, we could find a linearly independent subset and proceed likewise. For this reason it is common to assume that $X_1, X_2, \ldots, X_n$ are linearly independent to begin with, and we will often take this approach. However, the algorithm by which the optimal linear filter is found is different when the observations are linearly dependent, as opposed to being linearly independent. This difference has significant practical consequences regarding implementation and we will have more to say on the matter as we proceed through the chapter.

To avoid orthogonalization, we take a different approach to finding projections. We apply the orthogonality principle. In the framework of finite-dimensional inner product spaces, the orthogonality principle states that y_0 is the projection of y into the subspace $\mathcal{V}$ if and only if $\langle (y - y_0), v \rangle = 0$ for all $v \in \mathcal{V}$. In the context of optimal linear filtering, the orthogonality principle states that $\hat{Y}$ is the projection of the random variable Y into the subspace $\mathcal{S}_X$ spanned by $X_1, X_2,..., X_n$ if and only if for any other random variable $V \in \mathcal{S}_X$,

$$E[(Y - \hat{Y})V] = 0 \tag{4.33}$$

$V \in \mathcal{S}_X$ if and only if V is a linear combination of $X_1, X_2,..., X_n$, but this means that there exists a vector $\mathbf{A}$ of scalars such that $V = \mathbf{A}'\mathbf{X}$. It is such a linear combination that defines the linear filter ψ_A. We state the orthogonality principle as a theorem regarding finite collections of random variables.

Theorem 4.4. For the random vector $\mathbf{X} = (X_1, X_2,..., X_n)'$, there exists a set of constants that minimizes $MSE\langle\psi_A\rangle$ as an estimator of Y based on $X_1, X_2,...,$ X_n over all possible choices of constants. Moreover, $\hat{a}_1, \hat{a}_2,..., \hat{a}_n$ comprise a minimizing set if and only if, for any constants $a_1, a_2,..., a_n$,

$$E\left[\left(Y - \sum_{k=1}^{n} \hat{a}_k X_k\right)\left(\sum_{j=1}^{n} a_j X_j\right)\right] = 0 \tag{4.34}$$

meaning that $Y - \hat{\mathbf{A}}'\mathbf{X}$ is orthogonal to $\mathbf{A}'\mathbf{X}$. If $X_1, X_2,..., X_n$ are linearly independent, then the collection of optimizing constants is unique. ■

4.2.2. Design of the Optimal Linear Filter

According to the orthogonality principle, $\hat{Y}$ defined by Eq. 4.30 provides the best linear MSE estimator of Y based on $X_1, X_2,..., X_n$ if and only if

$$\begin{aligned} 0 &= E[(Y - \hat{Y})(a_1X_1 + a_2X_2 + \cdots + a_nX_n)] \\ &= \sum_{k=1}^{n} a_k E[(Y - \hat{Y})X_k] \end{aligned} \tag{4.35}$$

for any constants $a_1, a_2,..., a_n$. If $E[(Y - \hat{Y})X_k] = 0$ for all k, then Eq. 4.35 holds. Thus, for $k = 1, 2,..., n$, we need to solve the equation

$$0 = E[(Y - \hat{Y})X_k]$$

$$= E\left[\left(Y - \sum_{j=1}^{n} \hat{a}_j X_j\right) X_k\right]$$

$$= E[YX_k] - \sum_{j=1}^{n} \hat{a}_j E[X_j X_k] \tag{4.36}$$

For $j, k = 1, 2, \ldots, n$, let

$$R_{kj} = E[X_k X_j] = E[X_j X_k] = R_{jk} \tag{4.37}$$

and for $k = 1, 2, \ldots, n$, let

$$R_k = E[YX_k] \tag{4.38}$$

We are led to solving the system of equations

$$\begin{aligned} R_{11}\hat{a}_1 + R_{12}\hat{a}_2 + \cdots + R_{1n}\hat{a}_n &= R_1 \\ R_{21}\hat{a}_1 + R_{22}\hat{a}_2 + \cdots + R_{2n}\hat{a}_n &= R_2 \\ \vdots \quad\quad \vdots \quad\quad\quad \vdots \quad & \quad \vdots \\ R_{n1}\hat{a}_1 + R_{n2}\hat{a}_2 + \cdots + R_{nn}\hat{a}_n &= R_n \end{aligned} \tag{4.39}$$

Their solution provides the optimal homogeneous linear MSE estimator for Y.

Let $\mathbf{C} = (R_1, R_2, \ldots, R_n)'$, $\hat{\mathbf{A}} = (\hat{a}_1, \hat{a}_2, \ldots, \hat{a}_n)'$, and

$$\mathbf{R} = \begin{pmatrix} R_{11} & R_{12} & \cdots & R_{1n} \\ R_{21} & R_{22} & \cdots & R_{2n} \\ \vdots & \vdots & \ddots & \vdots \\ R_{n1} & R_{n2} & \cdots & R_{nn} \end{pmatrix} \tag{4.40}$$

$\mathbf{C}$ is the cross-correlation vector for $X_1, X_2, \ldots, X_n$ and Y, and $\mathbf{R}$ is the autocorrelation matrix for $X_1, X_2, \ldots, X_n$. The system of Eq. 4.39 has the matrix form

$$\mathbf{R}\hat{\mathbf{A}} = \mathbf{C} \tag{4.41}$$

If det[**R**] $\neq 0$, then the desired solution for the optimizing coefficient vector is

$$\hat{\mathbf{A}} = \mathbf{R}^{-1}\mathbf{C} \tag{4.42}$$

Regarding the determinant, if $x_1, x_2,..., x_n$ are vectors in an inner product space, then

$$\mathbf{G} = \begin{pmatrix} \langle x_1, x_1 \rangle & \langle x_1, x_2 \rangle & \cdots & \langle x_1, x_n \rangle \\ \langle x_2, x_1 \rangle & \langle x_2, x_2 \rangle & \cdots & \langle x_2, x_n \rangle \\ \vdots & \vdots & \ddots & \vdots \\ \langle x_n, x_1 \rangle & \langle x_n, x_2 \rangle & \cdots & \langle x_n, x_n \rangle \end{pmatrix} \tag{4.43}$$

is called the *Grammian* of $x_1, x_2,..., x_n$ and det[**G**] is the *Gram determinant*. The vectors are linearly independent if and only if det[**G**] $\neq 0$. The matrix **R** in Eq. 4.40 is the Grammian of $X_1, X_2,..., X_n$ and therefore det[**R**] $\neq 0$ if and only if $X_1, X_2,..., X_n$ are linearly independent. Assuming linear independence of the observations, the optimal linear filter is determined by Eq. 4.42.

Theorem 4.5. If $\mathbf{X} = (X_1, X_2,..., X_n)'$ is composed of linearly independent random variables, then the optimal homogeneous linear MSE filter for Y based on **X** is given by $\hat{Y} = \hat{\mathbf{A}}'\mathbf{X}$, where $\hat{\mathbf{A}} = \mathbf{R}^{-1}\mathbf{C}$. ■

To compute MSE for the optimal linear filter, note that $Y - \hat{Y}$ is orthogonal to any linear combination of the observation random variables, in particular, $\hat{Y}$. Thus,

$$\begin{aligned} E[|Y - \hat{Y}|^2] &= E[(Y - \hat{Y})Y - (Y - \hat{Y})\hat{Y}] \\ &= E[(Y - \hat{Y})Y] - E[(Y - \hat{Y})\hat{Y}] \\ &= E[(Y - \hat{Y})Y] \\ &= E[|Y|^2] - E\left[\left(\sum_{k=1}^{n} \hat{a}_k X_k\right) Y\right] \\ &= E[|Y|^2] - \sum_{k=1}^{n} \hat{a}_k R_k \end{aligned} \tag{4.44}$$

The optimal homogeneous linear filter depends only on the second-order moments of $X_1, X_2,..., X_n$, and Y, not on the actual distribution of the random variables. If $Z_1, Z_2,..., Z_n$ comprise a different set of observation variables and the joint distributions of $X_1, X_2,..., X_n, Y$ and of $Z_1, Z_2,..., Z_n, Y$ agree up to the second order, then the optimal linear filters for $X_1, X_2,..., X_n$ and for $Z_1, Z_2,..., Z_n$ are identical, even if the joint distributions are different. The optimal linear filter is a *second-order filter*: it is determined by second-order behavior of the random variables. The advantage of the filter is the elementary mathematical means by which the optimal parameter vector can be found, albeit at the cost of a constrained estimation rule and dependence on only second-order information.

To find the optimal nonhomogeneous linear MSE filter relative to the linearly independent random variables $X_1, X_2,..., X_n$, consider the set of $n + 1$ random variables consisting of $X_1, X_2,..., X_n$ together with the constant variable $X_0 = 1$. The filter is found by applying Theorem 4.5 to $X_0, X_1, X_2,..., X_n$. The resulting estimator is of the form

$$\hat{Y} = \hat{a}_0 + \sum_{k=1}^{n} \hat{a}_k X_k \tag{4.45}$$

***Example* 4.6.** For the optimal homogeneous linear estimator of Y in terms of a single random variable X_1, $\mathbf{R}\hat{\mathbf{A}} = \mathbf{C}$ reduces to the scalar equation $R_{11}\hat{a}_1 = R_1$, and

$$\hat{a}_1 = \frac{R_1}{R_{11}} = \frac{E[X_1 Y]}{E[X_1^2]}$$

For the nonhomogeneous case, we let $X_0 = 1$ and obtain the system

$$\begin{pmatrix} R_{00} & R_{01} \\ R_{10} & R_{11} \end{pmatrix} \begin{pmatrix} \hat{a}_0 \\ \hat{a}_1 \end{pmatrix} = \begin{pmatrix} R_0 \\ R_1 \end{pmatrix}$$

where $R_{00} = E[X_0^2] = E[1] = 1$, $R_{01} = R_{10} = E[X_0X_1] = E[X_1]$, $R_{11} = E[X_1^2]$, $R_0 = E[X_0Y] = E[Y]$, and $R_1 = E[X_1Y]$. Inverting $\mathbf{R}$ yields the solution vector according to Eq. 4.42:

$$\begin{pmatrix} \hat{a}_0 \\ \hat{a}_1 \end{pmatrix} = \frac{1}{\mathrm{Var}[X_1]} \begin{pmatrix} E[X_1^2] & -E[X_1] \\ -E[X_1] & 1 \end{pmatrix} \begin{pmatrix} E[Y] \\ E[X_1 Y] \end{pmatrix}$$ ■

***Example* 4.7.** For two observation random variables X_1 and X_2, $\mathbf{R}\hat{\mathbf{A}} = \mathbf{C}$ takes the form

$$\begin{pmatrix} R_{11} & R_{12} \\ R_{21} & R_{22} \end{pmatrix} \begin{pmatrix} \hat{a}_1 \\ \hat{a}_2 \end{pmatrix} = \begin{pmatrix} R_1 \\ R_2 \end{pmatrix}$$

In particular, suppose X_1, X_2, and Y are probabilistically independent and each is uniformly distributed on [0, 1]. Since X_2 is not identically a linear function of X_1, the two are linearly independent. The relevant moments are

$$R_{11} = E[X_1^2] = \int_0^1 x_1^2\, dx_1 = \frac{1}{3}$$

$$R_{12} = R_{21} = E[X_1 X_2] = \int_0^1 \int_0^1 x_1 x_2\, dx_2 dx_1 = \frac{1}{4}$$

$$R_{22} = E[X_2^2] = E[X_1^2] = \frac{1}{3}$$

$$R_1 = E[X_1 Y] = \int_0^1 \int_0^1 x_1 y\, dy dx_1 = \frac{1}{4}$$

$$R_2 = E[X_2 Y] = E[X_1 Y] = \frac{1}{4}$$

We need to solve

$$\begin{pmatrix} 1/3 & 1/4 \\ 1/4 & 1/3 \end{pmatrix} \begin{pmatrix} \hat{a}_1 \\ \hat{a}_2 \end{pmatrix} = \begin{pmatrix} 1/4 \\ 1/4 \end{pmatrix}$$

Inversion of the matrix yields the solution

$$\begin{pmatrix} \hat{a}_1 \\ \hat{a}_2 \end{pmatrix} = \begin{pmatrix} 48/7 & -36/7 \\ -36/7 & 48/7 \end{pmatrix} \begin{pmatrix} 1/4 \\ 1/4 \end{pmatrix} = \begin{pmatrix} 3/7 \\ 3/7 \end{pmatrix}$$

Thus, the optimal homogeneous linear MSE filter is

$$\hat{Y} = \frac{3}{7} X_1 + \frac{3}{7} X_2$$

and, since $E[Y^2] = 1/3$, the MSE is

$$E[(Y-\hat{Y})^2] = E[Y^2]-(\hat{a}_1R_1+\hat{a}_2R_2) = \frac{5}{42}$$

The estimator is biased relative to the mean of Y, since $E[\hat{Y}] = 3/7$ and $E[Y] = 1/2$.

For the nonhomogeneous case, let $X_0 = 1$. The five new components of **R** are $R_{00} = E[X_0^2] = E[1] = 1$, $R_{01} = R_{10} = R_{02} = R_{20} = 1/2$. Hence,

$$\mathbf{R} = \begin{pmatrix} 1 & 1/2 & 1/2 \\ 1/2 & 1/3 & 1/4 \\ 1/2 & 1/4 & 1/3 \end{pmatrix}$$

$$\mathbf{R}^{-1} = \begin{pmatrix} 7 & -6 & -6 \\ -6 & 12 & 0 \\ -6 & 0 & 12 \end{pmatrix}$$

Thus,

$$\begin{pmatrix} \hat{a}_0 \\ \hat{a}_1 \\ \hat{a}_2 \end{pmatrix} = \begin{pmatrix} 7 & -6 & -6 \\ -6 & 12 & 0 \\ -6 & 0 & 12 \end{pmatrix} \begin{pmatrix} 1/2 \\ 1/4 \\ 1/4 \end{pmatrix} = \begin{pmatrix} 1/2 \\ 0 \\ 0 \end{pmatrix}$$

The resulting optimal estimator is $\hat{Y} = \hat{a}_0X_0 = 1/2$. This result could have been predicted from the statement of the problem since the observation random variables are independent of Y, thereby making $E[Y]$ the best nonhomogeneous linear estimator of Y. Such a result did not occur in the homogeneous case since there was further constraint (homogeneity) and the estimator lacked a constant term. Since $R_0 = E[X_0Y] = E[Y] = 1/2$, the MSE is

$$E[(Y-\hat{Y})^2] = E[Y^2]-(\hat{a}_0R_0+\hat{a}_1R_1+\hat{a}_2R_2) = \frac{1}{12}$$

which is less than the MSE in the homogeneous case. ■

In practice, we are not concerned with optimal linear estimation when the variables are probabilistically independent; rather, it is when the observation random variables are correlated with the random variable to be estimated that the estimator is of importance. In the next example we examine the relationship of the optimal linear filter to correlations among the various random variables.

***Example* 4.8.** Suppose X_1, X_2, and Y each have unit variance, X_1 and X_2 possess mean 0, Y has mean μ, and the correlation coefficients of X_1 with X_2, X_1 with Y, and X_2 with Y are ρ, ρ_{1Y}, and ρ_{2Y}, respectively. For the optimal nonhomogeneous filter, the relevant second-order moments are found by straightforward calculation: $R_{00} = R_{11} = R_{22} = 1$, $R_{01} = R_{10} = R_{02} = R_{20} = 0$, $R_{12} = R_{21} = \rho$, $R_0 = \mu$, $R_1 = \rho_{1Y}$, $R_2 = \rho_{2Y}$. The second-moment matrix is

$$\mathbf{R} = \begin{pmatrix} 1 & 0 & 0 \\ 0 & 1 & \rho \\ 0 & \rho & 1 \end{pmatrix}$$

R is nonsingular if and only if $|\rho| \neq 1$; indeed, if $|\rho| = 1$, then X_1 is a linear function of X_2, so that X_1 and X_2 are not linearly independent. Stipulating that $|\rho| \neq 1$,

$$\mathbf{R}^{-1} = \frac{1}{1-\rho^2}\begin{pmatrix} 1-\rho^2 & 0 & 0 \\ 0 & 1 & -\rho \\ 0 & -\rho & 1 \end{pmatrix}$$

Thus, the optimizing coefficients are given by

$$\begin{pmatrix} \hat{a}_0 \\ \hat{a}_1 \\ \hat{a}_2 \end{pmatrix} = \frac{1}{1-\rho^2}\begin{pmatrix} 1-\rho^2 & 0 & 0 \\ 0 & 1 & -\rho \\ 0 & -\rho & 1 \end{pmatrix}\begin{pmatrix} \mu \\ \rho_{1Y} \\ \rho_{2Y} \end{pmatrix}$$

Hence,

$$\hat{a}_0 = \mu$$

$$\hat{a}_1 = \frac{\rho_{1Y} - \rho\rho_{2Y}}{1-\rho^2}$$

$$\hat{a}_2 = \frac{\rho_{2Y} - \rho\rho_{1Y}}{1-\rho^2}$$

To gain an appreciation of the optimal filter, we consider some special cases. If $\rho_{1Y} = \rho_{2Y} = 0$, then $\hat{a}_1 = \hat{a}_2 = 0$ and the optimal estimator is $\hat{Y} = \mu$, the mean of Y. This result is plausible since the lack of correlation between Y and either of the observation random variables means that Y is being estimated in the presence of no relevant observations.

Now suppose $\rho_{1Y} = 0$ and $\rho_{2Y} = 1$. Then there exist constants c and d, $c > 0$, such that $Y = cX_2 + d$. Since $\text{Var}[Y] = c^2\text{Var}[X_2]$ and both variances are 1,

$c = 1$. Moreover, $E[Y] = E[X_2] + d$, so that $d = \mu$. Consequently, $Y = X_2 + \mu$ and

$$\begin{aligned} 0 = \rho_{1Y} &= E[X_1 Y] \\ &= E[X_1(X_2 + \mu)] \\ &= E[X_1 X_2] + \mu E[X_1] = \rho \end{aligned}$$

Substituting these values into $\hat{a}_1$ and $\hat{a}_2$ yields the estimator $\hat{Y} = X_2 + \mu$, which makes sense since, in this case, $Y = X_2 + \mu$.

For the general result, the MSE is

$$\begin{aligned} E[(Y - \hat{Y})^2] &= E[Y^2] - (\hat{a}_0 R_0 + \hat{a}_1 R_1 + \hat{a}_2 R_2) \\ &= 1 - \frac{1}{1-\rho^2}[(\rho_{1Y} + \rho_{2Y}) - \rho(\rho_{1Y} + \rho_{2Y})] \\ &= 1 - \frac{\rho_{1Y} + \rho_{2Y}}{1+\rho} \end{aligned}$$

where we have used the relation

$$E[Y^2] = \mathrm{Var}[Y] + E[Y]^2 = 1 + \mu^2$$

In the case where $\rho_{1Y} = \rho_{2Y} = 0$, the MSE is 1, which is the variance of Y. As seen previously, if $\rho_{1Y} = 0$ and $\rho_{2Y} = 1$, then $\rho = 0$ and the MSE is 0. ■

4.2.3. Optimal Linear Filter in the Jointly Gaussian Case

In the previous section we observed that the optimal MSE filter for Y based on X is linear when X and Y are jointly Gaussian. This proposition extends to estimation based on more than one observation random variable.

Theorem 4.6. If $X_1, X_2, \ldots, X_n$, Y are jointly Gaussian, then the optimal linear MSE filter for Y in terms of $X_1, X_2, \ldots, X_n$ is also the optimal MSE filter. ■

To demonstrate the theorem, let $\mathbf{X} = (X_1, X_2, \ldots, X_n)'$ and first assume that $X_1, X_2, \ldots, X_n$, and Y all possess zero means. Let the optimal linear filter be given by Eq. 4.30. $X_1, X_2, \ldots, X_n$, and $Y - \hat{Y}$ are jointly Gaussian. Since $Y - \hat{Y}$ is uncorrelated with $X_1, X_2, \ldots, X_n$, joint normality implies that $Y - \hat{Y}$ is probabilistically independent of $X_1, X_2, \ldots, X_n$. Thus,

$$E[Y-\hat{Y}|\mathbf{X}] = E[Y-\hat{Y}] \tag{4.46}$$

Consequently,

$$\begin{aligned}
E[Y|\mathbf{X}]-\sum_{k=1}^{n}\hat{a}_k E[X_k] &= E[Y|\mathbf{X}]-\sum_{k=1}^{n}\hat{a}_k E[X_k|\mathbf{X}] \\
&= E[Y|\mathbf{X}]-E\left[\sum_{k=1}^{n}\hat{a}_k X_k \middle| \mathbf{X}\right] \\
&= E[Y|\mathbf{X}]-E[\hat{Y}|\mathbf{X}] \qquad (4.47) \\
&= E[Y-\hat{Y}|\mathbf{X}] \\
&= E[Y-\hat{Y}] \\
&= E\left[Y-\sum_{k=1}^{n}\hat{a}_k X_k\right] \\
&= E[Y] - \sum_{k=1}^{n}\hat{a}_k E[X_k]
\end{aligned}$$

which is 0 because all random variables have zero means. Therefore,

$$E[Y|\mathbf{X}] = \sum_{k=1}^{n}\hat{a}_k E[X_k] \tag{4.48}$$

and the globally optimal MSE filter is equal to the optimal linear MSE filter.

To finish the proof, we consider the nonzero-mean case. Let $\mu_X = (\mu_1, \mu_2, \ldots, \mu_n)'$ and μ_Y be the mean vector of $\mathbf{X}$ and mean of Y, respectively, and let $\mathbf{X}^0 = (X_1^0, X_2^0, \ldots, X_n^0)'$ and Y^0 be their centered versions. The preceding part of the proof applies to $\mathbf{X}^0$ and Y^0. Recognizing that the expectation of Y conditioned on $\mathbf{X}^0$ is the same as the expectation of Y conditioned on $\mathbf{X}$, and letting $\hat{\alpha}_1, \hat{\alpha}_2, \ldots, \hat{\alpha}_n$ be the optimizing coefficients for the optimal linear estimator of Y^0 based on $\mathbf{X}^0$, we obtain

$$E[Y|\mathbf{X}] = E[Y|\mathbf{X}^0]$$

$$= E[Y - \mu_Y + \mu_Y | \mathbf{X}^0]$$

$$= E[Y^0 | \mathbf{X}^0] + \mu_Y \tag{4.49}$$

$$= \sum_{k=1}^{n} \hat{\alpha}_k X_k^0 + \mu_Y$$

$$= \sum_{k=1}^{n} \hat{\alpha}_k (X_k - \mu_k) + \mu_Y$$

Hence, the conditional expectation is a linear operator and, since the MSE is minimal for the conditional expectation, the linear operator must be the optimal linear filter.

Theorem 4.6 is one reason why it is often assumed that the random variables are jointly Gaussian: this assumption makes the optimal linear filter globally optimal. In many situations the normality assumption is not needed because one need only obtain the optimal linear filter and then implicitly recognize that it is globally optimal if it happens that the random variables are jointly Gaussian.

4.2.4. Role of Wide-Sense Stationarity

For processing digital images, the optimal linear estimator for Y is applied as a window function of the n random variables $X_1, X_2,..., X_n$. There is a window $W = \{w_1, w_2,..., w_n\}$ consisting of n pixels, W is translated to a pixel z to form the observation window

$$W_z = \{z + w_1, z + w_2,..., z + w_n\} \tag{4.50}$$

(Fig. 4.1), and the filter is applied to the random variables in W_z. The optimal linear filter

$$\psi_{op,z} = \psi_{op,z}(X_1, X_2,..., X_n) \tag{4.51}$$

is pixel dependent because it depends on the second moments of $X_1, X_2,..., X_n$, and Y in W_z. In particular, the autocorrelation matrix and cross-correlation vector are functions of z, say $\mathbf{R}(z)$ and $\mathbf{C}(z)$. Relative to the ideal image process S_0 and the observed image process S, $Y = S_0(z)$ and $X_i = S(z + w_i)$ for $i = 1, 2,..., n$. Hence, the components of $\mathbf{R}(z)$ and $\mathbf{C}(z)$ are given by

$$R_{kj}(z) = E[S(z + w_k)S(z + w_j)] \tag{4.52}$$

$$R_k(z) = E[S_0(z)S(z + w_k)] \tag{4.53}$$

respectively. If S_0 and S are jointly WS stationary, then R_{kj} is a function of $w_k - w_j$ and R_k is a function of w_k, so that $\psi_{op,z}$ is independent of z and can be written ψ_{op}. The image filter Ψ_{op} is defined via ψ_{op} by

$$\Psi_{op}(S)(z) = \psi_{op}(S(z + w_1), S(z + w_2),..., S(z + w_n)) \tag{4.54}$$

Ψ_{op} is translation (spatially) invariant. Both Ψ_{op} and its window function ψ_{op} are referred to as the *optimal linear filter*.

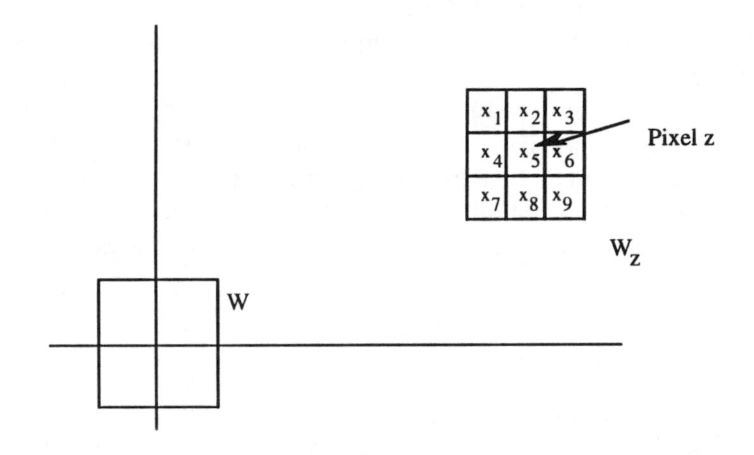

Figure 4.1 Window *W* translated to pixel *z*.

In the preceding examples, **R** and **C** were determined from a process model. In most practical applications they must be estimated. For discrete-time WS stationary signal processes $Y(j)$ and $X(j)$, we can estimate **R** and **C** from a single pair of realizations. Let the values of $X(j)$ in the window translated to z be $X_{z,-m}, X_{z,-m+1},..., X_{z,0}, ..., X_{z,m-1}, X_{z,m}$ and the value of $Y(j)$ at z be Y_z. For each z form the matrix $\hat{\mathbf{R}}_z$ and vector $\hat{\mathbf{C}}_z$ by

$$\hat{\mathbf{R}}_z = \begin{pmatrix} X_{z,-m}X_{z,-m} & X_{z,-m}X_{z,-m+1} & \cdots & X_{z,-m}X_{z,m} \\ X_{z,-m+1}X_{z,-m} & X_{z,-m+1}X_{z,-m+1} & \cdots & X_{z,-m+1}X_{z,m} \\ \vdots & \vdots & \ddots & \vdots \\ X_{z,m}X_{z,-m} & X_{z,m}X_{z,-m+1} & \cdots & X_{z,m}X_{z,m} \end{pmatrix} \tag{4.55}$$

$$\hat{\mathbf{C}}_z = \begin{pmatrix} X_{z,-m}Y_z \\ X_{z,-m+1}Y_z \\ \vdots \\ X_{z,m}Y_z \end{pmatrix} \tag{4.56}$$

If A is a set of points and N is the number of points in A, then $\mathbf{R}$ and $\mathbf{C}$ are estimated by

$$\hat{\mathbf{R}} = \frac{1}{N}\sum_{z \in A} \hat{\mathbf{R}}_z \tag{4.57}$$

$$\hat{\mathbf{C}} = \frac{1}{N}\sum_{z \in A} \hat{\mathbf{C}}_z \tag{4.58}$$

4.2.5. Signal-plus-Noise Model

A standard digital filtering problem is the observation of a number of random variables $X_1, X_2, \ldots, X_n$ in a window about a pixel when the underlying image has been corrupted by some degradation (noise) process. The true value at the pixel is Y but we can only observe corrupted values in the vicinity of the pixel. A commonly employed model is the *signal-plus-noise model*. For this model, there are n random variables $U_1, U_2, \ldots, U_n$ corresponding to the uncorrupted underlying random image (of which one might be Y), there are n noise random variables $N_1, N_2, \ldots, N_n$, and the n observed random variables $X_1, X_2, \ldots, X_n$ are modeled as having resulted from the noise-free signal variables plus noise,

$$X_k = U_k + N_k \tag{4.59}$$

for $k = 1, 2, \ldots, n$. Often, for the sake of mathematical tractability, it is assumed that the noise is uncorrelated to the noisefree image and is white.

***Example* 4.9.** We treat a simple case of the signal-plus-noise model: homogeneous linear estimation of a random variable Y given corrupted observations of itself and another random variable X. Let X and Y each have mean 0 and variance 1, and let their correlation coefficient be ρ. We estimate Y given the observations

$$X_1 = Y + N_1$$

$$X_2 = X + N_2$$

where the noise random variables N_1 and N_2 have zero means, possess common variance σ^2, are uncorrelated with X and Y, and are mutually uncorrelated. In short, the finite process (N_1, N_2) is white noise that is uncorrelated with (X, Y). The relevant moments are

$$R_{11} = E[(Y+N_1)^2] = E[Y^2]+2E[YN_1]+E[N^2] = 1+\sigma^2$$

$$R_{12} = E[(Y+N_1)(X+N_2)] = E[XY]+E[YN_2]+E[XN_1]+E[N_1N_2] = \rho$$

$$R_{21} = R_{12} = E[X_1X_2] = \rho$$

$$R_{22} = E[(X+N_2)^2] = 1+\sigma^2$$

$$R_1 = E[(Y+N_1)Y] = E[Y^2]+E[N_1Y] = 1$$

$$R_2 = E[(X+N_2)Y] = E[XY]+E[N_2Y] = \rho$$

The second-moment matrix is

$$\mathbf{R} = \begin{pmatrix} 1+\sigma^2 & \rho \\ \rho & 1+\sigma^2 \end{pmatrix}$$

and the inverse is

$$\mathbf{R}^{-1} = \frac{1}{(1+\sigma^2)^2-\rho^2}\begin{pmatrix} 1+\sigma^2 & -\rho \\ -\rho & 1+\sigma^2 \end{pmatrix}$$

Multiplying $\mathbf{R}^{-1}$ times the vector

$$\begin{pmatrix} R_1 \\ R_2 \end{pmatrix} = \begin{pmatrix} 1 \\ \rho \end{pmatrix}$$

yields the optimizing coefficients and the optimal homogeneous linear estimator for Y:

$$\hat{Y} = \frac{1+\sigma^2-\rho^2}{(1+\sigma^2)^2-\rho^2}X_1 + \frac{\rho\sigma^2}{(1+\sigma^2)^2-\rho^2}X_2$$

In the presence of no noise, $\sigma^2 = 0$, the estimator reduces to $\hat{Y} = X_1$, as expected. In the presence of extreme noise, σ^2 is essentially infinite and the estimator reduces to $\hat{Y} = 0 = E[Y]$. If $\rho = 0$, so that X and Y are uncorrelated, then

$$\hat{Y} = \frac{1}{1+\sigma^2} X_1$$

In general, the MSE is

$$E[(Y - \hat{Y})^2] = E[Y^2] - (\hat{a}_1 R_1 + \hat{a}_2 R_2)$$

$$= 1 - \frac{1+\sigma^2 - \rho^2 + \rho^2\sigma^2}{(1+\sigma^2)^2 - \rho^2}$$

For $\sigma^2 = 0$, the MSE is 0. Moreover, for fixed ρ,

$$\lim_{\sigma \to \infty} E[(Y - \hat{Y})^2] = 1$$ ■

***Example* 4.10.** The preceding window reasoning applies to a digital signal except that the window is over the discrete time domain. To apply it in the WS stationary setting, let the signal variables in the window be labeled Y_{-m}, $Y_{-m+1}, \ldots, Y_{-1}, Y_0, Y_1, \ldots, Y_{m-1}, Y_m$, the noise variables be labeled $N_{-m}, N_{-m+1}, \ldots$, $N_{-1}, N_0, N_1, \ldots, N_{m-1}, N_m$, and the observed variables be

$$X_j = Y_j + N_j$$

for $j = -m, \ldots, m$, where Y_0 is the value to be estimated at the window center. Consider the random telegraph signal of Example 2.18 and let the signal be the discrete-time process obtained by sampling the random telegraph signal at each integer. The covariance function of the resulting discrete sampled process is the same as the random telegraph signal except it is only evaluated at pairs of nonnegative integers. Let the noise process be white, uncorrelated to the signal, and have variance σ^2. The sampled random telegraph and noisy processes have mean 0 and variances 1 and $1 + \sigma^2$, respectively. Under the given labeling in the window,

$$R_{jj} = E[X_j^2] = E[Y_j^2] + E[N_j^2] = 1 + \sigma^2$$

for $j = -m, \ldots, m$,

$$R_{ij} = E[X_i X_j] = E[Y_i Y_j] = e^{-2\lambda|i-j|}$$

for $i \neq j$, $R_0 = 1$, and

$$R_j = E[X_j Y_0] = E[Y_j Y_0] = e^{-2\lambda|j|}$$

for $j \neq 0$. For instance, for a five-point window,

$$\mathbf{R} = \begin{pmatrix} 1+\sigma^2 & e^{-2\lambda} & e^{-4\lambda} & e^{-6\lambda} & e^{-8\lambda} \\ e^{-2\lambda} & 1+\sigma^2 & e^{-2\lambda} & e^{-4\lambda} & e^{-6\lambda} \\ e^{-4\lambda} & e^{-2\lambda} & 1+\sigma^2 & e^{-2\lambda} & e^{-4\lambda} \\ e^{-6\lambda} & e^{-4\lambda} & e^{-2\lambda} & 1+\sigma^2 & e^{-2\lambda} \\ e^{-8\lambda} & e^{-6\lambda} & e^{-4\lambda} & e^{-2\lambda} & 1+\sigma^2 \end{pmatrix}$$

$$\mathbf{C} = \begin{pmatrix} R_{-2} \\ R_{-1} \\ R_0 \\ R_1 \\ R_2 \end{pmatrix} = \begin{pmatrix} e^{-4\lambda} \\ e^{-2\lambda} \\ 1 \\ e^{-2\lambda} \\ e^{-4\lambda} \end{pmatrix}$$

For a numerical example, let $\lambda = 0.02145$ and $\sigma^2 = 0.2$. Then

$$\mathbf{R}^{-1} = \begin{pmatrix} 2.7290 & -1.1994 & -0.6448 & -0.3680 & -0.2498 \\ -1.1994 & 3.2333 & -0.9497 & -0.5421 & -0.3680 \\ -0.6448 & -0.9497 & 3.3360 & -0.9497 & -0.6448 \\ -0.3680 & -0.5421 & -0.9497 & 3.2333 & -1.1994 \\ -0.2498 & -0.3680 & -0.6448 & -1.1994 & 2.7290 \end{pmatrix}$$

$$\hat{\mathbf{A}} = \mathbf{R}^{-1}\mathbf{C} = \begin{pmatrix} 0.1290 \\ 0.1899 \\ 0.3328 \\ 0.1899 \\ 0.1290 \end{pmatrix}$$

The optimal linear filter is a weighted average, with contributions falling off for points away from the window center. This will have the effect of suppressing the additive point noise at the cost of smoothing the jumps in the underlying signal. ■

4.2.6. Edge Detection

Mathematical rigor aside, an edge image is a binary image purporting to mark visual edges in another image f, which here we take to be gray-scale. Many algorithms have been developed to construct edge images. One approach is to define a linear operator via a *gradient mask*. Given a mask $g = g(i, j)$, $i, j = -m, -m + 1, \ldots, m$, of numerical weights and an image f, define $h = \Gamma_g(f)$ by

$$h(i,j) = \sum_{r=-m}^{m} \sum_{s=-m}^{m} g(r,s) f(i+r, j+s) \tag{4.60}$$

The operator is linear and translation invariant. Typically there are two gradient masks to measure change in the horizontal and vertical directions. Various gradient masks have been employed, with the most popular pairs being of the form

$$k = \begin{pmatrix} -1 & 0 & 1 \\ -\lambda & 0 & \lambda \\ -1 & 0 & 1 \end{pmatrix} \qquad l = \begin{pmatrix} 1 & \lambda & 1 \\ 0 & 0 & 0 \\ -1 & -\lambda & -1 \end{pmatrix} \tag{4.61}$$

with $\lambda \geq 1$. The standard procedure is to compute $\Gamma_k(f)$ and $\Gamma_l(f)$, apply some norm to produce a gradient image, and then threshold the gradient image to obtain an edge image. A commonly employed norm is the maximum norm, $\max(|\Gamma_k(f)|, |\Gamma_l(f)|)$. The idea is that large gradients, either positive or negative, indicate that a pixel lies on the edge of some object in the image. The problem with this notion is that large gradients occur for many reasons besides the existence of edges — for instance, bright spikes, dark spikes, texture, and noise. Selection of an appropriate threshold is especially problematic: too low a threshold gives false edges and wide edges that need to be thinned; too high a threshold can result in broken or partial edges that need to be connected. Prefiltering to reduce noise is often necessary, but a smoothing filter flattens gradients, thereby making edges both wider and harder to detect by thresholding.

A similar approach to edge detection involves the *morphological gradient*. At each pixel (i, j) the morphological gradient is calculated as the maximum gray value minus the minimum gray value in a predefined window about (i, j). Once the gradient is computed, a binary edge image is found by thresholding.

While one can argue the merits of gradient edge detection, including threshold dependency, from our perspective the real issue is to carefully define the notion of edge within the context of random processes and discuss the manner in which a statistically meritorious edge operator can be synthesized.

The issue can be framed in the context of gray-scale-to-binary image operators. Consider binary and gray-scale image processes $Y(u, v)$ and $X(u, v)$, respectively. $Y(u, v)$ and $X(u, v)$ are the ideal and observed processes and we desire an operator Ψ to provide an estimator $\hat{Y}(u,v)$ of $Y(u, v)$. For locating targets, $Y(u, v)$ consists of random locations; for finding edges, $Y(u, v)$ is composed of random paths. The goodness of Ψ is judged by a probabilistic error measure between $\hat{Y}(u,v)$ and $Y(u, v)$. To illustrate the matter, we will give a detailed example involving a discrete one-dimensional model in which steps are considered to be edges and the goal is to find an optimal homogeneous MSE linear filter.

***Example* 4.11.** A model must be constructed in which there is an edge signal $Y(k)$ and an observed signal $X(k)$. For $k = 1, 2,\ldots,$ define the independent discrete-time process $Y(k)$, where, for each k, $Y(k)$ is a binomial random variable with $P(Y(k) = 1) = p$ and $P(Y(k) = 0) = q$, with $p > 0$, $q > 0$, and $p + q = 1$. Define the process $U(k)$ recursively by $P(U(0) = 1) = 1/2$, $P(U(0) = -1) = 1/2$, $U(k) = U(k - 1)$ if $Y(k) = 0$, and $U(k) = -U(k - 1)$ if $Y(k) = 1$. A realization of $U(k)$ randomly commences at -1 or 1, and then changes sign at points where $Y(k) = 1$. For $d = \pm 1$, $P(U(k) = d) = 1/2$. $U(k)$ behaves like the random telegraph signal; however, it is based on the binomial density and defined only at integers. $U(k)$ and $Y(k)$ form a discrete one-dimensional edge (step) model: the steps of $U(k)$ occur at points for which $Y(k) = 1$. Suppose the observed signal is corrupted by uncorrelated additive noise, so

$$X(k) = U(k) + N(k)$$

where $N(k)$ is white noise with variance σ^2. For the optimal linear edge-finding filter over a 3-point window, the observed signal is $X(k)$ and we estimate $Y(k)$. $X(k)$ is composed of noise plus the signal whose edges form the ideal image. The values of $X(k)$ observed in the window will be labeled X_{-1}, X_0, and X_1. Similar labeling will be used for $U(k)$, $N(k)$, and $Y(k)$. We compute the relevant moments. $E[X_{-1}^2] = E[X_0^2] = E[X_1^2] = 1 + \sigma^2$ and

$$\begin{aligned}
E[X_0X_1] &= E[U_0U_1] \\
&= P(U_0U_1 = 1) - P(U_0U_1 = -1) \\
&= P(U_0 = 1)P(U_1 = 1|U_0 = 1) + P(U_0 = -1)P(U_1 = -1|U_0 = -1) \\
&\quad - P(U_0 = 1)P(U_1 = -1|U_0 = 1) - P(U_0 = -1)P(U_1 = 1|U_0 = -1) \\
&= P(U_0 = 1)P(Y_1 = 0) + P(U_0 = -1)P(Y_1 = 0) \\
&\quad - P(U_0 = 1)P(Y_1 = 1) - P(U_0 = -1)P(Y_1 = 1)
\end{aligned}$$

$$= q - p$$

Similar calculations apply to moments of the form $E[X_jX_{j+1}]$. Next,

$$E[X_{-1}X_1] = P(U_{-1}U_1 = 1) - P(U_{-1}U_1 = -1)$$

$$= P(U_{-1} = 1)P(U_0 = 1|U_{-1} = 1)P(U_1 = 1|U_0 = 1)$$

$$+ P(U_{-1} = 1)P(U_0 = -1|U_{-1} = 1)P(U_1 = 1|U_0 = -1)$$

$$+ P(U_{-1} = -1)P(U_0 = -1|U_{-1} = -1)P(U_1 = -1|U_0 = -1)$$

$$+ P(U_{-1} = -1)P(U_0 = 1|U_{-1} = 1)P(U_1 = -1|U_0 = 1)$$

$$- P(U_{-1} = 1)P(U_0 = -1|U_{-1} = 1)P(U_1 = -1|U_0 = -1)$$

$$- P(U_{-1} = 1)P(U_0 = 1|U_{-1} = 1)P(U_1 = -1|U_0 = 1)$$

$$- P(U_{-1} = -1)P(U_0 = -1|U_{-1} = -1)P(U_1 = 1|U_0 = -1)$$

$$- P(U_{-1} = -1)P(U_0 = 1|U_{-1} = -1)P(U_1 = 1|U_0 = 1)$$

$$= (q - p)^2$$

Letting $a = q - p$,

$$\mathbf{R} = \begin{pmatrix} 1+\sigma^2 & a & a^2 \\ a & 1+\sigma^2 & a \\ a^2 & a & 1+\sigma^2 \end{pmatrix}$$

Since the noise is independent of Y_0 and $E[N_j] = 0$,

$$E[Y_0X_j] = E[Y_0(U_j + N_j)] = E[Y_0U_j]$$

for $j = -1, 0, 1$. Because $Y(k)$ is an independent process, Y_0 and U_{-1} are independent. Owing to their independence, for $d = \pm 1$,

$$P(Y_0 = 1, U_0 = d) = P(Y_0 = 1, U_{-1} = -d)$$

$$= P(Y_0 = 1)P(U_{-1} = -d)$$

$$= P(Y_0 = 1)P(U_0 = d)$$

Since a similar identity holds for $P(Y_0 = 0, U_0 = d)$, Y_0 and U_0 are independent. Their independence implies

$$P(Y_0 = d, U_1 = 1) = P(Y_0 = d, U_0 = 1, Y_1 = 0) + P(Y_0 = d, U_0 = 0, Y_1 = 1)$$

$$= P(Y_0 = d)P(U_0 = 1)P(Y_1 = 0) + P(Y_0 = d)P(U_0 = 0)P(Y_1 = 1)$$

$$= P(Y_0 = d)P(U_1 = 1)$$

for $d = \pm 1$. Since a similar identity holds for $P(Y_0 = d, U_0 = 0)$, Y_0 and U_1 are independent. Hence,

$$E[Y_0X_j] = E[Y_0U_j] = 0$$

for $j = -1, 0, 1$. Thus, $\mathbf{C} = (0, 0, 0)'$, $\hat{\mathbf{A}} = \mathbf{R}^{-1}\mathbf{C} = \mathbf{0}$, and the optimal linear edge-finder is the zero function, $\psi_{opt}(\mathbf{x}) = 0$. This result is unsatisfactory, but correct relative to the goal of estimating $Y(k)$. The problem is that, whereas steps of $U(k)$ are both up and down, the linear filter must estimate the process $Y(k)$ having only positive discrete impulses.

We can do better using the process $Z(k)$ that is 1 when $U(k)$ has a positive step and 0 otherwise. Estimating $Z(k)$ from $X(k)$ amounts to finding up-step edges. Using the fact that $P(Z_0 = 1 | U_{-1} = 1) = 0$ because there cannot be a positive step if the process is already at 1,

$$E[Z_0U_{-1}] = P(Z_0 = 1, U_{-1} = 1) - P(Z_0 = 1, U_{-1} = -1)$$

$$= P(U_{-1} = 1)P(Z_0 = 1|U_{-1} = 1) - P(U_{-1} = -1)P(Z_0 = 1|U_{-1} = -1)$$

$$= -P(U_{-1} = -1)P(Y_0 = 1)$$

$$= -\frac{p}{2}$$

$$E[Z_0U_0] = P(Z_0 = 1, U_0 = 1) - P(Z_0 = 1, U_0 = -1)$$

$$= P(Z_0 = 1, U_{-1} = -1)$$

$$= P(U_{-1} = -1)P(Z_0 = 1|U_{-1} = -1)$$

$$= \frac{p}{2}$$

$$E[Z_0U_1] = P(Z_0 = 1, U_1 = 1) - P(Z_0 = 1, U_1 = -1)$$

$$= P(U_{-1} = 0, Y_0 = 1, Y_1 = 0) - P(U_{-1} = 0, Y_0 = 1, Y_1 = 1)$$

$$= \frac{(q-p)p}{2}$$

$$\mathbf{C} = \frac{p}{2}\begin{pmatrix} -1 \\ 1 \\ a \end{pmatrix}$$

Weights for the optimal linear positive-step finder are given by $\mathbf{R}^{-1}\mathbf{C}$.

The case with no noise is instructive. $\mathbf{R}$ is unchanged except that $\sigma^2 = 0$, and

$$\mathbf{R}^{-1} = \frac{1}{(1-a^2)^2}\begin{pmatrix} 1-a^2 & a^3-a & 0 \\ a^3-a & 1-a^4 & a^3-a \\ 0 & a^3-a & 1-a^2 \end{pmatrix}$$

$$\hat{\mathbf{A}} = \frac{1}{1-a^2}\begin{pmatrix} -pq \\ pq \\ 0 \end{pmatrix}$$

$$\psi_{opt}(\mathbf{x}) = \frac{pq}{1-(q-p)^2}(x_0 - x_{-1}) = \frac{x_0 - x_{-1}}{4}$$

For the signal fragment,

$$s = (\ldots, 1, 1, 1, 1, -1, -1, -1, -1, 1, 1, 1, 1\ldots)$$

the signal filter corresponding to ψ_{opt} yields

$$\Psi_{opt}(s) = (\ldots, 0, 0, 0, 0, -1/2, 0, 0, 0, 1/2, 0, 0, 0,\ldots)$$

The filter has given $-1/2$ and $1/2$ at the negative and positive steps, respectively. This is not bad if our intent is to threshold the filter output to mark positive steps. Analogous reasoning applies to finding the optimal down-step filter, so that together the two filters provide edge detection. Owing to filter symmetry, one could simply apply the up-step filter and take absolute values to detect both positive and negative steps. ■

4.3. Steepest Descent

Filter weights for the optimal finite-observation linear filter can be found by multiplying the inverse of the autocorrelation matrix $\mathbf{R}$ times the cross-correlation vector **C**. For a large number of observations, matrix inversion can be computationally burdensome. Moreover, in practice **R** will likely be estimated from realizations, and estimation error can significantly affect inversion. Inversion can be avoided by applying a gradient-based iterative method to locate the minimum of the error surface.

4.3.1. Steepest Descent Iterative Algorithm

The optimal mean-square-error finite-observation, linear filter estimates a single random variable Y in terms of a linear combination of observation random variables $X_1, X_2,..., X_n$. The optimal filter minimizes the mean-square error between Y and estimators of the form

$$\psi_{\mathbf{A}}(\mathbf{X}) = \sum_{k=1}^{n} a_k X_k \tag{4.62}$$

where $\mathbf{A} = (a_1, a_2,..., a_n)'$ and $\mathbf{X} = (X_1, X_2,..., X_n)'$. Letting $\mathbf{C} = (R_1, R_2,..., R_n)'$ be the cross-correlation vector between Y and $\mathbf{X}$, letting $\mathbf{R}$ be the autocorrelation matrix as defined in Eq. 4.40, and assuming Y has zero mean, the mean-square error is given by

$$\begin{aligned} Q &= E[|Y - \psi_{\mathbf{A}}(\mathbf{X})|^2] \\ &= E[|Y|^2] - 2E[Y\psi_{\mathbf{A}}(\mathbf{X})] + E[\psi_{\mathbf{A}}(\mathbf{X})^2] \\ &= \mathrm{Var}[Y] - 2\sum_{k=1}^{n} a_k E[X_k Y] + \sum_{k=1}^{n}\sum_{j=1}^{n} a_k a_j E[X_k X_j] \\ &= \mathrm{Var}[Y] - 2\sum_{k=1}^{n} a_k R_k + \sum_{k=1}^{n}\sum_{j=1}^{n} a_k a_j R_{kj} \end{aligned} \tag{4.63}$$

Q is a quadratic error surface in the variables $a_1, a_1,..., a_n$ and has a unique minimum. At the unique minimum of Q, its gradient (as a function of the weights) vanishes: $\nabla Q = \mathbf{0}$. A vanishing gradient means that $\partial Q/\partial a_k = 0$ for $k = 1, 2,..., n$. From Eq. 4.63,

$$\frac{\partial Q}{\partial a_k} = -2R_k + 2\sum_{j=1}^{n} a_j R_{kj} \tag{4.64}$$

Combining these equations for $k = 1, 2,..., n$ yields the vector equation $\mathbf{C} = \mathbf{RA}$, the same system obtained via the orthogonality principle.

In vector form, Eq. 4.63 can be written as

$$Q = \text{Var}[Y] - 2\mathbf{A}'\mathbf{C} + \mathbf{A}'\mathbf{R}\mathbf{A} \tag{4.65}$$

For the optimal solution, $\mathbf{C} = \mathbf{R}\hat{\mathbf{A}}$ and this error becomes

$$\hat{Q} = \text{Var}[Y] - \hat{\mathbf{A}}'\mathbf{R}\hat{\mathbf{A}} \tag{4.66}$$

which agrees with Eq. 4.44. Subtraction yields

$$\begin{aligned} Q - \hat{Q} &= -2\mathbf{A}'\mathbf{R}\hat{\mathbf{A}} + \mathbf{A}'\mathbf{R}\mathbf{A} + \hat{\mathbf{A}}'\mathbf{R}\hat{\mathbf{A}} \\ &= \mathbf{A}'\mathbf{R}(\mathbf{A} - \hat{\mathbf{A}}) - (\mathbf{A} - \hat{\mathbf{A}})'\mathbf{R}\hat{\mathbf{A}} \\ &= (\mathbf{A} - \hat{\mathbf{A}})'\mathbf{R}(\mathbf{A} - \hat{\mathbf{A}}) \end{aligned} \tag{4.67}$$

This expression provides an explicit representation of the manner in which the optimizing coefficients provide minimum MSE. It expresses the increment in MSE owing to nonoptimality (as opposed to optimality).

Instead of finding the optimal coefficients by inverting **R**, one can proceed iteratively to an approximate solution for the optimal weight vector; that is, iteratively toward the minimum of the error surface defined by Q. As the error produced by the filter coefficients at the nth step of an iterative algorithm, Eq. 4.65 takes the form

$$Q(n) = \text{Var}[Y] - 2\mathbf{A}'(n)\mathbf{C} + \mathbf{A}'(n)\mathbf{R}\mathbf{A}(n) \tag{4.68}$$

where $\mathbf{A}(n)$ is the coefficient vector at step n. From Eq. 4.64, the gradient of $Q(n)$ is given by

$$\nabla Q(n) = -2\mathbf{C} + 2\mathbf{R}\mathbf{A}(n) \tag{4.69}$$

The method of steepest descent is based on the *correction*

$$\mathbf{A}(n+1) = \mathbf{A}(n) - \frac{\mu}{2}\nabla Q(n) \tag{4.70}$$

where μ is a positive constant and the factor 1/2 is inserted for subsequent simplification. Intuition behind the recursion is straightforward: minimum MSE occurs when the gradient of the error relative to the weights is zero; therefore, adjust the weight vector $\mathbf{A}(n)$ in the direction opposite to the error gradient to approach the minimum value of the error along the line of steepest descent. The magnitude of μ determines the magnitude of the adjustment. Insertion of $\nabla Q(n)$ from Eq. 4.69 into Eq. 4.70 yields the *steepest-descent algorithm*:

$$\mathbf{A}(n+1) = \mathbf{A}(n) + \mu[\mathbf{C} - \mathbf{R}\mathbf{A}(n)] \quad (4.71)$$

4.3.2. Convergence of the Steepest-Descent Algorithm

From the form of the steepest descent algorithm, it is evident that convergence of $\mathbf{A}(n)$ to the optimal coefficient vector $\hat{\mathbf{A}}$ depends on μ, $\mathbf{C}$, and $\mathbf{R}$. Since $\mathbf{C}$ and $\mathbf{R}$ are determined by the observed and ideal processes, we focus on μ. A large value of μ gives fast adjustment; a small value gives slow adjustment. Yet μ must not be set too large, since too large a value can cause *overcorrection* and lack of convergence.

To examine convergence, let $\mathbf{I}$ be the identity matrix and

$$\mathbf{D}(n) = \mathbf{A}(n) - \hat{\mathbf{A}} \quad (4.72)$$

Then

$$\begin{aligned}\mathbf{D}(n+1) &= \mathbf{A}(n+1) - \hat{\mathbf{A}} \\ &= \mathbf{A}(n) + \mu[\mathbf{C} - \mathbf{R}\mathbf{A}(n)] - \hat{\mathbf{A}} \\ &= \mathbf{D}(n) + \mu[\mathbf{R}\hat{\mathbf{A}} - \mathbf{R}\mathbf{A}(n)] \\ &= (\mathbf{I} - \mu\mathbf{R})\mathbf{D}(n) \end{aligned} \quad (4.73)$$

To characterize convergence, we use some properties of the autocorrelation matrix (see Section 1.6.5). $\mathbf{R}$ is real, symmetric, and nonnegative definite. If the observations are linearly independent, which is our assumption in solving the equation $\mathbf{RA} = \mathbf{C}$, then $\mathbf{R}$ is invertible and positive definite. Given that $\mathbf{R}$ is positive definite, its eigenvalues are real and positive. Because $\mathbf{R}$ (dimension $M + 1$) is real and symmetric, its eigenvalues satisfy two fundamental properties. First, there exist $M + 1$ mutually orthogonal eigenvectors corresponding to its $M + 1$ eigenvalues λ_0, $\lambda_1, \ldots, \lambda_M$. Second, $\mathbf{R}$ has the diagonalization $\mathbf{U}^{-1}\mathbf{R}\mathbf{U} = \mathbf{L}$, where $\mathbf{L}$ is the diagonal matrix whose diagonal is composed, in order, of the eigenvalues of $\mathbf{R}$, and $\mathbf{U}$ is the matrix whose columns are, in order, the orthonormalized eigenvectors corresponding to the eigenvalues. $\mathbf{U}$ is unitary $[\mathbf{U}' = \mathbf{U}^{-1}]$ so that the *unitary similarity transformation* $\mathbf{U}^{-1}\mathbf{R}\mathbf{U} = \mathbf{L}$ can be rewritten as $\mathbf{U}'\mathbf{R}\mathbf{U} = \mathbf{L}$.

Solving for $\mathbf{R}$ in the transformation and substituting into Eq. 4.73 yields

$$\mathbf{D}(n+1) = (\mathbf{I} - \mu\mathbf{U}\mathbf{L}\mathbf{U}')\mathbf{D}(n) \quad (4.74)$$

which, using $\mathbf{U}' = \mathbf{U}^{-1}$ and letting $\mathbf{B}(n) = \mathbf{U}'\mathbf{D}(n)$, can be written as

$$\begin{aligned}\mathbf{U}'\mathbf{D}(n+1) &= (\mathbf{I} - \mu\mathbf{L})\mathbf{U}'\mathbf{D}(n) \\ &= (\mathbf{I} - \mu\mathbf{L})\mathbf{B}(n)\end{aligned} \tag{4.75}$$

If we assume $\mathbf{A}(n)$ is initialized at $\mathbf{A}(0) = \mathbf{0}$, then

$$\mathbf{B}(0) = \mathbf{U}'\mathbf{D}(0) = -\mathbf{U}'\hat{\mathbf{A}} \tag{4.76}$$

According to the diagonal nature of $\mathbf{L}$, Eq. 4.75 is equivalent to the system

$$b_k(n+1) = (1 - \mu\lambda_k)b_k(n) \tag{4.77}$$

where

$$\mathbf{B}(n) = (b_0(n), b_1(n), \ldots, b_M(n))' \tag{4.78}$$

With $b_k(0)$ being the initial value of $\{b_k(n)\}$, Eq. 4.77 has the solution

$$b_k(n) = (1 - \mu\lambda_k)^n b_k(0) \tag{4.79}$$

Assuming $\mathbf{R}$ is positive definite, its eigenvalues are real and positive. Hence, the limit of $b_k(n)$ as n tends to infinity is 0 if and only if

$$|1 - \mu\lambda_k| < 1 \tag{4.80}$$

If $|1 - \mu\lambda_k| < 1$ for $k = 0, 1, \ldots, M$, then $\mathbf{B}(n) \to \mathbf{0}$; otherwise it does not. Since $\mathbf{D}(n) = \mathbf{U}\mathbf{B}(n)$, $\mathbf{A}(n)$ approaches the optimal solution $\hat{\mathbf{A}}$ as $n \to \infty$ if and only if

$$0 < \mu < \frac{2}{\lambda_{\max}} \tag{4.81}$$

where $\lambda_{\max}$ is the maximum eigenvalue of $\mathbf{R}$.

***Example* 4.12.** We apply the steepest-descent algorithm to the signal-plus-noise model of Example 4.9. To ease computations, let the noise variance be 1, so that

$$\mathbf{R} = \begin{pmatrix} 2 & \rho \\ \rho & 2 \end{pmatrix}$$

Since $\mathbf{C} = (\rho, 1)'$, the steepest-descent algorithm takes the form

$$\mathbf{A}(n+1) = \mathbf{A}(n) + \mu\left[\begin{pmatrix}\rho\\1\end{pmatrix} - \begin{pmatrix}2 & \rho\\ \rho & 2\end{pmatrix}\mathbf{A}(n)\right]$$

The eigenvalues of $\mathbf{R}$ are $2 + \rho$ and $2 - \rho$. Hence, μ must satisfy

$$0 < \mu < (2 + \rho)^{-1}$$

Letting $\mu = 1/4$ and $\mathbf{A}(0) = \mathbf{0}$, the algorithm proceeds in the following manner:

$$\mathbf{A}(1) = \begin{pmatrix}0\\0\end{pmatrix} + \frac{1}{4}\left[\begin{pmatrix}\rho\\1\end{pmatrix} - \begin{pmatrix}2 & \rho\\ \rho & 2\end{pmatrix}\begin{pmatrix}0\\0\end{pmatrix}\right] = \frac{1}{4}\begin{pmatrix}\rho\\1\end{pmatrix}$$

$$\mathbf{A}(2) = \frac{1}{4}\begin{pmatrix}\rho\\1\end{pmatrix} + \frac{1}{4}\left[\begin{pmatrix}\rho\\1\end{pmatrix} - \frac{1}{4}\begin{pmatrix}2 & \rho\\ \rho & 2\end{pmatrix}\begin{pmatrix}\rho\\1\end{pmatrix}\right] = \frac{1}{16}\begin{pmatrix}5\rho\\6-\rho^2\end{pmatrix}$$

$$\mathbf{A}(3) = \frac{1}{16}\begin{pmatrix}5\rho\\6-\rho^2\end{pmatrix} + \frac{1}{4}\left[\begin{pmatrix}\rho\\1\end{pmatrix} - \frac{1}{16}\begin{pmatrix}2 & \rho\\ \rho & 2\end{pmatrix}\begin{pmatrix}5\rho\\6-\rho^2\end{pmatrix}\right] = \frac{1}{64}\begin{pmatrix}20\rho+\rho^3\\28-7\rho^2\end{pmatrix}$$

$$\vdots$$

■

4.3.3. Least-Mean-Square Adaptive Algorithm

The steepest-descent algorithm avoids computation of $\mathbf{R}^{-1}$ but still requires that $\mathbf{R}$ and $\mathbf{C}$ have either been derived from some mathematical model of the random processes or have been estimated. In the absence of the kinds of mathematical models used to derive $\mathbf{R}$ in Section 4.2, $\mathbf{R}$ must be estimated from data. This can be accomplished from some number of realizations or, in the case of covariance egodicity, from a single realization. The least-mean-square (LMS) algorithm requires neither matrix inversion nor prior estimation of $\mathbf{R}$; instead, it is adaptive.

Consider applying a finite-observation linear filter across an observed signal $\{X(n)\}$ in a manner that is not time-invariant. The filter is of the form

$$\psi_{\mathbf{A}(n)}(n) = \sum_{k=0}^{M} a_k(n)X(n-M+k) \tag{4.82}$$

where $\mathbf{A}(n) = (a_0(n), a_1(n),..., a_M(n))'$ is time dependent and the filter provides estimation of the process $\{Y(n)\}$ at time n by an $(M + 1)$-point weighted sum over the process $\{X(n)\}$. Defining the moving average over the discrete time points $n - M, n - M + 1,..., n$ is notationally convenient and natural if we wish to estimate the current value by averaging over past values; however, any other set of $M + 1$ points would do. In the present framework, the observation vector is

$$\mathbf{X}(n) = \begin{pmatrix} X(n-M) \\ X(n-M+1) \\ \vdots \\ X(n) \end{pmatrix} \tag{4.83}$$

The situation we envision is one in which the observed and ideal processes are jointly WS stationary and the weight coefficients in Eq. 4.82 are to be adjusted in a way that in the long run (it is hoped) brings them closer to optimality. For each n, there is a transition $\mathbf{A}(n) \to \mathbf{A}(n + 1)$ based on the action of the filter (current state of the weight vector) at time n. The error is time-variant and is given by Eq. 4.68, with the recognition that $\mathbf{A}(n)$ is now indexed over the time points of the signal, not the step count in the iteration of the steepest-descent algorithm. With the same understanding, the gradient representation of Eq. 4.69 also continues to apply. If we know $\mathbf{R}$ and $\mathbf{C}$, then we can utilize the gradient correction of Eq. 4.70, which leads again to the steepest-descent algorithm of Eq. 4.71. Were we to take this approach, there would be nothing new except the extraneous indexing of the steepest-descent algorithm across the time points. Instead, we will use the gradient correction with estimates of $\mathbf{R}$ and $\mathbf{C}$ that depend on the signal at time n.

One estimator of $\mathbf{R}$, albeit a poor one, has entries

$$\hat{R}_{kj} = X(n - M + k)X(n - M + j) \tag{4.84}$$

derived from the mixed products of the observations within the window at n. This estimator can be compactly written in terms of the observation vector at n by

$$\hat{\mathbf{R}}(n) = \mathbf{X}(n)\mathbf{X}'(n) \tag{4.85}$$

A corresponding estimator for the cross-correlation vector has entries

$$\hat{R}_k = X(n - M + k)Y(n) \tag{4.86}$$

and is expressed in vector form by

$$\hat{\mathbf{C}}(n) = Y(n)\mathbf{X}(n) \tag{4.87}$$

Using these estimators, the steepest-descent algorithm of Eq. 4.71 becomes the recursive estimator equation

$$\begin{aligned}\mathbf{A}(n+1) &= \mathbf{A}(n) + \mu[Y(n)\mathbf{X}(n) - \mathbf{X}(n)\mathbf{X}'(n)\mathbf{A}(n)] \\ &= \mathbf{A}(n) + \mu[Y(n) - \mathbf{A}'(n)\mathbf{X}(n)]\mathbf{X}(n)\end{aligned} \tag{4.88}$$

The gradient expression of Eq. 4.69 becomes the estimator equation

$$\hat{\nabla} Q(n) = -2[Y(n) - \mathbf{A}'(n)\mathbf{X}(n)]\mathbf{X}(n) \tag{4.89}$$

Equation 4.88 characterizes the *least-mean-square* (*LMS*) *algorithm*. The filter output at n is $\mathbf{A}'(n)\mathbf{X}(n)$, which is an estimator of the optimal filter output $\hat{\mathbf{A}}'\mathbf{X}(n)$. The recursion of Eq. 4.88 adds a proportion of the error times the current observation vector to the current weight vector estimate $\mathbf{A}(n)$ to obtain an updated weight vector estimate $\mathbf{A}(n+1)$. The iteration commences with an initial weight estimate; for simplicity, we set $\mathbf{A}(0) = \mathbf{0}$.

4.3.4. Convergence of the LMS Algorithm

At each step of the iteration, sample moments are used to estimate the second-order moments

$$R_{kj} = E[X(n-M+k)X(n-M+j)] \tag{4.90}$$

Since these sample moments arise from a sample of one, any one estimate of $\mathbf{R}$ will be poor; however, the estimation procedure utilizes this information recursively, thereby recursively incorporating the sample observations into the estimation process. The issue is to what extent? We must consider convergence.

Since $\mathbf{A}(n)$ is used in the LMS algorithm as an estimator of the optimal linear weight vector $\hat{\mathbf{A}}$, error analysis can focus on the error vector

$$\mathbf{G}(n) = \mathbf{A}(n) - \hat{\mathbf{A}} \tag{4.91}$$

Substitution of Eq. 4.88 yields the recursive expression

$$\mathbf{G}(n+1) = \mathbf{A}(n+1) - \hat{\mathbf{A}}$$

$$= \mathbf{A}(n) + \mu[Y(n) - \hat{\mathbf{A}}'\mathbf{X}(n) + \hat{\mathbf{A}}'\mathbf{X}(n) - \mathbf{A}'(n)\mathbf{X}(n)]\mathbf{X}(n) - \hat{\mathbf{A}}$$

$$= [\mathbf{I} - \mu\mathbf{X}(n)\mathbf{X}'(n)]\mathbf{G}(n) + \mu[Y(n) - \hat{\mathbf{A}}'\mathbf{X}(n)]\mathbf{X}(n) \tag{4.92}$$

where $Y(n) - \hat{\mathbf{A}}'\mathbf{X}(n)$ is the error resulting from application of the optimal linear filter.

By the orthogonality principle, the error $Y(n) - \hat{\mathbf{A}}'\mathbf{X}(n)$ is orthogonal to the observation vector $\mathbf{X}(n)$, so that taking the expectation of $\mathbf{G}(n+1)$ yields

$$E[\mathbf{G}(n+1)] = E[(\mathbf{I} - \mu\mathbf{X}(n)\mathbf{X}'(n))\mathbf{G}(n)] \tag{4.93}$$

If we assume the observation vector $\mathbf{X}(n)$ at point n is independent of the weight vector $\mathbf{A}(n)$ at point n, then $\mathbf{X}(n)$ is independent of $\mathbf{G}(n)$ and

$$E[\mathbf{G}(n+1)] = (\mathbf{I} - \mu E[\mathbf{X}(n)\mathbf{X}'(n)])E[\mathbf{G}(n)]$$

$$= (\mathbf{I} - \mu\mathbf{R})E[\mathbf{G}(n)] \tag{4.94}$$

This equation is of the same form as Eq. 4.73 with $E[\mathbf{G}(n)]$ in place of $\mathbf{D}(n)$. Thus, the analysis following Eq. 4.73 applies and $E[\mathbf{G}(n)]$ converges to $\mathbf{0}$ so long as the eigenvalues of $\mathbf{R}$ satisfy the system defined by Eq. 4.80 or the single maximum inequality of Eq. 4.81. From the definition of $\mathbf{G}(n)$, we have the following theorem.

Theorem 4.7. If $\mathbf{X}(n)$ and $\mathbf{A}(n)$ are independent and $0 < \mu < 2/\lambda_{max}$, then the weight vectors generated by the LMS algorithm converge in the mean to the weight vector for the optimal linear filter, that is,

$$\lim_{n\to\infty} E[\mathbf{A}(n)] = \hat{\mathbf{A}} \quad \blacksquare \tag{4.95}$$

The independence assumption regarding $\mathbf{X}(n)$ and $\mathbf{A}(n)$ is problematic since $\mathbf{X}(n-1)$ is involved in the recursive expression of $\mathbf{A}(n)$; nevertheless, it is a modeling assumption needed to demonstrate convergence because it allows the expectations of $\mathbf{G}(n)$ and $\mathbf{X}(n)$ to be separated.

Mean convergence of $\mathbf{A}(n)$ to $\hat{\mathbf{A}}$ is important because it says that, for large n, the estimates of $\hat{\mathbf{A}}$ are, on average, close to $\hat{\mathbf{A}}$, which means that the distribution of $\mathbf{A}(n)$ is centered near $\hat{\mathbf{A}}$. However, mean convergence tells us nothing about the variation of $\mathbf{A}(n)$, nor does it characterize the mean-square error

$$Q(n) = E[|Y(n) - \mathbf{A}'(n)\mathbf{X}(n)|^2] \tag{4.96}$$

which is important if we wish to estimate $Y(n)$ by $\mathbf{A}'(n)\mathbf{X}(n)$.

Regarding convergence, beginning at the initialization, $\mathbf{A}(n)$ makes its way toward $\hat{\mathbf{A}}$ and once in the vicinity of $\hat{\mathbf{A}}$ (one hopes) stabilizes into a steady state. Convergence in the mean to $\hat{\mathbf{A}}$ ensures that $\mathbf{A}(n)$ will distribute about $\hat{\mathbf{A}}$ in the steady state. Even though $E[\mathbf{A}(n)] \to \hat{\mathbf{A}}$, the components of $\mathbf{A}(n)$ can have large variances. There are two possible approaches. We can take as the weight vector the estimate $\mathbf{A}(n)$ for some point n in the steady state. Then, although $E[\mathbf{A}(n)] \to \hat{\mathbf{A}}$, so that $\mathbf{A}(n)$ is an asymptotically unbiased estimator of $\hat{\mathbf{A}}$, the precision of the estimator is poor if some component variances are large, meaning the error $|\ \mathbf{A}(n) - \hat{\mathbf{A}}\ |$ can be significant. On the other hand, we can average $\mathbf{A}(n)$ over some time interval in the steady state to obtain an estimator

$$\hat{\mathbf{A}}_{av} = \frac{\mathbf{A}(n) + \mathbf{A}(n+1) + \cdots + \mathbf{A}(n+k)}{k+1} \tag{4.97}$$

of $\hat{\mathbf{A}}$. The component variances of $\hat{\mathbf{A}}_{av}$ decrease for increasing k.

Two questions arise concerning the mean-square error $Q(n)$: (1) Are there sufficient conditions under which $Q(n)$ converges? (2) If $Q(n)$ does converge, what is the relationship between its limit and that of the minimal MSE resulting for the optimal linear filter? We will state a theorem that addresses both of these concerns under very restrictive independence and distributional assumptions. Even under these highly restrictive conditions, $Q(n)$ does not converge to the minimal MSE resulting from the optimal linear filter.

Theorem 4.8. Suppose $\mathbf{X}(n)$ and $Y(n)$ satisfy the following conditions:
(i) $\mathbf{X}(1), \mathbf{X}(2), \ldots, \mathbf{X}(n)$ are independent;
(ii) $\mathbf{X}(n)$ is independent of $Y(1), Y(2), \ldots, Y(n-1)$;
(iii) $Y(n)$ is independent of $Y(1), Y(2), \ldots, Y(n-1)$;
(iv) $\mathbf{X}(n)$ and $Y(k)$ are jointly Gaussian for all n and k.
Then the MSE $Q(n)$ converges if and only if $0 < \mu < 2/\lambda_{\max}$ and

$$\sum_{k=0}^{M} \frac{\mu\lambda_k}{2(1-\mu\lambda_k)} < 1 \tag{4.98}$$

where $\lambda_0, \lambda_1, \ldots, \lambda_M$ are the eigenvalues of $\mathbf{R}$. If $\hat{Q}$ is the MSE for the optimal filter, then

$$\lim_{n\to\infty} Q(n) = \frac{\hat{Q}}{1 - \sum_{k=0}^{M} \frac{\mu\lambda_k}{2(1-\mu\lambda_k)}} \qquad \blacksquare \tag{4.99}$$

Since $\mathbf{A}(n+1)$ depends only on $\mathbf{X}(1)$, $\mathbf{X}(2)$,..., $\mathbf{X}(n)$, $Y(1)$, $Y(2)$,..., $Y(n)$, and $\mathbf{A}(0)$, under the first three assumptions, $\mathbf{A}(n+1)$ is independent of $\mathbf{X}(n+1)$ and $Y(n+1)$. In particular, independence of $\mathbf{A}(n+1)$ and $\mathbf{X}(n+1)$ are the conditions for Theorem 4.7, so that mean convergence is ensured if $0 < \mu < 2/\lambda_{max}$.

From the limit of Eq. 4.99, we see that, even when the MSE converges, it converges to an error greater than that of the optimal linear filter. Denote the summation of Eq. 4.98 by r. By choosing μ sufficiently small, r can be made arbitrarily small. The ratio of the error increase to the minimum error, known as the *misadjustment*, is

$$\chi = \frac{\lim_{n\to\infty} Q(n) - \hat{Q}}{\hat{Q}} = \frac{r}{1-r} \tag{4.100}$$

4.3.5. Nonstationary Processes

As thus far considered, the LMS algorithm has been applied under the assumption that the observation process and the process to be estimated are jointly WS stationary. This has ensured that $\mathbf{R}$, $\mathbf{C}$, and $\hat{\mathbf{A}}$ are time invariant. Suppose that the processes are not jointly WS stationary. Then the optimal weight vector is a function of n, say $\hat{\mathbf{A}}_n$, and the error between the optimal weight vector and its LMS estimate is given by

$$\begin{aligned} H(n) &= \mathbf{A}(n) - \hat{\mathbf{A}}_n \\ &= (\mathbf{A}(n) - E[\mathbf{A}(n)]) + (E[\mathbf{A}(n)] - \hat{\mathbf{A}}_n) \end{aligned} \tag{4.101}$$

The first summand in the error decomposition results from variation of $\mathbf{A}(n)$ about its mean $E[\mathbf{A}(n)]$ and the second (nonrandom) summand results from the difference in the mean of $\mathbf{A}(n)$ and the optimal weight vector at n. In the WS stationary case and under the assumption that $\mathbf{X}(n)$ and $\mathbf{A}(n)$ are independent, for large n, $E[\mathbf{A}(n)]$ is close to $\hat{\mathbf{A}} = \hat{\mathbf{A}}_n$, so that the second summand is small. In the nonstationary setting, the second summand need not be small for large n: it measures the *lag* in filter adaptation to changing statistics of the nonstationary process. Suppose adaptation takes place over a sequence of domains D_1, D_2,..., where the process is WS stationary when restricted to any domain D_j and each domain is sufficiently large that adaptation can be expected to reach a steady state. Then it can be expected that $\mathbf{A}(n)$ reaches a steady state while adaptation passes through D_1, re-adapts over D_2 to a second steady state, and so on. While such a property might be useful for filter design, to use the various adaptively discovered filters on a different image, one would have to segment the image and apply each adaptively designed filter over the appropriate domain.

A general question can be considered: What happens when the optimal linear filter is applied in a spatially invariant manner when the processes are not jointly WS stationary? WS stationarity makes the autocorrelation matrix and the cross-correlation vector constants, rather than functions of n. The problem of nonstationarity can be viewed in the following way: at n, the optimal weight vector is

$$\hat{\mathbf{A}}_n = \mathbf{R}^{-1}(n)\mathbf{C}(n) \tag{4.102}$$

but suppose it is instead taken to be $\mathbf{A} = \mathbf{R}^{-1}\mathbf{C}$, where $\mathbf{R}$ and $\mathbf{C}$ have been selected in some manner. For instance, $\mathbf{R}$ and $\mathbf{C}$ might have been estimated under the assumption of WS stationarity when the processes are not jointly WS stationary, thereby producing a *pooled* estimator. The MSE increase owing to estimation of $Y(n)$ using $\mathbf{A}$ instead of $\hat{\mathbf{A}}_n$ can be expressed via Eq. 4.67, where $Q(n)$ is the MSE resulting from estimation of $Y(n)$ by $\mathbf{A}'\mathbf{X}(n)$ and $\hat{Q}(n)$ is the MSE resulting from estimation of $Y(n)$ by $\hat{\mathbf{A}}_n'\mathbf{X}(n)$:

$$\begin{aligned} Q(n) - \hat{Q}(n) &= (\mathbf{A} - \hat{\mathbf{A}}_n)'\mathbf{R}(n)(\mathbf{A} - \hat{\mathbf{A}}_n) \\ &= (\mathbf{R}^{-1}\mathbf{C} - \mathbf{R}^{-1}(n)\mathbf{C}(n))'\mathbf{R}(n)(\mathbf{R}^{-1}\mathbf{C} - \mathbf{R}^{-1}(n)\mathbf{C}(n)) \end{aligned} \tag{4.103}$$

If $\mathbf{R}$ and $\mathbf{C}$ are close to $\mathbf{R}(n)$ and $\mathbf{C}(n)$, respectively, then the error from using $\mathbf{R}$ and $\mathbf{C}$ in place of $\mathbf{R}(n)$ and $\mathbf{C}(n)$ will be small. Using matrix and vector norms, this can be seen from $Q(n) - \hat{Q}(n)$ because

$$\begin{aligned} \|\mathbf{R}^{-1}\mathbf{C} - \mathbf{R}^{-1}(n)\mathbf{C}(n)\| &= \|\mathbf{R}^{-1}\mathbf{C} - \mathbf{R}^{-1}(n)\mathbf{C} + \mathbf{R}^{-1}(n)\mathbf{C} - \mathbf{R}^{-1}(n)\mathbf{C}(n)\| \\ &\le \|\mathbf{R}^{-1}\mathbf{C} - \mathbf{R}^{-1}(n)\mathbf{C}\| + \|\mathbf{R}^{-1}(n)\mathbf{C} - \mathbf{R}^{-1}(n)\mathbf{C}(n)\| \\ &= \|\mathbf{R}^{-1} - \mathbf{R}^{-1}(n)\|\,\|\mathbf{C}\| + \|\mathbf{R}^{-1}(n)\|\,\|\mathbf{C} - \mathbf{C}(n)\| \end{aligned} \tag{4.104}$$

A spatially invariant linear filter can be employed effectively in a nonstationary environment so long as the autocorrelation matrix and cross-correlation vector on which it is based do not vary too greatly from the actual varying autocorrelation matrices and cross-correlation vectors, respectively.

4.4. Least-Squares Estimation

For the next few sections we will consider vector estimation. This section concerns linear estimation of a deterministic vector $\mathbf{y}$ given an observation random m-vector $\mathbf{X}$ satisfying the model

$$\mathbf{X} = \mathbf{H}\mathbf{y} + \mathbf{N} \tag{4.105}$$

where the *design matrix* **H** is $m \times n$ with $m > n$, **y** is an n-vector, and **N** is a zero-mean random m-vector. This *finite linear model* occurs in a number of settings and the problem is to estimate **y** based on the observations constituting **X**. Estimation here differs from the kind of estimation previously discussed in this chapter. Heretofore, the task has been to estimate random variables, not deterministic variables. Because **y** is deterministic, the estimation discussed in this section can be categorized as *parametric estimation*.

If **Y** is a random vector with mean vector **y** and the means of **Y** and the random vector **X** are related by $E[\mathbf{X}] = \mathbf{H}E[\mathbf{Y}]$, then Eq. 4.105 is satisfied with

$$\mathbf{N} = \mathbf{X} - \mathbf{H}E[\mathbf{Y}] \tag{4.106}$$

being the centered version of **X** (the random displacement of **X** about its mean). The goal is to estimate $E[\mathbf{Y}]$ from **X**. A related view is that **y** is a realization of a vector process, the randomness of **X** results from observation noise **N** about **Hy**, and our desire is to estimate the realization **y** based on **X**. Another perspective is that **X** is a random vector having a parameterized mean $\mathbf{H}\boldsymbol{\theta}$ determined by a parameter vector $\boldsymbol{\theta} = (\theta_1, \theta_2, \ldots, \theta_n)'$. **N** is the dispersion of **X** about its mean $\mathbf{H}\boldsymbol{\theta}$, Eq. 4.105 holds with $\boldsymbol{\theta}$ in place of **y**, and the goal is to estimate $\boldsymbol{\theta}$ based on the observation **X**. While it is common to refer to **N** as noise, and we will also use the terminology, one can see from these model interpretations that **N** need not be noise, but rather the centered version of the observation process.

4.4.1. Pseudoinverse Estimator

If **H** is an invertible matrix, **x** and **y** are deterministic vectors, and

$$\mathbf{x} = \mathbf{H}\mathbf{y} \tag{4.107}$$

then $\mathbf{y} = \mathbf{H}^{-1}\mathbf{x}$ inverts the system, meaning

$$\mathbf{x} - \mathbf{H}(\mathbf{H}^{-1}\mathbf{x}) = \mathbf{0} \tag{4.108}$$

If **H** is not invertible, then $\mathbf{H}(\mathbf{H}^{-1}\mathbf{x})$ is not defined. For reasons to become apparent, we assume that the columns of **H** are linearly independent. If **H** were square, then it would be nonsingular. Therefore, we assume **H** is $m \times n$ with $m > n$, **x** is an m-vector, and **y** is an n-vector.

Even though **H** is not invertible, we can try to find a vector **y** to minimize the *sum-of-squares error*, given by

$$SSE = \|\mathbf{x} - \mathbf{H}\mathbf{y}\|^2$$

$$= \sum_{i=1}^{m} |x_i - u_i|^2$$

$$= \mathbf{x'x} - \mathbf{x'Hy} - \mathbf{y'H'x} + \mathbf{y'H'Hy} \tag{4.109}$$

where $\mathbf{x} = (x_1, x_2,..., x_m)'$ and $\mathbf{Hy} = (u_1, u_2,..., u_m)'$. To minimize *SSE* as a function of **y**, we need to take the partial derivative of *SSE* with respect to **y** and then set it equal to 0. Using the results of Exercise 4.26,

$$\frac{\partial\|\mathbf{x} - \mathbf{Hy}\|^2}{\partial \mathbf{y}} = 2\mathbf{H'Hy} - 2\mathbf{H'x} \tag{4.110}$$

Because the columns of **H** are linearly independent, **H'H** is invertible. Setting $\partial SSE/\partial \mathbf{y} = 0$ and solving yields the SSE-minimizing vector

$$\mathbf{y_x} = (\mathbf{H'H})^{-1}\mathbf{H'x} \tag{4.111}$$

$(\mathbf{H'H})^{-1}\mathbf{H'}$ is known as the *pseudoinverse* of **H** and is denoted by $\mathbf{H}^+$. Using this notation,

$$\mathbf{y_x} = \mathbf{H}^+\mathbf{x} \tag{4.112}$$

Were **H** invertible, then we would have $\mathbf{H}^+ = \mathbf{H}^{-1}$.

Now consider the finite linear model of Eq. 4.105 with

$$\begin{aligned} \mathbf{y} &= (y_1, y_2,..., y_n)' \\ \mathbf{X} &= (X_1, X_2,..., X_m)' \\ \mathbf{N} &= (N_1, N_2,..., N_m)' \end{aligned} \tag{4.113}$$

X is observed and we wish to estimate **y**. Given an observation **x**, $\mathbf{y_x}$ minimizes the SSE. Relative to the SSE, $\mathbf{Hy_x}$ provides a *best fit* for the data **x**. As **x** varies, so does $\mathbf{y_x}$. In this context, $\mathbf{y_x}$ becomes the estimator

$$\hat{\mathbf{y}} = \mathbf{H}^+\mathbf{X} \tag{4.114}$$

It is called the *least-squares estimator* for **y**. The estimation rule $\mathbf{H}^+\mathbf{x}$ has been developed to minimize a deterministic sum-of-squares error. Properties of $\hat{\mathbf{y}}$ as a statistical estimator of **y** need to be determined in the context of the finite linear model of Eq. 4.105.

Owing to the model, the estimator can be expressed as

$$\hat{\mathbf{y}} = (\mathbf{H}'\mathbf{H})^{-1}\mathbf{H}'(\mathbf{H}\mathbf{y} + \mathbf{N})$$
$$= (\mathbf{H}'\mathbf{H})^{-1}\mathbf{H}'\mathbf{H}\mathbf{y} + (\mathbf{H}'\mathbf{H})^{-1}\mathbf{H}'\mathbf{N}$$
$$= \mathbf{y} + (\mathbf{H}'\mathbf{H})^{-1}\mathbf{H}'\mathbf{N} \tag{4.115}$$

Accordingly,

$$E[\hat{\mathbf{y}}] = \mathbf{y} + (\mathbf{H}'\mathbf{H})^{-1}\mathbf{H}'E[\mathbf{N}] = \mathbf{y} \tag{4.116}$$

and $\hat{\mathbf{y}}$ is an unbiased estimator of y.

To this point we have only assumed that **N** is zero-mean. Now assume it is white, with covariance matrix $K = \sigma^2\mathbf{I}$, where **I** is the identity matrix. The covariance matrix of $\hat{\mathbf{y}}$ can be deduced from Eq. 4.115:

$$\mathbf{K}_{\hat{\mathbf{y}}} = E[(\hat{\mathbf{y}} - \mathbf{y})(\hat{\mathbf{y}} - \mathbf{y})']$$
$$= E[(\mathbf{H}'\mathbf{H})^{-1}\mathbf{H}'\mathbf{N}((\mathbf{H}'\mathbf{H})^{-1}\mathbf{H}'\mathbf{N})']$$
$$= E[(\mathbf{H}'\mathbf{H})^{-1}\mathbf{H}'\mathbf{N}\mathbf{N}'((\mathbf{H}'\mathbf{H})^{-1}\mathbf{H}')']$$
$$= (\mathbf{H}'\mathbf{H})^{-1}\mathbf{H}'E[\mathbf{N}\mathbf{N}']((\mathbf{H}'\mathbf{H})^{-1}\mathbf{H}')' \tag{4.117}$$
$$= \sigma^2(\mathbf{H}'\mathbf{H})^{-1}\mathbf{H}'\mathbf{H}((\mathbf{H}'\mathbf{H})^{-1})'$$
$$= \sigma^2(\mathbf{H}'\mathbf{H})^{-1}$$

An arbitrary linear estimator $\hat{\mathbf{y}} = \mathbf{DX}$ of y is unbiased if and only if

$$E[\mathbf{D}(\mathbf{H}\mathbf{y} + \mathbf{N})] = \mathbf{y} \tag{4.118}$$

which is equivalent to

$$\mathbf{DH} = \mathbf{I} \tag{4.119}$$

Hence, the covariance matrix of $\hat{\mathbf{y}}$ is

$$\mathbf{K}_{\hat{\mathbf{y}}} = E[(\hat{\mathbf{y}} - \mathbf{y})(\hat{\mathbf{y}} - \mathbf{y})']$$
$$= E[(\mathbf{D}(\mathbf{H}\mathbf{y} + \mathbf{N}) - \mathbf{D}\mathbf{H}\mathbf{y})(\mathbf{D}(\mathbf{H}\mathbf{y} + \mathbf{N}) - \mathbf{D}\mathbf{H}\mathbf{y})'] \tag{4.120}$$
$$= E[\mathbf{DN}(\mathbf{DN})']$$

$$= \sigma^2 \mathbf{DD}'$$

The *minimum-variance, unbiased, linear estimator* [or, *best linear, unbiased estimator* (*BLUE*)] is the one minimizing the diagonal elements of **DD**', since the diagonal elements of $\mathbf{K}_{\hat{\mathbf{y}}}$ are these multiplied by σ^2. Some matrix algebra shows that

$$\mathbf{DD}' = \mathbf{H}^+(\mathbf{H}^+)' + (\mathbf{D} - \mathbf{H}^+)(\mathbf{D} - \mathbf{H}^+)' \tag{4.121}$$

Diagonal terms of **DD**' are minimized when diagonal terms of the matrix product $(\mathbf{D} - \mathbf{H}^+)\,(\mathbf{D} - \mathbf{H}^+)'$ are minimized, but these are identically 0 if $\mathbf{D} = \mathbf{H}^+$, thereby proving that the variance terms of $\mathbf{K}_{\hat{\mathbf{y}}}$ are minimized when $\hat{\mathbf{y}} = \hat{\mathbf{y}}$. Thus, $\hat{\mathbf{y}}$ is the minimum-variance, unbiased, linear estimator (BLUE) for **y** in the finite linear model.

4.4.2. Least-Squares Estimation for Nonwhite Noise

The results for **N** white can be used to establish results for nonwhite noise having a positive-definite covariance matrix $K = E[\mathbf{NN}']$. K is positive-definite, real, and symmetric. Therefore there exists a matrix **Q** such that $K^{-1} = \mathbf{QQ}'$ and $\mathbf{Q}'K\mathbf{Q} = \mathbf{I}$. Premultiplying Eq. 4.105 by **Q**' yields

$$\mathbf{Q}'\mathbf{X} = (\mathbf{Q}'\mathbf{H})\mathbf{y} + \mathbf{Q}'\mathbf{N} \tag{4.122}$$

Since the covariance matrix $K_{\mathbf{Q}'\mathbf{N}}$ of **Q'N** satisfies

$$\begin{aligned} \mathbf{K}_{\mathbf{Q}'\mathbf{N}} &= E[\mathbf{Q}'\mathbf{N}(\mathbf{Q}'\mathbf{N})'] \\ &= \mathbf{Q}'E[\mathbf{NN}']\mathbf{Q} \\ &= \mathbf{Q}'\mathbf{KQ} \end{aligned} \tag{4.123}$$

which equals **I**, Eq. 4.122 is of the same form as the linear model of Eq. 4.105. Hence,

$$\hat{\mathbf{y}} = ((\mathbf{Q}'\mathbf{H})'\mathbf{Q}'\mathbf{H})^{-1}(\mathbf{Q}'\mathbf{H})'\mathbf{Q}'\mathbf{X} \tag{4.124}$$

is a BLUE for **y** in Eq. 4.122. Moreover,

$$\hat{\mathbf{y}} = (\mathbf{H}'\mathbf{QQ}'\mathbf{H})^{-1}\mathbf{H}'\mathbf{QQ}'\mathbf{X}$$

$$= (\mathbf{H}'\mathbf{K}^{-1}\mathbf{H})^{-1}\mathbf{H}'\mathbf{K}^{-1}\mathbf{X} \tag{4.125}$$

which reduces to the estimator of Eq. 4.114 when **N** is white.

In fact, $\hat{\mathbf{y}} = (\mathbf{H}'\mathbf{K}^{-1}\mathbf{H})^{-1}\mathbf{H}'\mathbf{K}^{-1}\mathbf{X}$ is a BLUE for **y** in the linear model of Eq. 4.105. It is unbiased because

$$E[\hat{\mathbf{y}}] = E[(\mathbf{H}'\mathbf{K}^{-1}\mathbf{H})^{-1}\mathbf{H}'\mathbf{K}^{-1}(\mathbf{H}\mathbf{y} + \mathbf{N})]$$

$$= (\mathbf{H}'\mathbf{K}^{-1}\mathbf{H})^{-1}\mathbf{H}'\mathbf{K}^{-1}\mathbf{H}\mathbf{y} + (\mathbf{H}'\mathbf{K}^{-1}\mathbf{H})^{-1}\mathbf{H}'\mathbf{K}^{-1}E[\mathbf{N}] \tag{4.126}$$

which reduces to **y** since $(\mathbf{H}'\mathbf{K}^{-1}\mathbf{H})^{-1}\mathbf{H}'\mathbf{K}^{-1}\mathbf{H} = \mathbf{I}$ and $E[\mathbf{N}] = \mathbf{0}$. Its covariance matrix is

$$\mathbf{K}_{\hat{\mathbf{y}}} = E[(\hat{\mathbf{y}} - \mathbf{y})(\hat{\mathbf{y}} - \mathbf{y})']$$

$$= E[(\mathbf{H}'\mathbf{K}^{-1}\mathbf{H})^{-1}\mathbf{H}'\mathbf{K}^{-1}\mathbf{N}((\mathbf{H}'\mathbf{K}^{-1}\mathbf{H})^{-1}\mathbf{H}'\mathbf{K}^{-1}\mathbf{N})']$$

$$= (\mathbf{H}'\mathbf{K}^{-1}\mathbf{H})^{-1}\mathbf{H}'\mathbf{K}^{-1}E[\mathbf{N}\mathbf{N}']\mathbf{K}^{-1}\mathbf{H}(\mathbf{H}'\mathbf{K}^{-1}\mathbf{H})^{-1} \tag{4.127}$$

$$= (\mathbf{H}'\mathbf{K}^{-1}\mathbf{H})^{-1}$$

To see that it has minimum variance among unbiased estimators for **y** in the linear model of Eq. 4.105, let **D** be any unbiased linear estimator for **y** in Eq. 4.105. Then

$$\mathbf{D}\mathbf{X} - \mathbf{D}(\mathbf{Q}')^{-1}\mathbf{Q}'\mathbf{X} \tag{4.128}$$

is an unbiased linear estimator for **y** in Eq. 4.122. By Eq. 4.125, $(\mathbf{H}'\mathbf{K}^{-1}\mathbf{H})^{-1}\mathbf{H}'\mathbf{K}^{-1}\mathbf{X}$ has minimum variance among such estimators. Thus, it is the BLUE for **y** in Eq. 4.105.

Theorem 4.9. In the linear model $\mathbf{X} = \mathbf{H}\mathbf{y} + \mathbf{N}$, where **H** is an $m \times n$ deterministic matrix with $m > n$, **H** has linearly independent columns, and **N** possesses zero mean and positive-definite covariance matrix **K**, the minimum-variance, unbiased, linear estimator for **y** is given by $(\mathbf{H}'\mathbf{K}^{-1}\mathbf{H})^{-1}\mathbf{H}'\mathbf{K}^{-1}\mathbf{X}$. ■

***Example* 4.13.** Consider discrete deterministic signals $z_1(n)$, $z_2(n)$,..., $z_m(n)$, $m > 3$, and $y(n)$, with $z_j(n)$ obtained from $y(n)$ via the moving average

$$z_j(n) = h_{j,-1}y(n-1) + h_{j,0}y(n) + h_{j,1}y(n+1)$$

If observation of $z_j(n)$ is obscured by additive white noise $N_j(n)$, then there are m observed random functions $X_j(n)$ defined by

$$X_j(n) = h_{j,-1}y(n-1) + h_{j,0}y(n) + h_{j,1}y(n+1) + N_j(n)$$

We wish to estimate $y(-1)$, $y(0)$, and $y(1)$ based on $X_1(0), X_2(0),..., X_m(0)$. The finite linear model is satisfied with

$$\mathbf{X} = (X_1(0), X_2(0),..., X_m(0))'$$

$$\mathbf{y} = (y(-1), y(0), y(1))'$$

$$\mathbf{N} = (N_1(0), N_2(0),..., N_m(0))'$$

and the $m \times 3$ design matrix

$$\mathbf{H} = \begin{pmatrix} h_{1,-1} & h_{1,0} & h_{1,1} \\ h_{2,-1} & h_{2,0} & h_{2,1} \\ \vdots & \vdots & \vdots \\ h_{m,-1} & h_{m,0} & h_{m,1} \end{pmatrix}$$

If the columns of $\mathbf{H}$ are linearly independent, then the BLUE is $\mathbf{H}^+\mathbf{X}$. When the noise is not white, then the BLUE is given by Theorem 4.9. ■

4.4.3. Multiple Linear Regression

The *multiple linear regression* model is a basic parametric model used in statistical inference. The model assumes there are k random variables $Z_1, Z_2,..., Z_k$ possessing means $\mu_1, \mu_2,..., \mu_k$, respectively, and indexed by n-vectors $\mathbf{z}_1, \mathbf{z}_2,..., \mathbf{z}_k$, where

$$\mathbf{z}_i = (z_{1i}, z_{2i},..., z_{ni})' \tag{4.129}$$

for $i = 1, 2,..., k$. Letting $N_i = Z_i - \mu_i$ be the random displacement between Z_i and its mean,

$$Z_i = \mu_i + N_i \tag{4.130}$$

For $i = 1, 2,..., k$, the mean is assumed to have the parametric form

$$\mu_i = \theta_0 + \theta_1 z_{1i} + \theta_2 z_{2i} + \cdots + \theta_n z_{ni} \tag{4.131}$$

We desire a least-squares estimator of

$$\boldsymbol{\theta} = (\theta_0, \theta_1,..., \theta_n)' \tag{4.132}$$

based on $m = m_1 + m_2 + \cdots + m_k$ independent observations from among Z_1, $Z_2,\ldots, Z_k$, where it is known prior to sampling that m_i observations are taken of Z_i. $N_1, N_2,\ldots, N_k$ are assumed to be identically distributed with a common variance σ^2.

Geometrically, the model can be viewed as having random variables Z_1, $Z_2,\ldots, Z_k$ defined at k points $\mathbf{z}_1, \mathbf{z}_2,\ldots, \mathbf{z}_k \in \Re^n$. The random point $(\mathbf{z}_i, Z_i)$ lies in $\Re^{n+1}$. If we view Z_i as a random variable on the line determined by $\mathbf{z}_i$ that extends into the $n + 1$ dimension of $\Re^{n+1}$, then its mean lies on the *regression hyperplane* determined by the linear equation

$$x = \theta_0 + \theta_1 z_1 + \theta_2 z_2 + \cdots + \theta_n z_n \tag{4.133}$$

where $\mathbf{z} = (z_1, z_2,\ldots, z_n)$ is a generic point in $\Re^n$. The least-squares estimator provides estimates of the hyperplane parameters, so that the estimated regression hyperplane is fit to the data. This is not the only interpretation of the model.

Suppose $X_{i1}, X_{i2},\ldots, X_{im_i}$ are the observations of (identically distributed to) Z_i for $i = 1, 2,\ldots, k$. For $j = 1, 2,\ldots, m_i$,

$$X_{ij} = \theta_0 + \theta_1 z_{1i} + \theta_2 z_{2i} + \cdots + \theta_n z_{ni} + N_{ij} \tag{4.134}$$

where N_{ij} is identically distributed to $N_1, N_2,\ldots, N_k$. The design matrix is $m \times (n + 1)$, with the first m_1 rows being $(1, \mathbf{z}_1')$, the second m_2 rows being $(1, \mathbf{z}_2')$, etc.:

$$\mathbf{H} = \begin{pmatrix} 1 & z_{11} & z_{21} & \cdots & z_{n1} \\ \vdots & \vdots & \vdots & \ddots & \vdots \\ 1 & z_{11} & z_{21} & \cdots & z_{n1} \\ 1 & z_{12} & z_{22} & \cdots & z_{n2} \\ \vdots & \vdots & \vdots & \ddots & \vdots \\ 1 & z_{1k} & z_{2k} & \cdots & z_{nk} \end{pmatrix} \tag{4.135}$$

Letting

$$\mathbf{X} = (X_{11}, X_{12},\ldots, X_{1m_1}, X_{21},\ldots, X_{km_k})' \tag{4.136}$$

$$\mathbf{N} = (N_{11}, N_{12},\ldots, N_{1m_1}, N_{21},\ldots, N_{km_k})' \tag{4.137}$$

results in the model

$$\mathbf{X} = \mathbf{H}\boldsymbol{\theta} + \mathbf{N} \tag{4.138}$$

corresponding to Eq. 4.105. Assuming **N** is white, $\mathbf{H}^+\mathbf{X}$ is the BLUE for $\boldsymbol{\theta}$.

***Example* 4.14.** The case for $n = 1$ is known as *simple linear regression*. $\boldsymbol{\theta} = (\theta_0, \theta_1)'$ and the estimated hyperplane is the straight line

$$x = \hat{\theta}_0 + \hat{\theta}_1 z$$

that best fits the observed points in the plane in the least-squares sense. The points $z_{11}, z_{12}, \ldots, z_{1k}$ are selected deterministically prior to data observation, there are m_i observations at each z_{1i}, and the estimated regression line fits the planar points (z_{1i}, X_{ij}), for $i = 1, 2, \ldots, k$ and $j = 1, 2, \ldots, m_i$ (see Fig. 4.2). The sum-of-squares error

$$SSE = \sum_{i=1}^{k} \sum_{j=1}^{m_i} |X_{ij} - (\theta_0 + \theta_1 z_{1i})|^2$$

is minimized. The design matrix transpose is

$$\mathbf{H}' = \begin{pmatrix} 1 & \cdots & 1 & 1 & \cdots & 1 \\ z_{11} & \cdots & z_{11} & z_{12} & \cdots & z_{1k} \end{pmatrix}$$

Let

$$\bar{z} = \frac{1}{m} \sum_{i=1}^{k} m_i z_{1i}$$

$$\overline{z^2} = \frac{1}{m} \sum_{i=1}^{k} m_i z_{1i}^2$$

$$\bar{X} = \frac{1}{m} \sum_{i=1}^{k} \sum_{j=1}^{m_i} X_{ij}$$

$$\overline{zX} = \frac{1}{m} \sum_{i=1}^{k} \sum_{j=1}^{m_i} z_{1i} X_{ij}$$

be the means of $\{z_{1i}\}$, $\{z_{1i}^2\}$, $\{X_{ij}\}$, and $\{z_{1i}X_{ij}\}$, respectively. Then

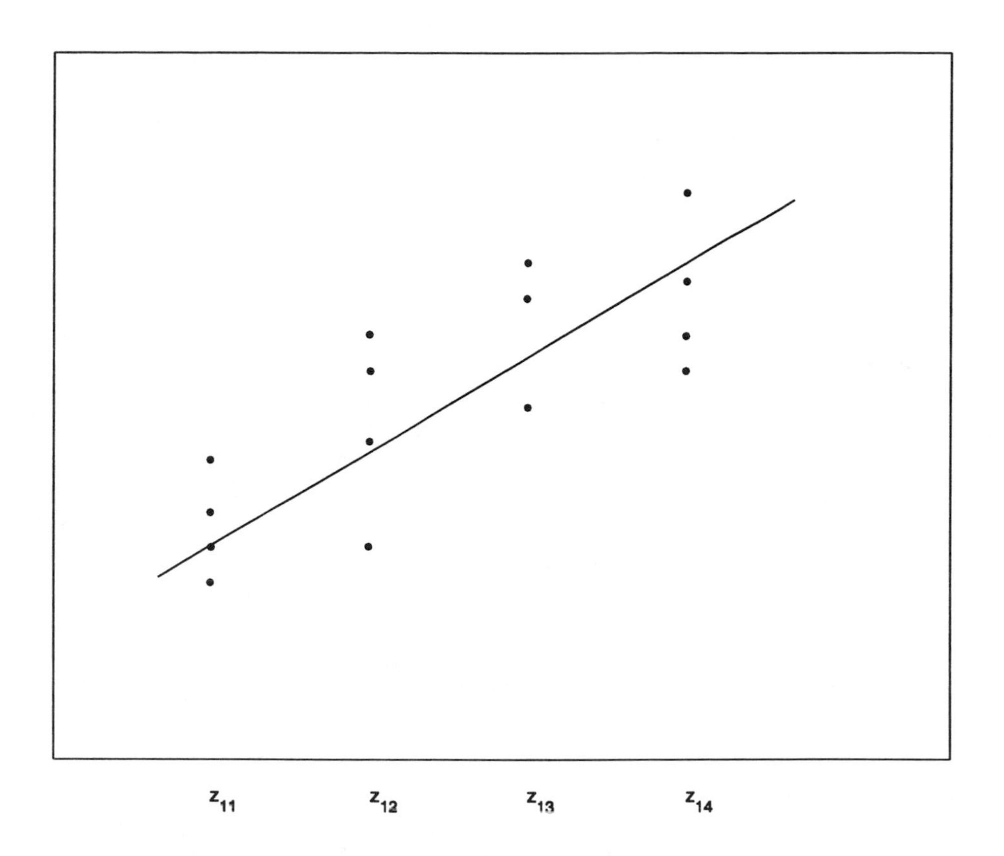

Figure 4.2 Simple linear regression.

$$\mathbf{H'H} = m\begin{pmatrix} 1 & \bar{z} \\ \bar{z} & \overline{z^2} \end{pmatrix}$$

$$(\mathbf{H'H})^{-1} = \frac{1}{m\left(\overline{z^2} - \bar{z}^2\right)}\begin{pmatrix} \overline{z^2} & -\bar{z} \\ -\bar{z} & 1 \end{pmatrix}$$

$$(\mathbf{H'H})^{-1}\mathbf{H'} = \frac{1}{m\left(\overline{z^2} - \bar{z}^2\right)}\begin{pmatrix} \overline{z^2} - z_{11}\bar{z} & \cdots & \overline{z^2} - z_{11}\bar{z} & \overline{z^2} - z_{12}\bar{z} & \cdots & \overline{z^2} - z_{1k}\bar{z} \\ z_{11} - \bar{z} & \cdots & z_{11} - \bar{z} & z_{12} - \bar{z} & \cdots & z_{1k} - \bar{z} \end{pmatrix}$$

$$(\mathbf{H'H})^{-1}\mathbf{H'X} = \frac{1}{\overline{z^2} - \bar{z}^2}\begin{pmatrix} \overline{z^2}\,\overline{X} - \bar{z}\overline{zX} \\ \overline{zX} - \bar{z}\overline{X} \end{pmatrix}$$

Consequently, the estimated regression line has the formulation

$$x = \frac{\overline{z^2}\,\overline{X} - \bar{z}\overline{zX}}{\overline{z^2} - \bar{z}^2} + \frac{\overline{zX} - \bar{z}\overline{X}}{\overline{z^2} - \bar{z}^2} z$$

According to Eq. 4.117,

$$\text{Var}[\hat{\theta}_0] = \frac{\sigma^2 \overline{z^2}}{m\left(\overline{z^2} - \bar{z}^2\right)}$$

$$\text{Var}[\hat{\theta}_1] = \frac{\sigma^2}{m\left(\overline{z^2} - \bar{z}^2\right)}$$

■

***Example* 4.15.** Viewing $\mathbf{z}_1$, $\mathbf{z}_2$,..., $\mathbf{z}_k$ as points in $\mathfrak{R}^n$ yields a multiple linear regression model that fits a regression hyperplane to the data points; instead, suppose

$$\mathbf{z}_i = (z_i, z_i^2,\ldots, z_i^n)'$$

for $i = 1, 2,\ldots, k$. Then Eq. 4.134 becomes

$$X_{ij} = \theta_0 + \theta_1 z_i + \theta_2 z_i^2 + \cdots + \theta_n z_i^n + N_{ij}$$

Geometrically, there are k points z_1, z_2,..., z_k on the real line and the least-squares estimator provides a best-fitting polynomial. The design matrix is

$$\mathbf{H} = \begin{pmatrix} 1 & z_1 & z_1^2 & \cdots & z_1^n \\ \vdots & \vdots & \vdots & \ddots & \vdots \\ 1 & z_1 & z_1^2 & \cdots & z_1^n \\ 1 & z_2 & z_2^2 & \cdots & z_2^n \\ \vdots & \vdots & \vdots & \ddots & \vdots \\ 1 & z_k & z_k^2 & \cdots & z_k^n \end{pmatrix}$$

■

4.4.4. Least-Squares Image Restoration

In Section 4.7 we will discuss a WS stationary linear model for discrete random images when the image process $Y(m, n)$ is degraded by blurring and additive noise to yield the observed image process $X(m, n)$. In that situation, the optimal linear filter will depend on knowledge relating to the joint distribution of the ideal and observed processes, namely, the covariance functions. For a given point (m, n) in the image domain, the filter will be an optimal linear estimator of the random variable $Y(m, n)$ based on observations $X(k, l)$ over all (k, l). The filter will restore the ideal image process from the observation process taken over all space.

Restoration based on the linear model of Eq. 4.105 is different. Here, the model involves a deterministic signal that is a realization of the ideal process (or its mean) and it will be estimated from an observed random signal that is a convolution of the deterministic signal plus additive noise. The signal model agrees with the linear model of Eq. 4.105 and Theorem 4.9 yields a BLUE for the deterministic signal.

In the present setting, the full-space degradation model takes the form

$$X(m, n) = \sum_{k=-\infty}^{\infty} \sum_{l=-\infty}^{\infty} h(m-k, n-l)y(k, l) + N(m, n) \tag{4.139}$$

(see Eq. 4.262 for the fully random model). For the moment, we will focus on the one-dimensional case, so that

$$X(m) = \sum_{k=-\infty}^{\infty} h(m-k)y(k) + N(m) \tag{4.140}$$

To formulate convolution over a finite domain, suppose $h(m)$ and $y(m)$ are periodic with period of length r. If one (or both) of their periods is less than r, then it can be extended with zeros to achieve a common period of length r. Define the finite convolution

$$v(m) = \sum_{k=0}^{r-1} h(m-k)y(k) \tag{4.141}$$

So defined, $v(m)$ has the same period as $h(m)$ and $y(m)$. Equation 4.141 becomes the vector equation $\mathbf{v} = \mathbf{H}\mathbf{y}$ by defining

$$\mathbf{y} = (y(0), y(1),..., y(r-1))' \tag{4.142}$$

$$\mathbf{v} = (v(0), v(1),..., v(r-1))' \tag{4.143}$$

$$\mathbf{H} = \begin{pmatrix} h(0) & h(-1) & h(-2) & \cdots & h(-r+1) \\ h(1) & h(0) & h(-1) & \cdots & h(-r+2) \\ h(2) & h(1) & h(0) & \cdots & h(-r+3) \\ \vdots & \vdots & \vdots & \ddots & \vdots \\ h(r-1) & h(r-2) & h(r-3) & \cdots & h(0) \end{pmatrix}$$

$$= \begin{pmatrix} h(0) & h(r-1) & h(r-2) & \cdots & h(1) \\ h(1) & h(0) & h(r-1) & \cdots & h(2) \\ h(2) & h(1) & h(0) & \cdots & h(3) \\ \vdots & \vdots & \vdots & \ddots & \vdots \\ h(r-1) & h(r-2) & h(r-3) & \cdots & h(0) \end{pmatrix} \tag{4.144}$$

where the preceding equality results from the periodicity of $h(m)$. In the latter form, $\mathbf{H}$ is *circulant*, meaning that the last entry of each row is the first entry in the next row, including the last entry of the last row being the first entry of the first row.

Returning to the image case of Eq. 4.139, we assume that $h(m, n)$ and $y(m, n)$ are periodic in two dimensions (having been extended with zeros if necessary) with period lengths r and s in the horizontal and vertical directions, respectively. Their finite convolution is defined by

$$v(m, n) = \sum_{k=0}^{r-1} \sum_{l=0}^{s-1} h(m-k, n-l) y(k, l) \tag{4.145}$$

In this setting, the model of Eq. 4.139 becomes the finite model

$$X(m, n) = \sum_{k=0}^{r-1} \sum_{l=0}^{s-1} h(m-k, n-l) y(k, l) + N(m, n) \tag{4.146}$$

To express the model in vector form, note that $X(m, n)$, $y(m, n)$, and $N(m, n)$ can be written as matrices over the discrete domain $\{0, 1, \ldots, r-1\} \times \{0, 1, \ldots, s-1\}$, with the functions defined elsewhere by periodic extension. Define $\mathbf{X}$ by taking its first r components to be the first row of the matrix corresponding to $X(m, n)$, its next r components to be the second row of the matrix, etc. Similarly define $\mathbf{y}$ and $\mathbf{N}$ corresponding to $y(m, n)$ and $N(m, n)$, respectively. Define $\mathbf{H}$ via the r^2 partitioning

$$\mathbf{H} = \begin{pmatrix} \mathbf{H}_0 & \mathbf{H}_{r-1} & \mathbf{H}_{r-2} & \cdots & \mathbf{H}_1 \\ \mathbf{H}_1 & \mathbf{H}_0 & \mathbf{H}_{r-1} & \cdots & \mathbf{H}_2 \\ \mathbf{H}_2 & \mathbf{H}_1 & \mathbf{H}_0 & \cdots & \mathbf{H}_3 \\ \vdots & \vdots & \vdots & \ddots & \vdots \\ \mathbf{H}_{r-1} & \mathbf{H}_{r-2} & \mathbf{H}_{r-3} & \cdots & \mathbf{H}_0 \end{pmatrix} \tag{4.147}$$

where, by the periodicity of $h(m, n)$, the blocks of $\mathbf{H}$ are defined by the circulant matrices

$$\mathbf{H}_j = \begin{pmatrix} h(j,0) & h(j,s-1) & h(j,s-2) & \cdots & h(j,1) \\ h(j,1) & h(j,0) & h(j,s-1) & \cdots & h(j,2) \\ h(j,2) & h(j,1) & h(j,0) & \cdots & h(j,3) \\ \vdots & \vdots & \vdots & \ddots & \vdots \\ h(j,s-1) & h(j,s-2) & h(j,s-3) & \cdots & h(j,0) \end{pmatrix} \tag{4.148}$$

for $j = 0, 1,..., r$. With these definitions, the model of Eq. 4.146 becomes the finite linear model of Eq. 4.105 and Theorem 4.9 applies.

4.5. Optimal Linear Estimation of Random Vectors

In this section we return to the general random setting and treat optimal estimation of random vectors based on a finite number of observations. It is straightforward to frame the theory in terms of the orthogonality principle as applied in Section 4.2; however, we do not want to be confined to linearly independent observations. Thus, we will begin by discussing the optimal linear filter for a single random variable when the observations are not linearly independent. The filter depends on a fundamental projection theorem. This approach will carry over into estimation of random vectors and recursive linear filtering.

4.5.1. Optimal Linear Filter for Linearly Dependent Observations

To this point, whenever we have invoked inner products to express a projection into a subspace spanned by n random variables, we have assumed that the random variables are linearly independent. The projection can be accomplished via a Fourier series relative to an orthogonal set of n vectors spanning the subspace. An orthogonal spanning set can always be obtained by the Gram-Schmidt orthogonalization procedure. In the case of optimal linear filters, we have bypassed the construction of orthogonal bases but have nevertheless assumed linearly independent observations when using the orthogonality principle. In a similar vein, even though the design matrix has not been assumed to be square in the method of least squares, the columns have been assumed to be linearly independent. In many instances, the assumption of linear independence is warranted. Moreover, it does not present a mathematical problem for finite-dimensional problems because we can always extract a linearly independent subset spanning the same subspace. But it does present a practical implementation problem. If a formula is derived assuming linearly independent observations, then there must be a reduction to linear independence before it is applied. We will now change directions and drop the linear independence assumption. This will require the introduction of some inner-product-space theory.

In the previous section we employed the pseudoinverse of an $m \times n$ matrix, $m \geq n$, under the condition that the columns of the matrix are linearly independent. Should the matrix be square, then the linearly independence assumption means that the matrix is nonsingular and the pseudoinverse reduces to the inverse. The pseudoinverse is actually more general, applying to linear operators between Hilbert spaces.

So as not to overly restrict our definition, we wish to define the pseudoinverse in the context of Hilbert spaces. This requires some definitions. If φ is a linear operator between Hilbert spaces, then φ is *bounded* if there exists a constant A such that

$$\|\varphi(x)\| \leq A\|x\| \tag{4.149}$$

for all x. For finite-dimensional spaces, matrix operators are bounded. A subspace $\mathcal{S}$ of a Hilbert space is *closed* if it contains all of its limit points. This means that if $x_1, x_2, \ldots \in \mathcal{S}$ and $\|x_n - x\| \to 0$ as $n \to \infty$, then $x \in \mathcal{S}$. All finite-dimensional subspaces are closed. If $\mathcal{S}$ is any subspace, then $\mathcal{S}^\perp$ denotes the subspace of elements that are orthogonal to all elements of $\mathcal{S}$. Finally, for any linear operator φ between Hilbert spaces, its *null space*, $\mathcal{N}_\varphi$, is the subspace consisting of all vectors y such that $\varphi(y)$ is the zero vector.

We continue to be concerned with filtering a finite number of observations, which means that the optimal linear filter is given by the unique projection into the finite-dimensional subspace spanned by the observations. In the more general setting, when subspaces can be infinite dimensional, the projection theorem requires that the subspace be closed. We state the general Hilbert-space *orthogonal projection theorem* here for the purposes of discussing the pseudoinverse of an operator. We will refer to it again when we discuss optimal linear filters for an infinite number of observation random variables.

Theorem 4.10. If $\mathcal{S}$ is a closed subspace of a Hilbert space $\mathcal{H}$, then there exists a unique pair of bounded linear operators π_S and ρ_S such that π_S maps $\mathcal{H}$ onto $\mathcal{S}$, ρ_S maps $\mathcal{H}$ onto $\mathcal{S}^\perp$, and each $y \in \mathcal{H}$ possesses a unique representation

$$y = \pi_S(y) + \rho_S(y) \quad \blacksquare \tag{4.150}$$

π_S and ρ_S are called the *orthogonal projections* of $\mathcal{H}$ onto $\mathcal{S}$ and $\mathcal{S}^\perp$, respectively. If $y \in \mathcal{S}$, then $\pi_S(y) = y$ and $\rho_S(y) = 0$; if $y \in \mathcal{S}^\perp$, then $\pi_S(y) = 0$ and $\rho_S(y) = y$. Hence, $\mathcal{S}^\perp$ is the null space of π_S and $\mathcal{S}$ is the null space of ρ_S. The norm of y is decomposed by the projections according to the relation

$$\|y\|^2 = \|\pi_S(y)\|^2 + \|\rho_S(y)\|^2 \tag{4.151}$$

The minimum over all distances $\|y - x\|$, for $x \in \mathcal{S}$, is achieved by $\|y - \pi_S(y)\|$. The properties of the orthogonal projections are familiar from the theory of finite-dimensional inner product spaces. Indeed, if $\mathcal{S}$ is finite-dimensional and $\{z_1, z_2,..., z_n\}$ is an orthogonal spanning set for $\mathcal{S}$, then π_S is realized as the Fourier series relative to $\{z_1, z_2,..., z_n\}$. The theorem is more interesting when trying to project into an infinite-dimensional subspace. The theorem then requires that the subspace be closed.

We now define the pseudoinverse of an operator φ between two Hilbert spaces. Denote the range of φ by $\mathcal{R}_\varphi$. Let $\varphi^\perp$ be the mapping defining the projection onto $\mathcal{R}_\varphi^{\perp}$. Let $\varphi^{\perp\perp} = (\varphi^\perp)^\perp$. $\varphi^{\perp\perp}$ provides the projection onto $(\mathcal{R}_{\varphi^\perp})^\perp = \mathcal{R}_\varphi^{\perp\perp} = \mathcal{R}_\varphi$ (the conjugate of the range of φ for complex spaces). Let φ_0 denote the restriction of φ to vectors lying in $\mathcal{N}_\varphi^{\perp}$. If $\mathcal{R}_\varphi$ is closed (which it is in finite-dimensional spaces), it can be shown that φ_0 is a one-to-one, linear mapping onto $\mathcal{R}_\varphi$. Hence, it has an inverse φ_0^{-1} mapping $\mathcal{R}_\varphi$ onto $\mathcal{N}_\varphi^{\perp}$. The *pseudoinverse* of φ is a bounded linear operator defined by the composition

$$\varphi^+ = \varphi_0^{-1}\varphi^{\perp\perp} \tag{4.152}$$

If φ has an inverse, then $\varphi^{\perp\perp}$ is the identity, $\varphi_0 = \varphi$ (because $\mathcal{N}_\varphi = \{0\}$), and $\varphi^+ = \varphi^{-1}$. Many other properties are satisfied by the pseudoinverse. We state a few: $\varphi^{++} = \varphi$, $\varphi\varphi^+ = \varphi^{\perp\perp}$, $\varphi^+\varphi^{\perp\perp} = \varphi^+$, $\varphi\varphi^+\varphi = \varphi$, and $\varphi^+\varphi\varphi^+ = \varphi^+$.

Our concern is with the pseudoinverse of a matrix operator. In the previous section, the design matrix $\mathbf{H}$ had the pseudoinverse $\mathbf{H}^+ = (\mathbf{H}'\mathbf{H})^{-1}\mathbf{H}'$. This representation holds when $\mathbf{H}$ is $m \times n$, $m \geq n$, and the columns of $\mathbf{H}$ are linearly independent. In this section and the next, we will be interested in square matrices when the matrix is singular (otherwise the pseudoinverse equals the inverse). For computational purposes we will state a single proposition concerning evaluation of pseudoinverses for symmetric matrices (that in the complex setting applies to hermitian matrices).

If $\mathbf{B}$ is a symmetric matrix and its unitary similarity transformation is $\mathbf{U}'\mathbf{B}\mathbf{U} = \mathbf{D}$, where $\mathbf{U}$ is unitary and $\mathbf{D}$ is diagonal, then the pseudoinverse of $\mathbf{B}$ is given by

$$\mathbf{B}^+ = \mathbf{U}\mathbf{D}^+\mathbf{U}' \tag{4.153}$$

To find the pseudoinverse, first find the unitary similarity transformation and then find the pseudoinverse of **D**. If the eigenvalues, λ_1, λ_2,..., λ_n, of **B** are nonnegative, then

$$\mathbf{D} = \begin{pmatrix} \lambda_1 & 0 & \cdots & 0 \\ 0 & \lambda_2 & \cdots & 0 \\ \vdots & \vdots & \ddots & \vdots \\ 0 & 0 & \cdots & \lambda_n \end{pmatrix} \tag{4.154}$$

where $\lambda_1 \geq \lambda_2 \geq \cdots \geq \lambda_n$. If $\lambda_r > 0$ and $\lambda_{r+1} = 0$, then the pseudoinverse of **D** is given by

$$\mathbf{D}^+ = \begin{pmatrix} \lambda_1^{-1} & 0 & \cdots & 0 & 0 & \cdots & 0 \\ 0 & \lambda_2^{-1} & \cdots & 0 & 0 & \cdots & 0 \\ \vdots & \vdots & \ddots & \vdots & \vdots & \ddots & \vdots \\ 0 & 0 & \cdots & \lambda_r^{-1} & 0 & \cdots & 0 \\ 0 & 0 & \cdots & 0 & 0 & \cdots & 0 \\ \vdots & \vdots & \ddots & \vdots & \vdots & \ddots & \vdots \\ 0 & 0 & \cdots & 0 & 0 & \cdots & 0 \end{pmatrix} \tag{4.155}$$

Having provided some brief background material on pseudoinverses, we can now state the fundamental theorem concerning projections into the span of a finite set of vectors in a Hilbert space that are not necessarily linearly independent. It involves the pseudoinverse of the Grammian (defined in Eq. 4.43).

Theorem 4.11. Let $\{x_1, x_2,..., x_n\}$ be a set of vectors in a Hilbert space with span S. Then the orthogonal projection of any vector y into S is given by

$$\pi_S(y) = \sum_{k=1}^{n} \alpha(y, x_k) x_k \tag{4.156}$$

where

$$\begin{pmatrix} \alpha(y,x_1) \\ \alpha(y,x_2) \\ \vdots \\ \alpha(y,x_n) \end{pmatrix} = \begin{pmatrix} \langle x_1,x_1\rangle & \langle x_1,x_2\rangle & \cdots & \langle x_1,x_n\rangle \\ \langle x_2,x_1\rangle & \langle x_2,x_2\rangle & \cdots & \langle x_2,x_n\rangle \\ \vdots & \vdots & \ddots & \vdots \\ \langle x_n,x_1\rangle & \langle x_n,x_2\rangle & \cdots & \langle x_n,x_n\rangle \end{pmatrix}^+ \begin{pmatrix} \langle y,x_1\rangle \\ \langle y,x_2\rangle \\ \vdots \\ \langle y,x_n\rangle \end{pmatrix} \quad \blacksquare \tag{4.157}$$

Letting **G** denote the Grammian, if $x_1, x_2,..., x_n$ form an orthonormal set, then **G** = **I** and $\pi_s(y)$ reduces to the Fourier series of y in terms of $x_1, x_2,..., x_n$.

If we let

$$\hat{\mathbf{A}} = (\alpha(y,x_1),\alpha(y,x_2),\ldots,\alpha(y,x_n))' \tag{4.158}$$

$$\mathbf{C} = (\langle y,x_1\rangle,\langle y,x_2\rangle,\ldots,\langle y,x_n\rangle)' \tag{4.159}$$

then Eq. 4.157 becomes

$$\hat{\mathbf{A}} = \mathbf{G}^{+}\mathbf{C} \tag{4.160}$$

If $x_1, x_2,..., x_n$ are linearly independent, then **G** is nonsingular, $\mathbf{G}^{+} = \mathbf{G}^{-1}$, and $\hat{\mathbf{A}} = \mathbf{G}^{-1}\mathbf{C}$.

In the case of optimal linear MSE filters, **G** = **R**, the autocorrelation matrix of the observations $X_1, X_2,..., X_n$, and **C** is the cross-correlation vector for the observations and the random variable Y to be estimated. When the observations are linearly independent, $\mathbf{R}^{+} = \mathbf{R}^{-1}$ and Eq. 4.160 becomes the defining equation for the optimal linear filter (Eq. 4.42). Should the observations not necessarily be linearly independent, then Eq. 4.42 is replaced by

$$\hat{\mathbf{A}} = \mathbf{R}^{+}\mathbf{C} \tag{4.161}$$

which yields a version of Theorem 4.5 without the assumption of linear independence.

4.5.2. Optimal Estimation of Random Vectors

An estimator of the random vector $\mathbf{Y} = (Y_1, Y_2,..., Y_m)'$ based on the observed random vector $\mathbf{X} = (X_1, X_2,..., X_n)'$ is a vector-valued function

$$\hat{\mathbf{Y}} = \psi(\mathbf{X}) \tag{4.162}$$

Componentwise,

$$\begin{pmatrix} \hat{Y}_1 \\ \hat{Y}_2 \\ \vdots \\ \hat{Y}_m \end{pmatrix} = \begin{pmatrix} \psi_1(X_1,X_2,\ldots,X_n) \\ \psi_2(X_1,X_2,\ldots,X_n) \\ \vdots \\ \psi_m(X_1,X_2,\ldots,X_n) \end{pmatrix} \tag{4.163}$$

Mean-square-error optimization is framed in terms of Eq. 3.49, so that

$$MSE\langle\psi\rangle = \sum_{i=1}^{m} E[|Y_i - \psi_i(\mathbf{X})|^2] \tag{4.164}$$

The MSE is minimized by minimizing each summand. These are minimized by letting

$$\psi_i(\mathbf{X}) = E[Y_i | \mathbf{X}] \tag{4.165}$$

Thus, the optimal MSE vector estimator is a vector of conditional expectations,

$$\hat{\mathbf{Y}} = \psi(\mathbf{X}) = \begin{pmatrix} E[Y_1 | X_1, X_2, \ldots, X_n] \\ E[Y_2 | X_1, X_2, \ldots, X_n] \\ \vdots \\ E[Y_m | X_1, X_2, \ldots, X_n] \end{pmatrix} \tag{4.166}$$

When there are r vectors, $\mathbf{X}_1, \mathbf{X}_2, \ldots, \mathbf{X}_r$, from which $\mathbf{Y}$ is to be estimated, the optimal estimator is the conditional expectation

$$\hat{\mathbf{Y}} = E[\mathbf{Y} | \mathbf{X}_1, \mathbf{X}_2, \ldots, \mathbf{X}_r] \tag{4.167}$$

where the notation means that the ith component of $\hat{\mathbf{Y}}$ is the conditional expectation of Y_i given all the components of $\mathbf{X}_1, \mathbf{X}_2, \ldots, \mathbf{X}_r$.

4.5.3. Optimal Linear Filters for Random Vectors

A linear estimator of $\mathbf{Y}$ based on $\mathbf{X} = (X_1, X_2, \ldots, X_n)'$ takes the form

$$\begin{pmatrix} \hat{Y}_1 \\ \hat{Y}_2 \\ \vdots \\ \hat{Y}_m \end{pmatrix} = \begin{pmatrix} \psi_1(\mathbf{X}) \\ \psi_2(\mathbf{X}) \\ \vdots \\ \psi_m(\mathbf{X}) \end{pmatrix} = \begin{pmatrix} a_{11} & a_{12} & \cdots & a_{1n} \\ a_{21} & a_{22} & \cdots & a_{2n} \\ \vdots & \vdots & \ddots & \vdots \\ a_{m1} & a_{m2} & \cdots & a_{mn} \end{pmatrix} \begin{pmatrix} X_1 \\ X_2 \\ \vdots \\ X_n \end{pmatrix} \tag{4.168}$$

Letting $\mathbf{A}$ denote the matrix, the estimator is given by

$$\psi(\mathbf{X}) = \mathbf{AX} \tag{4.169}$$

which represents the system of equations

$$\psi_i(\mathbf{X}) = \sum_{j=1}^{n} a_{ij} X_j \tag{4.170}$$

for $i = 1, 2, \ldots, m$. The MSE is given by

$$MSE\langle\psi\rangle = \sum_{i=1}^{m} E\left[\left|Y_i - \sum_{j=1}^{n} a_{ij} X_j\right|^2\right] \tag{4.171}$$

Minimization of the MSE results from finding the optimal linear filter for each component Y_i in terms of the components of $\mathbf{X}$.

$MSE\langle\psi\rangle$ is a squared distance in the space of all random m-vectors whose components possess finite second moments. The inner product is given by

$$\langle \mathbf{U}, \mathbf{V} \rangle = E[\mathbf{U}'\mathbf{V}] = \sum_{i=1}^{m} E[U_i V_i] \tag{4.172}$$

where $\mathbf{U} = (U_1, U_2, \ldots, U_m)'$ and $\mathbf{V} = (V_1, V_2, \ldots, V_m)'$. The norm is

$$\|\mathbf{U}\| = E[\mathbf{U}'\mathbf{U}]^{1/2} = \left(\sum_{i=1}^{m} E[|U_i|^2]\right)^{1/2} \tag{4.173}$$

The operator MSE is given by

$$\begin{aligned} MSE\langle\psi\rangle &= \|\mathbf{Y} - \psi(\mathbf{X})\|^2 \\ &= E[(\mathbf{Y} - \psi(\mathbf{X}))'(\mathbf{Y} - \psi(\mathbf{X}))] \end{aligned} \tag{4.174}$$

which expands to Eq. 4.171. The optimal linear MSE vector filter minimizes this squared norm.

The orthogonality principle extends directly by applying it componentwise. It states that $\hat{\mathbf{Y}}$ is the optimal linear estimator of $\mathbf{Y}$ based on the random vector $\mathbf{X}$ if and only if

$$E[(\mathbf{Y} - \hat{\mathbf{Y}})\mathbf{U}'] = \mathbf{0} \tag{4.175}$$

for any vector $\mathbf{U}$ whose components are linear combinations of the components of $\mathbf{X}$ (lie in the subspace spanned by the components of $\mathbf{X}$). Expanded, Eq. 4.175 becomes

$$\begin{pmatrix} E[(Y_1-\hat{Y}_1)U_1] & E[(Y_1-\hat{Y}_1)U_2] & \cdots & E[(Y_1-\hat{Y}_1)U_m] \\ E[(Y_2-\hat{Y}_2)U_1] & E[(Y_2-\hat{Y}_2)U_2] & \cdots & E[(Y_2-\hat{Y}_2)U_m] \\ \vdots & \vdots & \ddots & \vdots \\ E[(Y_m-\hat{Y}_m)U_1] & E[(Y_m-\hat{Y}_m)U_2] & \cdots & E[(Y_m-\hat{Y}_m)U_m] \end{pmatrix} = \mathbf{0} \tag{4.176}$$

Every component of $\mathbf{Y} - \hat{\mathbf{Y}}$ is orthogonal to every component of $\mathbf{U}$. Expressed in the form of Theorem 4.4, the orthogonality condition states that

$$E[(\mathbf{Y}-\hat{\mathbf{Y}})(\mathbf{BX})'] = \mathbf{0} \tag{4.177}$$

for any $m \times n$ matrix $\mathbf{B}$.

Theorem 4.5 also applies if the components of $\mathbf{X}$ are linearly independent. Let

$$\mathbf{C}_k = E[Y_k\mathbf{X}] \tag{4.178}$$

be the cross-correlation vector for Y_k with the components of $\mathbf{X}$ and $\mathbf{C}$ be the matrix with columns $\mathbf{C}_1$, $\mathbf{C}_2$,..., $\mathbf{C}_m$. According to Theorem 4.5, the optimal linear estimator for Y_k is

$$\hat{Y}_k = (\mathbf{R}^{-1}\mathbf{C}_k)'\mathbf{X} \tag{4.179}$$

Using the symmetry of $\mathbf{R}$, we obtain

$$\hat{\mathbf{Y}} = \begin{pmatrix} (\mathbf{R}^{-1}\mathbf{C}_1)'\mathbf{X} \\ (\mathbf{R}^{-1}\mathbf{C}_2)'\mathbf{X} \\ \vdots \\ (\mathbf{R}^{-1}\mathbf{C}_m)'\mathbf{X} \end{pmatrix} = \begin{pmatrix} \mathbf{C}_1'\mathbf{R}^{-1}\mathbf{X} \\ \mathbf{C}_2'\mathbf{R}^{-1}\mathbf{X} \\ \vdots \\ \mathbf{C}_m'\mathbf{R}^{-1}\mathbf{X} \end{pmatrix} = \mathbf{C}'\mathbf{R}^{-1}\mathbf{X} \tag{4.180}$$

Relative to equation Eq. 4.169, the optimal linear filter is defined by the matrix

$$\hat{\mathbf{A}} = \mathbf{C}'\mathbf{R}^{-1} \tag{4.181}$$

When the observations are not linearly independent, the same reasoning applies to Eq. 4.161 and the optimal linear filter is defined by the matrix

$$\hat{\mathbf{A}} = \mathbf{C}'\mathbf{R}^{+} \tag{4.182}$$

Writing **C**' and **R** as expectations, we have the following theorem, which is often called the *Gauss-Markov theorem*.

Theorem 4.12. If $\mathbf{X} = (X_1, X_2,..., X_n)'$ and $\mathbf{Y} = (Y_1, Y_2,..., Y_m)'$ are random vectors comprised of random variables possessing finite second moments, then the optimal linear MSE estimator of **Y** based on **X** is given by

$$\hat{\mathbf{Y}} = E[\mathbf{YX}']E[\mathbf{XX}']^{+}\mathbf{X} \qquad \blacksquare \qquad (4.183)$$

In the case of Theorem 4.5, we computed the MSE in Eq. 4.44. In the vector setting, we compute the *error-covariance matrix*:

$$\begin{aligned} E[(\hat{\mathbf{Y}}-\mathbf{Y})(\hat{\mathbf{Y}}-\mathbf{Y})'] &= E[(\hat{\mathbf{Y}}-\mathbf{Y})\hat{\mathbf{Y}}'] - E[(\hat{\mathbf{Y}}-\mathbf{Y})\mathbf{Y}'] \\ &= -E[(\hat{\mathbf{Y}}-\mathbf{Y})\mathbf{Y}'] \\ &= E[\mathbf{YY}'] - E[\hat{\mathbf{Y}}\mathbf{Y}'] \\ &= E[\mathbf{YY}'] - E[\mathbf{YX}']E[\mathbf{XX}']^{+}E[\mathbf{XY}'] \qquad (4.184) \end{aligned}$$

where the second equality follows from the orthogonality principle and the last from Theorem 4.12. The diagonal of the error-covariance matrix is composed of the MSEs for the components of **Y**, so that the trace of the error-covariance matrix gives the MSE.

***Example* 4.16.** Suppose **X** and **Y** satisfy the observation model

$$\mathbf{X} = \mathbf{HY} + \mathbf{N}$$

Theorem 4.12 gives the optimal linear estimator

$$\begin{aligned} \hat{\mathbf{Y}} &= E[\mathbf{Y}(\mathbf{HY}+\mathbf{N})']E[(\mathbf{HY}+\mathbf{N})(\mathbf{HY}+\mathbf{N})']^{+}\mathbf{X} \\ &= E[\mathbf{YY}'\mathbf{H}' + \mathbf{YN}']E[\mathbf{HYY}'\mathbf{H}' + \mathbf{HYN}' + \mathbf{NY}'\mathbf{H}' + \mathbf{NN}']^{+}\mathbf{X} \\ &= (E[\mathbf{YY}']\mathbf{H}' + E[\mathbf{YN}'])(\mathbf{H}E[\mathbf{YY}']\mathbf{H}' + \mathbf{H}E[\mathbf{YN}'] + E[\mathbf{NY}']\mathbf{H}' + E[\mathbf{NN}'])^{+}\mathbf{X} \end{aligned}$$

To proceed further, we assume that **N** is uncorrelated with **Y**. This means that every component of **N** is uncorrelated with every component of **Y**, so that their cross-correlation matrix, $E[\mathbf{NY}']$, is null. The autocorrelation matrices of **Y** and **N** are $\mathbf{R_Y} = E[\mathbf{YY}']$ and $\mathbf{R_N} = E[\mathbf{NN}']$, respectively. The optimal linear estimator of **Y** is

$$\hat{\mathbf{Y}} = \mathbf{R_Y H'(H R_Y H' + R_N)^+ X}$$ ■

4.6. Recursive Linear Filters

Rather then estimate a random vector **Y** based on all observations at once, it is often beneficial to estimate **Y** from a subset of the observations and then update the estimator with new observations. This is done recursively with observation random vectors $\mathbf{X}_0$, $\mathbf{X}_1$, $\mathbf{X}_2$,.... An initial linear estimator for **Y** is based on $\mathbf{X}_0$; this initial estimator is used in conjunction with $\mathbf{X}_1$ to obtain the optimal linear estimator based on $\mathbf{X}_0$ and $\mathbf{X}_1$; and this procedure is recursively performed to obtain linear estimators for **Y** based on $\mathbf{X}_0$, $\mathbf{X}_1$,..., $\mathbf{X}_j$, for $j = 1, 2,$.... The method can be extended to recursively estimate a vector random function $\mathbf{Y}(k)$ based on the random variables comprising a vector random function $\mathbf{X}(j)$ from time 0 up to time j. $\mathbf{Y}(k)$ is estimated based on $\mathbf{X}(0)$, $\mathbf{X}(1)$,..., $\mathbf{X}(j)$. It is assumed that there is a recursive relation between $\mathbf{Y}(k-1)$ and $\mathbf{Y}(k)$. This recursive relation is used in conjunction with prior estimators to derive an optimal linear estimator for $\mathbf{Y}(k)$.

4.6.1. Recursive Generation of Direct Sums

Optimal linear filtering is based on orthogonal projections onto subspaces spanned by observation random variables; recursive optimal linear filtering involves orthogonal projections onto direct sums of subspaces generated by a sequence of random vectors. As the recursion evolves, the filter incorporates the novel information in the new observations. The random variables in an observation sequence $\mathbf{X}_0$, $\mathbf{X}_1$, $\mathbf{X}_2$,..., $\mathbf{X}_j$,... span subspaces of increasing dimensionality as j increases. If $\mathcal{X}_j$ is the span of the random variables in $\mathbf{X}_0$, $\mathbf{X}_1$,,..., $\mathbf{X}_j$, then $\mathcal{X}_0 \subset \mathcal{X}_1 \subset \cdots$. The optimal linear filter for **Y** based on $\mathbf{X}_0$, $\mathbf{X}_1$,,..., $\mathbf{X}_j$ is the orthogonal projection (of the components) of **Y** into $\mathcal{X}_j$. The increase in demensionality between $\mathcal{X}_j$ and $\mathcal{X}_{j+1}$ depends on the linear dependence relations between the random variables comprising $\mathbf{X}_{j+1}$ and the random variables in $\mathbf{X}_0$, $\mathbf{X}_1$,,..., $\mathbf{X}_j$. Recursive optimal linear filtering is achieved by decomposing the subspaces $\mathcal{X}_0$, $\mathcal{X}_1$, $\mathcal{X}_2$,... into direct sums and then using the recursive projection procedure that generates these direct sums to obtain optimal linear estimators. Recursive discrete linear filtering follows directly from recursive generation of direct sums. In this subsection we discuss direct sums; in the next two, we use this theory to obtain recursive optimal linear filters.

If $\mathcal{W}_0$, $\mathcal{W}_1$,..., $\mathcal{W}_m$ are subspaces of a vector space, then their *sum*,

$$\mathcal{W} = \mathcal{W}_0 + \mathcal{W}_1 + \cdots + \mathcal{W}_m \tag{4.185}$$

is the span of the union of $\mathcal{W}_0$, $\mathcal{W}_1$,..., $\mathcal{W}_m$. Each vector $x \in \mathcal{W}$ can be expressed as

$$x = w_0 + w_1 + \cdots + w_m \tag{4.186}$$

where $w_k \in \mathcal{W}_k$ for $k = 0, 1,..., m$. If $\mathcal{B}_0, \mathcal{B}_1,..., \mathcal{B}_m$ are bases for $\mathcal{W}_0, \mathcal{W}_1,..., \mathcal{W}_m$, respectively, then $\mathcal{W}$ is the span of

$$\mathcal{B} = \bigcup_{k=0}^{m} \mathcal{B}_k \tag{4.187}$$

$\mathcal{B}$ is not necessarily a basis for $\mathcal{W}$ because $\mathcal{B}$ may not be linearly independent.

An added assumption makes $\mathcal{B}$ a basis. $\mathcal{W}_0, \mathcal{W}_1,..., \mathcal{W}_m$ are said to be *independent* if

$$w_0 + w_1 + \cdots + w_m = 0 \tag{4.188}$$

with $w_k \in \mathcal{W}_k$ for $k = 0, 1,..., m$, implies

$$w_0 = w_1 = \cdots = w_m = 0 \tag{4.189}$$

The following conditions are equivalent: (i) $\mathcal{W}_0, \mathcal{W}_1,..., \mathcal{W}_m$ are independent; (ii) $\mathcal{B}$ is a basis for $\mathcal{W}$; and (iii) for $k = 1, 2,..., m$,

$$\mathcal{W}_k \cap (\mathcal{W}_0 + \mathcal{W}_1 + \cdots + \mathcal{W}_{k-1}) = \{0\} \tag{4.190}$$

If $\mathcal{W}_0, \mathcal{W}_1,..., \mathcal{W}_m$ are independent, then we write $\mathcal{B} = (\mathcal{B}_0, \mathcal{B}_1,..., \mathcal{B}_m)$ to indicate that the order of the basis vectors is preserved when they are unioned to form a basis for $\mathcal{W}$, which is now called a *direct sum* and written

$$\mathcal{W} = \mathcal{W}_0 \oplus \mathcal{W}_1 \oplus \cdots \oplus \mathcal{W}_m \tag{4.191}$$

For inner product spaces, subspaces $\mathcal{W}_k$ and $\mathcal{W}_j$ are said to be *orthogonal* if $\langle w_k, w_j \rangle = 0$ for any $w_k \in \mathcal{W}_k$ and $w_j \in \mathcal{W}_j$. We write $\mathcal{W}_k \perp \mathcal{W}_j$. The subspaces $\mathcal{W}_0, \mathcal{W}_1,..., \mathcal{W}_m$ are orthogonal as a collection if they are mutually orthogonal. If $\mathcal{W}_0, \mathcal{W}_1,..., \mathcal{W}_m$ are orthogonal, then they are independent. Indeed, if Eq. 4.188 holds, then taking the inner product with $w_k \in \mathcal{W}_k$ yields $\langle w_k, w_k \rangle = 0$, implying that $w_k = 0$.

We are concerned with the situation in which there are collections of vectors,

$$\begin{aligned} C_0 &= \{x_{01}, x_{02},..., x_{0,n(0)}\} \\ C_1 &= \{x_{11}, x_{12},..., x_{1,n(1)}\} \\ &\vdots \end{aligned} \tag{4.192}$$

$$C_m = \{x_{m1}, x_{m2}, \ldots, x_{m,n(m)}\}$$
$$\vdots$$

where C_m has $n(m)$ vectors. The vectors in $C_0, C_1, \ldots, C_m$ are not necessarily linearly independent. Let $\mathcal{U}_0, \mathcal{U}_1, \ldots, \mathcal{U}_m$ be the spans of $C_0, C_1, \ldots, C_m$, respectively, $\mathcal{V}_0 = \mathcal{U}_0$, and

$$\mathcal{V}_m = \mathcal{U}_0 + \mathcal{U}_1 + \cdots + \mathcal{U}_m \tag{4.193}$$

We will express $\mathcal{V}_m$ as a direct sum,

$$\mathcal{V}_m = S_0 \oplus S_1 \oplus \cdots \oplus S_m \tag{4.194}$$

This will be done recursively.

To start the recursion, let $S_0 = \mathcal{U}_0$. Next, suppose we have the direct sum

$$\mathcal{V}_{m-1} = S_0 \oplus S_1 \oplus \cdots \oplus S_{m-1} \tag{4.195}$$

for $m - 1$. To obtain the direct-sum representation for m, let π_{m-1} be the orthogonal projection onto $\mathcal{V}_{m-1}$. For $j = 1, 2, \ldots, n(m)$, let

$$z_{mj} = x_{mj} - \pi_{m-1}(x_{mj}) \tag{4.196}$$

If $x_{mj} \in \mathcal{V}_{m-1}$, then $\pi_{m-1}(x_{mj}) = x_{mj}$ and $z_{mj} = 0$; if $x_{mj} \notin \mathcal{V}_{m-1}$, then $z_{mj} \neq 0$ and $z_{mj} \perp \mathcal{V}_{m-1}$. Let S_m be the span of $z_{m1}, z_{m2}, \ldots, z_{m,n(m)}$. These differences need not be linearly independent, but each is orthogonal to $\mathcal{V}_{m-1}$. Hence, $S_m \perp \mathcal{V}_{m-1}$. We form the direct sum $\mathcal{V}_{m-1} \oplus S_m$. Since vectors in $\mathcal{V}_{m-1} \oplus S_m$ are linear combinations of vectors in $\mathcal{V}_{m-1} + \mathcal{U}_m$,

$$\mathcal{V}_{m-1} \oplus S_m \subset \mathcal{V}_{m-1} + \mathcal{U}_m \tag{4.197}$$

Solving Eq. 4.196 for x_{mj} shows that $\mathcal{U}_m \subset \mathcal{V}_{m-1} \oplus S_m$. Therefore,

$$\mathcal{V}_{m-1} + \mathcal{U}_m \subset \mathcal{V}_{m-1} + (\mathcal{V}_{m-1} \oplus S_m) = \mathcal{V}_{m-1} \oplus S_m \tag{4.198}$$

Hence,

$$\begin{aligned} \mathcal{V}_m &= \mathcal{V}_{m-1} + \mathcal{U}_m \\ &= \mathcal{V}_{m-1} \oplus S_m \\ &= S_0 \oplus S_1 \oplus \cdots \oplus S_m \end{aligned} \tag{4.199}$$

We are interested in direct sums because of their relation to orthogonal projections. If y is a vector in a Hilbert space, then the orthogonal projection of y into a direct sum in which the subspaces are mutually orthogonal is the sum of the projections into the subspaces forming the direct sum. Let π_m and σ_m be the projections onto $\mathcal{V}_m$ and $\mathcal{S}_m$, respectively. The projections of y into $\mathcal{V}_0$, $\mathcal{V}_1$,... are found recursively by

$$\pi_0(y) = \sigma_0(y)$$

$$\pi_1(y) = \pi_0(y) + \sigma_1(y)$$

$$\vdots$$

$$\pi_m(y) = \pi_{m-1}(y) + \sigma_m(y) \tag{4.200}$$

$$\vdots$$

Referring to Eq. 4.196, let $\mathbf{z}_m = (z_{m1}, z_{m2},..., z_{m,n(m)})'$ for $m = 1, 2,...$. The components of $\mathbf{z}_m$ span $\mathcal{S}_m$, and according to Theorem 4.11 (see Eq. 4.160),

$$\sigma_m(y) = (\mathbf{G}_m^{+}\mathbf{C}_m)'\mathbf{z}_m \tag{4.201}$$

where $\mathbf{G}_m$ is the Grammian of $\mathbf{z}_m$ and

$$\mathbf{C}_m = (\langle y, z_{m1}\rangle, \langle y, z_{m2}\rangle,..., \langle y, z_{m,n(m)}\rangle)' \tag{4.202}$$

The recursion of Eq. 4.200 is initialized by

$$\pi_0(y) = \sigma_0(y) = (\mathbf{G}_0^{+}\mathbf{C}_0)'\mathbf{x}_0 \tag{4.203}$$

where $\mathbf{x}_0 = (x_{01}, x_{02},..., x_{0,n(0)})'$. Equations 4.200 and 4.201 are the basic equations for discrete recursive optimal linear filters.

4.6.2. Static Recursive Optimal Linear Filtering

The recursive projection theory applies at once to linear filtering. There is a sequence of observation random vectors, $\mathbf{X}_0$, $\mathbf{X}_1$, $\mathbf{X}_2$,..., and a random vector $\mathbf{Y}$ to be estimated. One can think of the observation vectors occurring at time points 0, 1, 2,... and the estimator of $\mathbf{Y}$ being updated at each time point. The estimation is static in the sense that new observations are included to get a better estimator of a fixed random vector $\mathbf{Y}$ (whereas in the next section the random vector to be estimated will also change with time). Applying Theorem 4.12 to the recursive system of Eq. 4.200 in conjunction with Eq. 4.201 yields an optimal recursive linear filter for $\mathbf{Y}$ based on $\mathbf{X}_0$, $\mathbf{X}_1$, $\mathbf{X}_2$,....

Theorem 4.13. Let $\mathbf{X}_0$, $\mathbf{X}_1$, $\mathbf{X}_2$,... and $\mathbf{Y}$ be random vectors possessing finite second moments. For $k = 0, 1,...$, let $\hat{\mathbf{Y}}_k$ be the optimal linear estimator of $\mathbf{Y}$ based on $\mathbf{X}_0, \mathbf{X}_1,..., \mathbf{X}_k$. For $k = 1, 2,...$, let $\hat{\mathbf{X}}_k$ be the optimal linear estimator of $\mathbf{X}_k$ based on $\mathbf{X}_0, \mathbf{X}_1,..., \mathbf{X}_{k-1}$. Then, for $k = 1, 2,...$,

$$\hat{\mathbf{Y}}_k = \hat{\mathbf{Y}}_{k-1} + \mathbf{M}_k(\mathbf{X}_k - \hat{\mathbf{X}}_k) \tag{4.204}$$

where the *gain matrix* $\mathbf{M}_k$ and the initialization are given by

$$\mathbf{M}_k = E[\mathbf{Y}(\mathbf{X}_k - \hat{\mathbf{X}}_k)']E[(\mathbf{X}_k - \hat{\mathbf{X}}_k)(\mathbf{X}_k - \hat{\mathbf{X}}_k)']^+ \tag{4.205}$$

$$\hat{\mathbf{Y}}_0 = E[\mathbf{Y}\mathbf{X}_0']E[\mathbf{X}_0\mathbf{X}_0']^+\mathbf{X}_0 \quad ■ \tag{4.206}$$

The estimator based on $\mathbf{X}_0, \mathbf{X}_1,..., \mathbf{X}_k$ is equal to the estimator based on $\mathbf{X}_0$, $\mathbf{X}_1,..., \mathbf{X}_{k-1}$ plus the gain matrix times the difference between $\mathbf{X}_k$ and the estimator of $\mathbf{X}_k$ based on $\mathbf{X}_0, \mathbf{X}_1,..., \mathbf{X}_{k-1}$. The differences $\mathbf{X}_k - \hat{\mathbf{X}}_k$ are called *innovations* because $\mathbf{X}_k - \hat{\mathbf{X}}_k$ provides the new information upon which the filter is to be updated. Should the components of $\mathbf{X}_k$ lie in the span of the components of $\mathbf{X}_0, \mathbf{X}_1,..., \mathbf{X}_{k-1}$, meaning that no observation dimensionality is added by $\mathbf{X}_k$, then $\mathbf{X}_k = \hat{\mathbf{X}}_k$ and $\hat{\mathbf{Y}}_k = \hat{\mathbf{Y}}_{k-1}$. At the other extreme, if all components of $\mathbf{X}_k$ are orthogonal to the span of the components of $\mathbf{X}_0, \mathbf{X}_1,...$, $\mathbf{X}_{k-1}$, then $\hat{\mathbf{X}}_k = \mathbf{0}$ and

$$\hat{\mathbf{Y}}_k = \hat{\mathbf{Y}}_{k-1} + E[\mathbf{Y}\mathbf{X}_k']E[\mathbf{X}_k\mathbf{X}_k']^+\mathbf{X}_k \tag{4.207}$$

The update is simply the optimal linear estimator for $\mathbf{Y}$ based on just $\mathbf{X}_k$.

The preceding theorem is quite general. Let us now specialize it to the situation in which the new observations are linear combinations of the random variables composing $\mathbf{Y}$ plus additive noise. The intent here is to model a linear measurement model $\mathbf{H}_k\mathbf{Y}$, defined at time k by the matrix $\mathbf{H}_k$. We assume that the measurements are degraded by "additive noise" $\mathbf{N}_k$, which is assumed to be uncorrelated with $\mathbf{X}_0, \mathbf{X}_1,..., \mathbf{X}_{k-1}$, and $\mathbf{Y}$. $\mathbf{N}_k$ is not presupposed to be white. These assumptions provide the observation model

$$\mathbf{X}_k = \mathbf{H}_k\mathbf{Y} + \mathbf{N}_k \tag{4.208}$$

$$E[\mathbf{Y}\mathbf{N}_k'] = E[\mathbf{X}_j\mathbf{N}_k'] = \mathbf{0} \quad (j < k) \tag{4.209}$$

Let $\mathbf{V}_k = (\mathbf{X}_0', \mathbf{X}_1',..., \mathbf{X}_k')'$ be the vector composed of the components of $\mathbf{X}_0, \mathbf{X}_1,..., \mathbf{X}_k$, taken in order. The preceding orthogonality conditions imply that $\mathbf{N}_k$ is uncorrelated with $\mathbf{V}_{k-1}$. $\hat{\mathbf{X}}_k$ is the optimal linear estimator of $\mathbf{X}_k$ based on $\mathbf{V}_{k-1}$. We let $\mathbf{E}_k$ denote the error-covariance matrix for estimation of $\mathbf{Y}$ by $\hat{\mathbf{Y}}_k$, namely,

$$\mathbf{E}_k = E[(\mathbf{Y}-\hat{\mathbf{Y}}_k)(\mathbf{Y}-\hat{\mathbf{Y}}_k)'] \tag{4.210}$$

and we denote the autocorrelation matrix for the noise by

$$\mathbf{R}_k = E[\mathbf{N}_k\mathbf{N}_k'] \tag{4.211}$$

By Theorem 4.12,

$$\begin{aligned}\hat{\mathbf{X}}_k &= E[\mathbf{X}_k\mathbf{V}_{k-1}']E[\mathbf{V}_{k-1}\mathbf{V}_{k-1}']^+\mathbf{V}_{k-1} \\ &= E[(\mathbf{H}_k\mathbf{Y}+\mathbf{N}_k)\mathbf{V}_{k-1}']E[\mathbf{V}_{k-1}\mathbf{V}_{k-1}']^+\mathbf{V}_{k-1} \\ &= (\mathbf{H}_k E[\mathbf{Y}\mathbf{V}_{k-1}']+E[\mathbf{N}_k\mathbf{V}_{k-1}'])E[\mathbf{V}_{k-1}\mathbf{V}_{k-1}']^+\mathbf{V}_{k-1} \\ &= \mathbf{H}_k E[\mathbf{Y}\mathbf{V}_{k-1}']E[\mathbf{V}_{k-1}\mathbf{V}_{k-1}']^+\mathbf{V}_{k-1} \\ &= \mathbf{H}_k\hat{\mathbf{Y}}_{k-1}\end{aligned} \tag{4.212}$$

Hence,

$$\begin{aligned}\mathbf{X}_k - \hat{\mathbf{X}}_k &= \mathbf{X}_k - \mathbf{H}_k\hat{\mathbf{Y}}_{k-1} \\ &= \mathbf{H}_k\mathbf{Y} + \mathbf{N}_k - \mathbf{H}_k\hat{\mathbf{Y}}_{k-1} \\ &= \mathbf{H}_k(\mathbf{Y}-\hat{\mathbf{Y}}_{k-1}) + \mathbf{N}_k\end{aligned} \tag{4.213}$$

We need to compute the gain matrix from Theorem 4.13. First,

$$\begin{aligned}E[\mathbf{Y}(\mathbf{X}_k - \hat{\mathbf{X}}_k)'] &= E[\mathbf{Y}(\mathbf{H}_k(\mathbf{Y}-\hat{\mathbf{Y}}_{k-1})+\mathbf{N}_k)'] \\ &= E[\mathbf{Y}(\mathbf{Y}-\hat{\mathbf{Y}}_{k-1})'\mathbf{H}_k']+E[\mathbf{Y}\mathbf{N}_k'] \\ &= E[\mathbf{Y}(\mathbf{Y}-\hat{\mathbf{Y}}_{k-1})']\mathbf{H}_k' \\ &= E[\mathbf{Y}(\mathbf{Y}-\hat{\mathbf{Y}}_{k-1})']\mathbf{H}_k' - E[\hat{\mathbf{Y}}_{k-1}(\mathbf{Y}-\hat{\mathbf{Y}}_{k-1})']\mathbf{H}_k'\end{aligned} \tag{4.214}$$

$$= E[(\mathbf{Y} - \hat{\mathbf{Y}}_{k-1})(\mathbf{Y} - \hat{\mathbf{Y}}_{k-1})']\mathbf{H}_k'$$

$$= \mathbf{E}_{k-1}\mathbf{H}_k'$$

where the fourth inequality follows from the orthogonality principle (the second summand being null). Next,

$$E[(\mathbf{X}_k - \hat{\mathbf{X}}_k)(\mathbf{X}_k - \hat{\mathbf{X}}_k)'] = E[(\mathbf{H}_k(\mathbf{Y} - \hat{\mathbf{Y}}_{k-1}) + \mathbf{N}_k)(\mathbf{H}_k(\mathbf{Y} - \hat{\mathbf{Y}}_{k-1}) + \mathbf{N}_k)']$$

$$= \mathbf{H}_k E[(\mathbf{Y} - \hat{\mathbf{Y}}_{k-1})(\mathbf{Y} - \hat{\mathbf{Y}}_{k-1})']\mathbf{H}_k' + E[\mathbf{N}_k\mathbf{N}_k']$$

$$+ \mathbf{H}_k E[(\mathbf{Y} - \hat{\mathbf{Y}}_{k-1})\mathbf{N}_k'] + E[\mathbf{N}_k(\mathbf{Y} - \hat{\mathbf{Y}}_{k-1})']\mathbf{H}_k' \tag{4.215}$$

Since the components of $\hat{\mathbf{Y}}_{k-1}$ lie in the span of the components of $\mathbf{X}_0, \mathbf{X}_1, \ldots, \mathbf{X}_{k-1}$, Eq. 4.209 implies that the last two summands in the preceding equation are null. Hence,

$$E[(\mathbf{X}_k - \hat{\mathbf{X}}_k)(\mathbf{X}_k - \hat{\mathbf{X}}_k)'] = \mathbf{H}_k\mathbf{E}_{k-1}\mathbf{H}_k' + \mathbf{R}_k \tag{4.216}$$

From Theorem 4.13, we deduce the following recursive updating proposition, in which the recursive expression of the error-covariance matrix is obtained by a direct (but tedious) matrix calculation.

Theorem 4.14. For the linear observation model $\mathbf{X}_k = \mathbf{H}_k\mathbf{Y} + \mathbf{N}_k$, subject to the orthogonality conditions $E[\mathbf{Y}\mathbf{N}_k'] = 0$ and $E[\mathbf{X}_j\mathbf{N}_k'] = 0$ for $j < k$, the optimal linear estimator for $\mathbf{Y}$ based on $\mathbf{X}_0, \mathbf{X}_1, \ldots, \mathbf{X}_k$ has the recursive formulation

$$\hat{\mathbf{Y}}_k = \hat{\mathbf{Y}}_{k-1} + \mathbf{M}_k(\mathbf{X}_k - \mathbf{H}_k\hat{\mathbf{Y}}_{k-1}) \tag{4.217}$$

where the gain matrix and error-covariance matrix are recursively given by

$$\mathbf{M}_k = \mathbf{E}_{k-1}\mathbf{H}_k'(\mathbf{H}_k\mathbf{E}_{k-1}\mathbf{H}_k' + \mathbf{R}_k)^+ \tag{4.218}$$

$$\mathbf{E}_k = \mathbf{E}_{k-1} - \mathbf{E}_{k-1}\mathbf{H}_k'(\mathbf{H}_k\mathbf{E}_{k-1}\mathbf{H}_k' + \mathbf{R}_k)^+\mathbf{H}_k\mathbf{E}_{k-1} \quad \blacksquare \tag{4.219}$$

4.6.3. Dynamic Recursive Optimal Linear Filtering

To this point we have considered recursive linear filtering for the situation in which new observations are used to estimate a fixed random vector. We now treat the dynamic case in which a vector random function over discrete time is estimated from an observed vector random function over discrete time. There are two vector random functions, $\mathbf{X}(j)$ and $\mathbf{Y}(k)$, and a linear filter is desired to estimate $\mathbf{Y}(k)$ from $\mathbf{X}(j)$. $\mathbf{Y}(k)$ satisfies a *linear dynamic model* composed of a state equation and a measurement equation.

The *state equation* has the form

$$\mathbf{Y}(k+1) = \mathbf{T}_k\mathbf{Y}(k) + \mathbf{U}(k) \tag{4.220}$$

for $k = 0, 1, \ldots$ $\mathbf{Y}(k)$ is called the *state vector*, $\mathbf{T}_k$ is an $n \times n$ matrix, $\mathbf{U}(k)$ is a discrete white-noise vector random function, called the *process noise*, and the cross-correlation between the state vector and the process noise is null for $j \leq k$, meaning that

$$\mathbf{R}_{\mathbf{UY}}(k, j) = E[\mathbf{U}(k)\mathbf{Y}(j)'] = \mathbf{0} \tag{4.221}$$

(see Eq. 3.237). In the vector setting, $\mathbf{U}(k)$ being white noise means that its autocorrelation function is given by

$$\mathbf{R}_{\mathbf{U}}(k, j) = E[\mathbf{U}(k)\mathbf{U}(j)'] = \mathbf{Q}_k\delta_{kj} \tag{4.222}$$

where

$$\mathbf{Q}_k = E[\mathbf{U}(k)\mathbf{U}(k)'] \tag{4.223}$$

The state equation describes the evolution of the random function $\mathbf{Y}(k)$ over time.

The *measurement equation* describes the observation of the state vector via a linear transformation together with additive measurement noise. It takes the form

$$\mathbf{X}(k) = \mathbf{H}_k\mathbf{Y}(k) + \mathbf{N}(k) \tag{4.224}$$

where $\mathbf{H}_k$ is an $m \times n$ *measurement matrix* and $\mathbf{N}(k)$ is discrete white noise possessing autocorrelation function

$$\mathbf{R}_{\mathbf{N}}(k, j) = E[\mathbf{N}(k)\mathbf{N}(j)'] = \mathbf{R}_k\delta_{kj} \tag{4.225}$$

where

$$\mathbf{R}_k = E[\mathbf{N}(k)\mathbf{N}(k)'] \tag{4.226}$$

In addition, for all k and j,

$$\mathbf{R}_{NU}(k, j) = E[\mathbf{N}(k)\mathbf{U}(j)'] = \mathbf{0} \tag{4.227}$$

$$\mathbf{R}_{NX}(k, j) = E[\mathbf{N}(k)\mathbf{X}(j)'] = \mathbf{0} \tag{4.228}$$

We desire the optimal linear estimator $\hat{\mathbf{Y}}(k|j)$ of $\mathbf{Y}(k)$, based on

$$\mathbf{V}(j) = (\mathbf{X}(0)', \mathbf{X}(1)',..., \mathbf{X}(j)')' \tag{4.229}$$

The components of $\mathbf{V}(j)$ consist of all observation random variables up to and including those at time j. The cross-correlation function of $\mathbf{U}$ and $\mathbf{V}$ is given by the blocked matrix

$$\begin{aligned}\mathbf{R}_{UV}(k, j) &= E[\mathbf{U}(k)\mathbf{V}(j)'] \\ &= \begin{pmatrix} E[\mathbf{U}(k)\mathbf{X}(0)'] & E[\mathbf{U}(k)\mathbf{X}(1)'] & \cdots & E[\mathbf{U}(k)\mathbf{X}(j)'] \end{pmatrix}\end{aligned} \tag{4.230}$$

Suppose $j \leq k$. For $i \leq j$, the ith block of the preceding matrix is given by

$$\begin{aligned}E[\mathbf{U}(k)\mathbf{X}(i)'] &= E[\mathbf{U}(k)(\mathbf{H}_i\mathbf{Y}(i) + \mathbf{N}(i))'] \\ &= E[\mathbf{U}(k)\mathbf{Y}(i)']\mathbf{H}_i' + E[\mathbf{U}(k)\mathbf{N}(i)']\end{aligned} \tag{4.231}$$

which, owing to the uncorrelatedness assumptions of the model, is null. Hence,

$$\mathbf{R}_{UV}(k, j) = \mathbf{0} \tag{4.232}$$

In the present context, $\hat{\mathbf{Y}}(k|k)$, $\hat{\mathbf{Y}}(k+1|k)$, and $\hat{\mathbf{Y}}(k|j)$, for $j > k$, are called the *filter*, *predictor*, and *smoothing filter* for $\mathbf{Y}(k)$, respectively. In recursive form and in the framework of a dynamical system, they are called *discrete Kalman filters*. We focus on the filter $\hat{\mathbf{Y}}(k+1|k+1)$.

From Theorem 4.12,

$$\hat{\mathbf{Y}}(k|k) = E[\mathbf{Y}(k)\mathbf{V}(k)']E[\mathbf{V}(k)\mathbf{V}(k)']^{+}\mathbf{V}(k) \tag{4.233}$$

From Theorem 4.12 and the state equation,

$$\begin{aligned}\hat{\mathbf{Y}}(k+1|k) &= E[(\mathbf{T}_k\mathbf{Y}(k) + \mathbf{U}(k))\mathbf{V}(k)']E[\mathbf{V}(k)\mathbf{V}(k)']^{+}\mathbf{V}(k) \\ &= (\mathbf{T}_k E[\mathbf{Y}(k)\mathbf{V}(k)'] + E[\mathbf{U}(k)\mathbf{V}(k)'])E[\mathbf{V}(k)\mathbf{V}(k)']^{+}\mathbf{V}(k) \\ &= \mathbf{T}_k E[\mathbf{Y}(k)\mathbf{V}(k)']E[\mathbf{V}(k)\mathbf{V}(k)']^{+}\mathbf{V}(k)\end{aligned} \tag{4.234}$$

$$= \mathbf{T}_k \hat{\mathbf{Y}}(k|k)$$

The recursive representation for the linear observation model of Theorem 4.14 applies one step at a time. Since the measurement model is the same for the linear dynamical system, the recursion and the computation of Eq. 4.212 apply in the present setting. Hence, from Theorem 4.14 and the preceding equation we deduce the Kalman filter recursive expression

$$\begin{aligned}\hat{\mathbf{Y}}(k+1|k+1) &= \hat{\mathbf{Y}}(k+1|k) + \mathbf{M}_{k+1}(\mathbf{X}(k+1) - \mathbf{H}_{k+1}\hat{\mathbf{Y}}(k+1|k)) \\ &= \mathbf{T}_k \hat{\mathbf{Y}}(k|k) + \mathbf{M}_{k+1}(\mathbf{X}(k+1) - \mathbf{H}_{k+1}\mathbf{T}_k \hat{\mathbf{Y}}(k|k))\end{aligned} \tag{4.235}$$

where the gain matrix is given by Eq. 4.218. It remains to find a recursive expression for the gain matrix.

In the present setting, the error-covariance matrix for the filter takes the form

$$\mathbf{E}_{k|j} = E[(\hat{\mathbf{Y}}(k|j) - \mathbf{Y}(k))(\hat{\mathbf{Y}}(k|j) - \mathbf{Y}(k))'] \tag{4.236}$$

Applying the state equation in conjunction with Eq. 4.234 yields the *covariance extrapolation*

$$\begin{aligned}\mathbf{E}_{k+1|k} &= E[(\hat{\mathbf{Y}}(k+1|k) - \mathbf{Y}(k+1))(\hat{\mathbf{Y}}(k+1|k) - \mathbf{Y}(k+1))'] \\ &= E[(\mathbf{T}_k \hat{\mathbf{Y}}(k|k) - (\mathbf{T}_k \mathbf{Y}(k) + \mathbf{U}(k)))(\mathbf{T}_k \hat{\mathbf{Y}}(k|k) - (\mathbf{T}_k \mathbf{Y}(k) + \mathbf{U}(k)))'] \\ &= E[(\mathbf{T}_k(\hat{\mathbf{Y}}(k|k) - \mathbf{Y}(k)) - \mathbf{U}(k))(\mathbf{T}_k(\hat{\mathbf{Y}}(k|k) - \mathbf{Y}(k)) - \mathbf{U}(k))'] \\ &= \mathbf{T}_k E[(\hat{\mathbf{Y}}(k|k) - \mathbf{Y}(k))(\hat{\mathbf{Y}}(k|k) - \mathbf{Y}(k))']\mathbf{T}_k' + E[\mathbf{U}(k)\mathbf{U}(k)'] \\ &\quad - \mathbf{T}_k E[(\hat{\mathbf{Y}}(k|k) - \mathbf{Y}(k))\mathbf{U}(k)'] - E[\mathbf{U}(k)(\hat{\mathbf{Y}}(k|k) - \mathbf{Y}(k))']\mathbf{T}_k' \\ &= \mathbf{T}_k \mathbf{E}_{k|k} \mathbf{T}_k' + \mathbf{Q}(k)\end{aligned} \tag{4.237}$$

where the last equality follows from the fact that the second two summands in the next to last expression are zero by Eqs. 4.221 and 4.232 [recognizing that $\hat{\mathbf{Y}}(k|k)$ lies in the span of the random variables constituting $\mathbf{V}(k)$]. The *Kalman gain matrix*,

$$\mathbf{M}_k = \mathbf{E}_{k|k-1}\mathbf{H}_k'(\mathbf{H}_k \mathbf{E}_{k|k-1}\mathbf{H}_k' + \mathbf{R}_k)^+ \tag{4.238}$$

follows from Eq. 4.218, and the *covariance update*,

$$\mathbf{E}_{k|k} = \mathbf{E}_{k|k-1} - \mathbf{M}_k\mathbf{H}_k\mathbf{E}_{k|k-1} \tag{4.239}$$

follows from Eq. 4.219.

Theorem 4.15. For the discrete linear dynamical system of Eqs. 4.220 and 4.224, the Kalman recursive filter is given by

$$\hat{\mathbf{Y}}(k+1|k+1) = \mathbf{T}_k\hat{\mathbf{Y}}(k|k) + \mathbf{M}_{k+1}(\mathbf{X}(k+1) - \mathbf{H}_{k+1}\mathbf{T}_k\hat{\mathbf{Y}}(k|k)) \tag{4.240}$$

where the covariance extrapolation, Kalman gain matrix, and covariance update are given by Eqs. 4.237, 4.238, and 4.239, respectively. ■

Theorem 4.15 addresses the Kalman filter. The Kalman predictor is obtained via Eq. 4.234 as

$$\begin{aligned}\hat{\mathbf{Y}}(k+1|k) &= \mathbf{T}_k(\mathbf{T}_{k-1}\hat{\mathbf{Y}}(k-1|k-1) + \mathbf{M}_k(\mathbf{X}(k) - \mathbf{H}_k\mathbf{T}_{k-1}\hat{\mathbf{Y}}(k-1|k-1)))\\ &= \mathbf{T}_k(\hat{\mathbf{Y}}(k|k-1) + \mathbf{M}_k(\mathbf{X}(k) - \mathbf{H}_k\hat{\mathbf{Y}}(k|k-1)))\end{aligned} \tag{4.241}$$

In theory, initialization of the Kalman filter involves application of Theorem 4.12 to obtain the estimator $\hat{\mathbf{Y}}(0|0)$ of $\mathbf{Y}(0)$, and then computation of $\mathbf{E}_{0|0}$ according to Eq. 4.236. Once $\hat{\mathbf{Y}}(0|0)$ and $\mathbf{E}_{0|0}$ are in hand, recursion can proceed by obtaining $\mathbf{E}_{1|0}$ via Eq. 4.237. The practical problem is that we likely lack sufficient distributional knowledge regarding $\mathbf{Y}(0)$ to obtain $\hat{\mathbf{Y}}(0|0)$. Equations 4.183 and 4.184 show that we would need the cross-correlation matrix of $\mathbf{Y}(0)$ with $\mathbf{X}(0)$ and the autocorrelation matrix of $\mathbf{Y}(0)$. There are different statistical ways to obtain initialization estimates without these, but we will not pursue them here. There is one easy situation: if initialization of the system is deterministic, then $\mathbf{Y}(0)$ is a known constant vector, $\hat{\mathbf{Y}}(0|0) = \mathbf{Y}(0)$, and $\mathbf{E}_{0|0} = \mathbf{0}$.

If $\mathbf{R}_k$, $\mathbf{E}_{k|k}$, $\mathbf{E}_{k|k-1}$, and $\mathbf{H}_k\mathbf{E}_{k|k-1}\mathbf{H}_k' + \mathbf{R}_k$ are nonsingular, then it can be shown via matrix algebra that

$$\mathbf{E}^{-1}_{k|k} = \mathbf{E}^{-1}_{k|k-1} + \mathbf{H}_k'\mathbf{R}_k^{-1}\mathbf{H}_k \tag{4.242}$$

$$\mathbf{M}_k = \mathbf{E}_{k|k}\mathbf{H}_k'\mathbf{R}_k^{-1} \tag{4.243}$$

If the measurement noise is large, then the gain is small. This agrees with our intuition that we should not significantly update an estimator based on very noisy observations.

The convergence of recursive estimators is a crucial issue. From the basic linear recursive system of Eq. 4.200, we see that the change in the estimator from step $k - 1$ to step k is given by $\sigma_k(y)$, the projection of y into S_k. There are two potential problems. First, the new vectors may produce little increase in dimension. The most degenerate dimensionality case occurs when all new vectors lie in the span of the previously incorporated vectors. Then, as noted following Theorem 4.13, the update is null.

A second issue concerns the correlation between y and the new observations. If $y \perp S_k$, then $\sigma_k(y) = 0$ and there is no update. Even if y is not orthogonal to S_k, its projection into S_k can be arbitrarily small. This potentiality is evident in Theorem 4.13, where the first factor of the gain matrix is the cross-correlation matrix of $\mathbf{Y}$ and $\mathbf{X}_k - \hat{\mathbf{X}}_k$. This problem also surfaces in the observation model of Theorem 4.14. The cross-correlation matrix for this model is computed in Eq. 4.214. Because $\mathbf{Y}$ and $\mathbf{N}_k$ are assumed to be uncorrelated, the noise plays no role. The first factor of the gain matrix in Eq. 4.218 is the cross-correlation matrix, expressed as $\mathbf{E}_{k-1}\mathbf{H}_k'$. In fact, Eq. 4.414 shows directly the dependence of convergence on the measurement matrix. Since the gain matrix for the Kalman filter is simply a restatement of the gain matrix for the model of Theorem 4.14, these comments apply directly to it. The difference with the Kalman filter is that the error-covariance matrix $\mathbf{E}_{k|k-1}$ is related to $\mathbf{E}_{k|k}$ by way of $\mathbf{T}_k$ (Eq. 4.237), but the import of this relationship is that it provides an error-covariance update. What is clear is that everything goes back to the basic linear recursive system of Eq. 4.200. The fact that in the dynamic linear system the random vector to be estimated is changing is of little consequence because of the uncorrelatedness assumptions of the model. This is evident in Eq. 4.234.

4.7. Optimal Infinite-Observation Linear Filters

The optimal MSE linear filter for a finite number of observations is found by projecting the random variable to be estimated into the subspace generated by the observations. As thus far developed, the solution does not apply to estimation based on an infinite number of random variables. For certain noise models, we wish to filter a random digital image (signal) based on observations over the entire grid (all discrete time) or a continuous random image (signal) based on observations over some domain (interval). In either case, there are infinitely many observations: in the first, one for each pixel in the grid; in the second, one for each point in the observation domain.

4.7.1. Wiener-Hopf Equation

For finite-observation linear filters, the orthogonality principle provides a necessary and sufficient condition for optimality. According to the projection theorem (Theorem 4.10), if a subspace is topologically closed, then the distance from a vector to the subspace is minimized by the orthogonal projection of the vector into the subspace. All finite-dimensional subspaces are closed, so therefore there always exists a minimizing vector in the subspace. This vector can be found by the orthogonality principle. The orthogonality principle also applies for infinite-dimensional subspaces; however, the existence of a minimizing vector is only assured if the subspace is closed. We state the orthogonality principle as it applies to random variables in this more general framework.

Theorem 4.16. If $\mathcal{S}$ is a subspace of finite-second-moment random variables and Y has a finite second moment, then $\hat{Y}$ is the optimal MSE estimator of Y lying in $\mathcal{S}$ if and only if $E[(Y - \hat{Y})U] = 0$ for any $U \in \mathcal{S}$. If it exists, then the best MSE estimator is unique. ■

The theorem states that $Y - \hat{Y}$ must be orthogonal to every element of $\mathcal{S}$, just as did Theorem 4.4; however, in the present context, where $\mathcal{S}$ is not spanned by a finite number of random variables, there is no uniform finite representation of the elements of $\mathcal{S}$ and therefore the methodology that resulted in the finite-observation solution cannot be directly applied. Moreover, the theorem does not assert the existence of an optimal MSE estimator in $\mathcal{S}$ (unless $\mathcal{S}$ is closed). Practically, however, if an estimator is found that satisfies the orthogonality principle, then existence is assured.

As conceived by Kolmogorov, the general theory of optimal linear filtering takes place in the framework of operators on Hilbert spaces, where subspaces are generated by operator classes. Here, we will specialize the theory to linear integral operators on random functions. Whereas the finite procedure is easily justifiable owing to finite representation of the subspace random variables, the application to infinite sums, or to integrals (which may be viewed as generalized sums), requires that certain stochastic integrals exist and that the random functions resulting from stochastic integrals possess finite second moments. When necessary we will appeal to Theorem 2.2.

The program is to estimate, for fixed s, the value $Y(s)$ of a random function based on observation of a random function $X(t)$ over some portion T of its domain. In the finite-observation case, we looked for an estimator represented as a linear combination of the observation random variables; now we look for an estimator of the form

$$W(s) = \int_T g(s,t)X(t)\,dt \tag{4.244}$$

Optimization is achieved by minimizing the MSE

$$MSE\langle W(s)\rangle = E\left[\left|Y(s) - \int_T g(s,t)X(t)\,dt\right|^2\right] \tag{4.245}$$

over all weighting functions $g(s, t)$. By the orthogonality principle, the optimal estimator

$$\hat{Y}(s) = \int_T \hat{g}(s,t)X(t)\,dt \qquad (4.246)$$

satisfies the relation

$$E[(Y(s) - \hat{Y}(s))W(s)] = 0 \qquad (4.247)$$

for any $W(s)$ of the form given in Eq. 4.244.

We need to be a bit careful here. In the finite-dimensional setting, the set of all linear combinations of a finite set of random variables generates a subspace, which is the span of the set. Here, we must be assured that the set of random variables $W(s)$ generated according to Eq. 4.244 by kernels $g(s, t)$ in some class $\mathcal{G}$ of kernels forms a subspace of the space of finite-second-moment random variables. First, a subspace must be linearly closed, meaning that if $W_1(s)$ and $W_2(s)$ are of the form given in Eq. 4.244, then, for arbitrary scalars c_1 and c_2,

$$W(s) = c_1W_1(s) + c_2W_2(s) \qquad (4.248)$$

has a similar representation. This is true so long as the kernel class $\mathcal{G}$ is linearly closed. If $\mathcal{G}$ is linearly closed and $W_1(s)$ and $W_2(s)$ result from Eq. 4.244 via the kernels $g_1(s, t)$ and $g_2(s, t)$, then $W(s)$ results from the kernel

$$g(s, t) = c_1g_1(s, t) + c_2g_2(s, t) \qquad (4.249)$$

If $\mathcal{G}$ is linearly closed, then it is a linear space under the usual addition and scalar operations on functions.

We must also be sure that every $W(s)$ generated according to Eq. 4.244 has a finite second moment. According to Theorem 2.2, a stochastic integral exists if the deterministic integral of Eq. 2.48 exists. The problem is similar to that encountered in Section 3.7.1, where we wished to construct coefficients for canonical expansions by using integral functions. If the covariance function of X is square-integrable over $T \times T$ and $g(s, t) \in L^2(T)$ in the variable t, then we can use the Cauchy-Schwarz inequality in the same way as in Eq. 3.145. Assuming that $\mu_X(t) = 0$ [else we could center $X(t)$] and applying Eq. 2.48, we obtain

$$E[|W(s)|^2] = K_W(s, s)$$

$$= \int_T \int_T g(s,t) g(s,t') K_X(t,t')\, dt dt'$$

$$\leq \int_T \int_T | g(s,t) g(s,t') K_X(t,t') | \, dt dt'$$

$$\leq \left(\int_T \int_T | g(s,t) g(s,t') |^2 \, dt dt' \right)^{1/2} \left(\int_T \int_T | K_X(t,t') |^2 \, dt dt' \right)^{1/2} \tag{4.250}$$

$$= \int_T | g(s,t) |^2 \, dt \left(\int_T \int_T | K_X(t,t') |^2 \, dt dt' \right)^{1/2}$$

Hence, if we assume that $K_X(t, t')$ is square-integrable, then Eq. 4.244 generates a subspace if weight functions are constrained to be square-integrable in t. Whether we make this constraint, or some other, we proceed under the assumption that the kernel class $\mathcal{G}$ is a linear space and that, for a given random function $X(t)$, integral operators having kernels in $\mathcal{G}$ yield random functions possessing finite second moments.

Expanding the orthogonality relation of Eq. 4.247 yields

$$E\left[\left(Y(s) - \int_T \hat{g}(s,u) X(u)\, du \right) \int_T g(s,t) X(t)\, dt \right] = 0 \tag{4.251}$$

Interchanging expectation and integration gives

$$\int_T g(s,t) \left(E[Y(s)X(t)] - \int_T \hat{g}(s,u) E[X(u)X(t)]\, du \right) dt = 0 \tag{4.252}$$

This relation is satisfied if and only if

$$E[Y(s)X(t)] = \int_T \hat{g}(s,u) E[X(u)X(t)]\, du \tag{4.253}$$

for (almost) all $t \in T$. Writing this equation in terms of moments yields the next theorem.

Theorem 4.17. Suppose $\mathcal{G}$ is a linear space of kernels $g(s, t)$ defined over the interval T and, for any $g(s, t) \in \mathcal{G}$, the stochastic integral of Eq. 4.244 gives a random variable possessing a finite second moment. Then $\hat{g}(s,t)$ yields the optimal MSE linear estimator of $Y(s)$ based on $X(t)$ according to Eq. 4.246 if and only if

$$R_{YX}(s,t) = \int_T \hat{g}(s,u) R_X(u,t)\, du \tag{4.254}$$

for (almost) all $t \in T$. ■

The preceding integral equation is called the *Wiener-Hopf equation* and is the integral form of the finite-dimensional system $\mathbf{R}\hat{\mathbf{A}} = \mathbf{C}$. The theorem does not assert the existence of a solution to the Wiener-Hopf equation; however, if a solution $\hat{g}(s,t)$ can be found, then there exists an optimal MSE linear filter and it is defined by Eq. 4.246.

For discrete signals, the Wiener-Hopf equation takes the summation form

$$R_{YX}(m,n) = \sum_{k=k_1}^{k_2} \hat{g}(m,k) R_X(k,n) \tag{4.255}$$

for all n between k_1 and k_2, where $-\infty \le k_1 \le k \le k_2 \le \infty$. For k_1 and k_2 finite, this reduces to the original finite system.

The system law of the optimal linear MSE filter is defined by the solution of the Wiener-Hopf equation. Optimization has been thrown into the domain of integral equations. Assuming that the random functions are sufficiently regular to justify interchanges of operations, no assumptions are made on the forms of $X(t)$ and $Y(s)$. Practically, however, the integral equation must often be solved by numerical methods. Although we have written single integrals, there is no restriction on the dimensions of the variables. Thus, the Wiener-Hopf equation applies to both signals and images.

Regarding the MSE for the optimal linear filter, if $Y(s)$ has zero mean, then

$$\begin{aligned} E[(Y(s) - \hat{Y}(s))^2] &= E[(Y(s) - \hat{Y}(s))Y(s)] \\ &= E[Y(s)^2] - \int_T \hat{g}(s,u) E[Y(s)X(u)]\, du \\ &= \mathrm{Var}[Y(s)] - \int_T \hat{g}(s,u) R_{YX}(s,u)\, du \end{aligned} \tag{4.256}$$

4.7.2. Wiener Filter

In general, the Wiener-Hopf integral equation determines a spatially variant filter; however, if $X(t)$ and $Y(s)$ are jointly WS stationary, then the optimal linear filter is spatially invariant with $\hat{g}(s,u) = \hat{g}(s-u)$. Moreover, if $X(t)$ is observed over all time, then the Wiener-Hopf equation becomes

$$r_{YX}(s-t) = \int_{-\infty}^{\infty} \hat{g}(s-u) r_X(u-t)\, du \tag{4.257}$$

Letting $\tau = u - t$ and $\xi = s - t$ yields the convolution integral

$$r_{YX}(\xi) = \int_{-\infty}^{\infty} \hat{g}(\xi-\tau) r_X(\tau)\, d\tau \tag{4.258}$$

Since the Fourier transform of the autocorrelation function is the power spectral density, applying the Fourier transform and dividing by $S_X(\omega)$ yields

$$\hat{G}(\omega) = \frac{S_{YX}(\omega)}{S_X(\omega)} \tag{4.259}$$

where $\hat{G}(\omega)$ is the Fourier transform of $\hat{g}(\tau)$. The filter is called the *Wiener filter*. For images, the Wiener filter involves the two-dimensional Fourier transform.

Assuming WS stationarity, the digital-image form of the Wiener-Hopf equation is

$$r_{YX}(m, n) = \sum_{k=-\infty}^{\infty} \sum_{l=-\infty}^{\infty} h(m-k, n-l) r_X(k-l) \tag{4.260}$$

for all m and n. Taking the two-dimensional discrete Fourier transform yields $\hat{G}(\omega_1, \omega_2)$ as in Eq. 4.259, where now $S_{YX}(\omega_1, \omega_2)$ and $S_X(\omega_1, \omega_2)$ are the two-dimensional discrete spectral densities corresponding to r_{YX} and r_X, respectively. In the WS stationary digital image setting, the filter representation becomes

$$\hat{Y}(m, n) = \sum_{k=-\infty}^{\infty} \sum_{l=-\infty}^{\infty} \hat{g}(m-k, n-l) X(k, l) \tag{4.261}$$

An important case arises when a digital image process $Y(m, n)$ is corrupted by both the blurring of a linear observation system and additive noise. Then the observed process is

$$X(m,n) = \sum_{k=-\infty}^{\infty} \sum_{l=-\infty}^{\infty} b(m-k, n-l)Y(k,l) + N(m,n) \tag{4.262}$$

where $b(m, n)$ is the blurring function and $N(m, n)$ is noise. Suppose $N(m, n)$ is uncorrelated with the ideal process $Y(m, n)$. Except for the additive noise, the observation process $X(m, n)$ is a convolution of $Y(m, n)$ with the impulse response of the imaging system. Rewriting Eqs. 3.219 and 3.220 with the roles of X and Y reversed, in the present case (and momentarily ignoring the additive noise) these equations become $S_{YX}(\omega) = \overline{S}_{XY}(\omega) = \overline{B}(\omega)\, S_Y(\omega)$ and $S_X(\omega) = |B(\omega)|^2 S_Y(\omega)$, respectively, now with a discrete interpretation. Including the additive noise with S_X and substituting S_{YX} and S_X directly into Eq. 4.259 yields

$$\hat{G}(\omega_1,\omega_2) = \frac{\overline{B}(\omega_1,\omega_2)S_Y(\omega_1,\omega_2)}{|B(\omega_1,\omega_2)|^2\, S_Y(\omega_1,\omega_2) + S_N(\omega_1,\omega_2)} \tag{4.263}$$

where $\overline{B}$ is the conjugate of B. When there is no additive noise, Eq. 4.263 reduces to the *inverse* filter, defined by

$$\hat{G}(\omega_1,\omega_2) = B(\omega_1, \omega_2)^{-1} \tag{4.264}$$

If there is no blur in Eq. 4.262, only additive noise, then Eq. 4.263 defines the *Wiener smoothing filter*,

$$\hat{G}(\omega_1,\omega_2) = \frac{S_Y(\omega_1,\omega_2)}{S_Y(\omega_1,\omega_2) + S_N(\omega_1,\omega_2)} \tag{4.265}$$

If we define the *signal-to-noise ratio* (*SNR*) at (ω_1, ω_2) by

$$SNR(\omega_1,\omega_2) = \frac{S_Y(\omega_1,\omega_2)}{S_N(\omega_1,\omega_2)} \tag{4.266}$$

then

$$\hat{G}(\omega_1,\omega_2) = \frac{SNR(\omega_1,\omega_2)}{SNR(\omega_1,\omega_2) + 1} \tag{4.267}$$

***Example* 4.17.** By inverting $\hat{G}(\omega_1,\omega_2)$, the Wiener filter can be applied via Eq. 4.261 to yield the estimator $\hat{Y}(m, n)$ from $X(m, n)$; however, $\hat{G}(\omega_1,\omega_2)$

can also be used directly. The double sum in Eq. 4.261 is an estimator of $Y(m, n)$. Therefore, for each realization $x(m, n)$ of $X(m, n)$,

$$\hat{y}(m, n) = \sum_{k=-\infty}^{\infty} \sum_{l=-\infty}^{\infty} \hat{g}(m-k, n-l)x(k,l)$$

is an estimate of $Y(m, n)$ given $x(m, n)$. In practice, a realization $x(m, n)$ is observed, $\hat{y}(m, n)$ is found for all m and n, and $\hat{y}(m, n)$ is taken as the filtered version of $x(m, n)$. The double sum defining $\hat{y}(m, n)$ is a deterministic convolution and taking the Fourier transform $\mathcal{F}$ yields

$$\mathcal{F}[\hat{y}](\omega_1, \omega_2) = \hat{G}(\omega_1, \omega_2)\, \mathcal{F}[x](\omega_1, \omega_2)$$

Thus, x can be filtered in the frequency domain and $\hat{y}$ recovered by inverting $\mathcal{F}$. For the Wiener smoothing filter,

$$\mathcal{F}[\hat{y}](\omega_1, \omega_2) = \frac{S_Y(\omega_1, \omega_2)}{S_Y(\omega_1, \omega_2) + S_N(\omega_1, \omega_2)}\, \mathcal{F}[x](\omega_1, \omega_2)$$

At one extreme, $S_N(\omega_1, \omega_2) \approx 0$ and the frequency component of x at (ω_1, ω_2) is passed in full; at the other extreme, $S_N(\omega_1, \omega_2) \approx \infty$ and the frequency component of x at (ω_1, ω_2) is suppressed. In terms of the signal-to-noise ratio,

$$\mathcal{F}[\hat{y}](\omega_1, \omega_2) = \frac{SNR(\omega_1, \omega_2)}{SNR(\omega_1, \omega_2) + 1}\, \mathcal{F}[x](\omega_1, \omega_2)$$

For (ω_1, ω_2) at which the SNR is large, $\hat{G}(\omega_1, \omega_2) \approx 1$ and these frequencies pass with little alteration. Frequencies at which the SNR is close to 0 are strongly attenuated. ■

***Example* 4.18.** A *Boolean random function* $Y(t)$ on the real line is defined by generating a set of Poisson points $\{t_i\}$, translating to each point a random signal $U_i(t)$ that is identically distributed to a given *primary signal* $U(t)$, and then taking the supremum of the translated signals as the signal $Y(t)$, which is SS stationary. This example concerns a digital approximation of a random Boolean signal for which the primary signal is triangular with random base and height. The observed signal $X(n)$ results from blurring $Y(n)$ and adding independent Gaussian white noise according to Eq. 4.262. Figure 4.3 shows (a) a realization $y(n)$ of the ideal signal; (b) the realization $x(n)$ of the observed process resulting from blurring and adding Gaussian white noise to $y(n)$; (c) $x(n)$ filtered by the 9-point optimal MSE linear filter; and (d) $x(n)$ filtered by the Wiener filter according to Eq. 4.263 using the discrete Fourier transform. Each part of the figure is labeled with the root-mean-square error, $RMS = MSE^{1/2}$. The required autocorrelation matrices and cross-correlation vectors have been estimated from realizations (as they will also be in the next example). The curve in Fig. 4.4 shows the decreasing RMS estimates of the optimal linear filter over ever larger windows in comparison with the Wiener filter (the horizontal axis being labeled with half the window length). ■

A fundamental aspect of any optimal filter is the degree to which its performance degrades when it is applied to random processes that are different than the one for which it has been designed. Qualitatively, a filter is said to be *robust* when its performance degradation is acceptable for processes close to the one for which it has been designed. Robustness is important for application because, once in practice, a filter will surely be applied in nondesign settings. For instance, consider the model of Eq. 4.262 and suppose the ideal process $Y(m, n)$ is controlled by a single parameter λ, the additive noise $N(m, n)$ is Gaussian noise with mean μ and variance σ^2, and the blur is fixed. If an optimal filter Ψ_0 is designed for parameter values λ_0, μ_0, and σ_0, but is applied to the model when the parameter values are λ_1, μ_1, and σ_1, then Ψ_0 can perform no better than the optimal filter Ψ_1 designed for λ_1, μ_1, and σ_1. Let $MSE_b\langle\Psi_a\rangle$ denote the MSE resulting from applying Ψ_a to the model determined by λ_b, μ_b, and σ_b. $MSE_1\langle\Psi_0\rangle$ can be no less than $MSE_1\langle\Psi_1\rangle$. The optimal filter Ψ_0 is robust if $MSE_1\langle\Psi_0\rangle$ is close to $MSE_1\langle\Psi_1\rangle$ when $(\lambda_1, \mu_1, \sigma_1)$ is close to $(\lambda_0, \mu_0, \sigma_0)$. This qualitative concept can be quantified by defining Ψ_0 to be *robust with degree* (ε, δ) $[\varepsilon, \delta > 0]$ if

$$MSE_1\langle\Psi_0\rangle - MSE_1\langle\Psi_1\rangle < \varepsilon \tag{4.268}$$

whenever $|\lambda_0 - \lambda_1| < \delta$, $|\mu_0 - \mu_1| < \delta$, and $|\sigma_0 - \sigma_1| < \delta$.

The optimal linear filter for Eq. 4.262 is determined by Eq. 4.263. $Y(m, n)$ contributes via its power spectral density and, for fixed blur, the degradation contributes via the power spectral density of $N(m, n)$. To indicate the dependencies of the optimal linear filter in this case, let us temporarily denote the Wiener filter for signal and noise power spectral densities S_{Y_0} and S_{N_0} by Ψ_0. We can define Ψ_0 to be robust with degree (ε, δ) [relative to Eq. 4.262] if the inequality of Eq. 4.268 holds whenever $\| S_{Y_0} - S_{Y_1} \| < \delta$ and $\| S_{N_0} - S_{N_1} \| < \delta$, where the distances are defined in some appropriate sense. The constrained model has resulted in a special form of the filter and this is reflected in the fact that robustness can be quantified by second-order information of the signal and noise, rather than their full distributions. With just the constraint of linearity and WS stationarity (not the specialized model of Eq. 4.262), the Wiener filter is determined by Eq. 4.259, so that robustness depends on S_{YX} and S_X.

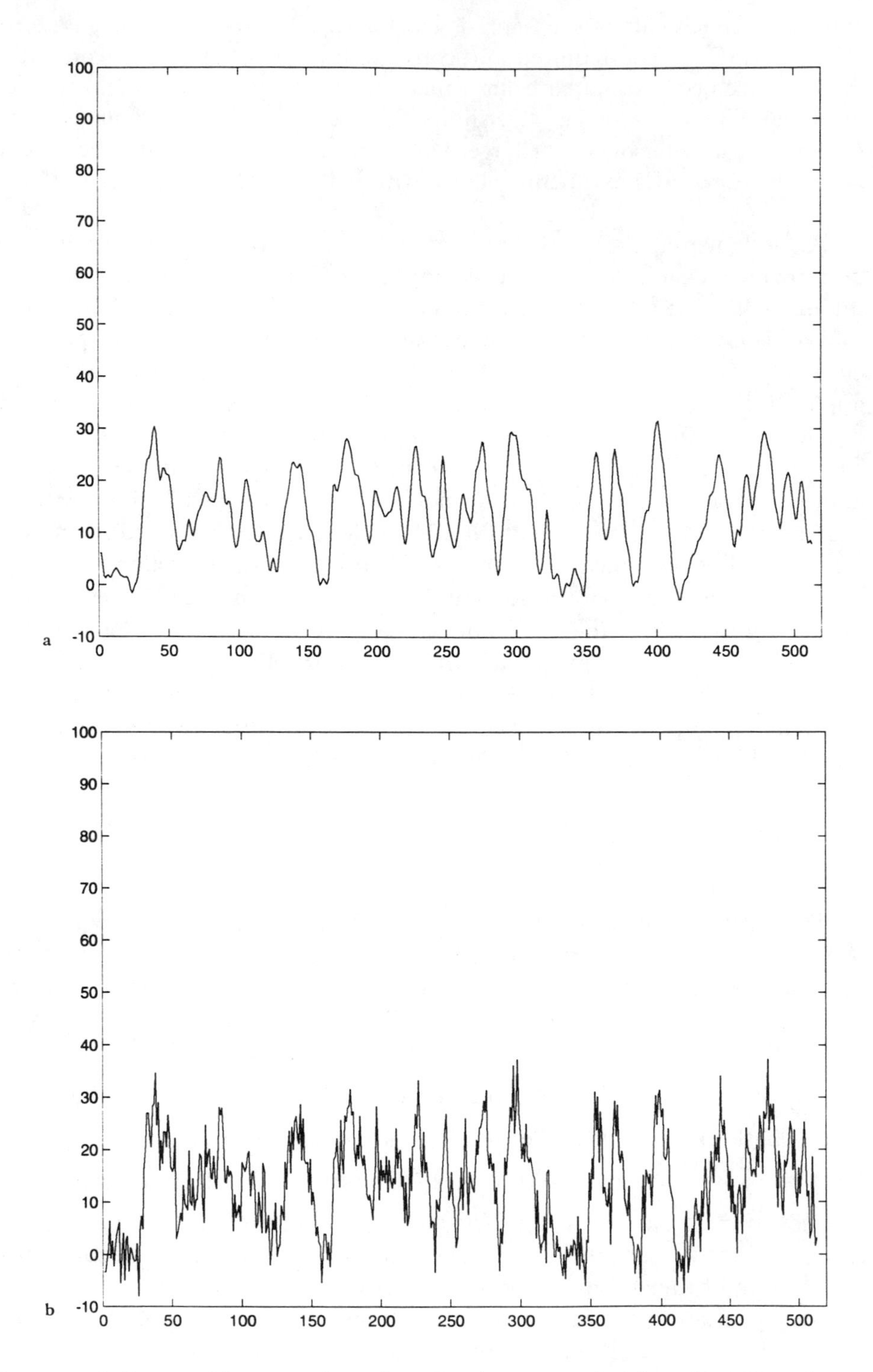

Figure 4.3 Linear filtering of random Boolean signal: (a) realization of ideal signal; (b) observed realization;

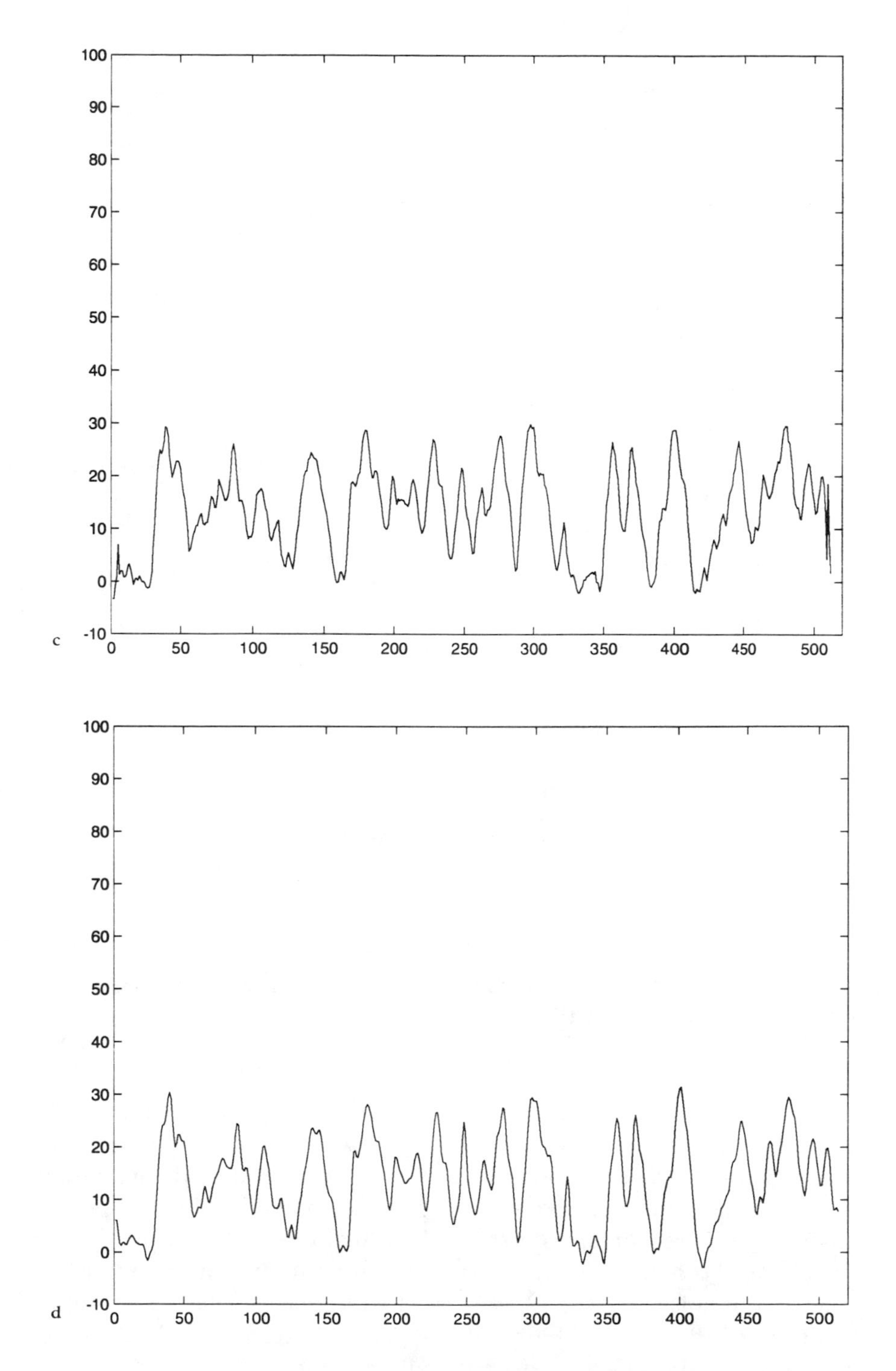

Figure 4.3 (cont): (c) observed realization filtered by 9-point optimal MSE linear filter; (d) observed realization filtered by Wiener filter.

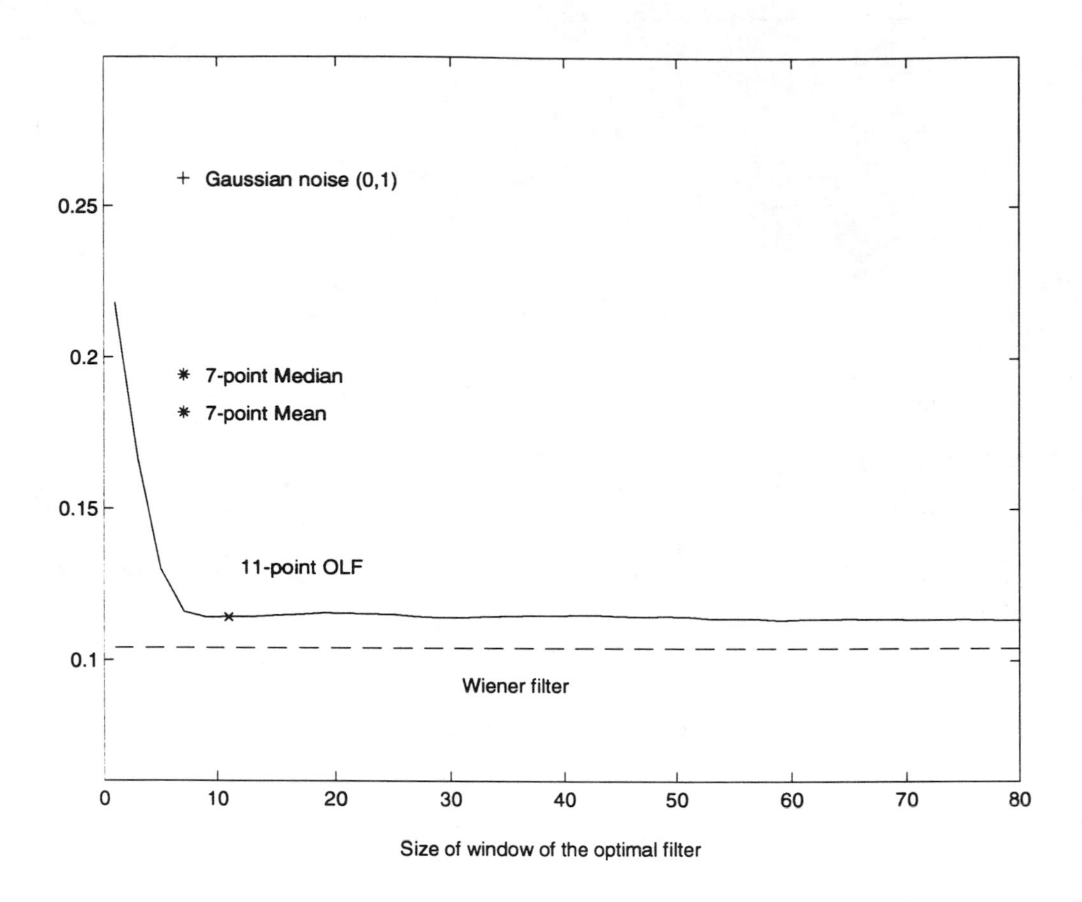

Figure 4.4 RMS estimates of optimal linear filter over ever larger windows in comparison with Wiener filter.

Example 4.19. A Boolean random function on $\Re^2$ is a random image. The Poisson points are spatial, the translations are two-dimensional, and the primary function is a random image. Consider a 4-bit gray-scale approximation of a random Boolean image for which the primary random signal is pyramidal and the observed image is degraded by blur and additive Gaussian white noise. Figure 4.5 shows (a) a realization $y(m, n)$ of the ideal image; (b) the observed realization $x(m, n)$ resulting from blurring and adding Gaussian white noise to $y(m, n)$; (c) $x(m, n)$ filtered by the 7×7 optimal linear filter; and (d) $x(m, n)$ filtered by the Wiener filter (to be denoted $\Psi_{b,n}$). Figure 4.6 shows the effect with only additive white noise: (a) the original realization $y(m, n)$ with additive Gaussian white noise; (b) the result of applying the Wiener filter Ψ_n designed for the white-noised image. Figure 4.7 shows (a) a blurred version of the original realization $y(m, n)$; (b) the inverse-filtered blurred image. An indication of the robustness problem can be seen in Figs. 4.8 and 4.9. Figure 4.8 shows (a) the Wiener filter $\Psi_{b,n}$ applied to the white-noised image of Fig. 4.6(a); (b) $\Psi_{b,n}$ applied to the blurred image of Fig. 4.7(a). Figure 4.9 shows (a) the Wiener filter Ψ_n applied to the blurred image of Fig. 4.7(a); (b) Ψ_n applied to the noisy, blurred realization $x(m, n)$.

Dividing the numerator and denominator of Eq. 4.263 by S_Y and then multiplying S_N/S_Y in the resulting denominator by the nonnegative parameter γ yields the *parametric Wiener filter*,

$$\hat{G}_\gamma(\omega_1,\omega_2) = \frac{\overline{B}(\omega_1,\omega_2)}{|B(\omega_1,\omega_2)|^2 + \gamma\dfrac{S_N(\omega_1,\omega_2)}{S_Y(\omega_1,\omega_2)}}$$

For $\gamma > 1$, the effect of the noise-to-signal ratio NSR is increased; for $\gamma < 1$, the NSR effect is diminished. In either case, the resulting filter is not MSE-optimal. Figure 4.10 shows the effect of applying the parametric filter to the original degradation $x(m, n)$ with: (a) $\gamma = 16$ and (b) $\gamma = 1/16$. For large γ, the filter behaves as though the NSR is very high and strongly suppresses high-frequency content by blurring the image; for small γ, it acts as though the NSR is very low and (as would the inverse filter for a blurred image) sharpens the image, thereby accentuating the point noise. ■

4.8. Optimal Linear Filter in the Context of a Linear Model

In this section, design of the optimal linear filter is facilitated by modeling the ideal and observed random functions via a pair of linear decompositions. No generality is lost by using the model. Its advantage lies in the fact that it can result in having to solve simpler equations than the Wiener-Hopf equation (as it appears in the general setting).

4.8.1. The Linear Signal Model

Throughout this section, we assume that the observed and ideal signals take the forms

$$X(t) = \sum_{k=1}^{n} W_k \varphi_k(t) + U(t) \tag{4.269}$$

$$Y(s) = \sum_{k=1}^{n} W_k \xi_k(s) + V(s)$$

a

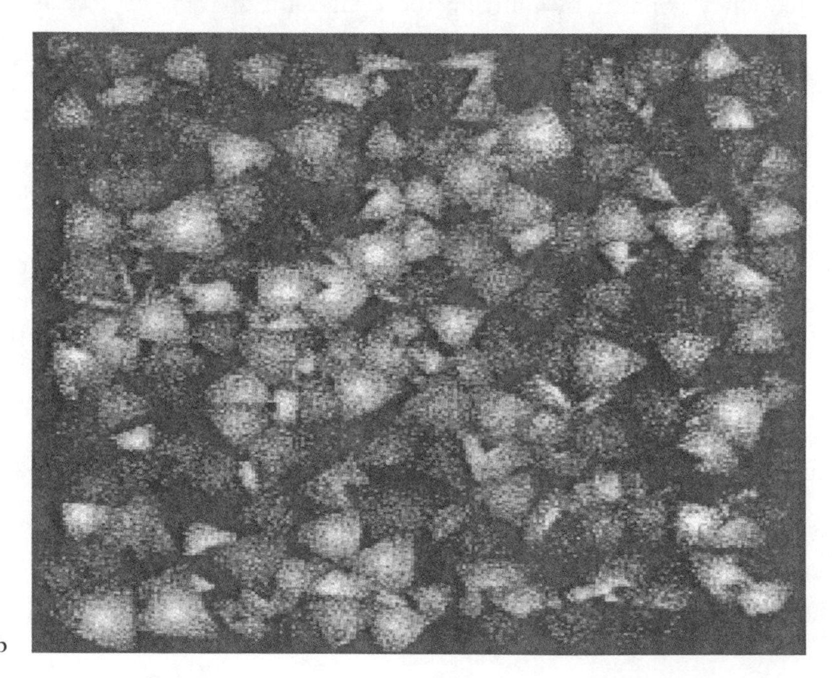

b

Figure 4.5 Linear filtering of random Boolean image degraded by blurring and additive white noise: (a) realization of ideal image; (b) observed realization;

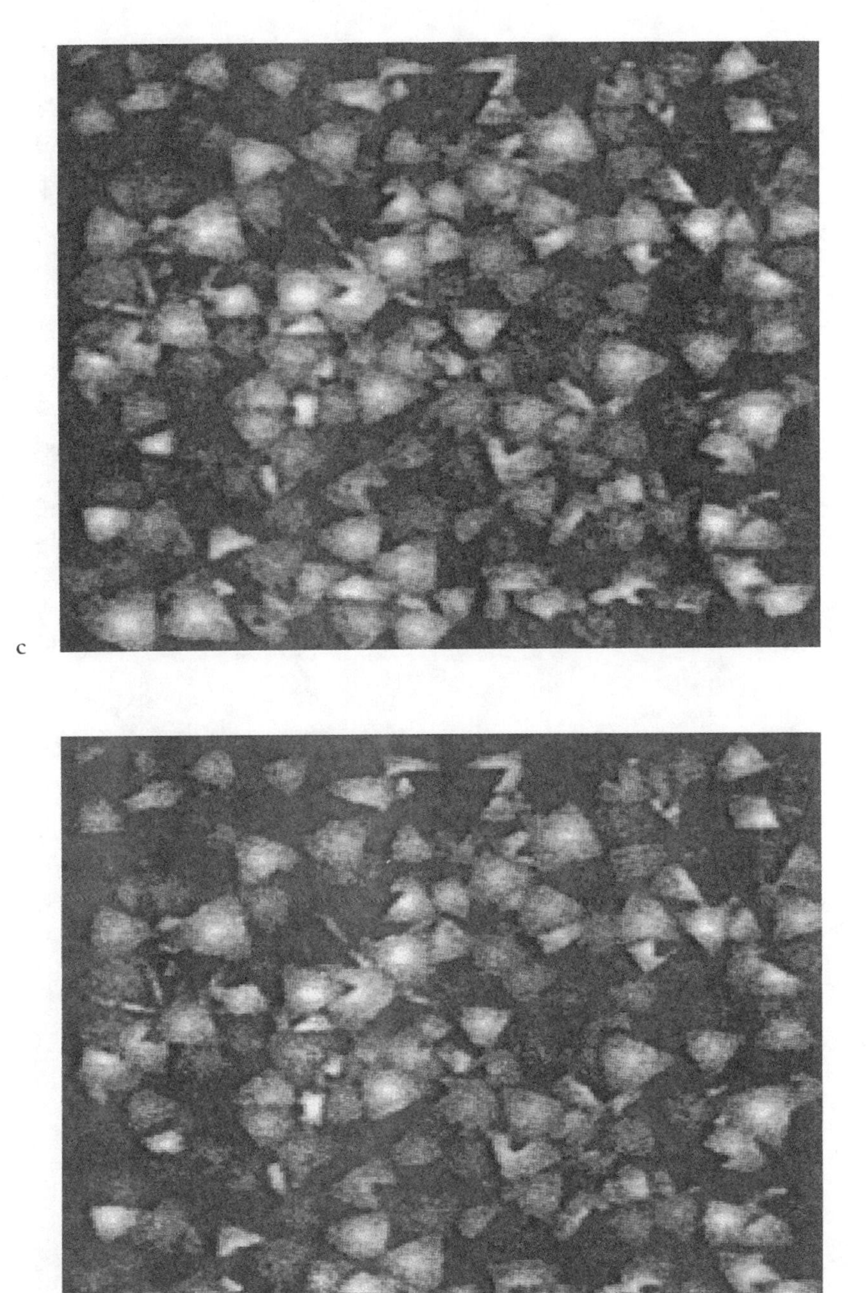

Figure 4.5 (cont): (c) observed realization filtered by the 7 × 7 optimal linear filter; (d) observed realization filtered by the Wiener filter $\Psi_{b,n}$ designed for blurred and white-noised image.

a

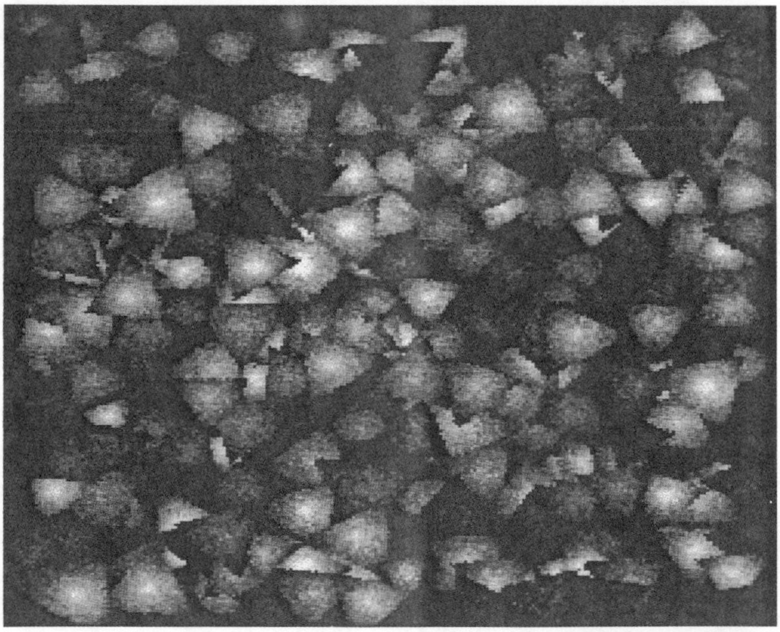

b

Figure 4.6 Linear filtering of random Boolean image degraded by additive white noise: (a) noisy realization; (b) noisy realization filtered by Wiener filter Ψ_n designed for additive white noise.

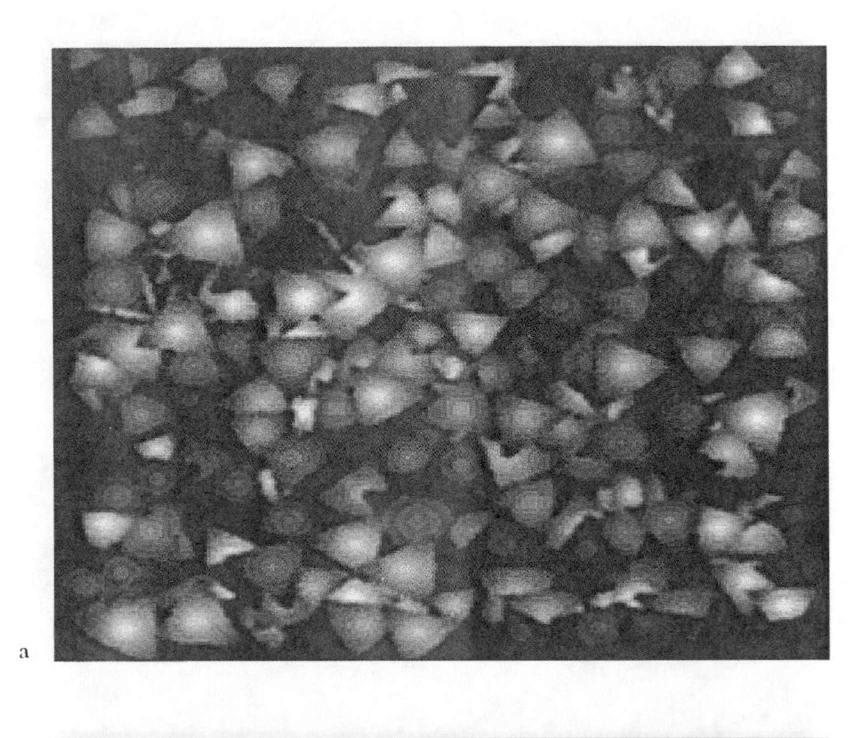
a

b

Figure 4.7 Linear filtering of blurred random Boolean image: (a) blurred realization; (b) inverse-filtered blurred realization.

a

b

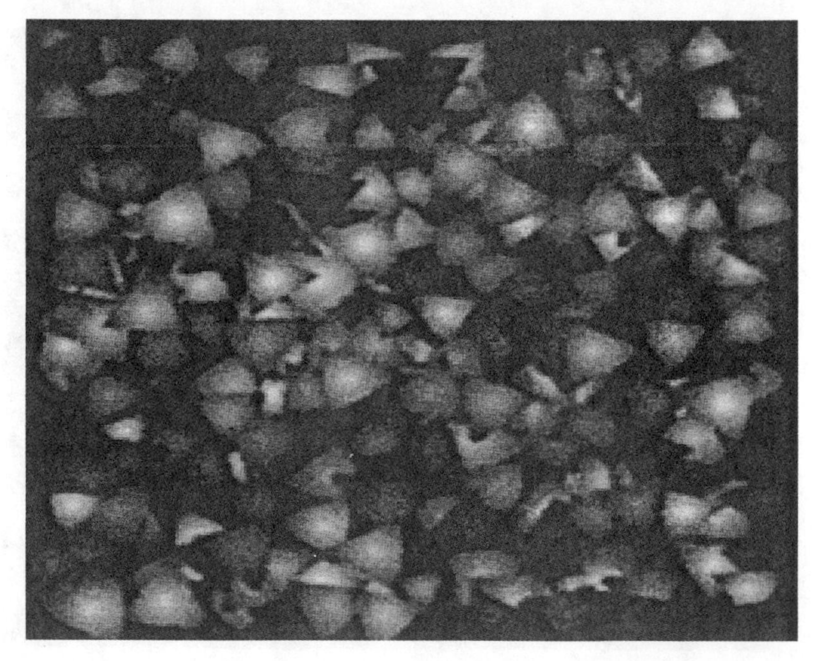

Figure 4.8 Robustness of Wiener filter: (a) $\Psi_{b,n}$ applied to white-noised realization; (b) $\Psi_{b,n}$ applied to blurred realization.

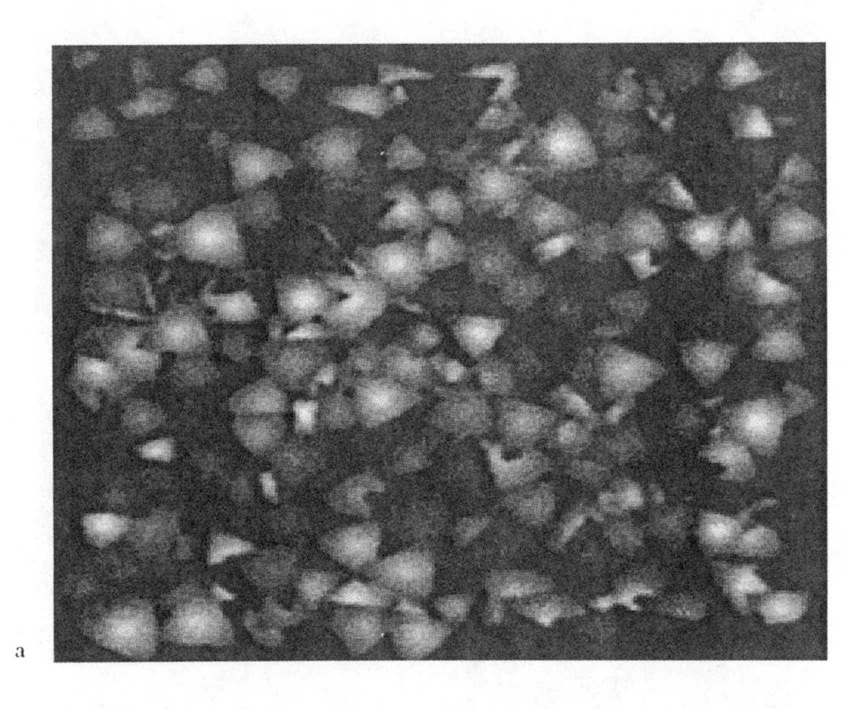

a

b

Figure 4.9 Robustness of Wiener filter: (a) Ψ_n applied to blurred realization; (b) Ψ_n applied to blurred-and-white-noised realization.

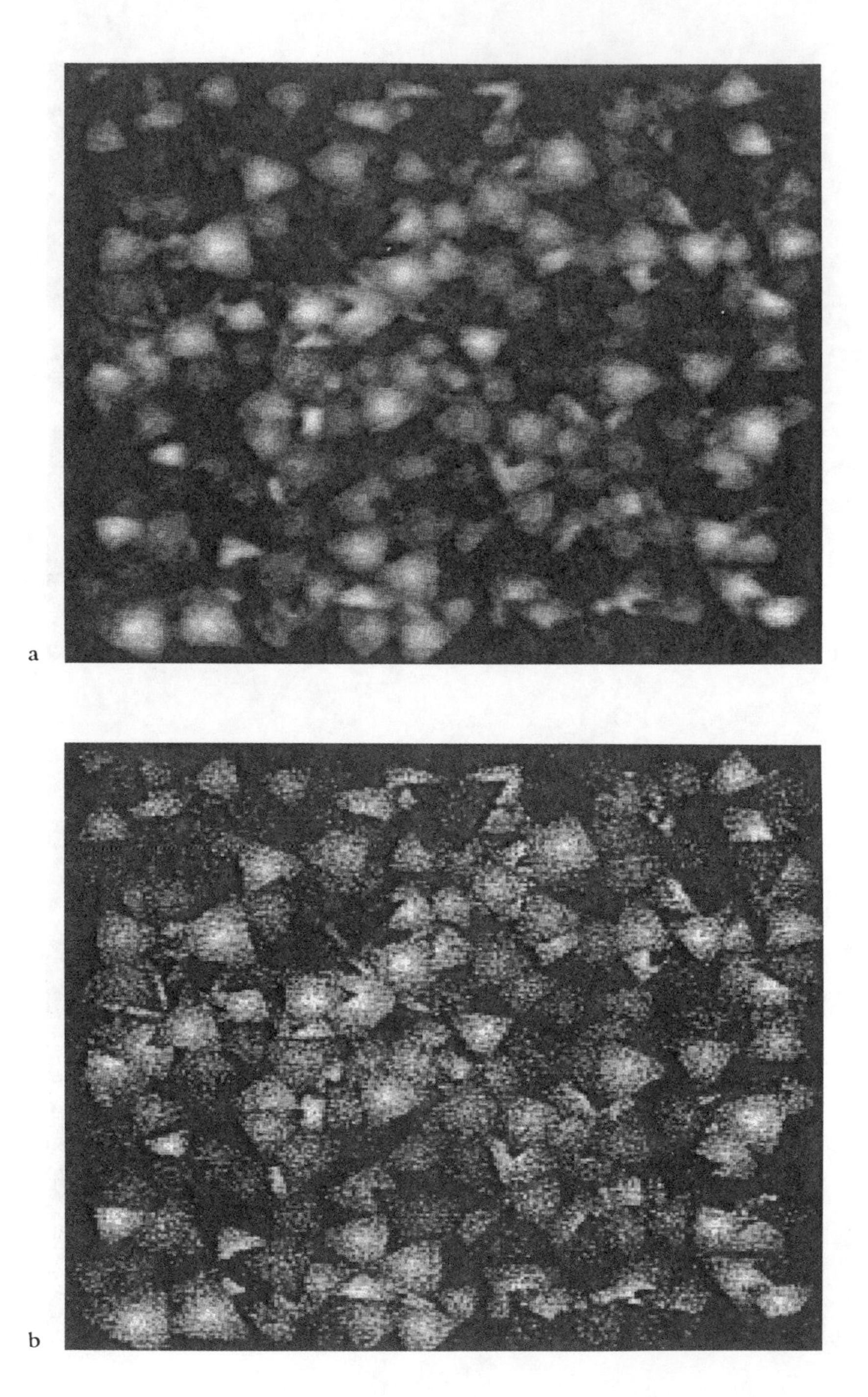

Figure 4.10 Application of parametric Wiener filter to blurred-and-white-noised realization with (a) $\gamma = 16$; (b) $\gamma = 1/16$.

respectively, where $W_1, W_2,..., W_n$ are linearly independent random variables that are uncorrelated with $U(t)$ and $V(s)$, $\varphi_k(t)$ and $\xi_k(s)$, $k = 1, 2,..., n$, are known deterministic functions, and t and s can be vector variables. We assume that both $U(t)$ and $V(s)$ possess zero mean. No generality is lost by assuming zero means, since otherwise we could write $X(t)$ and $Y(s)$ as

$$X(t) = \sum_{k=1}^{n} W_k \varphi_k(t) + \mu_U(t) + U_0(t) \tag{4.270}$$

$$Y(s) = \sum_{k=1}^{n} W_k \xi_k(s) + \mu_V(s) + V_0(s)$$

where $U_0(t)$ and $V_0(s)$ are the centered versions of $U(t)$ and $V(s)$, respectively. Letting $W_{n+1} = 1$, $\varphi_{n+1}(t) = \mu_U(t)$, and $\xi_{n+1}(s) = \mu_V(s)$, we obtain $X(t)$ and $Y(s)$ to be of the same form as in Eq. 4.269, except that W_{n+1} is not random. Subsequent to presenting a method for finding the weighting function $\hat{g}(s,t)$ defining the optimal linear filter when $W_1, W_2,..., W_n$ are random variables, we will discuss the case of constant coefficients.

If we let $W_1 = W_2 = \cdots = W_n = 0$ in Eq. 4.269, then the model reduces to the general case of estimating $V(s)$ by observing $U(t)$. Hence, the model subsumes the general problem, and conversely. This means that any method of solution obtained in the present setting provides a general method. The usefulness of the model, originally proposed by Semenov, is that the decomposition can facilitate discovery of an optimal filter when $U(t)$ and $V(s)$ are easy to describe. In particular, $U(t)$ is often additive noise and the linear decomposition for the underlying signal results in a simplified solution.

Letting

$$r_{ij} = E[W_i W_j] \tag{4.271}$$

and taking into account the uncorrelatedness between $W_1, W_2,..., W_n$ and both $U(t)$ and $V(s)$, from Eq. 4.269 we find the autocorrelation functions for $X(t)$ and $Y(s)$, as well as their cross-correlation function:

$$R_X(t, t') = E[X(t)X(t')]$$

$$= \sum_{i=1}^{n}\sum_{j=1}^{n} E[W_i W_j]\varphi_i(t)\varphi_j(t') + \sum_{k=1}^{n} E[U(t)W_k]\varphi_k(t')$$

$$+ \sum_{k=1}^{n} E[W_k U(t')]\varphi_k(t) + E[U(t)U(t')]$$

$$= \sum_{i=1}^{n}\sum_{j=1}^{n} r_{ij}\varphi_i(t)\varphi_j(t') + R_U(t, t') \tag{4.272}$$

$$R_Y(s, s') = \sum_{i=1}^{n}\sum_{j=1}^{n} r_{ij}\xi_i(s)\xi_j(s') + R_V(s, s') \tag{4.273}$$

$$R_{YX}(s, t) = \sum_{i=1}^{n}\sum_{j=1}^{n} r_{ij}\xi_i(s)\varphi_j(t) + R_{VU}(s, t) \tag{4.274}$$

4.8.2. Procedure for Finding the Optimal Linear Filter

Denoting the linear operator of Eq. 4.244 by $\Psi_{s,t}$, the equation takes the form

$$\Psi_{s,t}X(t) = \int_T g(s,t)X(t)\,dt \tag{4.275}$$

The first subscript denotes the point at which estimation is taking place and the second subscript denotes the variable of integration. It should be recognized that the domain T of integration can depend on s. Relative to $\Psi_{s,t}$, the Wiener-Hopf equation (Eq. 4.254) is

$$R_{YX}(s, t') = \Psi_{s,t}R_X(t, t') \tag{4.276}$$

for $t' \in T$. A solution $\hat{\Psi}_{s,t}$ of the Wiener-Hopf equation is defined by a kernel $\hat{g}(s,t)$ and

$$\hat{Y}(s) = \hat{\Psi}_{s,t}X(t) \tag{4.277}$$

is the optimal linear estimator of $Y(s)$.

Substituting in $R_X(t, t')$ and $R_{YX}(s, t')$ from Eqs. 4.272 and 4.274 for the autocorrelation and cross-correlation, the Wiener-Hopf equation becomes

$$\Psi_{s,t}R_U(t, t') = R_{VU}(s, t') + \sum_{i=1}^{n}\sum_{j=1}^{n} r_{ij}[\xi_i(s) - \Psi_{s,t}\varphi_i(t)]\varphi_j(t') \tag{4.278}$$

This single equation can be written as the system of $n + 1$ equations

$$\Psi_{s,t} R_U(t, t') = R_{VU}(s, t') + \sum_{j=1}^{n} d_j(s)\varphi_j(t') \tag{4.279}$$

$$d_j(s) = \sum_{i=1}^{n} r_{ij}[\xi_i(s) - \Psi_{s,t}\varphi_i(t)] \qquad (j = 1, 2, \ldots, n) \tag{4.280}$$

To put Eq. 4.280 into vector form, let $\mathbf{D}(s) = (d_1, d_2, \ldots, d_n)'$,

$$\mathbf{R} = \begin{pmatrix} r_{11} & r_{12} & \cdots & r_{1n} \\ r_{21} & r_{22} & \cdots & r_{2n} \\ \vdots & \vdots & \ddots & \vdots \\ r_{n1} & r_{n2} & \cdots & r_{nn} \end{pmatrix} \tag{4.281}$$

$$\mathbf{C} = \mathbf{R}^{-1} = \begin{pmatrix} c_{11} & c_{12} & \cdots & c_{1n} \\ c_{21} & c_{22} & \cdots & c_{2n} \\ \vdots & \vdots & \ddots & \vdots \\ c_{n1} & c_{n2} & \cdots & c_{nn} \end{pmatrix} \tag{4.282}$$

$$\mathbf{G}(s) = \begin{pmatrix} \xi_1(s) - \Psi_{s,t}\varphi_1(t) \\ \xi_2(s) - \Psi_{s,t}\varphi_2(t) \\ \vdots \\ \xi_n(s) - \Psi_{s,t}\varphi_n(t) \end{pmatrix} \tag{4.283}$$

R is the autocorrelation matrix of $\mathbf{W} = (W_1, W_2, \ldots, W_n)'$. **R** is nonsingular because $W_1, W_2, \ldots, W_n$ are linearly independent. Because $r_{ij} = r_{ji}$, the system of Eq. 4.280 and its solution are given by

$$\mathbf{D}(s) = \mathbf{R}\mathbf{G}(s) \tag{4.284}$$

$$\mathbf{G}(s) = \mathbf{R}^{-1}\mathbf{D}(s) = \mathbf{C}\mathbf{D}(s) \tag{4.285}$$

respectively. Termwise, Eq. 4.285 can be written as

$$\xi_i(s) - \Psi_{s,t}\varphi_i(t) = \sum_{j=1}^{n} c_{ij} d_j \qquad (i = 1, 2, \ldots, n) \tag{4.286}$$

The system of Eqs. 4.279 and 4.286 characterizes the optimal linear filter.

Owing to the linear structure of Eq. 4.279, the solution for $\Psi_{s,t}$ is of the form

$$\Psi_{s,t} = \Psi_{s,t}^{0} + \sum_{k=1}^{n} d_k \Psi_{s,t}^{k} \tag{4.287}$$

where the linear operators $\Psi_{s,t}^{0}$, $\Psi_{s,t}^{1}, \ldots,$ $\Psi_{s,t}^{n}$ satisfy the equations

$$\Psi_{s,t}^{0} R_U(t,t') = R_{VU}(s, t') \tag{4.288}$$

$$\Psi_{s,t}^{k} R_U(t,t') = \varphi_k(t') \tag{4.289}$$

for $k = 1, 2, \ldots, n$. Assuming that $\Psi_{s,t}^{0}$, $\Psi_{s,t}^{1}, \ldots,$ $\Psi_{s,t}^{n}$ solve this system, the optimal linear operator $\Psi_{s,t}$ is given by Eq. 4.287, where it is still necessary to solve for $d_1, d_2, \ldots, d_n$.

To solve for $d_1, d_2, \ldots, d_n$, let

$$b_{ij}(s) = \Psi_{s,t}^{j} \varphi_i(t) \tag{4.290}$$

for $i = 1, 2, \ldots, n$ and $j = 0, 1, \ldots, n$, and apply $\Psi_{s,t}$ to $\varphi_i(t)$ for $i = 1, 2, \ldots, n$ to yield

$$\Psi_{s,t}\varphi_i(t) = b_{i0} + \sum_{j=1}^{n} d_j b_{ij} \tag{4.291}$$

Substitution of this expression for $\Psi_{s,t}\varphi_i(t)$ into Eq. 4.286 yields

$$\xi_i(s) - b_{i0} = \sum_{j=1}^{n} (b_{ij} + c_{ij}) d_j \tag{4.292}$$

for $i = 1, 2, \ldots, n$

The preceding equations provide a procedure to determine $\hat{g}(s,t)$ for the optimal linear filter:

L1) Let $\mathbf{R} = (r_{ij})$ be the autocorrelation matrix of $(W_1, W_2,..., W_n)'$, and $\mathbf{C} = \mathbf{R}^{-1} = (c_{ij})$.

L2) For $t' \in T$, solve the following integral equation for $g_0(s, t)$:

$$\int_T R_U(t,t')g_0(s,t)\,dt = R_{VU}(s,t') \tag{4.293}$$

A solution always exists.

L3) For $k = 1, 2,..., n$, solve the n integral equations

$$\int_T R_U(t,t')g_k(s,t)\,dt = \varphi_k(t') \tag{4.294}$$

for $g_k(s, t)$. If there exist n solutions, then the remaining steps give the optimal linear filter. We will shortly describe an alternative method that gives the optimal filter should one or more of these equations not possess a solution.

L4) For $i = 1, 2,..., n$ and $j = 0, 1,..., n$,, let

$$b_{ij} = \int_T g_j(s,t)\varphi_i(t)\,dt \tag{4.295}$$

L5) Solve the following system of n equations for $d_1, d_2,..., d_n$:

$$\begin{aligned}
\sum_{k=1}^{n}(b_{1k}+c_{1k})d_k &= \xi_1(s)-b_{10}\\
\sum_{k=1}^{n}(b_{2k}+c_{2k})d_k &= \xi_2(s)-b_{20}\\
\vdots \qquad\qquad & \qquad \vdots\\
\sum_{k=1}^{n}(b_{nk}+c_{nk})d_k &= \xi_n(s)-b_{n0}
\end{aligned} \tag{4.296}$$

L6) The optimal MSE linear filter is determined by

$$\hat{g}(s,t) = g_0(s,t)+\sum_{k=1}^{n}d_k(s)g_k(s,t) \tag{4.297}$$

Several considerations apply. Some of the coefficients W_k may be unknown parameters rather than random variables. If W_k is a parameter, then it is treated as a random variable with infinite variance ($r_{kk} = \infty$), all c_{ij} in L1 for which either i or j equals k are set to 0, and the matrix $\mathbf{C}$ is the inverse of the autocorrelation matrix for the vector formed from $\mathbf{W}$ by deleting W_k. If m variables are nonrandom, then m deletions are made.

The matrix

$$\mathbf{B}(s) = \begin{pmatrix} b_{11} & b_{12} & \cdots & b_{1n} \\ b_{21} & b_{22} & \cdots & b_{2n} \\ \vdots & \vdots & \ddots & \vdots \\ b_{n1} & b_{n2} & \cdots & b_{nn} \end{pmatrix} \tag{4.298}$$

is nonnegative definite. Thus, $\det[\mathbf{B}(s)] \geq 0$. If $\varphi_1(t), \varphi_2(t),\ldots, \varphi_n(t)$ are linearly independent, then $\det[\mathbf{B}(s)] \neq 0$ and the system of equations in L5 has a unique solution. We can assume $\varphi_1(t), \varphi_2(t),\ldots, \varphi_n(t)$ are linearly independent; if not, they can be replaced by a linearly independent subset.

Sometimes the observed random function $X(t)$ represents the signal plus the noise $U(t)$. The c_{ij}, being elements of an inverse matrix of second moments of $W_1, W_2,\ldots, W_n$, vary inversely with respect to the signal variance. The b_{ij} vary inversely with respect to the noise variance. Thus, a large signal variance relative to that of the noise diminishes the effect that the probability characteristics of the signal have on the optimal linear filter.

Comparing the previous and present sections, and remembering that $U(t)$ and $V(s)$ possess zero means, we see that the integral equation of Eq. 4.293 is another form of the Wiener-Hopf equation. Thus, it appears that the previous orthogonality method is a special case of the present method, since letting $W_1 = W_2 = \cdots = W_n = 0$ gives the original orthogonality solution. In fact, the orthogonality solution also subsumes the current one because the decomposition assumes that the random functions satisfy a certain model.

The analysis goes through when $W_1, W_2,\ldots, W_n$ are dependent on the estimation point s, the change being that $\mathbf{R}$ is a function of s.

The MSE for the optimal linear filter is given by

$$MSE\langle \hat{Y}(s) \rangle = \mathrm{Var}[V(s)] - b_{00}(s) + \sum_{i=1}^{n}\sum_{j=1}^{n}(b_{ij}(s) + c_{ij})d_i(s)d_j(s) \tag{4.299}$$

where

$$b_{00}(s) = \int_T g_0(s,t)R_{VU}(s,t)\,dt \tag{4.300}$$

***Example* 4.20.** Consider the model

$$X(t) = W_1 + W_2 t + U(t)$$

$$Y(s) = W_1 + W_2 s + V(s)$$

where W_1 is a parameter and W_2 is a random variable. We wish to estimate $Y(s)$ by observing $X(t)$ over the interval $[a, s]$. Since W_1 is a parameter, we assume $r_{11} = \infty$ and $c_{11} = c_{12} = c_{21} = 0$. Consequently, $\mathbf{R} = (r_{22}) = \text{Var}[W_2]$ and $c_{22} = 1/r_{22}$. Applying step L2, $g_0(s, t)$ is determined by the integral equation

$$\int_a^s R_U(t,t')g_0(s,t)\,dt = R_{VU}(s,t')$$

for $a \leq t' \leq s$. Applying step L3, $g_1(s, t)$ and $g_2(s, t)$ are determined by the equations

$$\int_a^s R_U(t,t')g_1(s,t)\,dt = 1$$

$$\int_a^s R_U(t,t')g_2(s,t)\,dt = t'$$

By step L4, the b_{ij} are given by

$$b_{i0} = \int_a^s t^{i-1} g_0(s,t)\,dt \qquad (i = 1, 2)$$

$$b_{11} = \int_a^s g_1(s,t)\,dt$$

$$b_{i2} = \int_a^s t^{i-1} g_2(s,t)\,dt \qquad (i = 1, 2)$$

We obtain b_{21} by symmetry. According to step L5, d_1 and d_2 are found from the system

$$b_{11}d_1 + b_{12}d_2 = 1 - b_{10}$$

$$b_{21}d_1 + (b_{22} + 1/r_{22})d_2 = s - b_{20}$$

The optimal linear filter is determined by

$$\hat{g}(s,t) = g_0(s,t) + d_1(s)g_1(s,t) + d_2(s)g_2(s,t)$$

Its MSE is given by

$$MSE\langle \hat{Y} \rangle = \mathrm{Var}[V] - b_{00} + b_{11}d_1^2 + 2b_{12}d_1d_2 + (b_{22} + r_{22}^{-1})d_{22}^2 \quad \blacksquare$$

In image processing, it is common to estimate the ideal image by averaging over a window. In this case, the functions in Eq. 4.269 are of two variables, (t_1, t_2) and (s_1, s_2). The optimal filter and the integrals in steps L2 and L3 of the algorithm take the forms

$$\hat{Y}(s_1, s_2) = \int_{s_1-l}^{s_1+l} \int_{s_2-l}^{s_2+l} \hat{g}(s_1, s_2, t_1, t_2) X(t_1, t_2)\, dt_1 dt_2 \tag{4.301}$$

$$\int_{s_1-l}^{s_1+l} \int_{s_2-l}^{s_2+l} R_U((t_1, t_2), (t_1', t_2')) g_0(s_1, s_2, t_1, t_2)\, dt_1 dt_2 = R_{VU}(s_1, s_2, t_1', t_2') \tag{4.302}$$

$$\int_{s_1-l}^{s_1+l} \int_{s_2-l}^{s_2+l} R_U((t_1, t_2), (t_1', t_2')) g_k(s_1, s_2, t_1, t_2)\, dt_1 dt_2 = \varphi_k(t_1', t_2') \tag{4.303}$$

The integral equation in step L2 always has a solution, but the ones in L3 may not. As the method has so far been presented, we have assumed that all equations in L3 have solutions. Now assume that m of them have solutions but that $n - m$ of them do not. Without loss of generality, we assume there exist solutions for $\varphi_1(t)$, $\varphi_2(t)$,..., $\varphi_m(t)$, but not for $\varphi_{m+1}(t)$, $\varphi_{m+2}(t)$,..., $\varphi_n(t)$. In this case, it can be shown that there exist $n - m$ independent solutions, $h_{m+1}(s, t)$, $h_{m+2}(s, t)$,..., $h_n(s, t)$, of the homogeneous equation

$$\int_T R_U(t, t') h(s, t)\, dt = 0 \tag{4.304}$$

for $t' \in T$, for which

$$\int_T h_k(s,t)\varphi_j(t)\,dt = \delta_{kj} \tag{4.305}$$

for $j = 1, 2,..., n$ and $k = m + 1, m + 2,..., n$. The optimal linear filter is determined by

$$\hat{g}(s,t) = g_0(s,t) + \sum_{k=1}^{m} d_k g_k(s,t) + \sum_{k=m+1}^{n} e_k h_k(s,t) \tag{4.306}$$

where $d_1, d_2,..., d_m$ are solutions to the system

$$\sum_{i=1}^{m} (b_{ki} + c_{ki})d_i = \xi_k(s) - b_{k0} \qquad (k = 1, 2,..., m) \tag{4.307}$$

and, for $k = m + 1, m + 2,..., n$,

$$e_k = \xi_k(s) - b_{k0} - \sum_{i=1}^{m} (b_{ki} + c_{ki})d_i \tag{4.308}$$

4.8.3. Additive White Noise

The difficulty with the present method is that we are again confronted with solving integral equations; however, the model can result in much simpler integral equations, as exhibited in the preceding example. In fact, should $U(t)$ be white noise with covariance function

$$R_U(t, t') = I(t)\delta(t - t') \tag{4.309}$$

then, for $k = 0, 1,..., n$,

$$\int_T R_U(t,t')g_k(s,t)\,dt = I(t')g_k(s,t') \tag{4.310}$$

Thus, Eqs. 4.293 and 4.294 in L2 and L3 reduce to

$$g_0(s, t') = \frac{R_{VU}(s,t')}{I(t')} \tag{4.311}$$

$$g_k(s, t') = \frac{\varphi_k(t')}{I(t')} \tag{4.312}$$

If $R_{VU}(s, t)$ is known, then finding the optimal filter reduces to elementary algebra.

Consider the model

$$X(t) = W_1\varphi_1(t) + W_2\varphi_2(t) + U(t) \tag{4.313}$$

$$Y(s) = W_1\varphi_1(s) + W_2\varphi_2(s)$$

where $U(t)$ is white noise with intensity $I(t) = 1$ and both W_1 and W_2 are random variables. The c_{ij} are found according to L1 by inverting the autocorrelation matrix of $(W_1, W_2)'$. Since $V(s) = 0$, $R_{VU}(s, t) = 0$, and $g_0(s, t) = 0$ in L2. From Eq. 4.312, $g_k(s, t) = \varphi_k(t)$. Thus,

$$\hat{g}(s,t) = d_1\varphi_1(t) + d_2\varphi_2(t) \tag{4.314}$$

where d_1 and d_2 are obtained (step L5) by solving the algebraic equations

$$(b_{11} + c_{11})d_1 + (b_{12} + c_{12})d_2 = \varphi_1(s)$$

$$(b_{21} + c_{21})d_1 + (b_{22} + c_{22})d_2 = \varphi_2(s) \tag{4.315}$$

and where (step L4) the b_{ij} are given by

$$b_{11} = \int_T \varphi_1^2(t)\,dt$$

$$b_{12} = b_{21} = \int_T \varphi_1(t)\varphi_2(t)\,dt \tag{4.316}$$

$$b_{22} = \int_T \varphi_2^2(t)\,dt$$

Note that $g_0(s, t) = 0$ implies $b_{00} = b_{10} = b_{20} = 0$.

Examination of the equations yields two points: (1) for the given model, no integral equations need be solved, so that the entire procedure consists of solving linear equations; and (2) if $U(t)$ is white noise, then the approach works for any decomposition of the form

$$X(t) = \sum_{k=1}^{n} W_k\varphi_k(t) + U(t)$$

$$Y(s) = \sum_{k=1}^{n} W_k \varphi_k(s) \tag{4.317}$$

***Example* 4.21.** Consider the model

$$X(t) = W_1 + W_2 t + U(t)$$

$$Y(s) = W_1 + W_2 s$$

where $U(t)$ is white noise with unit intensity, and W_1 and W_2 are uncorrelated zero-mean, unit-variance random variables. Then $\mathbf{R}^{-1}$ is the identity matrix, so that $c_{11} = c_{22} = 1$ and $c_{12} = c_{21} = 0$. Assuming $X(t)$ is to be observed over the interval [0, 1], Eq. 4.316 yields $b_{11} = 1$, $b_{12} = b_{21} = 1/2$, and $b_{22} = 1/3$. The system of Eq. 4.315 is given by

$$2d_1 + \frac{d_2}{2} = 1$$

$$\frac{d_1}{2} + \frac{4d_2}{3} = s$$

These equations are solved by $d_1 = (13 - 6s)/29$ and $d_2 = (24s - 6)/29$. Hence,

$$\hat{g}(s,t) = \frac{1}{29}(13 - 6s - 6t + 24st)$$

Since $b_{00} = 0$, by Eq. 4.299 the MSE for the optimal linear estimator is

$$MSE\langle \hat{Y} \rangle = 2\left(\frac{13-6s}{29}\right)^2 + \left(\frac{13-6s}{29}\right)\left(\frac{24s-6}{29}\right) + \frac{4}{3}\left(\frac{24s-6}{29}\right)^2 \quad \blacksquare$$

4.8.4. Discrete Domains

Of particular importance to digital processing is the case where both random functions in the model are defined on discrete domains, the integers for signals and the grid points for images. For the discrete case, the decomposition takes the form

$$X(i) = \sum_{k=1}^{n} W_k \varphi_k(i) + U(i)$$

$$Y(j) = \sum_{k=1}^{n} W_k \xi_k(j) + V(j) \tag{4.318}$$

and the optimal MSE linear filter is of the form

$$\hat{Y}(j) = \sum_{i=1}^{\infty} \hat{g}(j,i)X(i) \tag{4.319}$$

The solution procedure is accomplished in the following manner:

L1') Perform step L1.

L2') For $l = 1, 2, \ldots$, solve the (possibly infinite) set of algebraic equations

$$\sum_{i=1}^{\infty} R_U(i,l)g_0(j,i) = R_{VU}(j,l) \tag{4.320}$$

L3') For $k = 1, 2, \ldots, n$, solve the equation system

$$\sum_{i=1}^{\infty} R_U(i,l)g_k(j,i) = \varphi_k(l) \qquad (l = 1, 2, \ldots) \tag{4.321}$$

L4') For $p = 1, 2, \ldots, n$ and $q = 0, 1, \ldots, n$, let

$$b_{pq} = \sum_{i=1}^{\infty} g_q(j,i)\varphi_p(i) \tag{4.322}$$

L5') Solve the same system as that given in L5, noting that j is used in place of s.

L6') The optimal linear filter is determined by

$$\hat{g}(j,i) = g_0(j,i) + \sum_{k=1}^{n} d_k(j)g_k(j,i) \tag{4.323}$$

and the MSE is calculated as in Eq. 4.299.

If $U(i)$ is discrete white noise, then $R_U(i, j) = \mathrm{Var}[U(i)]\delta_{ij}$ and L2' and L3' have the solutions

$$g_0(j,i) = \frac{R_{VU}(j,i)}{\mathrm{Var}[X(i)]} \tag{4.324}$$

$$g_k(j,i) = \frac{\varphi_k(i)}{\mathrm{Var}[X(i)]} \tag{4.325}$$

As in the continuous case, the problem is reduced to one of solving algebraic equations.

4.9. Optimal Linear Filters via Canonical Expansions

A canonical expansion of a random function provides a decomposition into uncorrelated random variables. Estimation of a random function $Y(s)$ in terms of an observed random function $X(t)$ can be facilitated by first canonically decomposing $X(t)$ into white noise $Z(\xi)$ and then estimating $Y(s)$ by means of $Z(\xi)$. The advantage of this approach results from the simplified form of the Wiener-Hopf integral equation for white noise input. Application of canonical representations for the determination of optimal linear filters is largely due to Pugachev.

4.9.1. Integral Decomposition into White Noise

If the zero-mean random function $X(t)$ possesses an integral canonical expansion in terms of white noise $Z(\xi)$, then according to the theory of Section 3.9,

$$X(t) = \int_{\Xi} Z(\xi)x(t,\xi)\,d\xi \tag{4.326}$$

$$Z(\xi) = \int_{T} X(t)\overline{a(t,\xi)}\,dt \tag{4.327}$$

Rather than estimate $Y(s)$ directly by a linear filter acting on $X(t)$, suppose we can find the optimal linear filter that estimates $Y(s)$ from $Z(\xi)$,

$$\hat{Y}(s) = \int_{\Xi} \hat{w}(s,\xi)Z(\xi)\,d\xi \tag{4.328}$$

where $\hat{w}(s,\xi)$ is the optimal weighting function. Then

$$\hat{Y}(s) = \int_{\Xi}\int_{T}\hat{w}(s,\xi)\overline{a(t,\xi)}X(t)\,dt\,d\xi$$

$$= \int_{T}\hat{g}(s,t)X(t)\,dt \tag{4.329}$$

where

$$\hat{g}(s,t) = \int_{\Xi}\hat{w}(s,\xi)\overline{a(t,\xi)}\,d\xi \tag{4.330}$$

To see that $\hat{g}(s,t)$ gives the optimal linear filter for $X(t)$, suppose $g(s, t)$ is a different weighting function. Because $\hat{w}(s,\xi)$ is the optimal kernel for $Z(\xi)$,

$$E\left[\left|Y(s) - \int_{T}g(s,t)X(t)\right|^2\right] = E\left[\left|Y(s) - \int_{T}g(s,t)\left(\int_{\Xi}Z(\xi)x(t,\xi)\,d\xi\right)dt\right|^2\right]$$

$$= E\left[\left|Y(s) - \int_{\Xi}\left(\int_{T}g(s,t)x(t,\xi)\,dt\right)Z(\xi)\,d\xi\right|^2\right] \tag{4.331}$$

$$\geq E\left[\left|Y(s) - \int_{\Xi}\hat{w}(s,\xi)Z(\xi)\,d\xi\right|^2\right]$$

$$= E\left[\left|Y(s) - \int_{T}\hat{g}(s,t)X(t)\,dt\right|^2\right]$$

To summarize the methodology: (1) $X(t)$ is transformed to white noise $Z(\xi)$; (2) the optimal linear filter for $Z(\xi)$ is found; and (3) the optimal linear filter for $X(t)$ is found by concatenation.

It remains to find an expression for $\hat{g}(s,t)$. According to the Wiener-Hopf equation,

$$R_{YZ}(s,t) = \int_{T}\hat{w}(s,u)R_{Z}(u,t)\,du$$

$$= \int_T I(t)\delta(u-t)\hat{w}(s,u)\,du \tag{4.332}$$

$$= I(t)\,\hat{w}(s,t)$$

Hence,

$$\hat{w}(s,t) = \frac{R_{YZ}(s,t)}{I(t)} \tag{4.333}$$

From Eq. 4.330 and the cross-covariance expansion

$$R_{YZ}(s,\xi) = E\left[Y(s)\overline{\int_T X(u)\overline{a(u,\xi)}\,du}\right]$$

$$= \int_T R_{YX}(s,u)a(u,\xi)\,du \tag{4.334}$$

we deduce the desired expression for the weighting function for the optimal linear filter with input $X(t)$:

$$\hat{g}(s,t) = \int_\Xi \frac{R_{YZ}(s,\xi)}{I(\xi)}\overline{a(t,\xi)}\,d\xi$$

$$= \int_\Xi \frac{\overline{a(t,\xi)}}{I(\xi)}\int_T R_{YX}(s,u)a(u,\xi)\,du\,d\xi \tag{4.335}$$

In Section 4.7.2 it has been seen that, when $X(t)$ and $Y(s)$ are jointly WS stationary and $X(t)$ is observed over all time, the Wiener-Hopf equation reduces to a convolution integral (Eq. 4.258) and the Fourier transform of the optimal weighting function is thereby determined (Eq. 4.259). According to the results of Section 3.10.3, the solution of Eq. 4.259 is a special case of the general solution given in Eq. 4.335 with $a(t,\xi) = e^{j\xi t}$ and $T = (-\infty, \infty)$. From the perspective of representation, WS stationarity and observation over all time are not central issues.

4.9.2. Integral Equations Involving the Autocorrelation Function

Whether one approaches the optimal linear filter directly through the Wiener-Hopf integral equation or via the linear model of the previous section, the problem ultimately comes down to solving integral equations of the form

$$\int_T R_X(t,\tau) g(s,t)\, dt = f(s,\tau) \tag{4.336}$$

for $\tau \in T$. The Wiener-Hopf equation is of this form, as are the integral equations (Eqs. 4.293 and 4.294) in the procedure for the linear model. We will call this integral equation the *extended Wiener-Hopf equation* (and avoid another use of the term "generalized"). Given an integral canonical expansion for $X(t)$, based on the preceding method involving white-noise decomposition (in particular, Eq. 4.330), we are motivated to look for a solution of the form

$$g(s,t) = \int_\Xi h(s,\xi)\overline{a(t,\xi)}\, d\xi \tag{4.337}$$

We need to find the function $h(s, \xi)$.

Substitution of the proposed form into the integral equation and application of Eq. 3.194 yields

$$\begin{aligned} f(s,\tau) &= \int_T R_X(t,\tau) \int_\Xi h(s,\xi)\overline{a(t,\xi)}\, d\xi\, dt \\ &= \int_\Xi h(s,\xi) \int_T \overline{R_X(\tau,t)a(t,\xi)}\, dt\, d\xi \\ &= \int_\Xi I(\xi)h(s,\xi)\overline{x(\tau,\xi)}\, d\xi \end{aligned} \tag{4.338}$$

This equation holds for all $\tau \in T$. Multiplying it by $a(\tau, t)$, integrating over T, and applying the bi-orthogonality relation of Eq. 3.196 yields

$$\int_T a(\tau,t) f(s,\tau)\, d\tau = \int_T a(\tau,t) \int_\Xi I(\xi)h(s,\xi)\overline{x(\tau,\xi)}\, d\xi\, d\tau$$

$$= \int_{\Xi} I(\xi)h(s,\xi) \int_{T} a(\tau,t)\overline{x(\tau,\xi)}\, d\tau\, d\xi \tag{4.339}$$

$$= \int_{\Xi} I(\xi)h(s,\xi)\delta(t-\xi)\, d\xi$$

$$= I(t)h(s,t)$$

Hence,

$$h(s,t) = \frac{1}{I(t)} \int_{T} a(\tau,t) f(s,\tau)\, d\tau \tag{4.340}$$

From Eq. 4.337,

$$g(s,t) = \int_{\Xi} \frac{\overline{a(t,\xi)}}{I(\xi)} \int_{T} f(s,\tau)a(\tau,\xi)\, d\tau\, d\xi \tag{4.341}$$

Equation 4.341 yields the optimal filter of Eq. 4.335 if $f(s, \tau) = R_{YX}(s, \tau)$, which is the case when the extended Wiener-Hopf equation reduces to the Wiener-Hopf equation.

Besides the fact that the solution of Eq. 4.341 is more general, there are differences between Eqs. 4.335 and 4.341. In deriving the former, it is assumed that $\hat{w}(s,t)$ provides an optimal linear filter and Eq. 4.335 follows because it is necessary that $\hat{w}(s,t)$ satisfy the Wiener-Hopf equation. The more general approach postulates a form for the solution and then proceeds to find the form of the kernel $g(s, t)$. It does not assure us that $g(s, t)$ provides an optimal weighting function. Moreover, whereas Eq. 4.340 necessarily follows from Eq. 4.338, the converse is not true. Omitting the theoretical details, it can be shown that if the extended Wiener-Hopf equation has a solution, then the solution must be given by Eq. 4.341.

Not only does Eq. 4.341 give the optimal linear filter (if one exists) for the direct approach, it also provides solutions for steps L2 and L3 in the procedure of Section 4.8. For Eq. 4.293, $f(s, \tau) = R_{VU}(s, \tau)$, and for Eq. 4.294, $f(s, \tau) = \varphi_k(\tau)$.

4.9.3. Solution via Discrete Canonical Expansions

There are two drawbacks with using integral canonical expansions. First, they are theoretically more difficult than discrete canonical expansions;

second, discrete canonical representation is more readily achievable. We now assume that $X(t)$ has the canonical expansion

$$X(t) = \sum_{k=1}^{\infty} Z_k x_k(t) \tag{4.342}$$

where Z_k and $x_k(t)$ are given by 3.143 and 3.148, respectively, and $a_j(t)$ and $x_k(t)$ satisfy the bi-orthogonality relation of Eq. 3.151. In analogy to the solution by integral canonical expansions, we look for an optimal kernel of the form

$$g(s, t) = \sum_{k=1}^{\infty} h_k(s)\overline{a_k(t)} \tag{4.343}$$

Substitution of the proposed kernel into the extended Wiener-Hopf equation and application of Eq. 3.148 yields

$$f(s, \tau) = \int_T R_X(t, \tau)\left(\sum_{k=1}^{\infty} h_k(s)\overline{a_k(t)}\right) dt$$

$$= \sum_{k=1}^{\infty} h_k(s) \int_T \overline{R_X(\tau, t) a_k(t)}\, dt \tag{4.344}$$

$$= \sum_{k=1}^{\infty} V_k h_k(s)\overline{x_k(\tau)}$$

where $V_k = \text{Var}[Z_k]$. Two points should be noted. First, the preceding calculation assumes that integration and summation can be interchanged. We will return to this point shortly. Second, the resulting representation of $f(s, \tau)$ is only a necessary condition for $g(s, t)$, as defined in Eq. 4.343, to be a solution of the extended Wiener-Hopf equation. It can, in fact, be proven that for the extended Wiener-Hopf equation to possess a solution that results in an operator that transforms $X(t)$ into a random function possessing a finite variance (which is a requirement in the context of optimal linear filtering), it is necessary that $f(s, \tau)$ be representable in terms of the coordinate functions of $X(t)$. If $f(s, \tau)$ has such a representation, we will say that it is *coordinate-function representable*.

Multiplying by $a_j(\tau)$, integrating over T, and applying the bi-orthogonality relation of Eq. 3.151 yields

$$\int_T a_j(\tau) f(s,\tau)\, d\tau = \sum_{k=1}^{\infty} V_k h_k(s) \int_T a_j(\tau) \overline{x_k(\tau)}\, d\tau$$

$$= V_j h_j(s) \tag{4.345}$$

Solving for $h_j(s)$ yields

$$h_j(s) = \frac{1}{V_j} \int_T a_j(\tau) f(s,\tau)\, d\tau \tag{4.346}$$

Putting this expression into Eq. 4.343 yields

$$g(s,t) = \sum_{k=1}^{\infty} \frac{\overline{a_k(t)}}{V_k} \int_T a_k(\tau) f(s,\tau)\, d\tau \tag{4.347}$$

If there exists a solution of the extended Wiener-Hopf equation, then it is given by Eq. 4.347.

We now wish to apply this solution to the Wiener-Hopf equation and the linear model of Section 4.8. First, the cross-covariance function of the random functions $Y(s)$ and $X(t)$ is coordinate-function representable. Letting $f(s, \tau) = R_{YX}(s, \tau)$,

$$g(s,t) = \sum_{k=1}^{\infty} \frac{\overline{a_k(t)}}{V_k} \int_T a_k(\tau) R_{YX}(s,\tau)\, d\tau \tag{4.348}$$

For the linear model of Eq. 4.269, a similar expression holds for $U(t)$ and $V(s)$:

$$g_0(s,t) = \sum_{k=1}^{\infty} \frac{\overline{a_k(t)}}{V_k} \int_T a_k(\tau) R_{VU}(s,\tau)\, d\tau \tag{4.349}$$

If the functions $\varphi_1(t)$, $\varphi_2(t)$,..., $\varphi_n(t)$ are coordinate-function representable, then

$$g_j(s,t) = \sum_{k=1}^{\infty} \frac{\overline{a_k(t)}}{V_k} \int_T a_k(\tau) \varphi_j(\tau)\, d\tau \tag{4.350}$$

for $j = 1, 2,..., n$. For $i, j = 1, 2,..., n$, the quantities b_{ij} in step L4 of the procedure become

$$b_{ij} = \int_T \varphi_i(t)\left(\sum_{k=1}^{\infty} \frac{\overline{a_k(t)}}{V_k} \int_T a_k(\tau)\varphi_j(\tau)\,d\tau\right) dt$$

$$= \sum_{k=1}^{\infty} \frac{1}{V_k} \int_T \varphi_i(t)\overline{a_k(t)}\,dt \int_T \varphi_j(\tau)a_k(\tau)\,d\tau \tag{4.351}$$

For $i = 1, 2,..., n$,

$$b_{i0} = \int_T \varphi_i(t)\left(\sum_{k=1}^{\infty} \frac{\overline{a_k(t)}}{V_k} \int_T a_k(\tau)R_{VU}(s,\tau)\,d\tau\right) dt$$

$$= \sum_{k=1}^{\infty} \frac{1}{V_k} \int_T \varphi_i(t)\overline{a_k(t)}\,dt \int_T a_k(\tau)R_{VU}(s,\tau)\,d\tau \tag{4.352}$$

The procedure is completed using these values for b_{ij}. For computation of the MSE, Eq. 4.300 becomes

$$b_{00} = \int_T R_{VU}(s,t)\left(\sum_{k=1}^{\infty} \frac{\overline{a_k(t)}}{V_k} \int_T a_k(\tau)R_{VU}(s,\tau)\,d\tau\right) dt$$

$$= \sum_{k=1}^{\infty} \frac{1}{V_k} \int_T \overline{R_{VU}(s,t)a_k(t)}\,dt \int_T R_{VU}(s,\tau)a_k(\tau)\,d\tau$$

$$= \sum_{k=1}^{\infty} \frac{1}{V_k} \left| \int_T R_{VU}(s,\tau)a_k(\tau)\,d\tau \right|^2 \tag{4.353}$$

For the method to be legitimate, Eqs. 4.349 and 4.350 (or Eq. 4.348 when the Wiener-Hopf equation is used directly without the linear model) must converge. Moreover, if they converge in the mean over T, then the aforementioned integral-sum interchange in Eq. 4.344 is justified.

***Example* 4.22.** Let $Y(t)$ be a zero-mean signal, $\{u_k(t)\}$ be the Karhunen-Loeve coordinate-function system for $K_Y(t, \tau)$, $\{\lambda_k\}$ be the corresponding eigenvalue sequence, $N(t)$ be white noise with constant intensity κ that is uncorrelated with the signal, and

$$X(t) = Y(t) + N(t)$$

be observed. Then

$$K_X(t, \tau) = K_Y(t, \tau) + \kappa\delta(t - \tau)$$

$$K_{YX}(t, \tau) = K_Y(t, \tau)$$

$$\begin{aligned}\int_T K_X(t,\tau)u_k(\tau)\,d\tau &= \int_T (K_Y(t,\tau)+\kappa\delta(t-\tau))u_k(\tau)\,d\tau \\ &= (\lambda_k + \kappa)u_k(t)\end{aligned}$$

$X(t)$ has the same coordinate functions as $Y(t)$, but with eigenvalues $\lambda_k + \kappa$. According to Eq. 4.348,

$$\begin{aligned}\hat{g}(s,t) &= \sum_{k=1}^{\infty} \frac{\overline{u_k(t)}}{\lambda_k + \kappa} \int_T u_k(\tau) R_Y(s,\tau)\,d\tau \\ &= \sum_{k=1}^{\infty} \frac{\lambda_k}{\lambda_k + \kappa} u_k(s)\overline{u_k(t)}\end{aligned}$$

The optimal linear filter is given by

$$\begin{aligned}\hat{Y}(s) &= \int_T \left(\sum_{k=1}^{\infty} \frac{\lambda_k}{\lambda_k + \kappa} u_k(s)\overline{u_k(t)} \right) X(t)\,dt \\ &= \sum_{k=1}^{\infty} \frac{\lambda_k}{\lambda_k + \kappa} Z_k u_k(s)\end{aligned}$$

where $Z_1, Z_2, \ldots$ are the Karhunen-Loeve coefficients for $X(t)$. ■

4.10. Optimal Binary Filters

According to Theorem 4.3, the conditional expectation is the overall optimal MSE filter. Unfortunately, it is impractical for many applications because it does not usually lead to a mathematically tractable form involving a relatively small number of parameters that can be estimated from observations. Although they are generally suboptimal [except when the observation and ideal random functions are jointly Gaussian (Theorem 4.6)], optimal linear MSE filters can be derived analytically owing to their representation as orthogonal projections in Hilbert spaces. In this section we return to the conditional expectation and consider the discrete binary setting, in which both the observation random variables and the random variable to be estimated are binary valued. For the binary situation, it is computationally feasible to form the joint distribution of the random variables as a probability table so long as their number is not too great. This table can be estimated by the histogram of a large number of observations. Discrete binary filters are important in digital image processing, where the input random vector represents a random pattern.

4.10.1. Binary Conditional Expectation

For n observation binary random variables $X_1, X_2,\ldots, X_n$, a binary random variable Y to be estimated, and a function $\psi(X_1, X_2,\ldots, X_n)$, the mean-square error for ψ as an estimator of Y is equivalent to the *mean-absolute error* (*MAE*) of ψ, namely,

$$MAE\langle\psi\rangle = E[|Y - \psi(X_1, X_2,\ldots, X_n)|] \tag{4.354}$$

Were the optimal MAE estimator not constrained by a need to be binary, it would be given by the conditional expectation

$$E[Y \mid X_1, X_2,\ldots, X_n] = P(Y = 1 \mid X_1, X_2,.., X_n) \tag{4.355}$$

To be a binary filter on $X_1, X_2,\ldots, X_n$, the conditional expectation requires quantization. Letting $\mathbf{X} = (X_1, X_2,\ldots, X_n)'$, the optimal binary MAE filter is the binary conditional expectation, defined by

$$E_b[Y|\mathbf{X}] = \begin{cases} 1, & \text{if } P(Y = 1|\mathbf{X}) > 0.5 \\ 0, & \text{if } P(Y = 1|\mathbf{X}) \leq 0.5 \end{cases} \tag{4.356}$$

where the subscript b denotes the binary-operator form of the conditional expectation.

Binary-conditional-expectation estimation can be viewed via tabulation of the relevant probabilities. For each outcome of **X**, Y can take on the value 0 or 1. We can form a table consisting of 2^n rows of the form

$$[\mathbf{x},\ P(\mathbf{X}=\mathbf{x}),\ P(Y=0|\,\mathbf{x}),\ P(Y=1|\,\mathbf{x})] \tag{4.357}$$

where $\mathbf{x} = (x_1, x_2,..., x_n)'$ is one of the 2^n possible outcomes of **X** and

$$P(Y=y\,|\,\mathbf{x}) = \frac{P(Y=y, \mathbf{X}=\mathbf{x})}{P(\mathbf{X}=\mathbf{x})} \tag{4.358}$$

If **x** is observed ($\mathbf{X} = \mathbf{x}$), then $E_b[Y \mid \mathbf{X}] = 1$ if

$$P(Y=1\,|\,\mathbf{x}) > P(Y=0\,|\,\mathbf{x}) \tag{4.359}$$

Otherwise, $E_b[Y \mid \mathbf{X}] = 0$. The value of the optimal filter can be determined by finding **x** in the table and checking Eq. 4.359.

Let ψ_{op} denote the binary conditional expectation. From Eq. 4.354,

$$\begin{aligned} MAE\langle\psi_{op}\rangle &= E[|Y - E_b[Y|\mathbf{X}]|] \\ &= \sum_{\mathbf{x},y} |y - E_b[Y|\mathbf{x}]|\,P(\mathbf{X}=\mathbf{x}, Y=y) \\ &= \sum_{\mathbf{x}} |0 - E_b[Y|\mathbf{x}]|\,P(\mathbf{X}=\mathbf{x}, Y=0) + \sum_{\mathbf{x}} |1 - E_b[Y|\mathbf{x}]|\,P(\mathbf{X}=\mathbf{x}, Y=1) \end{aligned} \tag{4.360}$$

where the sums are over all possible observation vectors. For a given **x**, $E_b[Y \mid \mathbf{x}]$ is either 0 or 1. Therefore the two summations have 2^n null terms and 2^n nonnull terms. The term for **x** in the first sum is null if $E_b[Y \mid \mathbf{x}] = 0$; the term for **x** in the second sum is null if $E_b[Y \mid \mathbf{x}] = 1$. Hence, $MAE\langle\psi_{op}\rangle$ can be rewritten as

$$\begin{aligned} MAE\langle\psi_{op}\rangle &= \sum_{\{\mathbf{x}:E_b[Y|\mathbf{x}]=1\}} P(\mathbf{X}=\mathbf{x}, Y=0) + \sum_{\{\mathbf{x}:E_b[Y|\mathbf{x}]=0\}} P(\mathbf{X}=\mathbf{x}, Y=1) \\ &= \sum_{\{\mathbf{x}:P(Y=1|\mathbf{x})>0.5\}} P(\mathbf{X}=\mathbf{x})P(Y=0|\mathbf{x}) + \sum_{\{\mathbf{x}:P(Y=1|\mathbf{x})\leq 0.5\}} P(\mathbf{X}=\mathbf{x})P(Y=1|\mathbf{x}) \end{aligned} \tag{4.361}$$

The last expression has a simple interpretation relative to the probability table. For the pair of entries corresponding to **x**, take $P(\mathbf{X} = \mathbf{x})P(Y = y \mid \mathbf{x})$ corresponding to the y (0 or 1) not chosen for ψ_{op} and then sum up these probabilities to find $MAE\langle\psi_{op}\rangle$.

ψ_{op} is defined via the pairs $(\mathbf{x}, y)$, $y = 0$ or $y = 1$, as determined by the inequality of Eq. 4.359. As a logical operator, ψ_{op} is defined by the 2^n rows of a truth table, where each row is of the form $[x_1, x_2,..., x_n, y]$. ψ_{op} also possesses a logical-variable representation.

4.10.2. Boolean Functions and Optimal Translation-Invariant Filters

If $x_1, x_2,..., x_n$ are binary variables taking on the logical values 0 or 1, then an n-variable binary operator $\psi(x_1, x_2,..., x_n)$ is called a *Boolean function* and possesses a sum-of-products (disjunction-of-conjunction) representation

$$\psi(x_1, x_2,..., x_n) = \sum_i x_1^{p(i,1)} x_2^{p(i,2)} \cdots x_n^{p(i,n)} \tag{4.362}$$

where $p(i, j)$ is either null or "c," the former indicating the presence of the jth variable in the ith product (*minterm*) and the latter indicating the presence of the complemented jth variable in the ith product. There are at most 2^n minterms and each corresponds to a 1-valued row in the truth table defining $\psi(x_1, x_2,..., x_n)$. The minterm expansion is called the *disjunctive normal form* of ψ. Reduction of the minterm expansion can be accomplished in accordance with the laws of Boolean algebra. The expansion then takes the form

$$\psi(x_1, x_2,..., x_n) = \sum_i x_{i,1}^{p(i,1)} x_{i,2}^{p(i,2)} \cdots x_{i,n(i)}^{p(i,n(i))} \tag{4.363}$$

where $x_{i,1}, x_{i,2},..., x_{i,n(i)}$ denote the $n(i)$ distinct variables in the ith product. There exist various algorithms for logic reduction, which, in general, is not unique.

Suppose we wish to filter an observed binary-valued discrete random image $S(z)$ to estimate the ideal random image $S_0(z)$ by applying a translation-invariant window operator Ψ. Then Ψ is defined via a window $W = \{w_1, w_2,..., w_n\}$ and a Boolean function ψ on n binary random variables $X_1, X_2,..., X_n$. Ψ is defined at a pixel z by translating the window to z and applying the Boolean function ψ to the binary random values in the translated window W_z (see Eq. 4.50). Ψ is translation (spatially) invariant because we are assuming that the Boolean function ψ is not dependent on the pixel at which Ψ is defined.

Considering the matter probabilistically, for any pixel z, let $Y_z = S_0(z)$ and $\mathbf{X}_z$ be the vector of variables for the observed image $S(z)$ in the window W_z. The binary conditional expectation,

$$\psi_{op,z}(\mathbf{X}_z) = E_{b,z}[Y_z \mid \mathbf{X}_z] \tag{4.364}$$

is the optimal MAE filter for estimating Y_z based on the observations $\mathbf{X}_z$. If $S_0(z)$ and $S(z)$ are jointly SS stationary, then the estimator $\psi_{op,z}$ is independent of z and we can write $\psi_{op}(\mathbf{X}) = E_b[Y \mid \mathbf{X}]$. The optimal windowed MAE filter Ψ_{op} is translation invariant and is defined by

$$\Psi_{op}(S)(z) = \psi_{op}(\mathbf{X}_z) \tag{4.365}$$

which is expressed componentwise by Eq. 4.54.

***Example* 4.23.** Consider a three-point window centered at the origin for filtering digital signals and suppose the following conditional probabilities, but no others, exceed 0.5: $P(Y = 1|\mathbf{x} = 001)$, $P(Y = 1|\mathbf{x} = 011)$, $P(Y = 1|\mathbf{x} = 100)$, $P(Y = 1|\mathbf{x} = 110)$, $P(Y = 1|\mathbf{x} = 111)$. Then ψ_{op} is defined by the truth table

x	000	001	010	011	100	101	110	111
y	0	1	0	1	1	0	1	1

ψ_{op} has the disjunctive normal form

$$\psi_{op}(x_1, x_2, x_3) = x_1^c x_2^c x_3 + x_1^c x_2 x_3 + x_1 x_2^c x_3^c + x_1 x_2 x_3^c + x_1 x_2 x_3$$

Logic reduction yields

$$\psi_{op}(x_1, x_2, x_3) = x_1^c x_3 + x_1 x_3^c + x_1 x_2 x_3$$

For a second illustration, suppose the following conditional probabilities exceed 0.5: $P(Y = 1|\mathbf{x} = 011)$, $P(Y = 1|\mathbf{x} = 101)$, $P(Y = 1|\mathbf{x} = 110)$, $P(Y = 1|\mathbf{x} = 111)$. Then ψ_{op} is defined by the truth table

x	000	001	010	011	100	101	110	111
y	0	0	0	1	0	1	1	1

ψ_{op} has the disjunctive normal form

$$\psi_{op}(x_1, x_2, x_3) = x_1^c x_2 x_3 + x_1 x_2^c x_3 + x_1 x_2 x_3^c + x_1 x_2 x_3$$

Logic reduction yields

$$\psi_{op}(x_1, x_2, x_3) = x_1x_2 + x_1x_3 + x_2x_3$$

Looking at the table, we see that $\psi_{op}(x_1, x_2, x_3) = 1$ if and only if at least two pixels are 1-valued. Hence, the preceding logical expression for ψ_{op} is the logical expression for a median. Expressed differently, Ψ_{op} is the moving three-point binary median. ■

In practice, filters are designed from pairs of realizations S and S_0 of the observed and ideal images, respectively. A direct approach is to apply Eq. 4.359 using estimates of the probabilities. Conditional probabilities $P(Y = 1 \mid \mathbf{x})$ are estimated by scanning a window across realizations and recording the fraction of times that a co-located pixel in the ideal image realization has value 1 for a given vector $\mathbf{x}$ in the window about that pixel in the observed image. For successful application, image realizations used in the design procedure must agree well with real-world images on which the filter will act. Both ideal- and degraded-image realizations can be simulated according to image and noise models, and probability estimates can be based on these simulations. Design via estimates of all conditional probabilities is limited owing to the number of observation vectors: 2^n for an n-point window. For large windows, only a small fraction of the 2^n vectors will be observed during training (sampling), and many of these will be observed too few times to obtain precise estimates of $P(Y = 1 \mid \mathbf{x})$. Various techniques have been developed to facilitate a satisfactory estimate of the optimal filter.

Once designed, a filter must be implemented either in software or directly in logic hardware. In either case, requiring a large number of product terms to form the filter can be prohibitive. Since ψ_{op} is a logical sum of products, once its disjunctive normal form has been found, logic reduction can be employed to reduce the expansion; nonetheless, owing to practical demands, further reduction may be necessary.

To mitigate the amount of logic, we would like to constrain the number of terms in the logic expansion for ψ_{op}. To analyze the problem, we assume ψ_{op} is expressed in disjunctive normal (nonreduced) form. According to the correspondence between the truth-table and disjunctive-normal-form expansion for ψ_{op}, constraint means creating a new filter ψ_{con} whose truth table is derived from the defining truth table of ψ_{op} by changing some 1-valued rows to 0-valued rows, in the process increasing MAE. For each $\mathbf{x}$ whose row is changed from y-value 1 to y-value 0, MAE increases

$$\delta_{\mathbf{x}} = [P(Y = 1 \mid \mathbf{x}) - P(Y = 0 \mid \mathbf{x})]P(\mathbf{X} = \mathbf{x}) \tag{4.366}$$

$\delta_{\mathbf{x}}$ is called the *restoration effect* of $\mathbf{x}$. The first factor corresponds to the decision to estimate Y as 0, rather than 1, when $\mathbf{x}$ is observed. The second factor is the prior probability of observing $\mathbf{x}$. For any filter ψ, the set of its 1-valued observation vectors is called its *kernel* (or 1-set) and is denoted by

$\mathcal{K}[\psi]$. Under the present constraint, $\mathcal{K}[\psi_{con}] \subset \mathcal{K}[\psi_{op}]$ and the increase in MAE from applying ψ_{con} instead of ψ_{op} is

$$MAE\langle\psi_{con}\rangle - MAE\langle\psi_{op}\rangle = \sum_{\mathbf{x}\in\mathcal{K}[\psi_{op}]-\mathcal{K}[\psi_{con}]} \delta_{\mathbf{x}} \tag{4.367}$$

The best size-constrained approximation to the binary-conditional-expectation filter is achieved by forming a new truth table by changing from 1 to 0 the truth values of those vectors in $\mathcal{K}[\psi_{op}]$ having the smallest restoration effects, forming the corresponding disjunctive normal form, and then applying logic reduction.

The matter can be viewed from a more general perspective by defining the *restoration effect* of any observation $\mathbf{x}$ according to Eq. 4.366. Then $\mathcal{K}[\psi_{op}]$, which is composed of observations for which Eq. 4.359 holds, is equivalently defined as the set of $\mathbf{x}$ for which $\delta_{\mathbf{x}} > 0$, where we only consider $\mathbf{x}$ for which $P(\mathbf{X} = \mathbf{x}) > 0$ (all other observations being removed from consideration). If $\mathcal{K}[\psi]$ is the kernel of any other filter ψ, then the cost (increase in MAE) resulting from applying a nonoptimal filter ψ instead of ψ_{op} is given in terms of the symmetric difference of their kernels by

$$MAE\langle\psi\rangle - MAE\langle\psi_{op}\rangle = \sum_{\mathbf{x}\in\mathcal{K}[\psi_{op}]\Delta\mathcal{K}[\psi]} |\delta_{\mathbf{x}}| \tag{4.368}$$

Since in practice filters are designed from realization statistics, a designed "optimal" filter is likely not optimal and Eq. 4.368 plays a central role in analyzing the loss of optimality owing to realization-based design. Briefly, if ψ_{des} is a filter designed from realization statistics, then the increase in error owing to using ψ_{des} instead of ψ_{op} is given by Eq. 4.368 with ψ_{des} in place of ψ. Full analysis of design error is more complicated because a designed filter is a function of the realizations and therefore the cost of design is a random variable and must be analyzed as such.

For images, an observation vector is of the form $\mathbf{x} = (x_1, x_2,..., x_n)'$ and $\mathbf{x}$ results from scanning across the translated window horizontally row by row. Each observation is usually written as a matrix of 0s and 1s. Moreover, each minterm in the disjunctive normal form of a Boolean function corresponds to a 0-1 matrix (*template*) by placing 0 or 1 in each matrix component corresponding to a complemented or uncomplemented variable, respectively. A reduced expansion is treated similarly, except that for each variable not appearing in a product term we place an $\times$ (*don't care* symbol) in the corresponding component. A template is said to be *canonical* if it consists only of 0s and 1s (and therefore corresponds to a full minterm).

If a Boolean function is defined by a single product and template T corresponds to the product, then we denote the corresponding image operator over the window W by $\Lambda_T(S)$. If ψ is defined by the sum of products of Eq. 4.363 containing m terms, Ψ is the translation-invariant image operator defined by ψ, and $T_1, T_2, \ldots, T_m$ are the templates corresponding to the m products, then

$$\Psi(S) = \bigcup_{i=1}^{m} \Lambda_{T_i}(S) \tag{4.369}$$

because each logical OR in the sum of products corresponds to a union when considering the individual products as image operators.

***Example* 4.24.** Figures 4.11(a) and 4.11(b) show realizations of a text process and the process degraded by edge noise. Two methods are used to design an estimate ψ_{des} of the optimal filter for a 5×5 window. The *direct* approach estimates $P(Y = 1 \mid \mathbf{x})$ and uses the estimate in place of $P(Y = 1 \mid \mathbf{x})$ in Eq. 4.359 to determine $\psi_{des}(\mathbf{x})$. Many vectors will be observed rarely or not at all during training. For these, there will be poor probability estimates, or none. If $\mathbf{x}$ is not observed during training, then $\psi_{des}(\mathbf{x})$ must be determined in some manner. Figure 4.11(c) shows the effect of a directly designed filter on the degraded image of Fig. 4.11(b). Figure 4.11(d) shows the effect of a filter designed using a method called *differencing*. Differencing design takes $\psi_{des}(\mathbf{x})$ to be the observed binary value at the window center unless $P(Y = 1 \mid \mathbf{x})$ has been estimated with sufficient precision during training. Differencing is commonly used in digital document restoration and enhancement, and often provides better filter estimation than the direct approach (although performance comparisons depend on the image model). ■

***Example* 4.25.** Devices comprising a digital document system may operate at different sampling resolutions and therefore spatial resampling must often be performed to allow porting of a digital image among the various devices. If a document has been captured at 300 spi (spots per inch) and is to be printed on a 600-spi printer, then resolution conversion forms an estimate of the underlying image spatially sampled at 600 spi from its observed 300 spi sampling. Resolution conversion is characterized by the ratio of the input and output sampling resolutions. A conversion ratio is defined in terms of the pagewise horizontal and vertical sampling resolutions, u and v (in spi), of the input image and the corresponding sampling resolutions, r and s, of the output image. This example is concerned with integer conversion to a higher resolution, meaning $r/u > 1$, $s/v > 1$, and both r/u and s/v are integer ratios. Let $A_{u,v}$ and $A_{r,s}$ be the low-resolution input and high-resolution output images, respectively. If each pixel i in $A_{u,v}$ is mapped into L pixels, then each pixel value $A_{u,v}(i)$ corresponds to a vector defining the ith block of $A_{r,s}$,

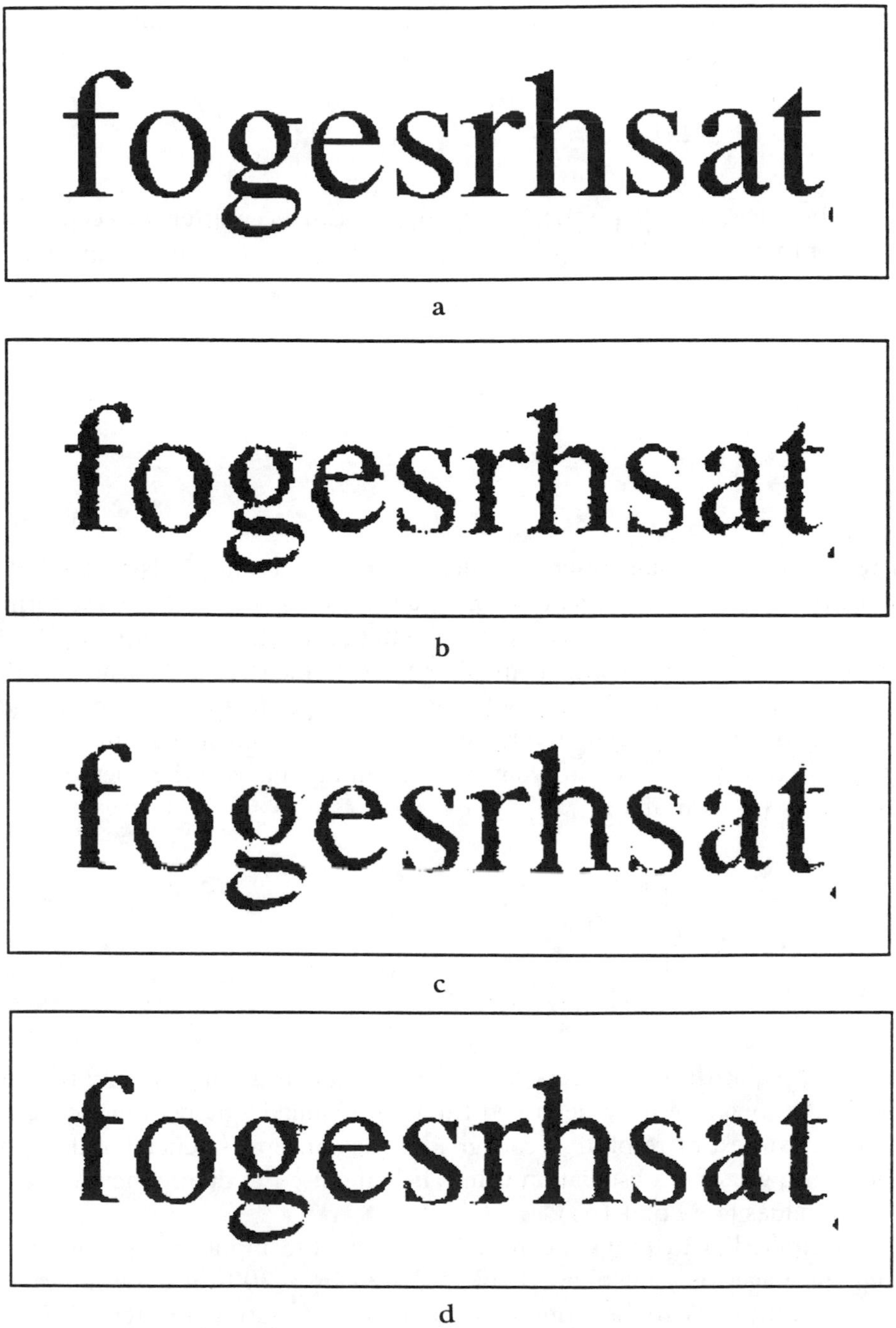

fogesrhsat

a

fogesrhsat

b

fogesrhsat

c

fogesrhsat

d

Figure 4.11 Restoration of text process: (a) realization of ideal process; (b) realization degraded by edge noise; (c) restoration by directly designed filter; (d) restoration by filter designed by differencing method.

$$\mathbf{A}_{r,s}[i] = \begin{pmatrix} A_{r,s}(i,1) \\ A_{r,s}(i,2) \\ \vdots \\ A_{r,s}(i,L) \end{pmatrix}$$

Assuming there are N pixels forming $A_{u,v}$, it can be written in vector form with components $A_{u,v}(1)$, $A_{u,v}(2),\ldots, A_{u,v}(N)$. $A_{r,s}$ can be written in block form as

$$A_{r,s} = \begin{pmatrix} \mathbf{A}_{r,s}[1] \\ \mathbf{A}_{r,s}[2] \\ \vdots \\ \mathbf{A}_{r,s}[N] \end{pmatrix}$$

A resolution-conversion filter $\mathbf{\Psi}$ is defined by observing pixels in a window about pixel i in $A_{u,v}$ to estimate the L values forming $\mathbf{A}_{r,s}[i]$. $\mathbf{\Psi}$ has the form $\mathbf{\Psi} = (\Psi_1, \Psi_2,\ldots, \Psi_L)'$, where $\Psi_j(A_{u,v})(i)$ estimates $A_{r,s}(i, j)$. Blockwise, $\mathbf{\Psi}(A_{u,v})(i)$ estimates $\mathbf{A}_{r,s}[i]$. Each component filter Ψ_j is defined via a Boolean function ψ_j defined over an n-pixel window W. $\Psi_j(A_{u,v})(i)$ is evaluated by translating W to i and computing ψ_j using the binary values in the translated window. The structure can be viewed differently by defining the vector-valued Boolean function $\boldsymbol{\psi}$ on n variables $x_1, x_2,\ldots, x_n$ by

$$\boldsymbol{\psi}(x_1, x_2,\ldots, x_n) = \begin{pmatrix} \psi_1(x_1, x_2,\ldots, x_n) \\ \psi_2(x_1, x_2,\ldots, x_n) \\ \vdots \\ \psi_L(x_1, x_2,\ldots, x_n) \end{pmatrix}$$

$\boldsymbol{\psi}$ is an L-valued function on n variables and serves as the window function for $\mathbf{\Psi}$. Although we have described the vector-valued window function $\boldsymbol{\psi}$ for $\mathbf{\Psi}$ in the context of resolution conversion, the concept is general and applies wherever a vector of observation values is being used to determine a vector of output values (see Eq. 4.163).

Consider 300 to 600-spi conversion, where the input image class arises from the pages of a telephone book digitized by a 300-spi desktop scanner and the output is to be printed on a higher resolution printer. For this conversion, $L = 4$. The window is 5×5 and $\boldsymbol{\psi}$ is estimated from realizations by the differencing method. An estimate of $\mathbf{\Psi}$ has been applied to the low-resolution realization of Fig. 4.12(a) to obtain the high-resolution realization of Fig. 4.12(b). For the designed filter, the canonical expansions of ψ_1, ψ_2, ψ_3, and ψ_4 have 1914, 1694, 1659, and 1941 templates, respectively. ■

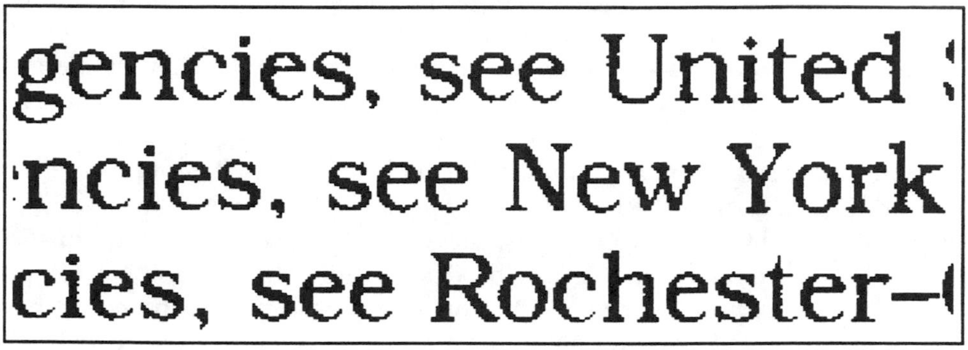

gencies, see United

ncies, see New York

cies, see Rochester–

a

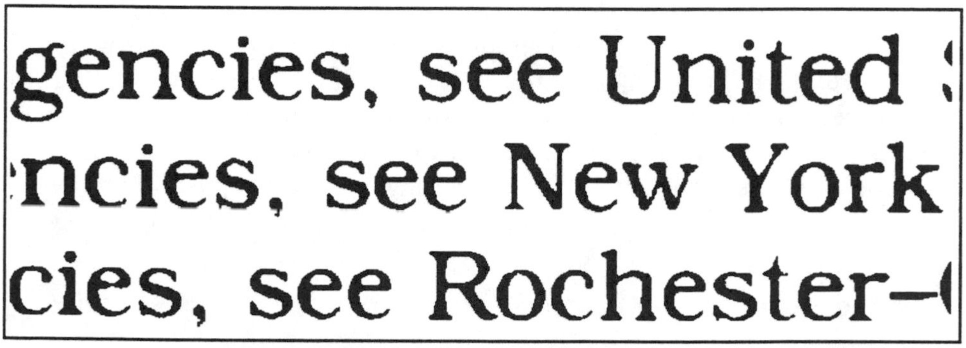

gencies, see United

ncies, see New York

cies, see Rochester–

b

Figure 4.12 Resolution conversion: (a) low-resolution realization; (b) high-resolution realization.

4.10.3. Optimal Increasing Filters

A Boolean function ψ is *increasing* if $\mathbf{x} \le \mathbf{y}$ implies $\psi(\mathbf{x}) \le \psi(\mathbf{y})$, where, by definition,

$$\mathbf{x} = (x_1, x_2, \ldots, x_n)' \le \mathbf{y} = (y_1, y_2, \ldots, y_n)' \tag{4.370}$$

if and only if $x_i \leq y_i$ for $i = 1, 2,.., n$. ψ is increasing if and only if it possesses a *positive* (complement-free) sum-of-products expansion, meaning that ψ can be expressed as

$$\psi(x_1, x_2,..., x_n) = \sum_i x_{i,1} x_{i,2} \cdots x_{i,n(i)} \tag{4.371}$$

A positive (increasing) Boolean function must have a positive representation, but it also has nonpositive representations, including its disjunctive normal form. Unless otherwise stated, we assume that a positive Boolean function is represented by a positive expansion. For the second filter of Example 4.23, logic reduction yielded an expansion having no complemented variables.

If a set of variables in any product of a positive expansion contains the set of variables in a distinct product, then, whenever the former product has a value of 1, so does the latter. Thus, inclusion of the former product in the expansion is redundant and it can be deleted from the expansion without changing the function defined by the expansion. No product whose variable set does not contain the variable set of a distinct product can be deleted without changing the function. Performing the permitted deletions produces a unique *minimal representation* of the positive Boolean function.

From a set perspective, if S is a binary digital image, then pixel $z \in S$ if and only if $S(z) = 1$, where $S(z)$ denotes the logical value of S at z, and $z \notin S$ if and only if $S(z) = 0$. A binary filter Ψ is *increasing* if and only if $S_1 \subset S_2$ implies $\Psi(S_1) \subset \Psi(S_2)$. If Ψ is defined by the Boolean function ψ defined over window W, then Ψ is increasing if and only if ψ is positive.

For $B = \{b_1, b_2,..., b_r\} \subset W$, define the single-product Boolean function ε_B by

$$\varepsilon_B(x_1, x_2,..., x_n) = x_{B,1} x_{B,2} \cdots x_{B,r} \tag{4.372}$$

where the variables $x_{B,1}, x_{B,2},..., x_{B,r}$ correspond to $b_1, b_2,..., b_r$. Let E_B be the image operator defined by ε_B. $\mathrm{E}_B(S)(z) = 1$ $[z \in \mathrm{E}_B(S)]$ if and only if $S(z + b_i) = 1$ $[z + b_i \in S]$ for $i = 1, 2,..., r$. Relative to B, $z \in \mathrm{E}_B(S)$ if and only if $B_z \subset S$. E_B and ε_B are both called *erosion*, B is called a *structuring element*, and E_B and ε_B are related according to Eq. 4.365.

More generally, suppose the Boolean function ψ corresponding to Ψ is defined via the positive sum of products of Eq. 4.371. Then Ψ can be expressed as

$$\Psi(S) = \bigcup_{i=1}^{m} \mathrm{E}_{B_i}(S) \tag{4.373}$$

where B_i corresponds to the ith product in the logical expansion and $B_i \subset W$ for all i. Because each positive Boolean expansion has a minimal form, so too does any union of erosions formed from structuring elements in a finite window. These structuring elements correspond to minimal products and compose the *basis*, $\mathcal{B}[\Psi]$, of Ψ. This basis is unique and we can always assume the expansion giving Ψ is taken over its basis.

There is a relation between the union expansions of Eqs. 4.369 and 4.373. If T is a (not necessarily canonical) template defining the image operator Λ_T, and B and C are the subsets of the window W corresponding to the 1-valued and 0-valued elements of T, respectively, then

$$\Lambda_T(S) = \mathrm{E}_B(S) \cap \mathrm{E}_C(S^c) \tag{4.374}$$

meaning that $\Lambda_T(S)(z) = 1$ if and only if $B_z \subset S$ and $C_z \subset S^c$. In this context, Λ_T is known as the *hit-or-miss transform*. Using the erosion representation of Λ_T yields an equivalent form of the general representation of Eq. 4.369, namely,

$$\Psi(S) = \bigcup_{i=1}^{m} \mathrm{E}_{B_i}(S) \cap \mathrm{E}_{C_i}(S^c) \tag{4.375}$$

which reduces to the erosion expansion of Eq. 4.373 when there are no 0-valued elements in any of the templates (meaning the corresponding sum of products is complement-free).

If for a particular image-degradation model, the optimal binary filter is increasing, then reduction of the logical expansion for the Boolean function will yield a positive expansion. From a purely probabilistic standpoint, derivation of an increasing optimal filter via the conditional expectation and logical reduction is a valid approach; however, practical matters can be quite different. If the image processes are such that the optimal filter is increasing, then it may be beneficial to take a direct approach to designing an increasing filter. Indeed, there are a number of reasons why it might be beneficial to apply the best increasing filter even when it is not fully optimal. Minimal positive representations can significantly reduce the logic cost in comparison to nonpositive representations. Also, one often needs to employ only a fraction of the optimal-basis elements to achieve satisfactory performance. More relevant to our concerns in the present context is that there can be great statistical advantage in directly designing an increasing filter. A filter (nonlinear or linear) designed from realization-based statistics is an estimate of the actual optimal filter. The precision in estimating a nonincreasing filter from realization-based statistics is dependent on the precision of estimates of the restoration effects. Goodness of the designed filter (as opposed to the theoretically best filter) depends on these estimates, and a very large number of realizations may be required to obtain good estimates. Typically, many fewer realizations are required to obtain good estimates of optimal increasing filters than are required for nonincreasing filters. We now discuss the design of increasing filters.

Consider a single-erosion filter. At each pixel the observed image values in the translated window comprise a binary random vector $\mathbf{X} = (X_1, X_2,\ldots, X_n)'$ and, as an operator on random vectors, erosion takes the form

$$\varepsilon_B(\mathbf{X}) = \min \{X_i\colon b_i = 1\} \tag{4.376}$$

where $B = \{b_1, b_2,\ldots, b_r\} \subset W$. Filter MAE, denoted by $MAE\langle B\rangle$, is given by

$$\begin{aligned} MAE\langle B\rangle &= E[|Y - \varepsilon_B(\mathbf{X})|] \\ &= P(Y \neq \varepsilon_B(\mathbf{X})) \\ &= \sum_{\{(\mathbf{x},y):y\neq\varepsilon_B(\mathbf{x})\}} f(x_1, x_2, \ldots, x_n, y) \end{aligned} \tag{4.377}$$

where Y is the value of the ideal image at the pixel and $f(x_1, x_2,\ldots, x_n, y)$ is the joint density of $X_1, X_2,\ldots, X_n, Y$. Owing to SS stationarity, reference to the pixel has been suppressed.

In practice, $MAE\langle B\rangle$ is estimated from realizations of the ideal- and observed-image processes. In the case of restoration, we take a realization of the uncorrupted process and transform it by the degradation transformation to produce a realization of the degraded process. The degraded realization is eroded pixelwise and compared to the ideal-image realization. An estimate of $MAE\langle B\rangle$ is obtained by dividing the number of pixels at which the eroded observation disagrees with the ideal by the total number of pixels considered. Estimation precision increases with the number of pixels observed.

For a general m-erosion filter Ψ defined by Eq. 4.373 with basis $\{B_1, B_2,\ldots, B_m\}$, the filter error is given by

$$MAE\langle\Psi\rangle = \sum_{\{(\mathbf{x},y):y\neq\max_i \varepsilon_{B_i}(\mathbf{x})\}} f(x_1, x_2, \ldots, x_n, y) \tag{4.378}$$

The optimal filter can be estimated by estimating $MAE\langle\Psi\rangle$ for all possible bases and choosing Ψ_{op} as the filter corresponding to the basis having minimal MAE. If the window has n pixels, then it has 2^n subsets, but many of these are eliminated from consideration owing to the minimality condition for a basis. Nonetheless, except for relatively small windows, constraints must be imposed, thereby (hopefully slightly) increasing MAE. Constraints include limiting the basis size and also constraining the search to a subclass of all possible structuring elements.

As expressed in Eq. 4.378, it would appear that filter design must include obtaining realization-based statistics for every basis, a prohibitive task. In fact, the following theorem shows that one need only obtain MAE estimates for single-erosion filters and then recursively obtain MAE estimates for multiple-erosion filters.

Theorem 4.18. The MAE of an m-erosion filter Ψ_m can be expressed in terms of a single-erosion filter with structuring element B_m and two $(m-1)$-erosion filters Ψ_{m-1} and Φ_{m-1}:

$$MAE\langle\Psi_m\rangle = MAE\langle\Psi_{m-1}\rangle - MAE\langle\Phi_{m-1}\rangle + MAE\langle B_m\rangle \tag{4.379}$$

where the relevant filter bases are given by

$$\mathcal{B}[\Psi_{m-1}] = \{B_1, B_2,\ldots, B_{m-1}\}$$

$$\mathcal{B}[\Psi_m] = \mathcal{B}[\Psi_{m-1}] \cup \{B_m\} = \{B_1, B_2,\ldots, B_m\} \tag{4.380}$$

$$\mathcal{B}[\Phi_{m-1}] = \{B_1 \cup B_m, B_2 \cup B_m,\ldots, B_{m-1} \cup B_m\}$$ ■

The recursive MAE representation of Theorem 4.18 is a corollary of a direct error representation which we now derive. Letting S_0 and S be the ideal and observed images, respectively, for any structuring element B, the MAE of erosion defined by B can be expressed in terms of the symmetric difference of the events $[B_z \subset S]$ and $[z \in S_0]$, namely,

$$MAE\langle B\rangle = P([B_z \subset S] \,\Delta\, [z \in S_0]) \tag{4.381}$$

where z is an arbitrary point in S. More generally,

$$MAE\langle\Psi_m\rangle = P\left(\left(\bigcup_{k=1}^{m}[(B_k)_z \subset S]\right) \Delta\, [z \in S_0]\right) \tag{4.382}$$

Letting $C_k = [(B_k)_z \subset S]$ and $D = [z \in S_0]$,

$$
\begin{aligned}
MAE\langle\Psi_m\rangle &= P\left(\bigcup_{k=1}^{m} C_k\right) + P(D) - 2P\left(\left(\bigcup_{k=1}^{m} C_k\right) \cap D\right) \\
&= P\left(\bigcup_{k=1}^{m} C_k\right) + P(D) - 2P\left(\bigcup_{k=1}^{m} (C_k \cap D)\right) \\
&= P(D) + \sum_{j=1}^{m} (-1)^{j+1} \sum_{1 \le i_1 < i_2 < \cdots < i_j \le m} \left[P\left(\bigcap_{k=1}^{j} C_{i_k}\right) - 2P\left(\bigcap_{k=1}^{j} (C_{i_k} \cap D)\right)\right] \\
&= \sum_{j=1}^{m} (-1)^{j+1} \sum_{1 \le i_1 < i_2 < \cdots < i_j \le m} \left[P\left(\bigcap_{k=1}^{j} C_{i_k}\right) + P(D) - 2P\left(\bigcap_{k=1}^{j} (C_{i_k} \cap D)\right)\right] \\
&= \sum_{j=1}^{m} (-1)^{j+1} \sum_{1 \le i_1 < i_2 < \cdots < i_j \le m} P\left(\left(\bigcap_{k=1}^{j} C_{i_k}\right) \Delta D\right) \qquad (4.383)
\end{aligned}
$$

where, in turn, the equalities result from the probability of a symmetric difference, distribution of intersection, the probability addition theorem (Theorem 1.1), the identity

$$
\sum_{j=1}^{m} (-1)^{j+1} \binom{m}{j} = 1 \qquad (4.384)
$$

and again the probability of a symmetric difference. Substituting the event equivalence

$$
\bigcap_{k=1}^{j} [(B_{i_k})_z \subset S] = \left[\left(\bigcup_{k=1}^{j} B_{i_k}\right)_z \subset S\right] \qquad (4.385)
$$

into Eq. 4.383 in place of the intersection of the C_k yields

$$
MAE\langle\Psi_m\rangle = \sum_{j=1}^{m} (-1)^{j+1} \sum_{1 \le i_1 < i_2 < \cdots < i_j \le m} P\left(\left[\left(\bigcup_{k=1}^{j} B_{i_k}\right)_z \subset S\right] \Delta \left[z \in S_0\right]\right) \qquad (4.386)
$$

which upon application of Eq. 4.381 yields the direct error representation

$$MAE\langle\Psi_m\rangle = \sum_{j=1}^{m}(-1)^{j+1} \sum_{1\le i_1<i_2<\cdots<i_j\le m} MAE\left\langle \bigcup_{k=1}^{j} B_{i_k} \right\rangle \tag{4.387}$$

The recursive form given in Theorem 4.18 results from rewriting Eq. 4.387 as

$$MAE\langle\Psi_m\rangle = MAE\langle B_m\rangle + \sum_{j=1}^{m-1}(-1)^{j+1} \sum_{1\le i_1<i_2<\cdots<i_j\le m-1} MAE\left\langle \bigcup_{k=1}^{j} B_{i_k} \right\rangle$$

$$- \sum_{j=1}^{m-1}(-1)^{j+1} \sum_{1\le i_1<i_2<\cdots<i_j\le m-1} MAE\left\langle \left(\bigcup_{k=1}^{j} B_{i_k}\right) \cup B_m \right\rangle \tag{4.388}$$

Relative to Eq. 4.380, the three summands correspond to the MAEs for erosion by B_m, Ψ_{m-1}, and Φ_{m-1}, respectively.

To illustrate the theorem, we write it out for three erosions. Direct representation is

$$MAE\langle B_1, B_2, B_3\rangle = MAE\langle B_1\rangle + MAE\langle B_2\rangle + MAE\langle B_3\rangle$$

$$- MAE\langle B_1 \cup B_2\rangle - MAE\langle B_1 \cup B_3\rangle - MAE\langle B_2 \cup B_3\rangle$$

$$+ MAE\langle B_1 \cup B_2 \cup B_3\rangle \tag{4.389}$$

In recursive form,

$$MAE\langle B_1, B_2, B_3\rangle = MAE\langle B_1, B_2\rangle + MAE\langle B_3\rangle - MAE\langle B_1 \cup B_3, B_2 \cup B_3\rangle \tag{4.390}$$

***Example* 4.26.** A filter Ψ is *antiextensive* if, for any image S, $\Psi(S) \subset S$; it is *extensive* if, for any S, $S \subset \Psi(S)$. Every binary-to-binary image filter Ψ possesses a decomposition

$$\Psi(S) = (\Psi^{\cup}(S) - S) \cup \Psi^{\cap}(S)$$

where $\Psi^{\cup}(S) = \Psi(S) \cup S$ and $\Psi^{\cap}(S) = \Psi(S) \cap S$. Since $\Psi^{\cup}$ and $\Psi^{\cap}$ are extensive and antiextensive, respectively, Ψ is decomposed into extensive and antiextensive parts. Often an optimal filter (not necessarily increasing) is well approximated by an optimal increasing filter when the optimal filter is extensive or antiextensive. The decomposition is useful because the extensive and antiextensive parts can be designed separately, each being approximated by an optimal increasing filter. To illustrate *paired-representation* resolution conversion, we consider a 300 to 600-spi conversion using the window

$$W = \begin{pmatrix} 0 & 0 & 1 & 0 & 0 \\ 0 & 1 & 1 & 1 & 0 \\ 1 & 1 & 1 & 1 & 1 \\ 0 & 1 & 1 & 1 & 0 \\ 0 & 0 & 1 & 0 & 0 \end{pmatrix}$$

Increasing filters have been designed using Theorem 4.18 to approximate the extensive and antiextensive parts of the optimal filter. Figures 4.13(a) through 4.13(d) show a 600-spi image, the original image at 300 spi, the paired-representation-resolution-conversion output of the 300-spi image, and the extensive and antiextensive filter bases for each of the four filters (phases) composing ψ. ■

4.11. Pattern Classification

Pattern classification involves a decision rule to assign an observed pattern to one of a number of classes, $C_0, C_1,..., C_m$, where the term "pattern" is used generically, so that it might be an image, subimage, signal, etc. $\mathfrak{R}^n$ is partitioned into $m + 1$ regions $R_0, R_1,..., R_m$ associated with $C_0, C_1,..., C_m$, respectively, an n-dimensional feature vector $\mathbf{x}$ is associated with each observed pattern, and the classification decision is made according to which of $m + 1$ regions $\mathbf{x}$ belongs.

4.11.1. Optimal Classifiers

The theory of optimal MAE binary filters depends on reduction of the conditional expectation to discrete binary representation, specifically, the binary quantization of $E[Y|\mathbf{X}] = P(Y = 1|\mathbf{X})$. The binary theory extends to the case where $\mathbf{X}$ is a continuous random vector and, if we view Y as a classification variable, $Y = 0$ or $Y = 1$ according to whether some observation is to be placed into class C_0 or C_1, then the optimal binary filter can be interpreted in terms of optimal classification.

If a binary random variable Y is to be estimated based on observation of a continuous random vector $\mathbf{X} = (X_1, X_2,..., X_n)'$, then the conditional expectation of Eq. 4.20 reduces to Eq. 4.23 just as in the discrete case. The difference is that, for continuous observations, $P(Y = 1| x_1, x_2,..., x_n)$ is defined by a quotient of densities (Eq. 4.21) rather than a quotient of discrete probabilities, so that the conditional expectation is a function of a continuous random vector. The optimal MAE binary estimator, $\xi_{op}(\mathbf{X})$, for Y in terms of $\mathbf{X}$ is the binary (quantized) conditional expectation defined in

basrc

stote

a

basrc

stote

b

basrc

stote

c

Figure 4.13 Paired-representation resolution conversion: (a) 600 spi realization; (b) original 300 spi realization; (c) filter output;

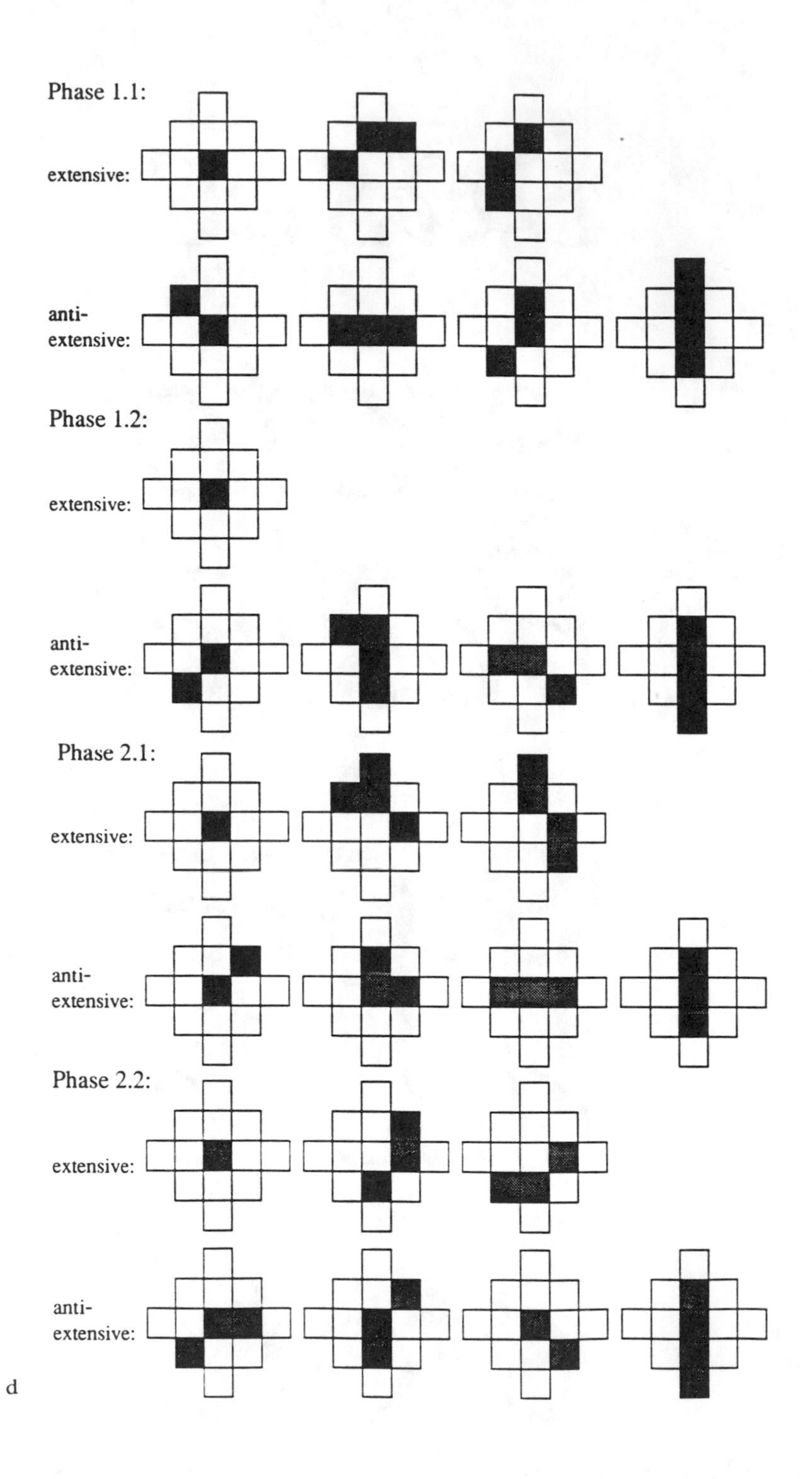

Figure 4.13 (cont): (d) bases for four filter phases.

Eq. 4.356. Its error is given by the density form of Eq. 4.361,

$$MAE\langle \xi_{op} \rangle = \int_{R_1} f(x_1, x_2, \ldots, x_n, 0)\, dx_1 dx_2 \cdots dx_n + \int_{R_0} f(x_1, x_2, \ldots, x_n, 1)\, dx_1 dx_2 \cdots dx_n \tag{4.391}$$

where

$$R_1 = \{\mathbf{x}\text{: } P(Y = 1|\mathbf{x}) > 0.5\} \tag{4.392}$$

$$R_0 = \{\mathbf{x}\text{: } P(Y = 1|\mathbf{x}) \leq 0.5\}$$

From the perspective of pattern classification, Y represents the occurrence of a pattern in class C_0 ($Y = 0$) or class C_1 ($Y = 1$). A classifier ξ classifies a pattern based on a related observation $\mathbf{x}$, which may be the readings over some pixel set or, more often, a vector of numerical features computed from the observed pattern, for instance, Fourier descriptors. The misclassification probability of ξ is given by

$$M\langle \xi \rangle = P(\xi(\mathbf{X}) = 1,\ Y = 0) + P(\xi(\mathbf{X}) = 0,\ Y = 1) \tag{4.393}$$

which equals $MAE\langle \xi \rangle$. The optimal binary filter ξ_{op} minimizes the misclassification probability and is therefore the optimal classifier. $M\langle \xi_{op} \rangle$ is given by Eq. 4.391. An observation $\mathbf{x}$ is classified by ξ_{op} into C_0 or C_1 according to the conditional-density relations $f(0 \mid \mathbf{x}) \geq f(1 \mid \mathbf{x})$ or $f(1 \mid \mathbf{x}) > f(0 \mid \mathbf{x})$, respectively.

To generalize the analysis to $m + 1$ pattern classes $C_0, C_1, \ldots, C_m$, assume that the random variable Y has the range $\{0, 1, \ldots, m\}$. For classifier ξ, the probability of misclassification is

$$\begin{aligned} M\langle \xi \rangle &= \sum_{k=0}^{m} P(Y = k, \xi(\mathbf{X}) \neq k) \\ &= 1 - \sum_{k=0}^{m} P(Y = k, \xi(\mathbf{X}) = k) \\ &= 1 - \sum_{k=0}^{m} \int_{R_k} f(x_1, x_2 \ldots, x_n, k)\, dx_1 dx_2 \cdots dx_n \end{aligned} \tag{4.394}$$

where $R_k = \{\mathbf{x}\text{: } \xi(\mathbf{x}) = k\}$. $M\langle \xi \rangle$ is minimized by choosing ξ to maximize the latter sum and this is accomplished if

$$R_k = \{\mathbf{x}: f(\mathbf{x}, k) \geq f(\mathbf{x}, j) \text{ for } j = 0, 1,\ldots, m\} \tag{4.395}$$

Hence, an optimal classifier is defined by $\xi_{op}(\mathbf{x}) = k$ if $f(\mathbf{x}, k) \geq f(\mathbf{x}, j)$ for $j = 0, 1,\ldots, m$. Relative to conditional probabilities, $\xi_{op}(\mathbf{x}) = k$ if $f(k \mid \mathbf{x}) \geq f(j \mid \mathbf{x})$ for $j = 0, 1,\ldots, m$ because

$$f(j \mid \mathbf{x}) = \frac{f(\mathbf{x}, j)}{f(\mathbf{x})} \tag{4.396}$$

[assuming $f(\mathbf{x}) \neq 0$]. Vectors for which $\xi_{op}(\mathbf{x})$ is optionally defined form *decision boundaries* in $\Re^n$. $M\langle\xi_{op}\rangle$ is given by Eq. 4.394 using R_k in Eq. 4.395.

It is common to view classification via *discriminant* functions: there are $m + 1$ functions $d_0, d_1,\ldots, d_m$ and $\mathbf{x}$ is classified into C_k if $d_k(\mathbf{x}) \geq d_j(\mathbf{x})$ for $j = 0, 1,\ldots, m$. If we let

$$f(\mathbf{x}, k) = f(\mathbf{x} \mid k) f(k) \tag{4.397}$$

be the discriminant, then misclassification error is minimized. Since the order relation among discriminant functions is unaffected by any strictly increasing function, taking the logarithm yields the equivalent optimal discriminant

$$d_k(\mathbf{x}) = \log f(\mathbf{x} \mid k) + \log f(k) \tag{4.398}$$

4.11.2. Gaussian Maximum-Likelihood Classification

Now assume that the conditional densities $f(\mathbf{x} \mid 0), f(\mathbf{x} \mid 1),\ldots, f(\mathbf{x} \mid m)$ are normal. For $k = 0, 1,\ldots, m$, $f(\mathbf{x} \mid k)$ is of the form given in Eq. 1.190, where $\mathbf{K}$ and $\boldsymbol{\mu}$, the covariance matrix and mean vector, are replaced by $\mathbf{K}_k$ and $\mathbf{u}_k$, respectively. Dropping the constant terms and multiplying by the factor 2 (which has no effect on discrimination), the discriminant of Eq. 4.398 becomes

$$d_k(\mathbf{x}) = -(\mathbf{x} - \mathbf{u}_k)'\mathbf{K}_k^{-1}(\mathbf{x} - \mathbf{u}_k) - \log(\det[\mathbf{K}_k]) + 2\log f(k) \tag{4.399}$$

The form of Eq. 4.399 shows that decision boundaries $d_k(\mathbf{x}) = d_j(\mathbf{x})$ are quadratic. The corresponding classifier is called the *Gaussian maximum-likelihood classifier* because of the Gaussian assumption and the fact that it is based on maximizing the likelihood of pattern C_k given observation $\mathbf{x}$.

If all conditional densities possess the same covariance matrix $\mathbf{K}$, then $\log(\det[\mathbf{K}_k])$ can be dropped from $d_k(\mathbf{x})$ and, since the symmetry of $\mathbf{K}^{-1}$ implies $\mathbf{x}'\,\mathbf{K}^{-1}\mathbf{u}_k = \mathbf{u}_k'\,\mathbf{K}^{-1}\mathbf{x}$,

$$d_k(\mathbf{x}) = -(\mathbf{x} - \mathbf{u}_k)'\mathbf{K}^{-1}(\mathbf{x} - \mathbf{u}_k) + 2\log f(k)$$

$$= -(\mathbf{x} - \mathbf{u}_k)'\mathbf{K}^{-1}\mathbf{x} + (\mathbf{x} - \mathbf{u}_k)'\mathbf{K}^{-1}\mathbf{u}_k + 2\log f(k) \tag{4.400}$$

$$= -\mathbf{x}'\mathbf{K}^{-1}\mathbf{x} + 2\mathbf{u}_k'\mathbf{K}^{-1}\mathbf{x} - \mathbf{u}_k'\mathbf{K}^{-1}\mathbf{u}_k + 2\log f(k)$$

The first summand can be dropped because it is independent of k. Thus, we arrive at the equivalent discriminant

$$d_k(\mathbf{x}) = 2\mathbf{u}_k'\mathbf{K}^{-1}\mathbf{x} - \mathbf{u}_k'\mathbf{K}^{-1}\mathbf{u}_k + 2\log f(k) \tag{4.401}$$

which is a linear function of $\mathbf{x}$ and produces decision boundaries that are hyperplanes.

If, besides a common covariance matrix, we assume that the conditional random vectors are uncorrelated with covariance matrix $\mathbf{K} = \sigma^2\mathbf{I}$, then Eq. 4.400 reduces to

$$d_k(\mathbf{x}) = -\sigma^{-2}(\mathbf{x} - \mathbf{u}_k)'(\mathbf{x} - \mathbf{u}_k) + 2\log f(k)$$

$$= -\sigma^{-2}\|\mathbf{x} - \mathbf{u}_k\|^2 + 2\log f(k) \tag{4.402}$$

If we go one step further and assume that the classes possess equal prior probabilities $f(k)$, then the logarithm term and the square can be dropped to yield the discriminant

$$d_k(\mathbf{x}) = -\sigma^{-2}\|\mathbf{x} - \mathbf{u}_k\| \tag{4.403}$$

In this simplified situation, $\mathbf{x}$ is assigned to class C_k if its distance to the mean vector for C_k is minimal. Equations 4.402 and 4.403 are convenient because they do not require a theoretical model or estimation of the covariance matrix; however, uncorrelatedness is a very restrictive assumption, and equal class probabilities are even more so.

Rather than view Eq. 4.399 as a discriminant to be maximized, we can view its negation as the distance from $\mathbf{x}$ to C_k, which is the viewpoint mentioned in Section 3.5.2. Equation 4.399 is then rewritten as a distance function

$$D(\mathbf{x}, C_k) = (\mathbf{x} - \mathbf{u}_k)'\mathbf{K}_k^{-1}(\mathbf{x} - \mathbf{u}_k) + \log\frac{\det[\mathbf{K}_k]}{f(k)^2} \tag{4.404}$$

The observed vector **x** is classified into C_k if $D(\mathbf{x}, C_k) \leq D(\mathbf{x}, C_j)$ for $j = 0, 1,..., m$.

The observed vector is random when viewed across the entire pattern collection and each pattern class C_k is characterized by the distribution of the random vector $\mathbf{X}_k$ derived from the patterns in C_k. $\mathbf{X}_k$ has density $f(\mathbf{x} \mid k)$ or, as sometimes loosely written, $f(\mathbf{x} \mid C_k)$. Each vector component is called a *feature*; $\mathbf{X}_k$ is called a *feature vector*. We write $D(\mathbf{x}, \mathbf{X}_k)$ in place of $D(\mathbf{x}, C_k)$, and $D(\mathbf{x}, \mathbf{X}_k)$ provides a distance measurement between the observed realization and the distribution of $\mathbf{X}_k$.

An observation **x** is an estimate of the mean of the random vector **X** for which it is a realization. Hence $\mathbf{x} - \mathbf{u}_k$ in Eq. 4.404 estimates the difference between the means of **X** and $\mathbf{X}_k$. When all feature vectors possess a common covariance matrix and all features are uncorrelated with common variance σ^2, then the determinant is dropped and Eq. 4.404 becomes

$$D(\mathbf{x}, \mathbf{X}_k) = \sigma^{-2} \|\mathbf{x} - \mathbf{u}_k\|^2 + \frac{1}{\log f(k)^2} \tag{4.405}$$

If σ is large, then the prior probabilities play a significant role; if σ is small, then the distance between **x** and $\mathbf{u}_k$ is dominant. Note also the effects of small and large class probabilities.

Gaussian maximum-likelihood classification is relatively robust with respect to the assumption of normally distributed conditional densities; however, successful application of the classifier depends on having reasonably good estimates of the covariance matrices.

4.11.3. Linear Discriminants

Even if the conditional densities are not Gaussian or the feature vectors do not possess a common covariance matrix, one can still desire a linear classifier. In this case, there are $m + 1$ discriminant functions of the form

$$d_k(\mathbf{x}) = \sum_{i=0}^{n} a_{ki} x_i \tag{4.406}$$

where $\mathbf{x} = (x_0, x_1,..., x_n)'$ and nonhomogeneity results from letting $x_0 = 1$. Letting

$$\mathbf{A} = \begin{pmatrix} a_{00} & a_{01} & \cdots & a_{0n} \\ a_{10} & a_{11} & \cdots & a_{1n} \\ \vdots & \vdots & \ddots & \vdots \\ a_{m0} & a_{m1} & \cdots & a_{mn} \end{pmatrix} \tag{4.407}$$

and $\mathbf{d}(\mathbf{x}) = (d_0(\mathbf{x}), d_1(\mathbf{x}),..., d_m(\mathbf{x}))'$, the discriminant functions can be expressed vectorially by

$$\mathbf{d}(\mathbf{x}) = \mathbf{A}\mathbf{x} \tag{4.408}$$

An observed feature vector $\mathbf{x}$ is assigned to class C_k if $d_k(\mathbf{x})$ is maximum among discriminant values. For linear discriminant functions, the maximum criterion yields convex decision regions having hyperplane boundaries. We would like to find a matrix $\mathbf{A}$ to minimize misclassification error among all possible weight matrices; however, except in special circumstances, such as Gaussian maximum-likelihood with a common covariance matrix, this task is mathematically prohibitive. Even when theoretically possible, it requires a good estimate of the covariance matrix.

Another way to proceed is to define an error function based on target values for the discriminant functions and then select $\mathbf{A}$ to minimize this target error. Suppose $r + 1$ vectors $\mathbf{x}_0, \mathbf{x}_1,..., \mathbf{x}_r$ are observed and it is known to which class each should be assigned. For any observed vector $\mathbf{x}_j$ and discriminant d_k, let b_{kj} be a target value for $d_k(\mathbf{x}_j)$. Since we know the class to which a particular vector $\mathbf{x}_j$ belongs, say C_q, the target values should be chosen so that b_{kj} is maximum for $k = q$. Then, should $d_k(\mathbf{x}_j)$ equal its target value for all k, $\mathbf{x}_j$ will be properly placed into C_q.

Having chosen target values, weights need to be determined so that discriminant values are close to their target values. To this end, define a sum-of-squares error by

$$\begin{aligned} SSE &= \sum_{k=0}^{m}\sum_{j=0}^{r} |d_k(\mathbf{x}_j) - b_{kj}|^2 \\ &= \sum_{k=0}^{m}\sum_{j=0}^{r}\left(\sum_{i=0}^{n} a_{ki}x_{ji} - b_{kj}\right)^2 \end{aligned} \tag{4.409}$$

where $\mathbf{x}_j = (x_{j0}, x_{j1},..., x_{jn})'$. To minimize SSE with respect to the weights, take partial derivatives with respect to a_{ki} and set $\partial SSE/\partial a_{ki} = 0$ to obtain the equations

$$\sum_{j=0}^{r}\left(\sum_{l=0}^{n} a_{kl}x_{jl} - b_{kj}\right)x_{ji} = 0 \tag{4.410}$$

for $k = 0, 1,..., m$ and $i = 0, 1,..., n$. Letting

$$\mathbf{H} = \begin{pmatrix} x_{00} & x_{01} & \cdots & x_{0n} \\ x_{10} & x_{11} & \cdots & x_{1n} \\ \vdots & \vdots & \ddots & \vdots \\ x_{r0} & x_{r1} & \cdots & x_{rn} \end{pmatrix} \tag{4.411}$$

$$\mathbf{B} = \begin{pmatrix} b_{00} & b_{01} & \cdots & b_{0r} \\ b_{10} & b_{11} & \cdots & b_{1r} \\ \vdots & \vdots & \ddots & \vdots \\ b_{m0} & b_{m1} & \cdots & b_{mr} \end{pmatrix} \tag{4.412}$$

the system of Eq. 4.410 can be expressed as

$$\mathbf{H'HA'} = \mathbf{H'B'} \tag{4.413}$$

so that the least-squares weights are determined via the pseudoinverse of **H** by

$$\mathbf{A'} = (\mathbf{H'H})^{-1}\mathbf{H'B'} \tag{4.414}$$

Existence of the pseudoinverse is ensured if **H'H** is invertible, which is itself ensured if the columns of **H** are linearly independent. Since it is desirable to use a large number of observations for estimation, typically r significantly exceeds n and the columns of **H** can be expected to be linearly independent. If **H** were square, then $(\mathbf{H'H})^{-1}\mathbf{H'} = \mathbf{H}^{-1}$ and Eq. 4.414 would reduce to $\mathbf{A'} = \mathbf{H}^{-1}\mathbf{B'}$, or $\mathbf{B'} = \mathbf{HA'}$. Componentwise, this means that

$$b_{kj} = (x_{j0}, x_{j1},..., x_{jn})(a_{k0}, a_{k1},..., a_{kn})' = \sum_{i=0}^{n} a_{ki}x_{ji} \tag{4.415}$$

which means that $SSE = 0$. While such a result might appear appealing, it is based on a small number of training observations and therefore the chosen weights cannot be expected to perform well on feature vectors outside the training set. If **H** is square and singular, or if $r < n$, then the least-squares system is underdetermined and there exists a class of weight matrices all giving zero *SSE*. As in the invertible case, these solutions are not consequential owing to the paucity of training observations.

4.12. Neural Networks

Optimization over the class of nonhomogeneous linear operators can represent a severe constraint on optimality. Optimization can be improved by considering function classes containing the class of linear filters as a subclass. Consider a linear filter

$$\psi_{\mathbf{A}}(\mathbf{X}) = a_0 + \sum_{k=1}^{m} a_k X_k = \sum_{k=0}^{m} a_k X_k \tag{4.416}$$

where $\mathbf{X} = (X_1, X_2, ..., X_m)'$, $\mathbf{A} = (a_0, a_1, ..., a_m)'$ is the weight vector, and (for notational simplicity) $X_0 \equiv 1$. In the context of neural networks, a_0 is called the *bias*. One way to introduce nonlinearity is to apply functions $h_0, h_1, ..., h_m$ (called *basis* functions) to the input vector $\mathbf{X}$ and then form a linear combination to arrive at the functional form

$$\psi_{\mathbf{A}}(\mathbf{X}) = \sum_{k=0}^{m} a_k h_k(\mathbf{X}) \tag{4.417}$$

A bias term remains by setting $h_0(\mathbf{X}) \equiv 1$. Finding an optimal weight vector $\mathbf{A}$ is a linear optimization problem with the observation random variables X_1, $X_2, ..., X_m$ replaced by the random variables $h_1(\mathbf{X}), h_2(\mathbf{X}), ..., h_m(\mathbf{X})$. Improved estimation over using $X_1, X_2, ..., X_m$ requires a judicious choice of basis functions. A further extension is to follow the linear combination with the application of some nonlinear function g to obtain

$$\psi_{\mathbf{A}}(\mathbf{X}) = g\left(\sum_{k=0}^{m} a_k h_k(\mathbf{X})\right) \tag{4.418}$$

$\psi_{\mathbf{A}}$ is called a *single-layer neural network* (or *single-layer perceptron*) with *activation function g*. A special case occurs when $h_k(\mathbf{X}) = X_k$ for $k = 1, 2, ..., m$, in which case a single-layer network is a linear operator followed by an activation function.

4.12.1. Two-Layer Neural Networks

Rather than employ fixed basis functions, greater flexibility in network design is achieved by letting each function h_k be a single-layer network composed of a linear filter followed by an activation function. The resulting *two-layer neural network* (*two-layer perceptron*) is of the form

$$Z = \psi_{\mathbf{A}}(\mathbf{X}) = g_2\left(\sum_{j=0}^{r} a_j g_1\left(\sum_{k=0}^{m} a_{jk} X_k\right)\right) \tag{4.419}$$

where $\mathbf{A} = (a_0, a_1,..., a_r, a_{00}, a_{01},...., a_{rm})'$, $X_0 \equiv 1$, and g_1 and g_2 are the activation functions for the first and second layers, respectively. A bias term can be inserted into the second layer by changing Eq. 4.419 slightly to become

$$Z = \psi_{\mathbf{A}}(\mathbf{X}) = g_2\left(a_0 W_0 + \sum_{j=1}^{r} a_j g_1\left(\sum_{k=0}^{m} a_{jk} X_k\right)\right) \tag{4.420}$$

where $W_0 \equiv 1$. In either case, there are intermediate outputs

$$W_j = g_1\left(\sum_{k=0}^{m} a_{jk} X_k\right) \tag{4.421}$$

$j = 0, 1,..., r$. Further generality can be introduced by allowing the neural network to have more than a single output. This extension will be addressed subsequently.

The graph structure of the network of Eq. 4.420 is illustrated in Fig. 4.14. Except for X_0, each node at the extreme left of the graph is a component of the input vector; all other nodes are called *processing units* or *neurons*. Neurons in the first processing layer are called *hidden units* because their outputs are not directly observable at the completion of network processing; neurons in the second processing layer are called *output units*. The network is *feed-forward*, meaning there are no feedback loops sending the output of a neuron to an input of a neuron in a preceding layer. The network is *fully connected* because each output in a layer feeds into all units in the succeeding layer. A neural network need not be feed-forward nor fully connected, and it can have more than two layers of processing units. We will consider only feed-forward, fully connected, two-layer networks.

Three functions are commonly employed as activation functions. *Linear* activation refers to the identity function $g(t) = t$; *threshold* activation involves the threshold function $g(t) = 0$ if $t < 0$ and $g(t) = 1$ if $t \geq 0$; and *sigmoid* activation involves the logistic sigmoid function $g(t) = (1 + e^{-t})^{-1}$. The logistic sigmoid function g is a strictly increasing function of t, $g(0) = 0.5$, $g(t) \to 1$ as $t \to \infty$, $g(t) \to 0$ as $t \to -\infty$, and g is approximately linear for small $|t|$. If the

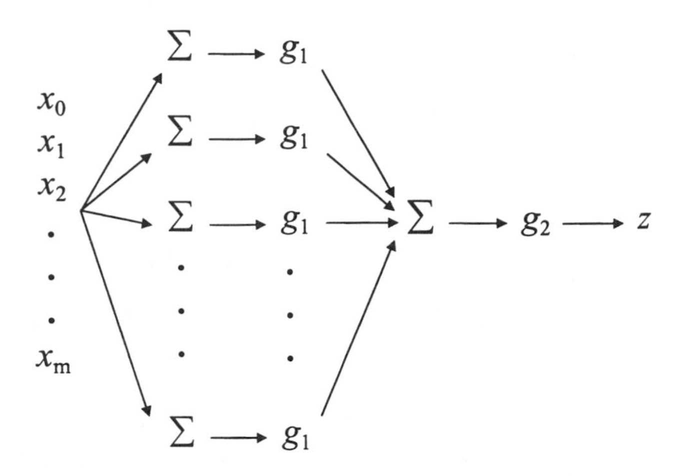

Figure 4.14 Single-output, two-layer neural network.

output activation function is linear, then the two-layer network becomes a linear function of the hidden outputs W_j. If the output unit is a threshold function, then the network produces a decision region $D \subset \Re^m$: $Z = 1$ for $\mathbf{X} \in D$ and $Z = 0$ for $\mathbf{X} \notin D$.

The degree to which a neural network can minimize error depends on the function class possessing such representation. Relative to mean-square error, the conditional expectation $E[Y|\mathbf{X}]$ provides optimal filtering but, except in special cases, such as binary digital images or small numbers of observation variables for low-bit gray-scale images, it is computationally impractical. Error minimization over a subclass of filters results in increased error. For linear filters, this error increase is offset (to some degree) by the design advantage of parameterized optimization and straightforward solution via the orthogonality principle. A neural network also provides a parameterized framework but optimization is more difficult. Error increase over the conditional expectation depends on the extent of the function class defined by the neural network $\psi_{\mathbf{A}}$ as the model parameters forming $\mathbf{A}$ run over all possible choices. If there exists a set of model parameters forming a weight vector $\mathbf{A}^*$ such that $\psi_{\mathbf{A}^*}(\mathbf{x}) = E[Y|\mathbf{x}]$ for any realization $\mathbf{x}$, then, in principle, parametrically optimizing the neural network $\psi_{\mathbf{A}}$ imposes no constraint on optimality. The following theorem addresses the existence of such a neural network in the sense that the conditional expectation can be approximated to any desired degree so long as it is a continuous function of $\mathbf{x}$.

Theorem 4.19. If $\xi(x_1, x_2,..., x_m)$ is a continuous function defined on $[0, 1]^m$, g is a bounded, nonconstant, montonically increasing continuous function, and $\varepsilon > 0$, then there exists an integer r and constants $c_1, c_2,..., c_r, a_1, a_2,..., a_r, a_{11}, a_{12},..., a_{1m}, a_{21},..., a_{rm}$ such that

$$\left| \xi(x_1, x_2, \ldots, x_m) - \sum_{i=1}^{r} a_i g\left(c_i + \sum_{j=1}^{m} a_{ij} x_j \right) \right| < \varepsilon \tag{4.422}$$

for all $(x_1, x_2, \ldots, x_m) \in [0, 1]^m$. ■

Since the approximating double sum in Eq. 4.422 partakes of the same form as the neural network of Eq. 4.419 with linear output and logistic-sigmoid hidden-layer activation function, it guarantees that a two-layer network can uniformly approximate any continuous function on $[0, 1]^m$ to any desirable tolerance so long as the number of units in the hidden layer is sufficiently large. Consequently, if the joint distribution of **X** has compact support and the conditional expectation is continuous, then there exists a neural network uniformly approximating the conditional expectation to any arbitrarily desired degree. While such a proposition is encouraging, it is deficient from a practical perspective. Not only does one not know the function whose approximation is desired, but, even were we to know the function and how to find the necessary coefficients, a close approximation can require an extremely large number of hidden units (and therefore model parameters). As the number of model parameters grows, use of the model for filter design becomes increasingly intractable from the perspectives of both computation and statistical estimation of the model parameters. For optimal linear filters, the twin problems of computation and estimation appear in the inversion and estimation of the autocorrelation matrix, respectively, or, for steepest descent, in convergence and autocorrelation estimation; for neural networks, they appear in convergence for iteratively finding an error-surface minimum and estimation of the error surface.

In practice, the task is to estimate the optimal weights for a given form of neural network; that is, the number of hidden units is prescribed beforehand. Since the number of hidden units must be kept relatively small, there is no assurance that the optimal neural network of the prescribed form approximates the conditional expectation, nor do we necessarily know the degree to which it is suboptimal. Employing a two-layer neural network with an insufficient number of hidden units to approximate the conditional expectation represents a constraint on optimization. Using multiple-layer and nonfully connected networks allows greater constraint flexibility. The problem is akin to the one found in binary filtering where, in practice, various constraints are imposed in order to use the disjunctive-normal-form representational model of Eq. 4.362 to estimate the binary conditional expectation or to use Eq. 4.373 when using Theorem 4.18 to estimate the optimal increasing filter.

4.12.2. Steepest Descent for Nonquadratic Error Surfaces

The method of steepest descent to find the optimal linear filter, discussed in Section 4.3, can be applied to find local minima for any error function possessing continuous second derivatives relative to the weights. In general (and in particular for neural networks), the algorithm is more problematic than in the case of linear filters, where the mean-square error is quadratic and the resulting error surface has a single local minimum that is also the global minimum. The error surface for a neural network can have many local minima, so that a gradient method such as steepest descent can get stuck at a local minimum.

Consider a weight vector $\mathbf{W} = (w_0, w_1,\ldots, w_M)'$; let $Q = Q(\mathbf{W})$ be an error function possessing continuous second derivatives relative to $w_0, w_1,\ldots, w_M$; and let $\mathbf{W}_0$ be a fixed set of weights. The Taylor expansion of Q around $\mathbf{W}_0$ yields the approximation

$$Q(\mathbf{W}) \approx Q(\mathbf{W}_0) + (\mathbf{W} - \mathbf{W}_0)'\nabla Q(\mathbf{W}_0) + \frac{1}{2}(\mathbf{W} - \mathbf{W}_0)'\mathbf{H}(\mathbf{W} - \mathbf{W}_0) \quad (4.423)$$

where $\nabla Q(\mathbf{W}_0)$ is the gradient of Q evaluated at $\mathbf{W}_0$ and $\mathbf{H}$ is the $(M + 1) \times (M + 1)$ *Hessian* matrix, whose (i, j) component is $(\partial^2 Q/\partial w_i \partial w_j)(\mathbf{W}_0)$. The Taylor series expansion of ∇Q around $\mathbf{W}_0$ yields the first-order approximation

$$\nabla Q(\mathbf{W}) \approx \nabla Q(\mathbf{W}_0) + \mathbf{H}(\mathbf{W} - \mathbf{W}_0) \quad (4.424)$$

If $\mathbf{W}_0$ is a minimum of Q, then $\nabla Q(\mathbf{W}_0) = 0$ and Eq. 4.423 reduces to an approximate representation of the error increase resulting from not using optimal weights, namely,

$$Q(\mathbf{W}) - Q(\mathbf{W}_0) \approx \frac{1}{2}(\mathbf{W} - \mathbf{W}_0)'\mathbf{H}(\mathbf{W} - \mathbf{W}_0) \quad (4.425)$$

For the linear-filter setting (Section 4.3), Q is given by the MSE. From Eq. 4.63 we see that the Hessian is the autocorrelation matrix and Eq. 4.425 corresponds to Eq. 4.67. The present situation is more general, so that even for MSE, Q does not generally reduce to such a simple form and the Hessian does not reduce to the autocorrelation matrix.

In the present setting, the steepest descent algorithm is given by

$$\mathbf{W}(n + 1) = \mathbf{W}(n) - \mu\nabla Q(n) \quad (4.426)$$

where $\nabla Q(n)$ is the gradient of the error function at the nth weight vector in the iteration. Equation 4.426 corresponds to Eq. 4.70; however, the reduction to Eq. 4.71 does not apply because that reduction depends on the specific form of the linear filter.

Convergence can be analyzed via the Hessian matrix. Since **H** is real and symmetric, there are $M + 1$ eigenvectors $\mathbf{E}_0, \mathbf{E}_1, \ldots, \mathbf{E}_M$ corresponding to $M + 1$ nonnegative, real eigenvalues $\lambda_0, \lambda_1, \ldots, \lambda_M$ such that $\mathbf{E}_0, \mathbf{E}_1, \ldots, \mathbf{E}_M$ form an orthonormal system. Assuming that **H** is positive definite, all eigenvalues are positive. $\mathbf{W} - \mathbf{W}_0$ has the representation

$$\mathbf{W} - \mathbf{W}_0 = \sum_{k=0}^{M} v_k \mathbf{E}_k \tag{4.427}$$

For **W** close to $\mathbf{W}_0$, $\nabla Q(\mathbf{W}_0) = 0$ together with Eq. 4.424 shows that the gradient of Q at **W** possesses the approximation

$$\nabla Q(\mathbf{W}) \approx \mathbf{H}(\mathbf{W} - \mathbf{W}_0) \tag{4.428}$$

Inserting Eq. 4.427 into Eq. 4.428 and using the fact that the eigenvalues of **H** are $\lambda_0, \lambda_1, \ldots, \lambda_M$ yields a local approximation of the gradient,

$$\nabla Q(\mathbf{W}) \approx \sum_{k=0}^{M} \lambda_k v_k \mathbf{E}_k \tag{4.429}$$

which improves as **W** approaches $\mathbf{W}_0$. If we now consider the weight vector $\mathbf{W}(n)$ occurring in an iterative procedure using the gradient descent of Eq. 4.426, then

$$\mathbf{W}(n + 1) - \mathbf{W}(n) \approx -\mu \sum_{k=0}^{M} \lambda_k v_k(n) \mathbf{E}_k \tag{4.430}$$

Componentwise,

$$v_k(n + 1) \approx (1 - \mu\lambda_k) v_k(n) \tag{4.431}$$

so that

$$v_k(n) \approx (1 - \mu\lambda_k)^n v_k(0) \tag{4.432}$$

(which is reminiscent of Eq. 4.79). Since the eigenvectors form an orthonormal system, multiplication by $\mathbf{E}_k'$ in Eq. 4.427 yields

$$\mathbf{E}_k'(\mathbf{W}(n) - \mathbf{W}_0) = v_k(n) \tag{4.433}$$

Thus, $v_k(n)$ is the distance from $\mathbf{W}(n)$ to $\mathbf{W}_0$ along $\mathbf{E}_k$. From Eq. 4.432, $v_k(n) \to 0$ as $n \to \infty$ if and only if Eq. 4.80 holds, so that $\mathbf{W}(n) \to \mathbf{W}_0$ if and only if Eq. 4.81 holds, with λ_{max} being the largest eigenvalue of $\mathbf{H}$.

The inefficiency of gradient descent can be seen in Eq. 4.432, which expresses the distance from $\mathbf{W}(n)$ to $\mathbf{W}_0$ along $\mathbf{E}_k$. If we choose μ as large as possible, $\mu = 2/\lambda_{max}$, and select k so that $\lambda_k = \lambda_{min}$, the smallest eigenvalue of $\mathbf{H}$, then

$$v_k(n) \approx (1 - 2\lambda_{min}/\lambda_{max})^n v_k(0) \tag{4.434}$$

If $\lambda_{min}/\lambda_{max}$ is very small, meaning the eigenvalues of $\mathbf{H}$ are widely spread, then convergence will be slow. There are a number of enhanced gradient-descent algorithms and other minimization algorithms. We leave these to a text on nonlinear optimization.

4.12.3. Sum-of-Squares Error

The form of linear filters facilitates optimization directly in terms of mean-square error. Nonlinear activation functions in neural networks make such a direct MSE approach problematic. To circumvent error representation, one can use an error estimate in place of the error. An error estimate is found that is a function of the model parameters, and optimizing parameters are found to minimize the error estimate. Since optimization is relative to the estimate, the minimizing parameters are estimates of the true optimizing parameters (which is also the case with linear filters if the autocorrelation matrix and cross-correlation vector are estimated from realizations). For a sufficient sample size, the error estimate (as a function of the parameters) is likely to be close to the error (as a function of the parameters) and the estimated parameters will likely be close to optimal. The situation is akin to the method of least-squares in Section 4.4, where the pseudoinverse estimate $(\mathbf{H}'\mathbf{H})^{-1}\mathbf{H}'\mathbf{x}$ is deduced from the error sum of squares for a single observation $\mathbf{x}$, but statistical properties are deduced for the estimator $(\mathbf{H}'\mathbf{H})^{-1}\mathbf{H}'\mathbf{X}$.

For estimation of random variable Y by means of the parameterized function $\psi_{\mathbf{A}}$ of the random vector $\mathbf{X} = (X_1, X_2, \ldots, X_m)'$, the MSE is given by

$$MSE\langle\psi_{\mathbf{A}}\rangle = E[|Y - \psi_{\mathbf{A}}(\mathbf{X})|^2] \tag{4.435}$$

If $(\mathbf{X}_1, Y_1), (\mathbf{X}_2, Y_2), \ldots, (\mathbf{X}_N, Y_N)$ comprise a random sample S from the joint distribution of Y and $\mathbf{X}$, then, relative to the sample, we define the *sum-of-squares error* by

$$SSE\langle\psi_A\rangle = \sum_{n=1}^{N} |Y_n - \psi_{\mathbf{A}}(\mathbf{X}_n)|^2 \tag{4.436}$$

$SSE\langle\psi_A\rangle/N$ is the sample-mean estimator of $MSE\langle\psi_A\rangle$ for S, so that

$$E[SSE\langle\psi_A\rangle/N] = MSE\langle\psi_A\rangle \tag{4.437}$$

$SSE\langle\psi_A\rangle/N$ is an unbiased, consistent estimator of $MSE\langle\psi_A\rangle$.

The precision with which $SSE\langle\psi_A\rangle/N$ estimates $MSE\langle\psi_A\rangle$ depends on N. In particular,

$$\mathrm{Var}\left[\frac{SSE\langle\psi_{\mathbf{A}}\rangle}{N}\right] = \frac{\mathrm{Var}[|Y - \psi_{\mathbf{A}}(\mathbf{X})|^2]}{N} \tag{4.438}$$

and, from Chebyshev's inequality, for any $\eta > 0$,

$$P\left(\left|\frac{SSE\langle\psi_{\mathbf{A}}\rangle}{N} - MSE\langle\psi_{\mathbf{A}}\rangle\right| < \frac{\eta}{2}\right) \geq 1 - \frac{\mathrm{Var}[SSE\langle\psi_{\mathbf{A}}\rangle/N]}{(\eta/2)^2}$$

$$= 1 - \frac{\mathrm{Var}[|Y - \psi_{\mathbf{A}}(\mathbf{X})|^2]}{N(\eta/2)^2} \tag{4.439}$$

Minimizing $SSE\langle\psi_A\rangle/N$ as a function of the model parameters is equivalent to minimizing $SSE\langle\psi_A\rangle$. The bound in Eq. 4.439 depends on the model parameters. If $\mathrm{Var}[|Y - \psi_A(\mathbf{X})|^2]$ is uniformly bounded by γ for all model-parameter vectors $\mathbf{A}$ in some region R of the parameter space, then Eq. 4.439 implies that, for the parameter vector $\mathbf{A}^*$ minimizing $SSE\langle\psi_A\rangle$ in R,

$$P\left(\min_{\mathbf{A}\in R} MSE\langle\psi_{\mathbf{A}}\rangle \geq MSE\langle\psi_{\mathbf{A}^*}\rangle - \eta\right) \geq 1 - 4\gamma\eta^{-2}N^{-1} \tag{4.440}$$

Assuming that the minimum MSE is achieved for a vector $\mathbf{A} \in R$, this implies that, for any $\eta > 0$, a sufficiently large sample will ensure that minimizing the SSE will provide a parameter vector for which the probability of the minimum MSE being less than $MSE\langle\psi_{A^*}\rangle - \eta$ is as small as desired. When $MSE\langle\psi_A\rangle$ does not have a tractable expression, optimization can be performed relative to $SSE\langle\psi_A\rangle$ and is a form of least-squares estimation. As quantified by Eq. 4.440, the cost of using the SSE in place of the MSE depends on the sample size.

If ψ_A is the two-layer neural network of Eq. 4.419, consider the termwise errors

$$Q_n = |Y_n - \psi_A(\mathbf{X}_n)|^2 \tag{4.441}$$

contributing to $SSE\langle\psi_A\rangle$ in Eq. 4.436. Total error $Q = SSE\langle\psi_A\rangle$ is given by

$$Q = \sum_{n=1}^{N} Q_n \tag{4.442}$$

If

$$S = \{(\mathbf{x}_1, y_1), (\mathbf{x}_2, y_2), \ldots, (\mathbf{x}_N, y_N)\} \tag{4.443}$$

is a specific data set from the random sample $\mathcal{S}$, then the numerical values Q_n and Q formed by replacing the random variable Y_n and random vector $\mathbf{X}_n$ by y_n and $\mathbf{x}_n$, respectively, are estimates of the random errors defined by Eqs. 4.441 and 4.442. Relative to S, $SSE\langle\psi_A\rangle/N$ is an estimate of $MSE\langle\psi_A\rangle$; relative to $\mathcal{S}$, $SSE\langle\psi_A\rangle/N$ is an estimator of $MSE\langle\psi_A\rangle$. As is common in statistics, we will employ the same notation (Q_n, Q, SSE) for estimators and estimates.

4.12.4. Error Back-Propagation

Application of steepest descent to Q according to Eq. 4.426 requires knowledge of the partial derivatives of Q with respect to the weights. Since Q is a sum of the termwise errors, the partial derivatives of Q are known once the partial derivatives of the termwise errors are known. The method of error back-propagation estimates these termwise derivatives from a sample data set. If $\psi_A(\mathbf{x}_1)$, $\psi_A(\mathbf{x}_2), \ldots, \psi_A(\mathbf{x}_N)$ are the network outputs, then we wish to assign network weights based on the termwise errors $Q_1, Q_2, \ldots, Q_N$. The difficulty to be overcome is that observation of an error Q_n does not explicitly provide information regarding the contribution of the hidden units to the error. In error back-propagation, partial derivatives of the error function Q_n relative to weights of the second layer are estimated from observable training errors and partial derivatives relative to weights of the first layer are found by back propagating these errors through the network. Because the method only depends on a termwise decomposition of the error function, not on SSE in particular, we develop it for any error function satisfying Eq. 4.442. Although we only consider error back-propagation for two-layer networks, the method applies to feed-forward networks having more than two layers.

We consider a single term of Eq. 4.442. To avoid cumbersome notation, we omit the subscript n from the corresponding data pair and simply write $\mathbf{x}$ and z. For the two-layer network of Eq. 4.419, let

$$A_j = \sum_{k=0}^{m} a_{jk} x_k$$

$$w_j = g_1(A_j) \tag{4.444}$$

$$B = \sum_{j=0}^{r} a_j w_j$$

$$z = g_2(B)$$

Application of the chain rule (and writing all derivatives using the ∂ notation) yields

$$\frac{\partial Q_n}{\partial a_j} = \frac{\partial Q_n}{\partial z}\frac{\partial z}{\partial B}\frac{\partial B}{\partial a_j} = \frac{\partial Q_n}{\partial z} g_2'(B) w_j \tag{4.445}$$

$$\begin{aligned}\frac{\partial Q_n}{\partial a_{jk}} &= \frac{\partial Q_n}{\partial z}\frac{\partial z}{\partial B}\frac{\partial B}{\partial a_{jk}} \\ &= \frac{\partial Q_n}{\partial z}\frac{\partial z}{\partial B}\left(\sum_{q=0}^{r}\frac{\partial B}{\partial w_q}\frac{\partial w_q}{\partial A_q}\frac{\partial A_q}{\partial a_{jk}}\right) \\ &= \frac{\partial Q_n}{\partial z} g_2'(B) g_1'(A_j) a_j x_k\end{aligned} \tag{4.446}$$

Letting

$$\delta_2 = \frac{\partial Q_n}{\partial B} = \frac{\partial Q_n}{\partial z}\frac{\partial z}{\partial B} = g_2'(B)\frac{\partial Q_n}{\partial z} \tag{4.447}$$

$$\delta_{1,j} = \frac{\partial Q_n}{\partial A_j} = \frac{\partial Q_n}{\partial B}\frac{\partial B}{\partial w_j}\frac{\partial w_j}{\partial A_j} = g_1'(A_j) a_j \delta_2 \tag{4.448}$$

the desired partial derivatives take the forms

$$\frac{\partial Q_n}{\partial a_j} = \delta_2 w_j \tag{4.449}$$

$$\frac{\partial Q_n}{\partial a_{jk}} = \delta_{1,j} x_k \tag{4.450}$$

δ_2 and $\delta_{1,j}$ are called *error terms*.

The method is called *error back-propagation* because the output error δ_2 is observable from the network output, the hidden-unit errors $\delta_{1,j}$ are not observable, and $\delta_{1,j}$ is obtained via Eq. 4.448 by propagating the output error back through the network. Computation of A_j and B from the sample data is called *forward propagation*. Equations 4.449 and 4.450 give the partial derivatives for the terms of Eq. 4.442; the full partial derivatives, $\partial Q/\partial a_j$ and $\partial Q/\partial a_{jk}$, are found by summing from $n = 1, 2,..., N$. To apply error back-propagation to sample data, compute the output and hidden linear combinations from the sample data, use Eqs. 4.447 through 4.450 to find the termwise partial derivatives, and then sum.

***Example* 4.27.** Consider error back-propagation for least-squares estimation using the logistic-sigmoid activation function g for both layers, noting that

$$g'(t) = g(t)(1 - g(t))$$

The error term for the data pair $(\mathbf{x}_n, y_n)$ is $Q_n = (y - z)^2$. Since $g(B) = z$ and $w_j = g(A_j)$,

$$\delta_2 = g(B)(1 - g(B))\frac{\partial Q_n}{\partial z} = 2z(1 - z)(z - y)$$

$$\delta_{1,j} = g'(A_j)a_j\delta_2 = g(A_j)(1 - g(A_j))a_j\delta_2 = w_j(1 - w_j)a_j\delta_2$$

$\partial Q_n/\partial a_j$ and $\partial Q_n/\partial a_{jk}$ are found from Eqs. 4.449 and 4.450, respectively, and $\partial Q/\partial a_j$ and $\partial Q/\partial a_{jk}$ are found by summing over n. ■

For application of a neural-network filter across an image (or signal), the observation variables are taken in a window about a pixel and the output value at the pixel is computed by the neural network. Spatial invariance of the network depends on joint stationarity of the processes. If a neural network is designed using windows across a realization of a nonstationary process, then it suffers from lack of estimation accuracy, just as would a realization-based estimate of an optimal binary filter, or an estimate of an optimal linear filter designed across a realization of a non-WS-stationary process.

We have been examining neural networks from the standpoint of optimal filtering based on a finite collection of values from an observed image (signal). This perspective has been taken to provide a unifying theme to filter design in terms of filter form, error estimation, and error-surface minimization. In practice, neural networks are often used to estimate optimal discriminants based on feature-vector inputs. A single-layer, single-output, linear-activation neural network can be viewed as either a finite-window linear filter or a discriminant function. Two-class linear discrimination results from letting g be threshold activation in a single-layer network (Eq. 4.418). More generally, for pattern discrimination, a neural network has multiple outputs comprising a discriminant vector.

4.12.5. Error Back-Propagation for Multiple Outputs

Only slight modifications are needed when applying error back-propagation to multiple-output, two-layer networks. In this case, the network is a vector-valued filter serving as an estimator of a random vector $\mathbf{Y} = (Y_1, Y_2,..., Y_M)'$, the output of the network is a random vector $\mathbf{Z} = (Z_1, Z_2,..., Z_M)'$, and the filter is defined by

$$Z_i = \psi_i(\mathbf{X}) = h_i\left(\sum_{j=0}^{r} b_{ij} g_j\left(\sum_{k=0}^{m} a_{jk} X_k\right)\right) \tag{4.451}$$

for $i = 1, 2,..., M$, where h_i and g_j are the activation functions for the ith output and jth hidden node, respectively (Fig. 4.15), and $X_0 \equiv 1$.

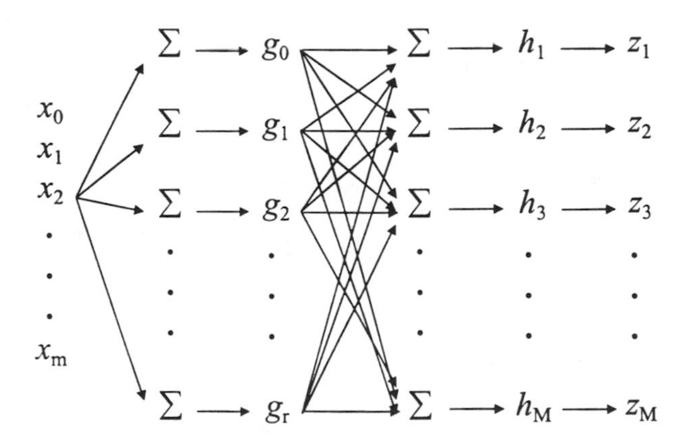

Figure 4.15 Multiple-output, two-layer neural network.

If $(\mathbf{X}_1, \mathbf{Y}_1), (\mathbf{X}_2, \mathbf{Y}_2),..., (\mathbf{X}_N, \mathbf{Y}_N)$ comprise a random sample, then, relative to the sample, SSE is defined by Eq. 4.442 with

$$Q_n = \sum_{i=1}^{M} (Y_i(n) - Z_i(n))^2 \tag{4.452}$$

where

$$\mathbf{Y}_n = (Y_1(n), Y_2(n), ..., Y_M(n))' \tag{4.453}$$

$$\mathbf{Z}_n = (Z_1(n), Z_2(n), ..., Z_M(n))' \tag{4.454}$$

Error back-propagation only requires a summation error expression according to Eq. 4.442. Error minimization is accomplished for a data set $\{(\mathbf{x}_n, \mathbf{y}_n)\}$ from the random sample.

The following four substitutions (corresponding to Eq. 4.444) are used to express the back-propagation algorithm:

$$A_j = \sum_{k=0}^{m} a_{jk} x_k$$

$$w_j = g_j(A_j) \tag{4.455}$$

$$B_i = \sum_{j=0}^{r} b_{ij} w_j$$

$$z_i = h_i(B_i)$$

The error terms for the outer and hidden layers are defined by $\delta_{2,i} = \partial Q_n/\partial B_i$ and $\delta_{1,j} = \partial Q_n/\partial A_j$, respectively, and the chain-rule yields

$$\delta_{2,i} = \frac{\partial Q_n}{\partial B_i} = \frac{\partial Q_n}{\partial z_i}\frac{\partial z_i}{\partial B_i} = h_i'(B_i)\frac{\partial Q_n}{\partial z_i} \tag{4.456}$$

$$\delta_{1,j} = \frac{\partial Q_n}{\partial A_j} = \sum_{i=0}^{M}\frac{\partial Q_n}{\partial B_i}\frac{\partial B_i}{\partial w_j}\frac{\partial w_j}{\partial A_j} = g_j'(A_j)\sum_{i=0}^{M}\delta_{2,i}b_{ij} \tag{4.457}$$

Partial derivatives of Q_n with respect to the parameters are derived by the chain-rule:

$$\frac{\partial Q_n}{\partial b_{ij}} = \frac{\partial Q_n}{\partial B_i}\frac{\partial B_i}{\partial b_{ij}} = \delta_{2,i} w_j \tag{4.458}$$

$$\frac{\partial Q_n}{\partial a_{jk}} = \frac{\partial Q_n}{\partial A_j}\frac{\partial A_j}{\partial a_{jk}} = \delta_{1,j}x_k \tag{4.459}$$

4.12.6. Adaptive Network Design

To the extent that estimates of the autocorrelation matrix and cross-correlation vector are used for linear-filter optimization via steepest descent, they correspond to error-back-propagation estimation of partial derivatives for gradient descent. In analogy to the LMS algorithm, error back-propagation can be used for adaptive design. Weights are adapted based on the error when a data vector is processed through the neural network. Each weight is adjusted according to the rate of change of the error with respect to the weight.

Adaptation is defined recursively by assuming the network to be in state n when the nth training pair $(\mathbf{x}_n, \mathbf{y}_n)$ is presented to the network to produce the output $\mathbf{z}_n$. This means that model parameters are of the form $a_{jk}(n)$ and $b_{ij}(n)$, and comprise a parameter vector $\mathbf{A}(n)$ defining the state of the system. It also means that Q_n is now the error resulting from the training pair $(\mathbf{x}_n, \mathbf{y}_n)$ when the network is defined by $\mathbf{A}(n)$, and that the variables defined by Eq. 4.455 are functions of n: $A_j(n)$, $w_j(n)$, $B_i(n)$, $z_i(n)$.

An instantaneous gradient is defined for $a_{jk}(n)$ and $b_{ij}(n)$ by $\partial Q_n/\partial a_{jk}(n)$ and $\partial Q_n/\partial b_{ij}(n)$. As with the LMS algorithm, a rate parameter μ is selected and the corrections

$$\Delta b_{ij}(n) = -\mu\frac{\partial Q_n}{\partial b_{ij}(n)} \tag{4.460}$$

$$\Delta a_{jk}(n) = -\mu\frac{\partial Q_n}{\partial a_{jk}(n)} \tag{4.461}$$

are applied to the weights by

$$b_{ij}(n+1) = b_{ij}(n) + \Delta b_{ij}(n) \tag{4.462}$$

and similarly for $a_{jk}(n+1)$. According to Eqs. 4.458 and 4.459,

$$\begin{aligned}\Delta b_{ij}(n) &= -\mu\delta_{2,i}(n)w_j(n)\\ &= -\mu h_i'(B_i(n))\frac{\partial Q_n}{\partial z_i(n)}w_j(n)\end{aligned} \tag{4.463}$$

$$\Delta a_{jk}(n) = -\mu\delta_{1,j}(n)x_k(n)$$

$$= -\mu g_j'(A_j(n))\left(\sum_{i=0}^{M}\delta_{2,i}(n)b_{ij}(n)\right)x_k(n) \tag{4.464}$$

There are many important considerations regarding application of adaptive error back-propagation; we briefly comment on a few. The network parameters need to be initialized. The performance of a neural network can be greatly enhanced by a good initialization. If there is prior information, then it can be used to initialize the weights; otherwise, it is common to choose the weights randomly from a uniform distribution having a small range. As with the LMS algorithm, choice of the parameter μ is important. If μ is too small, then adaptation will be very slow; if μ is chosen too large, then the state vector $\mathbf{A}(n)$ can oscillate (so that the network becomes unstable during training). Lastly, the training data should be randomly ordered and passed through the algorithm. The data set can be reused, with the training pairs being randomly ordered each time. As with all estimation problems, if the data set is too small, then the resulting estimator (network) will have a high variance, thereby resulting in designed networks that perform poorly. While reusing the data set may help to train the network, a small data set is not likely to be representative of the joint distribution of $\mathbf{X}$ and $\mathbf{Y}$.

There is no general theorem regarding convergence of the algorithm. As to choosing a stopping criterion, the optimal weight vector minimizes the error surface and therefore has zero gradient. Therefore, a reasonable (but not always successful) choice is to halt adaptive back-propagation when the gradient-vector norm falls below a predetermined threshold. Whenever the algorithm is stopped, the final state vector $\mathbf{A}(n)$ defines a filter that will be used to estimate the random vector $\mathbf{Y}$ from the observed random vector $\mathbf{X}$. The goodness of the network depends on the properties of this estimator.

Exercises for Chapter 4

1. Let X and Y be independent binomial random variables, both having parameters n and p. Show that the conditional probability mass function for X given $X + Y = m \le n$ is

$$P(X = k | X + Y = m) = \frac{\binom{n}{k}\binom{n}{m-k}}{\binom{2n}{m}}$$

for $k \le m$. Note that $X + Y$ possesses a binomial distribution with parameters $2n$ and p.

2. Let X and Y be uniformly distributed over the region bounded by the lines $x = 0$, $y = 0$, and $y = 1 - x$. Find the conditional density for Y given X and $E[Y|X]$.

3. Suppose a document imaging algorithm is expected to count the number of vowels on a set of randomly selected pages from some page archive, the number of pages to be selected is a normal random variable with mean μ and variance σ^2, the number of vowels on any page is normally distributed with mean ξ and variance η^2, and the numbers of vowels on pages are independent of the number of pages selected. Let T denote the total number of vowels selected, X_i denote the number of vowels on the ith selected page, and Y denote the number of pages selected. Use Theorem 4.1 to show that $E[T] = \mu\xi$.

4. Let $X_1, X_2,\ldots$ be a sequence of independent random variables identically distributed to X and N be a random variable independent of $X_1, X_2,\ldots$. For

$$Y = \sum_{k=1}^{N} X_k$$

show that

$$\text{Var}[Y] = E[N]\text{Var}[X] + E[X]^2\,\text{Var}[N]$$

Apply the result to the situation where N is Poisson distributed with mean λ and X is gamma distributed with parameters α and β.

5. Let $Y_1, Y_2,\ldots$ be a sequence of independent random variables identically distributed with Y and let $N(t)$ be a Poisson process with parameter λ. The random process

$$X(t) = \sum_{n=1}^{N(t)} Y_n$$

is called a *compound Poisson process*. Show that

(i) $E[X(t)] = \lambda t E[Y]$

(ii) $K_X(t,t') = \lambda E[Y^2]\min(t,t')$

6. Suppose X is a random analog signal and we desire to quantize it for digital processing. Quantization requires sequences of real numbers ..., z_{-2}, z_{-1}, z_0, z_1, z_2,... and ..., y_{-2}, y_{-1}, y_0, y_1, y_2,... such that the quantized signal Y is defined by $Y = y_i$ if $z_i < X \leq z_{i+1}$. Find the optimal values ..., $y_{-2}, y_{-1}, y_0, y_1, y_2,...$, those being the ones that minimize the mean-square error $E[(X - Y)^2]$. Show that, for the optimal quantizer, $E[Y] = E[X]$ and

$$\text{Var}[Y] = \text{Var}[X] - E[(X - Y)^2]$$

Hint: Note that for any quantizer Y,

$$E[(X - Y)^2] = \sum_{i=-\infty}^{\infty} E[(X - y_i)^2 | z_i < X \leq z_{i+1}] P(z_i < X \leq z_{i+1})$$

Define the random variable W by $W = i$ if $z_i < X \leq z_{i+1}$ and use Theorem 4.3 to find the optimal y_i.

7. A continuous-time random function $X(t)$, $t \geq 0$, with finite mean is called a *continuous-parameter martingale* if the conditional expectation of the process at a point in time given values of the process at n prior times is equal to the most recent observed value. Analytically, for any $t_1 < t_2 < ... < t_n < t_{n+1}$,

$$E[X(t_{n+1}) | X(t_1), X(t_2), \ldots, X(t_n)] = X(t_n)$$

Show that the Wiener process is a martingale.

8. A *discrete-parameter martingale* is defined in the same way as a continuous-parameter martingale, except that the time points are restricted to integers. Show that, if Y_1, Y_2,... are independent zero-mean random variables, then the discrete-time process

$$X(k)=\sum_{i=1}^{k} Y_i$$

is a discrete-time martingale.

9. Suppose jobs arrive for service in a Poisson stream with an arrival rate of q per hour. Then the number X of jobs arriving in an hour possesses a Poisson distribution with parameter q. Now, suppose the arrival rate is not constant, but is a gamma-distributed random variable Q possessing a prior density with parameters α and β. Show that the posterior density relative to a single observation is gamma distributed. Based on a single observation of the process, find the Bayes estimator for Q.
10. Suppose X possesses an exponential distribution with a random parameter B that is uniformly distributed over the interval [0, 1]. Find the posterior distribution for a random sample of size n and find the Bayes estimator for B.
11. Redo Exercise 4.10 for the situation in which B is itself exponentially distributed with parameter c.
12. Suppose X, Y, and Z are uniformly distributed over the region bounded by the xy, xz, and yz planes and the plane passing through the points (1, 0, 0), (0, 1, 0), and (0, 0, 1). Find the optimal homogeneous linear estimator of Z in terms of X and Y. Find the MSE for the filter.
13. Reconsider Example 4. 8 for the optimal homogeneous linear filter.
14. Reconsider Example 4.9 for three observations: $X_1 = Y_1 + N_1$, $X_2 = Y_2 + N_2$, and $X_3 = Y_3 + N_3$. The signal variables Y_1, Y_2, and Y_3 are zero-mean and possess covariance matrix

$$\mathbf{K}=\begin{pmatrix} 1 & 0.80 & 0.64 \\ 0.80 & 1 & 0.80 \\ 0.64 & 0.80 & 1 \end{pmatrix}$$

The noise variables N_1, N_2, and N_3 are zero-mean and possess the common variance 0.50. Find the optimal linear filter for Y_3 relative to the three observations.

15. For the random variables X_1, X_2, and Y, find the optimal estimator of Y of the form

$$\hat{Y}=a_1X_1+a_2X_2+a_{12}X_1X_2+a_{11}X_1^2+a_{22}X_2^2$$

where the coefficients a_1, a_2, a_{12}, a_{11}, and a_{22} are to be determined. Hint: let Z_1, Z_2, Z_3, Z_4, and Z_5 equal X_1, X_2, X_1X_2, X_1^2, and X_2^2 respectively, so that estimation is linear in terms of Z_1, Z_2, Z_3, Z_4, and Z_5. More generally, show that optimal estimation of Y by a linear combination of products of the random variables $X_1, X_2, \ldots, X_n$ can be reduced to a linear filtering problem involving a matrix of mixed moments of $X_1, X_2, \ldots, X_n$.

16. Formulate the optimal discrete linear filter for *optimal linear forward prediction*. In this case a stochastic process $X(k)$ is observed at n time points $k - n, k - n + 1, \ldots, k - 1$ to give the best MSE prediction of the process at time point k. Assuming $X(k)$ is WS stationary, express the relevant quantities in terms of the appropriate moments of $X(k)$. Also express the MSE.
17. Formulate the optimal discrete linear filter for *optimal linear backward prediction*. In this case a stochastic process $X(k)$ is observed at n time points $k - n + 1, \ldots, k - 1, k$ to give the best MSE prediction of the process at time point $k - n$. Assuming that $X(k)$ is WS stationary, express the relevant quantities in terms of the appropriate moments of $X(k)$. Also express the MSE.
18. For a zero-mean n-vector $\mathbf{X}$, the Hadamard matrix transform achieves compression via the compression-decompression scheme

$$\mathbf{X} \to \mathbf{Y} = \mathbf{HX} \to \mathbf{Y}_m \to \mathbf{X}_m = \mathbf{H}^{-1}\mathbf{Y}_m = \mathbf{HY}_m$$

where $\mathbf{Y}_m$ is the truncation of $\mathbf{Y}$ to the first m components. From the perspective of filtering, $\mathbf{HY}_m$ is an estimator of the random vector $\mathbf{X}$. If

$$\mathbf{H} = \begin{pmatrix} h_{11} & h_{12} & \cdots & h_{1n} \\ h_{21} & h_{22} & \cdots & h_{2n} \\ \vdots & \vdots & \ddots & \vdots \\ h_{n1} & h_{n2} & \cdots & h_{nn} \end{pmatrix}$$

then the random-variable component X_k of vector $\mathbf{X}$ is estimated by

$$\hat{X}_k = \sum_{i=1}^{m} Y_i h_{ki}$$

which is a linear filter. For $n = 2$, show that this linear filter is not necessarily an optimal linear filter from the perspective that the original vector $\mathbf{X}$ is to be linearly estimated from the compressed vector $\mathbf{Y}_1$; in fact, the estimate from decompression by $\mathbf{H}$ is the optimal linear filter if and only if X_1 and X_2 have equal variances. Specifically, show that $\hat{X}_1 = (X_1 + X_2)/2$ and that the optimal linear estimator of X_1 based on $\mathbf{Y}_1$ is given by

$$\hat{X}_1 = \frac{\sigma_1^2 + \mathrm{Cov}[X_1, X_2]}{\sigma_1^2 + \sigma_2^2 + 2\mathrm{Cov}[X_1, X_2]}(X_1 + X_2)$$

where σ_k is the variance of X_k (and similarly for X_2).

19. Reconsider Exercise 4.18 for the Hadamard transform with $n = 4$.
20. Reconsider Exercise 4.18 for the DCT with $n = 3$.
21. Continuing in the spirit of Exercise 4.18, in Section 3.6.6 it was shown that the Karhunen-Loeve transform is optimal relative to the error criterion of Eq. 3.41. This optimality was of a particular form. Letting $\mathbf{Y} = \mathbf{UX}$ be a matrix transform of the zero-mean n-vector $\mathbf{X}$, optimality is with respect to the compression-decompression scheme

$$\mathbf{X} \rightarrow \mathbf{Y} = \mathbf{UX} \rightarrow \mathbf{Y}_m \rightarrow \mathbf{X}_m = \mathbf{U}^{-1}\mathbf{Y}_m$$

Minimal error occurs when $\mathbf{U}$ is the Karhunen-Loeve matrix. From the perspective of filtering, $\mathbf{U}^{-1}\mathbf{Y}_m$ is an estimator of the random vector $\mathbf{X}$. Letting $\mathbf{U}$ be of the form given in Eq. 3.116 and letting $\mathbf{Z} = \mathbf{UX}$ be the Karhunen-Loeve transform, the random-variable component X_k of vector $\mathbf{X}$ is estimated by the linear filter

$$\hat{X}_k = \sum_{i=1}^{m} Z_i u_{ki}$$

For $n = 2$ and letting

$$\begin{pmatrix} w_{11} & w_{12} \\ w_{21} & w_{22} \end{pmatrix} = \begin{pmatrix} u_{11} & u_{12} \\ u_{21} & u_{22} \end{pmatrix}^{-1}$$

show that

$$\hat{X}_1 = w_{11}u_{11}X_1 + w_{11}u_{12}X_2$$

Show that the optimal homogeneous linear estimator of X_1 based on the single random variable $u_{11}X_1 + u_{12}X_2$ is given by

$$\hat{X}_1 = \frac{u_{11}\sigma_1^2 + u_{12}\mathrm{Cov}[X_1, X_2]}{u_{11}^2\sigma_1^2 + u_{12}^2\sigma_2^2 + 2u_{11}u_{12}\mathrm{Cov}[X_1, X_2]}(u_{11}X_1 + u_{12}X_2)$$

Hence, the estimate of X_1 from decompression by $\mathbf{U}^{-1}$ is the optimal linear filter if and only if the coefficient in the preceding expression equals w_{11}. Demonstrate a similar result for estimation of X_2.

22. Find $\mathbf{A}(3)$ for the steepest descent algorithm applied to the homogeneous case of Example 4.7 with $\mu = 1/\lambda_{max}$. Initialize the algorithm with $\mathbf{A}(0) = \mathbf{0}$.
23. Find $\mathbf{A}(3)$ for the steepest descent algorithm applied to Exercise 4.14 with $\mu = 1/\lambda_{max}$. Initialize the algorithm with $\mathbf{A}(0) = \mathbf{0}$.
24. Write a program to implement the steepest descent algorithm and apply it to Exercise 4.14 with $\mu = 1/\lambda_{max}$. Initialize the algorithm with $\mathbf{A}(0) = \mathbf{0}$.
25. Write a program to implement the LMS algorithm. Let $Y(n)$ be the discrete random signal resulting from sampling at each positive integer the random telegraph signal of Example 2.18 with $\lambda = 0.10$, $Z(n)$ be discrete white noise with variance $\sigma_Z^2 = 0.5$ that is independent of the random telegraph signal, and $X(n) = Y(n) + Z(n)$ be the observed process. Apply the LMS algorithm to adaptively find a 5-point linear filter to restore $Y(n)$ from $X(n)$ by simulating the processes. Run the algorithm for $\mu = 0.05$, $\mu = 0.10$, $\mu = 0.25$, and $\mu = 0.50$. Initialize the algorithm with $\mathbf{A}(0) = \mathbf{0}$.
26. If $\xi(\mathbf{x})$ is a scalar function of the vector $\mathbf{x} = (x_1, x_2, \ldots, x_m)'$, then the derivative of $\xi(\mathbf{x})$ with respect to $\mathbf{x}$ is defined by

$$\frac{d\xi(\mathbf{x})}{d\mathbf{x}} = \begin{pmatrix} \partial\xi(\mathbf{x})/\partial x_1 \\ \partial\xi(\mathbf{x})/\partial x_2 \\ \vdots \\ \partial\xi(\mathbf{x})/\partial x_m \end{pmatrix}$$

Show the following:
(a) for a real symmetric $m \times m$ matrix $\mathbf{A}$, if $\xi(\mathbf{x}) = \mathbf{x}'\mathbf{A}\mathbf{x}$, then $d\xi(\mathbf{x})/d\mathbf{x} = 2\mathbf{A}\mathbf{x}$. Hint: Write out $\mathbf{x}'\mathbf{A}\mathbf{x}$ in terms of the components of $\mathbf{A}$ and $\mathbf{x}$.
(b) for an m-vector $\mathbf{b}$, if $\xi(\mathbf{x}) = \mathbf{b}'\mathbf{x}$, then $d\xi(\mathbf{x})/d\mathbf{x} = \mathbf{b}$.
(c) for m-vector $\mathbf{b}$ and $m \times m$ matrix $\mathbf{A}$, if $\xi(\mathbf{x}) = \mathbf{b}'\mathbf{A}\mathbf{x}$, then $d\xi(\mathbf{x})/d\mathbf{x} = \mathbf{A}'\mathbf{b}$.
27. For Walsh-Hadamard transform coding, suppose the transformed vector is compressed by changing the bottom six components to 0. For the Hadamard matrix $\mathbf{H}_8$ of dimension 8, if $\mathbf{y}_0$ is the compressed vector, then $\mathbf{H}_8\mathbf{y}_0$ is the decompressed vector. Suppose the decompressed vector is degraded by additive, zero-mean noise $\mathbf{N}$, so that the observed vector is $\mathbf{X} = \mathbf{H}_8\mathbf{y}_0 + \mathbf{N}$. Rather than apply $\mathbf{H}_8$ to the vector $\mathbf{y}_0$, one could instead multiply the 2-vector $\mathbf{y}$ defined by the first two components of $\mathbf{y}_0$ by the design matrix $\mathbf{H}$ composed of the first two columns of $\mathbf{H}_8$, thereby resulting in the observation equation $\mathbf{X} = \mathbf{H}\mathbf{y} + \mathbf{N}$. The receiver, knowing that $\mathbf{X}$ is a noisy observation, finds the least-squares

estimator $\hat{\mathbf{y}}$ of $\mathbf{y}$ and then takes $\mathbf{H}\hat{\mathbf{y}}$ as an estimator of the correct compressed signal. Compute $\hat{\mathbf{y}}$ and $\mathbf{H}\hat{\mathbf{y}}$, and interpret the results.

28. For simple linear regression, show that $E[SSE] = (m - 2)\sigma^2$. Hint: First show that

$$SSE = \sum_{i=1}^{k}\sum_{j=1}^{m_i} X_{ij}^2 - m\overline{X}^2 - \hat{\theta}_1\left(\sum_{i=1}^{k}\sum_{j=1}^{m_i} z_{ij}X_{ij} - m\overline{z}\overline{X}\right)$$

29. Derive the expressions for $\mathbf{H'H}$, $(\mathbf{H'H})^{-1}$, $(\mathbf{H'H})^{-1}\mathbf{H'}$, and $(\mathbf{H'H})^{-1}\mathbf{H'X}$ in Example 4.14.
30. What condition guarantees that the columns of the design matrix are linearly independent in multiple linear regression?
31. Reconsider Example 4.13 for images when the observed processes are

$$X_j(m, n) = h_{j,0,0}y(m, n) + h_{j,0,-1}y(m, n-1) + h_{j,0,1}y(m, n+1)$$

$$+ h_{j,1,0}y(m+1, n) + h_{j,-1,0}y(m-1, n) + N_j(m, n)$$

for $j = 1, 2,\ldots, r$.

32. Consider Eq. 4.105 with white noise of variance σ^2 and let $\mathbf{B}$ be a known matrix. Show that the BLUE for the vector $\mathbf{By}$ is given by $\mathbf{B}(\mathbf{H'H})^{-1}\mathbf{H'X}$. Hint: if $\mathbf{DX}$ is any linear unbiased estimator of $\mathbf{By}$, then $\mathbf{DH} = \mathbf{B}$.
33. Find the pseudoinverse for the matrix

$$\mathbf{B} = \begin{pmatrix} 1 & 2 & 1 \\ 2 & 4 & 2 \\ 1 & 2 & 1 \end{pmatrix}$$

34. Let X_1 and X_2 be zero-mean random variables with $X_2 = aX_1$ $(a \neq 0)$ and $E[|X_1|^2] = 1$. Let the random variable Y also have zero mean and let $E[YX_1] = c \neq 0$. Apply Theorem 4.11 to find the optimal linear estimator of Y based on X_1 and X_2.
35. Redo Exercise 4.34 when Y has mean $\mu_Y = d \neq 0$. Compare the results.
36. Write out Eq. 4.177 without matrix notation. Show all expectations and sums.
37. If $\mathbf{X}$ and $\mathbf{Y}$ possess mean vector $\mathbf{0}$, then $\hat{\mathbf{Y}}$ is unbiased, with $E[\hat{\mathbf{Y}}] = \mathbf{0}$. If $\mathbf{Y}$ has nonnull mean vector $E[\mathbf{Y}] = \mathbf{b}$, then $\hat{\mathbf{Y}}$ is biased. Show what happens when we adjoin the constant random variable $X_0 = 1$ to the observation vector, which now becomes $\mathbf{X}_0 = (1, \mathbf{X}')'$. Inclusion of X_0 makes the filter nonhomogeneous. Go one step further and show what happens when $\mathbf{X}$ has nonnull mean vector $E[\mathbf{X}] = \mathbf{a}$.

38. Go through three steps of the recursion of Theorem 4.14 for the scalar observation model $X_k = hY + N_k$. Assume that the second moment of the noise is r and that the procedure is initialized by $\hat{Y}_0 = 0$ and $e_0 = 0$.
39. Under what conditions does Theorem 4.14 yield unbiased estimators? If $\mathbf{Y}$ has nonnull mean vector $\mathbf{b}$, what is the effect of including the constant vector $X_0 = 1$ as part of the initial observation vector $\mathbf{X}_0$?
40. Demonstrate Eq. 4.219. Hint: Let

$$\mathbf{C}_k = (\mathbf{H}_k\mathbf{E}_{k-1}\mathbf{H}_k' + \mathbf{R}_k)^+$$

$\mathbf{C}$ is symmetric and

$$\hat{\mathbf{Y}}_k = \hat{\mathbf{Y}}_{k-1} + \mathbf{E}_{k-1}\mathbf{H}_k'\mathbf{C}_k(\mathbf{X}_k - \mathbf{H}_k\hat{\mathbf{Y}}_{k-1})$$

Now proceed to evaluate $\mathbf{E}_k$ using this representation of $\hat{\mathbf{Y}}_k$.
41. Initialize at zero and compute the recursion of Theorem 4.15 for the scalar system

$$Y(k+1) = tY(k) + U(k)$$

$$X(k) = hY(k) + N(k)$$

where $U(k)$ and $N(k)$ have second moments q and r, respectively. What happens as $r \to \infty$?
42. For the observation model of Theorem 4.14, show how to construct a measurement matrix $\mathbf{H}_k$ so that the update is null, even though the span of the components of $\mathbf{X}_k$ is orthogonal to the span of all previous observation random variables.
43. Derive Eq. 4.243 from Eq. 4.242. A more difficult problem is to derive Eq. 4.242.
44. A key proposition for recursive linear filtering is that the orthogonal projection onto a direct sum of mutually orthogonal subspaces is the sum of the projections onto the subspaces. Demonstrate this proposition. Hint: Let y be a vector, $\mathcal{B}_1, \mathcal{B}_2, \ldots, \mathcal{B}_m$ be bases for $S_1, S_2, \ldots, S_m$, respectively, $\mathcal{B}_k = \{z_{k1}, z_{k2}, \ldots, z_{k,q(k)}\}$, $\sigma_k(y)$ be the projection of y into S_k, $\pi(y)$ be the projection of y into the direct sum $\mathcal{V}$ of $S_1, S_2, \ldots, S_m$, $\mathbf{G}$ be the Grammian of the vectors comprising the basis $\mathcal{B} = (\mathcal{B}_1, \mathcal{B}_2, \ldots, \mathcal{B}_m)$ for $\mathcal{V}$,

$$\mathbf{x}_k = (z_{k1}, z_{k2}, \ldots, z_{k,q(k)})'$$

$$\mathbf{C}_k = (\langle y, z_{k1}\rangle, \langle y, z_{k2}\rangle, \ldots, \langle y, z_{k,q(k)}\rangle)'$$

$$\mathbf{x} = (\mathbf{x}_1', \mathbf{x}_2', \ldots, \mathbf{x}_m')'$$

$\mathbf{C} = (\mathbf{C}_1', \mathbf{C}_2',..., \mathbf{C}_m')'$

Show that

$$\mathbf{G}^{-1}\mathbf{C} = \begin{pmatrix} \mathbf{G}_1^{-1} & \mathbf{0} & \cdots & \mathbf{0} \\ \mathbf{0} & \mathbf{G}_2^{-1} & \cdots & \mathbf{0} \\ \vdots & \vdots & \ddots & \vdots \\ \mathbf{0} & \mathbf{0} & \cdots & \mathbf{G}_m^{-1} \end{pmatrix} \begin{pmatrix} \mathbf{C}_1 \\ \mathbf{C}_2 \\ \vdots \\ \mathbf{C}_m \end{pmatrix} = \begin{pmatrix} \mathbf{G}_1^{-1}\mathbf{C}_1 \\ \mathbf{G}_2^{-1}\mathbf{C}_2 \\ \vdots \\ \mathbf{G}_m^{-1}\mathbf{C}_m \end{pmatrix}$$

where $\mathbf{G}_k$ is the Grammian of the vectors in $\mathcal{B}_k$, and then show that

$$\pi(y) = (\mathbf{G}^{-1}\mathbf{C})'\mathbf{x} = \sum_{k=1}^{m} (\mathbf{G}_k^{-1}\mathbf{C}_k)'\mathbf{x}_k$$

45. In this exercise we derive an alternative form of Theorem 4.13 for the case in which there are a number of observation vectors and the observation random variables are linearly independent. First, suppose we wish to estimate a single random variable Y based on the observed random n-vectors, $\mathbf{X}_0, \mathbf{X}_1,..., \mathbf{X}_k$, where $\mathbf{X}_i = (X_{i1}, X_{i2},..., X_{in})'$ for $i = 0, 1,..., k$, and all random variables in $\mathbf{X}_0, \mathbf{X}_1,..., \mathbf{X}_k$ are linearly independent. Let $\mathcal{S}_i$ be the subspace spanned by the components of $\mathbf{X}_i$. Apply the Gram-Schmidt orthogonalization procedure to find a set of orthogonal random variables, $Z_{i1}, Z_{i2},..., Z_{in}$, spanning $\mathcal{S}_i$ and let $\mathbf{Z}_i = (Z_{i1}, Z_{i2},..., Z_{in})'$. Let $\mathcal{S}$ denote the subspace spanned by the full set of observations. Show that $\mathcal{S} = \mathcal{S}_0 \oplus \mathcal{S}_1 \oplus \cdots \oplus \mathcal{S}_k$ and

$$\hat{Y} = \sum_{i=0}^{k} \sum_{j=1}^{n} \frac{E[YZ_{ij}]}{E[|Z_{ij}|^2]} Z_{ij}$$

Based on this single-random-variable expression, show that the optimal linear estimator of a random vector $\mathbf{Y} = (Y_1, Y_2,..., Y_n)'$ based on $\mathbf{X}_0, \mathbf{X}_1,..., \mathbf{X}_k$ is given by

$$\hat{\mathbf{Y}} = \sum_{i=0}^{k} E[\mathbf{Y}\mathbf{Z}_i']E[\mathbf{Z}_i\mathbf{Z}_i']^{-1}\mathbf{Z}_i$$

For one observation vector, this expression provides a special form of Eq. 4.183.

46. We continue the development of the preceding exercise to find an alternative form of Theorem 4.15 when the observation random variables are linearly independent. We switch to random-function

notation. According to the final equation of Exercise 4.44, the predictor (prior to recursive expression) is given by

$$\hat{\mathbf{Y}}(k+1|k) = \sum_{i=0}^{k} \mathbf{R}_{\mathbf{YZ}}(k+1,i)\mathbf{R}_{\mathbf{Z}}^{-1}(i)\mathbf{Z}(i)$$

where $\mathbf{R}_{\mathbf{Z}}(i) = E[\mathbf{Z}(i)\mathbf{Z}(i)']$ and $\mathbf{R}_{\mathbf{YZ}}$ is the cross-correlation function of $\mathbf{Y}(k)$ and $\mathbf{Z}(i)$. Now consider the linear dynamic model of Eqs. 4.220 and 4.224. Show that

$$\mathbf{R}_{\mathbf{YZ}}(k+1, i) = \mathbf{T}_k\mathbf{R}_{\mathbf{YZ}}(k, i)$$

Substitute this expression into the predictor equation, apply $\mathbf{T}_k$ to the original predictor equation at time k, and subtract the latter from the former. Show that these operations yield the recursive predictor

$$\hat{\mathbf{Y}}(k+1|k) = \mathbf{T}_k(\hat{\mathbf{Y}}(k|k-1) + \mathbf{M}_k\mathbf{Z}(k))$$

where the gain matrix is given by

$$\mathbf{M}_k = \mathbf{R}_{\mathbf{YZ}}(k,k)\mathbf{R}_{\mathbf{Z}}^{-1}(k)$$

Show that Eq. 4.241 holds (with this gain matrix). What is the corresponding filter form?

47. Express the model of Eq. 4.262 for analog images. State the Wiener filter, inverse filter, and Wiener smoothing filter for the model.
48. Find the Wiener smoothing filter when the signal is the random telegraph signal of Example 2.18 and the noise is white of intensity κ. What happens to the filter as $\kappa \to \infty$?
49. Find the inverse filter when an analog signal is blurred by the function $b(t) = e^{-|t|}$.
50. Consider the ideal situation in which the power spectral density for image noise has constant value κ on the disk of radius ρ and constant value 0 outside the disk. What happens to the Wiener smoothing filter as $\kappa \to \infty$?
51. Implement the model of Example 4.19 for just additive, zero-mean Gaussian noise. Design an experiment to study filter robustness as the noise variance varies. Analyze the effect of applying a filter designed for variance σ_1^2 when the variance is actually σ_2^2.
52. Apply the method of Section 4.8.2 with $\varphi_1(t) = 1$, $\varphi_2(t) = t^2$, $\xi_1(s) = 1$, and $\xi_2(s) = s^2$.

53. In Section 4.8.3 it was stated that the white-noise approach is applicable for any finite number of terms. Find the general expressions when $n = 3$.
54. Work out in detail the discrete-domain model of steps L1' through L7' for the model

$$X(i) = Z_1\varphi_1(i) + Z_2\varphi_2(i) + U(i)$$

$$Y(j) = Z_1\xi_1(j) + Z_2\xi_2(j)$$

where $U(i)$ is white noise with constant intensity σ^2, and Z_1 and Z_2 are zero-mean jointly normally distributed random variables possessing the covariance matrix of Eq. 1.191.
55. Express the linear model and the algorithm L1' through L7' for a finite discrete domain in matrix format. This means that $g_k(j, i)$ is treated as a vector

$$\mathbf{g}_k(j) = (g_k(j, 1), g_k(j, 2), \ldots, g_k(j, m))'$$

For instance, the equations of step L2' form the matrix

$$\begin{pmatrix} E[U_1U_1] & E[U_1U_2] & \cdots & E[U_1U_m] \\ E[U_2U_1] & E[U_2U_2] & \cdots & E[U_2U_m]] \\ \vdots & \vdots & \ddots & \vdots \\ E[U_mU_1] & E[U_mU_2]] & \cdots & E[U_mU_m] \end{pmatrix} \begin{pmatrix} g_0(j,1) \\ g_0(j,2) \\ \\ g_0(j,m) \end{pmatrix} = \begin{pmatrix} E[V_jU_1] \\ E[V_jU_2] \\ \\ E[V_jU_m] \end{pmatrix}$$

where we have employed the symmetry of the autocorrelation matrix. Hence, the desired matrix expression is

$$\mathbf{R}_U\mathbf{g}_0(j) = \mathbf{R}_{VU}'\mathbf{d}_j$$

where $\mathbf{R}_U$ is the autocorrelation matrix of $\mathbf{U}$, $\mathbf{R}_{VU}$ is the cross-correlation matrix for $\mathbf{V}$ and $\mathbf{U}$, and $\mathbf{d}_j$ is the vector whose jth component is 1 and all other components are 0.
56. Suppose that based on observing $X(t)$ over the interval $(-\infty, t_0)$ we wish to predict the value of $X(t)$ at time $t_1 > t_0$. Assuming WS stationarity, use the Karhunen-Loeve expansion to find the optimal linear predictor of $X(t_1)$.
57. Reconsider the previous exercise when the observed process is corrupted by uncorrelated additive white noise.
58. Express Eqs. 4.347 through 4.353 when the random functions are over discrete time.
59. State the image versions of Eqs. 4.347 through 4.353.

60. Let X_1, X_2, X_3, X_4 be independent binary random variables with $P(X_i = 1) = p$, $0 < p < 1$. Define the binary random variable Y by $Y = 1$ if and only if $X_1 + X_2 + X_3 + X_4 \geq 2$. Find the binary conditional expectation $E[Y|X_1, X_2]$ for $p = 1/4$ and $p = 1/2$. Find the MSE in both cases.
61. Consider a five-point cross window W centered at the origin in the two-dimensional grid and label the binary values in the window by $x_1, x_2,..., x_5$. Suppose $P(Y = 1|\mathbf{x}) > 0.5$ if and only if at least three of the components are 1-valued. Explain why ψ_{op} is a positive Boolean function, give the disjunctive normal form for ψ_{op}, and give the minimal positive sum-of-products expansion for ψ_{op}.
62. Consider a three-point horizontal window centered at the origin. Suppose the relevant probabilities and restoration effects are given in the following table:

$\mathbf{x}$	$P(\mathbf{X} = \mathbf{x})$	$P(Y = 1\vert \mathbf{x})$	$P(Y = 0\vert \mathbf{x})$	$\delta_{\mathbf{x}}$
000	0.30	0.10	0.90	−0.24
001	0.05	0.10	0.90	−0.04
010	0.20	0.60	0.40	0.04
011	0.05	0.70	0.30	0.02
100	0.05	0.10	0.90	−0.04
101	0.05	0.30	0.70	−0.02
110	0.10	0.70	0.30	0.04
111	0.20	0.90	0.10	0.16

Find the optimal filter and the MAE for the optimal filter. What is the cost of applying the median instead of the optimal filter?

63. The desired vector process $\mathbf{Y}$ over a two-point window is defined by the table:

$\mathbf{y}$	$P(\mathbf{Y} = \mathbf{y})$
00	0.10
01	0.20
10	0.30
11	0.40

Let the noise vector $\mathbf{N}$ consist of two independent binary random variables, each being equiprobable between 0 and 1. The observed process is $\mathbf{X} = \mathbf{Y} \vee \mathbf{N}$, where the maximum is applied componentwise. Find the optimal MAE filter to estimate the first component of $\mathbf{Y}$ based upon observation of $\mathbf{X}$.

64. Suppose the desired vector process $\mathbf{Y}$ over a two-point window is the same one given in Exercise 4.63 and the observed vector $\mathbf{X}$ is defined in a manner dependent on $\mathbf{Y}$ according to the following table:

$\mathbf{y}$	$P(\mathbf{X} = 00 \mid \mathbf{y})$	$P(\mathbf{X} = 01 \mid \mathbf{y})$	$P(\mathbf{X} = 10 \mid \mathbf{y})$	$P(\mathbf{X} = 11 \mid \mathbf{y})$
00	0.10	0.25	0.25	0.40
01	0.20	0.25	0.25	0.30
10	0.30	0.25	0.25	0.20
11	0.40	0.25	0.25	0.10

Find the optimal MAE filter.

65. For a two-point window with the origin being the left pixel, we can consider four structuring elements: (**0**, 0), (**0**, 1), (**1**, 0), (**1**, 1), with bold indicating the origin. Geometrically, these contain 0, 1, 1, and 2 pixels, respectively. The structuring element (**0**, 0) is null and produces the filter having output 1 for any input: $\varepsilon_{(0,0)}(x_1, x_2) = 1$ for any (x_1, x_2). It is usually not considered a legitimate structuring element but is necessary probabilistically in case the optimal filter requires an output of 1 for any input vector. The structuring element (**1**, 0) produces the identity map: $\varepsilon_{(1,0)}(x_1, x_2) = x_1$. Using Theorem 4.18, find the optimal increasing filter for the processes of Exercise 4.64.

66. Using a 512 × 512 image frame, generate a random Boolean process in the following manner: generate a set of pixels by randomly selecting each pixel with probability $p = 0.02$, at each selected pixel place a centered square of uniformly random edge length between 6 and 14 pixels, inclusively, and take the union of the squares. The ideal image process S_0 consists of such realizations. The observed process S consists of degraded realizations of S_0 for which a pixel value is flipped with probability $q = 0.02$. Using a 3 × 3 square (centered) window, find the optimal filter to restore S_0 from S. Design the optimal filter using 20 ideal realizations and associated degraded observations. Estimate MAE for the optimal filter. Generate a new image pair (ideal and degraded) not used during the design procedure, apply the filter to the new observed image, and compute MAE.

67. Using a 512 × 512 image frame, generate S_0 and the observed image process S as in the preceding example. Now, however, the ideal image S_1 consists of all pixels that have a neighbor (horizontal, vertical, or diagonal) in S_0 but are not in S_0 themselves. S_1 is an edge process for S_0. Design the optimal filter Ψ to estimate S_1 from S using 20 ideal

realizations of S_1 and associated observations from S. The goal of Ψ is to optimally find edges in noisy images. Generate a new image pair (ideal edge and degraded) not used during the design procedure, apply Ψ to the new observed image, and compute MAE.

68. Show that the decision regions for linear discriminants are convex. Set A is convex if for any vectors $\mathbf{x}$, $\mathbf{z} \in A$ and nonnegative scalars a and b such that $a + b = 1$, $a\mathbf{x} + b\mathbf{z} \in A$.
69. Consider a two-class, single-feature classifier in which the features possess uniform distributions over [0, 4] and [3, 9], respectively. Find the optimal decision boundary, the misclassification error, and the discriminant functions.
70. Consider a two-class, single-feature classifier in which the features possess Gaussian distributions with means μ_0 and μ_1 ($\mu_0 < \mu_1$), respectively, and common variance σ^2. Find the optimal decision boundary, the misclassification error, and the discriminant functions.
71. Repeat Exercise 4.70 when the standard deviations are different, with $\sigma_0 < \sigma_1$.
72. Consider a two-class, two-feature Gaussian maximum-likelihood classifier. Assume (a) the conditional density $f(\mathbf{x} \mid 0)$ has marginal means μ_{00} and μ_{01}, marginal standard deviations σ_{00} and σ_{01}, and correlation coefficient ρ_0; (b) the conditional density $f(\mathbf{x} \mid 1)$ has marginal means μ_{10} and μ_{11}, marginal standard deviations σ_{10} and σ_{11}, and correlation coefficient ρ_1; and (c) the class probabilities are equal. Find the discriminant functions and the equation of the decision boundary. Assuming equal covariance matrices, find the discriminant functions and the equation of the decision boundary.
73. To give more flexibility to linear discrimination, the form of Eq. 4.406 can be changed to

$$d_k(\mathbf{x}) = \sum_{i=0}^{n} a_{ki} h_i(\mathbf{x})$$

where $h_1, h_2, \ldots, h_n$ are predetermined functions of $\mathbf{x}$ and $h_0(\mathbf{x}) \equiv 1$. Redefine the design matrix of Eq. 4.411 according to this new form of linear discriminant and show that the pseudoinverse solution of Eq. 4.414 still applies.

74. Generate a 512×512 random texture $\mathcal{T}_1$ in the following matter: select a pixel with probability $p_1 = 0.008$; if a pixel is selected, place at it the lower-left corner of a square of edge length 10. Generate a second texture $\mathcal{T}_2$ in the same manner, except use only the lower half of the square. Generate a third texture $\mathcal{T}_3$ like the first except use only the left half of the square. Two features will be used for classification. For the first feature, select every fiftieth pixel in the bottom row of the image (starting with the twentieth pixel), count the number of 0-to-1 changes

in the image while traversing the vertical pixel column above the pixel, and then sum the counts. The second feature is generated similarly except that horizontal 0-to-1 changes are counted. Using numerical methods, find the misclassification error of the optimal classifier according to Eq. 4.394. Next, estimate the means and covariance matrices of the feature vectors for the three textures by generating 50 images of each texture. Using these estimates, construct the Gaussian maximum-likelihood classifier. Finally, generate 50 new images of each texture and apply each classifier to each image to see how they perform on independent data.

75. Instead of all misclassifications being equally costly, one can define a cost c_{kj} incurred by misclassifying a class-C_k pattern into class C_j. The expected misclassification error associated with patterns in C_k is then

$$\rho_k = \sum_{j=0}^{m} c_{kj} \int_{R_j} f(x_1, x_2, \ldots, x_m | k)\, dx_1 dx_2 \cdots dx_m$$

where R_j is the decision region for class C_j. The total expected loss is given by

$$\rho = \sum_{k=0}^{m} \rho_k P(C_k)$$

Find the classifier that minimizes ρ.

76. Consider two-class, n-feature classification in which the conditional densities are Gaussian with common covariance matrix $\mathbf{K}$ and mean vectors $\mathbf{u}_0$ and $\mathbf{u}_1$ for classes C_0 and C_1, respectively. Let Y denote the class number. Show that $P(Y = 1 | \mathbf{x})$ can be expressed as a single-layer neural network with the logistic sigmoid activation function g. Hint: First use Bayes' theorem to show

$$P(Y = 1 | \mathbf{x}) = g(v)$$

$$v = \log \frac{P(\mathbf{x}| Y = 0)\, P(Y = 0)}{P(\mathbf{x}| Y = 1)\, P(Y = 1)}$$

Find $P(\mathbf{x} | Y = k)$, $k = 0$ or 1, from the form of the Gaussian distribution, and then use these conditional probabilities in the expression for v to show that v is of the form given in Eq. 4.416 by finding $a_0, a_1, \ldots, a_n$.

77. Consider two-class, n-feature classification in which the feature vectors are binary, with x_i equal to 0 or 1, the probability of x_i being 1 is p_{ki} for class C_k ($k = 0, 1$), and Y denotes class number. Show that, if the

component random variables are independent, then $P(Y = 1 | \mathbf{x})$ can be expressed as a single-layer neural network with logistic sigmoid activation function g. Hint: Show that

$$P(\mathbf{x}|Y=k) = \sum_{i=1}^{m} p_{ki}^{x_i}(1-p_{ki})^{1-x_i}$$

for $k = 0, 1$, and use Bayes' theorem as in the preceding exercise.

78. Consider a 0-1 binary vector $\mathbf{x} = (x_1, x_2, ..., x_n)'$. Let $a_{\mathbf{x},i} = 1$ if $x_i = 1$ and $a_{\mathbf{x},i} = -1$ if $x_i = 0$ for $i = 1, 2, ..., n$. Let b be the number of 1-valued components of $\mathbf{x}$, $a_{\mathbf{x},0} = -b$, and g be the threshold activation function. Show that, for an arbitrary binary vector $\mathbf{z} = (z_1, z_2, ..., z_n)'$,

$$g\left(a_{\mathbf{x},0} + \sum_{i=1}^{n} a_{\mathbf{x},i} z_i\right) = \begin{cases} 1, & \text{if } \mathbf{z} = \mathbf{x} \\ 0, & \text{if } \mathbf{z} \neq \mathbf{x} \end{cases}$$

Using such functions for the first layer, show that any Boolean function can be represented by a neural network.

79. Derive error back-propagation formulas for a three-layer network using nonadaptive design. Generalize to networks containing an arbitrary number of layers.
80. Extend the conclusions of the preceding exercise to adaptive design.
81. Interpret error back-propagation for a two-layer, multiple-output network that is not fully connected.
82. Interpret Eq. 4.439 in terms of convergence in probability.
83. The SSE analysis of Section 4.12.3 depends on the assumption that the observation pairs comprise a random sample. Suppose they are identically distributed but are not uncorrelated. In this case, Eq. 4.438 does not apply; instead we must appeal to Eq. 1.182. Reinterpret Eqs. 4.438 through 4.440 when the observation pairs are correlated. What do you conclude about using correlated data when training a neural network?

Chapter 5
Random Models

5.1. Markov Chains

In many applications, we observe a system transitioning through various possible states and the probability of observing the system in any given state is conditioned by states occupied at previous times. A key example is a queue, or waiting line, where jobs arrive for service and the state of the system is the number of jobs in the system. This number is random and, at any point in time, depends on arrivals to the system and service to jobs in the system prior to that point. Since arrivals and service times are random, so is the state of the system. In the next three sections, we study systems for which conditional state probabilities depend only on the most recent conditioning event. The first two sections treat mainly discrete-time processes; the third treats continuous-time processes.

5.1.1. Chapman-Kolmogorov Equations

Given values of a one-dimensional, discrete-valued random process at times prior to some time t', (excepting independence) $X(t)$ will depend on one or more of the preceding observations. For instance, for $t < t'$ we consider the *transition probability*

$$p_{x,x'}(t, t') = P(X(t') = x' \mid X(t) = x) \tag{5.1}$$

In the current context, the set of all possible process values is called the *state space* and the transition probability $p_{x,x'}(t, t')$ gives the conditional probability of being in state x' at time t' given the process is in state x at time t. More generally, we consider conditional probabilities $P(X(t') = x' \mid V)$, where V is a set of values for $X(t)$ at some number of points prior in time to t'. In many important situations, conditioning the process on its values at some collection of previous time points reduces to conditioning only at the last time point at which it is conditioned. A discrete-state-space stochastic process is said to be a *Markov chain* if, given any set of time points $t_1 < t_2 < \cdots < t_n$, it satisfies the *Markov property*

$$P(X(t_n) = x_n \mid X(t_1) = x_1, X(t_2) = x_2,\ldots, X(t_{n-1}) = x_{n-1}) = P(X(t_n) = x_n \mid X(t_{n-1}) = x_{n-1}) \tag{5.2}$$

A Markov chain is said to be *homogeneous* if the transition probabilities depend only on the time difference. Letting $\tau = t' - t$, the transition probabilities of a homogeneous Markov chain take the form

$$p_{x,x'}(\tau) = P(X(t+\tau) = x' \mid X(t) = x) \tag{5.3}$$

Markov chains are often defined over discrete time with state space $S = \{1, 2, \ldots\}$. For any time t and state n we define the *state probability*

$$p_n(t) = P(X(t) = n) \tag{5.4}$$

giving the probability that the Markov chain is in state n at time t. The *initial state probabilities* are $p_n(0) = P(X(0) = n)$. At any time t, $\{p_n(t)\colon n \in S\}$ provides a probability distribution, the initial distribution being $\{p_n(0)\colon n \in S\}$.

Example **5.1.** A Poisson process $X(t)$ with parameter λ is a continuous-time homogeneous Markov chain with transition probabilities given by

$$p_{m,n}(t, t') = p_{m,n}(t' - t) = P(X(t') - X(t) = n - m) = \frac{e^{-\lambda(t'-t)}[\lambda(t'-t)]^{n-m}}{(n-m)!} \quad \blacksquare$$

Theorem 5.1. (*Chapman-Kolmogorov equations*) For a discrete-time Markov chain, times $i < r < j$, and states m and n,

$$p_{m,n}(i,j) = \sum_{l=1}^{\infty} p_{m,l}(i,r)\, p_{l,n}(r,j) \quad \blacksquare \tag{5.5}$$

The theorem expands the transition probability in terms of intermediate states. To avoid going into rigorous mathematical conditioning expressions, we give an intuitive argument. For any state k, the transition $X(i) = m$ to $X(j) = n$ can be accomplished by a concatenation of transitions, $X(i) = m$ to $X(r) = l$ followed by $X(r) = l$ to $X(j) = n$. Under the Markovian assumption, the transitions are independent. Moreover, the transition $X(i) = m$ to $X(j) = n$ is the mutually exclusive union of all (over l) such transitions. Therefore the probability of the transition $X(i) = m$ to $X(j) = n$ can be expressed as a sum of products over all l, which is what the theorem states.

5.1.2. Transition Probability Matrix

Now consider a finite state space $S = \{1, 2,..., N\}$. For times i and j, the *transition probability matrix* of a discrete-time Markov chain with state space S is defined to be

$$\mathbf{P}(i,j) = \begin{pmatrix} p_{1,1}(i,j) & p_{1,2}(i,j) & \cdots & p_{1,N}(i,j) \\ p_{2,\mathrm{i}}(i,j) & p_{2,2}(i,j) & \cdots & p_{2,N}(i,j) \\ \vdots & \vdots & \ddots & \vdots \\ p_{N,1}(i,j) & p_{N,2}(i,j) & \cdots & p_{N,N}(i,j) \end{pmatrix} \tag{5.6}$$

The sum of the probabilities in any row of $\mathbf{P}(i, j)$ is 1 because the events from which the row probabilities are derived consist of all possible transitions from a particular state in the time frame i to j. Any matrix for which each row sums to 1 is called a *stochastic matrix*. According to the definition of matrix multiplication, the Chapman-Kolmogorov equations reduce to the matrix equation

$$\mathbf{P}(i,j) = \mathbf{P}(i,r)\mathbf{P}(r,j) \tag{5.7}$$

Proceeding recursively, we see that $\mathbf{P}(i, j)$ can be expressed as a matrix product of one-step transition probability matrices:

$$\begin{aligned} \mathbf{P}(i,j) &= \mathbf{P}(i,j-1)\mathbf{P}(j-1,j) \\ &= \mathbf{P}(i,j-2)\mathbf{P}(j-2,j-1)\mathbf{P}(j-1,j) \\ &\vdots \\ &= \mathbf{P}(i,i+1)\mathbf{P}(i+1,i+2)\cdots\mathbf{P}(j-1,j) \end{aligned} \tag{5.8}$$

For $j = 0, 1, 2,...$, let

$$\mathbf{p}(j) = \begin{pmatrix} p_1(j) \\ p_2(j) \\ \vdots \\ p_N(j) \end{pmatrix} \tag{5.9}$$

be the *j-state probability vector*, $\mathbf{p}(0)$ being the *initial-state probability vector*, and let $\mathbf{p}_k(0, j)$ be the kth column of $\mathbf{P}(0, j)$. Then

$$\mathbf{p}_k(0,j)'\mathbf{p}(0) = \sum_{n=1}^{N} P(X(j)=k \mid X(0)=n)P(X(0)=n)$$

$$= \sum_{n=1}^{N} P(X(j) = k, X(0) = n) \qquad (5.10)$$

$$= P(X(j) = k)$$

$$= p_k(j)$$

From the definition of matrix multiplication, it follows that $\mathbf{p}(j) = \mathbf{P}(0, j)'\mathbf{p}(0)$. Putting this result together with Eq. 5.8 (for $i = 0$) yields an expression for the j-state vector in terms of the initial vector and the one-step transition probabilities:

$$\mathbf{p}(j) = [\mathbf{P}(0, 1)\mathbf{P}(1, 2)...\mathbf{P}(j - 1, j)]'\mathbf{p}(0) \qquad (5.11)$$

For a discrete-time, homogeneous Markov chain, the transition probabilities depend only on the time difference (number of steps) $j - i$. Letting $r = j - i$ be the number of steps, the transition probabilities take the form $p_{m,n}(r) = p_{m,n}(j - i) = p_{m,n}(i, j)$ and these are called the *r-step probabilities*. The *one-step probabilities*,

$$p_{m,n} = P(X(i + 1) = n \mid X(i) = m) \qquad (5.12)$$

are key. In the homogeneous case, the Chapman-Kolmogorov equations can be written as

$$p_{m,n}(r + s) = \sum_{k=1}^{\infty} p_{m,k}(r) p_{k,n}(s) \qquad (5.13)$$

The probability of transitioning from state m to state n in $r + s$ steps is the sum over all states k of the products of the probabilities of transitioning from m to k in r steps and from k to n in s steps.

***Example* 5.2.** In the homogeneous form, the Chapman-Kolmogorov equations have a useful graphical interpretation in terms of one-step probabilities. Figure 5.1 shows the Markov diagram for a three-state homogeneous Markov chain. Each circle represents a state, arrows between states represent possible transitions, and labels on arrows denote transition probabilities. Although all possible transition arrows are drawn, those with a transition probability of zero can be omitted. The arrows having no *tail states* represent the initial distribution. Referring to the figure, we can see the meaning of the homogeneous form of the Chapman-Kolmogorov equations in terms of "arrow chasing." For instance, to transition from state 1 to state 3 in two steps, the process can take any one of three paths: 1–1–3, 1–2–3, or 1–3–3. The strings represent mutually exclusive events and therefore

$$p_{1,3}(2) = P(1-1-3) + P(1-2-3) + P(1-3-3)$$

$$= p_{1,1}p_{1,3} + p_{1,2}p_{2,3} + p_{1,3}p_{3,3}$$

which is the Chapman-Kolmogorov equation (Eq. 5.13) for $r = s = m = 1$ and $n = 3$. ■

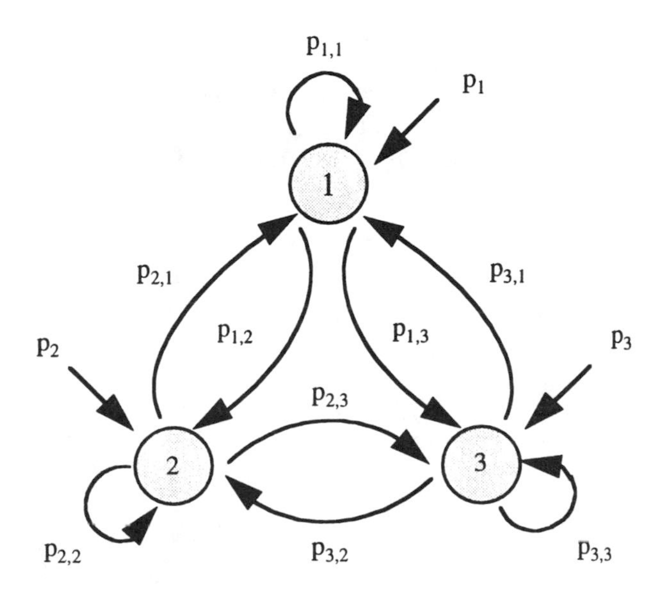

Figure 5.1 Markov diagram for a three-state homogeneous Markov chain.

If a discrete-time, finite-state Markov chain is homogeneous, then we need only consider its *one-step transition probability matrix*

$$\mathbf{P} = \begin{pmatrix} p_{1,1} & p_{1,2} & \cdots & p_{1,N} \\ p_{2,1} & p_{2,2} & \cdots & p_{2,N} \\ \vdots & \vdots & \ddots & \vdots \\ p_{N,1} & p_{N,2} & \cdots & p_{N,N} \end{pmatrix} \tag{5.14}$$

Owing to homogeneity, $\mathbf{P} = \mathbf{P}(i, i+1)$ for $i = 1, 2, \ldots, N-1$. Hence Eq. 5.11 takes the form given in the next theorem.

Theorem 5.2. For a homogeneous, discrete-time Markov chain with state space $S = \{1, 2,..., N\}$, the k-state vector is given by $\mathbf{p}(k) = [\mathbf{P}^k]'\mathbf{p}(0)$. ■

***Example* 5.3.** In Section 2.7.2 we considered a one-dimensional random walk. The process is a discrete-time homogeneous Markov chain. According to Eq. 2.95, if $X(i)$ denotes the state of the process after i steps, then the state probabilities are given by

$$p_n(i) = P(X(i) = n) = \binom{i}{\frac{i+n}{2}} p^{(i+n)/2} q^{(i-n)/2}$$

where p and $q = 1 - p$ are the probabilities of steps to the right and left, respectively. The state space becomes finite if there are *barriers* to constrain movement. For instance, if there are barriers at 0 and 3, then the state space is $S = \{0, 1, 2, 3\}$ and the one-step transition probability matrix is

$$\mathbf{P} = \begin{pmatrix} p_{0,0} & p_{0,1} & p_{0,2} & p_{0,3} \\ p_{1,0} & p_{1,1} & p_{1,2} & p_{1,3} \\ p_{2,0} & p_{2,1} & p_{2,2} & p_{2,3} \\ p_{3,0} & p_{3,1} & p_{3,2} & p_{3,3} \end{pmatrix} = \begin{pmatrix} 0 & 1 & 0 & 0 \\ q & 0 & p & 0 \\ 0 & q & 0 & p \\ 0 & 0 & 1 & 0 \end{pmatrix}$$

Assuming the walk starts at the origin, the initial vector is $\mathbf{p}(0) = (1, 0, 0, 0)'$. According to Theorem 5.2, the state vector at time $j = 2$ is

$$\mathbf{p}(2) = [\mathbf{P}^2]'\mathbf{p}(0) = \begin{pmatrix} q & 0 & q^2 & 0 \\ 0 & q(1+p) & 0 & q \\ p & 0 & p(1+q) & 0 \\ 0 & p^2 & 0 & p \end{pmatrix} \begin{pmatrix} 1 \\ 0 \\ 0 \\ 0 \end{pmatrix} = \begin{pmatrix} q \\ 0 \\ p \\ 0 \end{pmatrix}$$ ■

5.1.3. Markov Processes

Thus far in this section, and as will be the case in the next two sections, we have assumed a discrete state space. This allowed us in Eq. 5.2 to give the Markov property in terms of the conditional probability that $X(t_n) = x_n$. When the process does not have a discrete state space, then the Markov property takes the following form: if $t_1 < t_2 < \cdots < t_n$, then

$$P(X(t_n) \le x_n \mid X(t_1) = x_1, X(t_2) = x_2,..., X(t_{n-1}) = x_{n-1}) = P(X(t_n) \le x_n \mid X(t_{n-1}) = x_{n-1}) \tag{5.15}$$

If $X(t)$ possesses the Markov property, then it is called a *Markov process*. In terms of conditional densities, the Markov property becomes

$$f(x_n; t_n \mid x_1, x_2,\ldots, x_{n-1}; t_1, t_2,\ldots, t_{n-1}) = f(x_n; t_n \mid x_{n-1}; t_{n-1}) \tag{5.16}$$

By repeatedly employing the multiplication principle for conditional densities together with the Markovian conditioning property, we can express the nth-order joint density of a Markov process in terms of second-order densities:

$$\begin{aligned}
&f(x_1, x_2,\ldots, x_n; t_1, t_2,\ldots, t_n)\\
&= f(x_1, x_2,\ldots, x_{n-1}; t_1, t_2,\ldots, t_{n-1}) f(x_n; t_n \mid x_1, x_2,\ldots, x_{n-1}; t_1, t_2,\ldots, t_{n-1})\\
&= f(x_1, x_2,\ldots, x_{n-1}; t_1, t_2,\ldots, t_{n-1}) f(x_n; t_n \mid x_{n-1}; t_{n-1})\\
&= f(x_1, x_2,\ldots, x_{n-2}; t_1, t_2,\ldots, t_{n-2}) f(x_{n-1}; t_{n-1} \mid x_1, x_2,\ldots, x_{n-2}; t_1, t_2,\ldots, t_{n-2}) f(x_n; t_n \mid x_{n-1}; t_{n-1})\\
&= f(x_1, x_2,\ldots, x_{n-2}; t_1, t_2,\ldots, t_{n-2}) f(x_{n-1}; t_{n-1} \mid x_{n-2}; t_{n-2}) f(x_n; t_n \mid x_{n-1}; t_{n-1})\\
&\quad\vdots\\
&= f(x_1; t_1) \prod_{k=2}^{n} f(x_k; t_k \mid x_{k-1}; t_{k-1})\\
&= f(x_1; t_1) \prod_{k=2}^{n} \frac{f(x_{k-1}, x_k; t_{k-1}, t_k)}{f(x_{k-1}; t_{k-1})}\\
&= \frac{\prod_{k=2}^{n} f(x_{k-1}, x_k; t_{k-1}, t_k)}{\prod_{k=2}^{n-1} f(x_k; t_k)}
\end{aligned} \tag{5.17}$$

Like Gaussian processes, Markov processes are completely characterized by their second-order distributions.

5.2. Steady-State Distributions for Discrete-Time Markov Chains

Perhaps the most important aspect of Markov chains in terms of applications is the analysis of their long-run behavior. Does there exist a probability distribution giving limiting probabilities of the Markov chain being in the various states, independent of a given initial state? And if such a limiting distribution exists, how can it be found? We consider the long-run behavior of homogeneous discrete-time Markov chains.

5.2.1. Long-Run Behavior of a Two-State Markov Chain

To appreciate long-run behavior, consider a two-state Markov chain with state space {0, 1} and one-step transition probability matrix

$$\mathbf{P} = \begin{pmatrix} p_{0,0} & p_{0,1} \\ p_{1,0} & p_{1,1} \end{pmatrix} = \begin{pmatrix} a & 1-a \\ 1-d & d \end{pmatrix} \tag{5.18}$$

where $a = p_{0,0}$ and $d = p_{1,1}$. Squaring $\mathbf{P}$ yields

$$\mathbf{P}^2 = \begin{pmatrix} a^2 + (1-a)(1-d) & a(1-a) + d(1-a) \\ a(1-d) + d(1-d) & (1-a)(1-d) + d^2 \end{pmatrix} \tag{5.19}$$

More generally, if $u = 2 - a - d$ and $|1 - u| < 1$, then mathematical induction shows that

$$\mathbf{P}^k = u^{-1}\begin{pmatrix} 1-d & 1-a \\ 1-d & 1-a \end{pmatrix} + u^{-1}(1-u)^k \begin{pmatrix} 1-a & a-1 \\ d-1 & 1-d \end{pmatrix} \tag{5.20}$$

Thus,

$$\lim_{k\to\infty} \mathbf{P}^k = u^{-1}\begin{pmatrix} 1-d & 1-a \\ 1-d & 1-a \end{pmatrix} \tag{5.21}$$

In the homogeneous case, $\mathbf{P}(0, k) = \mathbf{P}^k$ and therefore the matrix limit can be expressed in terms of the individual components of $\mathbf{P}(0, k)$:

$$\pi_0 = \lim_{k\to\infty} p_{0,0}(k) = \lim_{k\to\infty} p_{1,0}(k) = \frac{1 - p_{1,1}}{2 - p_{0,0} - p_{1,1}} \tag{5.22}$$

$$\pi_1 = \lim_{k\to\infty} p_{0,1}(k) = \lim_{k\to\infty} p_{1,1}(k) = \frac{1 - p_{0,0}}{2 - p_{0,0} - p_{1,1}} \tag{5.23}$$

The pair (π_0, π_1), for which $\pi_0 + \pi_1 = 1$, is called the *steady-state (long-run) distribution* of the Markov chain. From a practical perspective, the existence of steady-state probabilities means that, for large k, we can treat the transition probabilities as constants, independent of k. For large k, the probability of transitioning from state 1 to state 0 in k steps is approximately π_0, regardless of the precise value of k. Moreover, for large k, $p_{0,0}(k) \approx \pi_0 \approx p_{1,0}(k)$, so that after k steps the probability of the process being in state 0 is essentially independent of the state of the process prior to commencing the k steps. Similarly, for large k, $p_{0,1}(k) \approx \pi_1 \approx p_{1,1}(k)$.

5.2.2. Classification of States

Analysis of steady-state behavior for homogeneous Markov chains possessing more than two states is facilitated by classification of states. State n is said to be *accessible* from state m if state n can be reached from state m in a finite number of steps: probabilistically, there exists r such that $p_{m,n}(r) > 0$. State n is not accessible from state m if $p_{m,n}(r) = 0$ for all $r \geq 0$. States m and n *communicate* if each is accessible from the other. We write $m \leftrightarrow n$ to denote that m and n communicate. Communication is an equivalence relation: it is *reflexive* ($m \leftrightarrow m$), *symmetric* ($m \leftrightarrow n$ implies $n \leftrightarrow m$), and *transitive*, ($m \leftrightarrow n$ and $n \leftrightarrow l$ imply $m \leftrightarrow l$). Consequently, the set of states is partitioned into the union of equivalence classes, where each equivalence class consists of states that communicate with each other but communicate with no other states. Note that state n may be accessible from state m but that state m might not be accessible from state n. A Markov chain is said to be *irreducible* if all states belong to a single equivalence class, that is, if they all communicate with each other. Often it is feasible to break the analysis of a Markov chain into individual analyses of its equivalence classes.

To characterize the recurrence of a Markov chain with respect to a given state n, let

$$f_{nn}(i) = P(X(i) = n, X(j) \neq n \text{ for } j = 1, 2, \ldots, i-1 \mid X(0) = n) \tag{5.24}$$

$$f_{nn} = \sum_{i=1}^{\infty} f_{nn}(i) \tag{5.25}$$

$$\mu_{nn} = \sum_{i=1}^{\infty} i f_{nn}(i) \tag{5.26}$$

Then $f_{nn}(i)$ is the probability that, starting from state n, the process returns to state n for the first time in i steps; f_{nn} is the probability that, starting from state n, the process returns to state n in a finite number of steps; and μ_{nn} is the expected number of steps for the first return of the process to state n. The number of steps for the first return to state n is called the *recurrence time* and μ_{nn} is the *mean recurrence time* of state n. State n is said to be *recurrent* if, starting from state n, the probability of return is 1, that is, if $f_{nn} = 1$; state n is *nonrecurrent* if $f_{nn} < 1$. A recurrent state n is called *null recurrent* if $\mu_{nn} = \infty$ and *positive recurrent* if $\mu_{nn} < \infty$. As defined, state n is recurrent or nonrecurrent according to whether $f_{nn} = 1$ or $f_{nn} < 1$. It can be shown that recurrence can be equivalently characterized via the transition probabilities for the state: state n is recurrent if and only if

$$\sum_{j=1}^{\infty} p_{n,n}(j) = \infty \tag{5.27}$$

Employing the characterization of recurrence by Eq. 5.27, we can show that, if m and n communicate and m is recurrent, then n is recurrent. To see this, note that $m \leftrightarrow n$ implies there exist times r and s such that $p_{m,n}(r) > 0$ and $p_{n,m}(s) > 0$. For any $j \geq 1$,

$$p_{n,n}(r + j + s) \geq p_{m,n}(r) p_{m,m}(j) p_{n,m}(s) \tag{5.28}$$

Hence,

$$\sum_{j=1}^{\infty} p_{n,n}(r + j + s) \geq p_{m,n}(r) p_{n,m}(s) \sum_{j=1}^{\infty} p_{m,m}(j) \tag{5.29}$$

The sum on the right diverges to infinity because m is recurrent. Therefore the sum on the left also diverges to infinity, which forces the summation in Eq. 5.27 to diverge to infinity, thereby implying that state n is recurrent. Since all states communicating with a given recurrent state are recurrent, recurrence is a property of state equivalence classes: in any equivalence class, either all states are recurrent or none are. Hence, we characterize each equivalence class as being recurrent or nonrecurrent.

The *period* of state n is the greatest common divisor of all integers $i \geq 1$ for which $p_{n,n}(i) > 0$. If the period is 1, then the state is said to be *aperiodic*. In analogy to recurrence and state communication, if states m and n communicate, then m and n have the same period. Hence, periodicity is also a property of equivalence classes, so that we can speak of the period of an equivalence class or of an equivalence class being aperiodic. If a state n in an equivalence class can be found for which $p_{n,n} > 0$, then the state, and therefore the equivalence class, is aperiodic.

5.2.3. Steady-State and Stationary Distributions

Turning to long-run behavior, a homogeneous, discrete-time Markov chain with state space S is said to possess a *steady-state* (*long-run*) *distribution* if there exists a probability distribution $\{\pi_n : n \in S\}$ such that, for all states $m, n \in S$,

$$\lim_{k \to \infty} p_{m,n}(k) = \pi_n \tag{5.30}$$

If there exists a steady-state distribution, then, regardless of the state m, the probability of the Markov chain being in state n in the long run is π_n. In particular, for any initial distribution $\{p_n(0): n \in S\}$, the state probability $p_n(k)$ approaches π_n as $k \to \infty$. Indeed,

$$\begin{aligned}\lim_{k\to\infty} p_n(k) &= \lim_{k\to\infty} \sum_{m\in S} p_m(0)p_{m,n}(k) \\ &= \sum_{m\in S}\left(p_m(0)\lim_{k\to\infty} p_{m,n}(k)\right) \qquad (5.31) \\ &= \left(\sum_{m\in S} p_m(0)\right)\pi_n\end{aligned}$$

which, because the total initial-state probability is 1, equals π_n.

A homogeneous Markov chain with state space S is said to possess a *stationary distribution* if there exists a probability distribution $\{\pi_n: n \in S\}$ such that, for any $n \in S$,

$$\pi_n = \sum_{m\in S} \pi_m p_{m,n} \qquad (5.32)$$

If there exists a stationary distribution $\{\pi_n: n \in S\}$, then, for any step time r,

$$\pi_n = \sum_{m\in S} \pi_m p_{m,n}(r) \qquad (5.33)$$

which can be demonstrated by mathematical induction in conjunction with the Chapman-Kolmogorov equations. By definition it holds for $r = 1$. Assuming it holds for r, then

$$\begin{aligned}\sum_{m=1}^{\infty} \pi_m p_{m,n}(r+1) &= \sum_{m=1}^{\infty} \pi_m \sum_{l=1}^{\infty} p_{m,l}p_{l,n}(r) \\ &= \sum_{l=1}^{\infty}\left(\sum_{m=1}^{\infty} \pi_m p_{m,l}\right)p_{l,n}(r) \qquad (5.34) \\ &= \sum_{l=1}^{\infty} \pi_l p_{l,n}(r)\end{aligned}$$

which, by the induction hypothesis, reduces to π_n. According to Eq. 5.33, if the initial distribution is $\{\pi_n: n \in S\}$, then $p_n(i) = \pi_n$ for all i. Hence, the Markov chain is a stationary random process. More generally, Eq. 5.33 shows that once the state probability distribution assumes the stationary distribution, henceforth the state probability distribution remains unchanged, thereby rendering the Markov chain stationary from that point forward.

In general, for an irreducible Markov chain, there exists a set $\{x_n, n \in S\}$ of nonnegative values such that, for any $m, n \in S$,

$$\lim_{k\to\infty} \frac{1}{k}\sum_{j=1}^{k} p_{m,n}(j) = x_n \tag{5.35}$$

Even if $x_1 + x_2 + \cdots = 1$, this does not mean that $\{x_n: n \in S\}$ forms a steady-state distribution because convergence of the mean in Eq. 5.35 does not imply convergence of the terms. Moreover, while for a finite state space, $x_1 + x_2 + \cdots = 1$, the sum is not necessarily 1 for an infinite state space.

The matter of whether or not a Markov chain possesses a steady-state distribution is not simply one of whether or not $p_{m,n}(k)$ has a limit as $k \to \infty$. According to Eq. 5.27, state n is nonrecurrent if and only if

$$\sum_{j=1}^{\infty} p_{n,n}(j) < \infty \tag{5.36}$$

Hence, if state n is nonrecurrent, then

$$\lim_{k\to\infty} p_{n,n}(k) = 0 \tag{5.37}$$

Since recurrence is a class property, this implies that an irreducible Markov chain having a nonrecurrent state cannot possess a steady-state distribution since, for any steady-state distribution, the sum of the state probabilities must be 1.

For a recurrent state n there need not be a limit for $p_{m,n}(k)$ as $k \to \infty$ and, if limits exist for all states $n \in S$, these limits need not form a steady-state distribution. We summarize the basic limiting properties for recurrent states. Let state n belong to an aperiodic recurrent equivalence class. If m is any state in the equivalence class containing n, then

$$\lim_{k\to\infty} p_{m,n}(k) = \frac{1}{\mu_{nn}} \tag{5.38}$$

where μ_{nn} is the mean recurrence time for state n. If n is positive recurrent, then

$$\lim_{k\to\infty} p_{m,n}(k) > 0 \tag{5.39}$$

If n is null recurrent, then

$$\lim_{k\to\infty} p_{m,n}(k) = 0 \tag{5.40}$$

***Example* 5.4.** The probability transition matrix for the homogeneous Markov chain described by the diagram of Fig. 5.2 is

$$\mathbf{P} = \begin{pmatrix} 0 & a & 1-a & 0 \\ b & c & 0 & 1-b-c \\ 0 & 0 & 0 & 1 \\ 0 & 0 & 1 & 0 \end{pmatrix}$$

where we assume that $0 < a, b, c < 1$. From the diagram we can see that state 1 is nonrecurrent. In fact, $f_{11}(1) = 0$, $f_{11}(2) = ab$, $f_{11}(3) = abc, \ldots, f_{11}(i) = abc^{i-2}, \ldots$, and

$$f_{11} = \sum_{k=0}^{\infty} abc^k = \frac{ab}{1-c} < 1$$

There are two irreducible classes, {1, 2} and {3, 4}. Class {1, 2} is nonrecurrent and aperiodic. Class {3, 4} is recurrent $[f_{33}(2) = f_{44}(2) = 1]$ and has period 2. If we consider the class {3, 4} in isolation, then it has the transition probability matrix

$$\mathbf{P} = \begin{pmatrix} 0 & 1 \\ 1 & 0 \end{pmatrix}$$

Also $p_{3,3}(k) = 0$ if k is odd and $p_{3,3}(k) = 1$ if k is even; on the other hand, $p_{4,3}(k) = 1$ if k is odd and $p_{4,3}(k) = 1$ if k is even. As a two-state Markov chain, the chain does not possess a steady-state distribution. For future reference, note that

$$\mathbf{P}'\begin{pmatrix} 1/2 \\ 1/2 \end{pmatrix} = \begin{pmatrix} 1/2 \\ 1/2 \end{pmatrix}$$

■

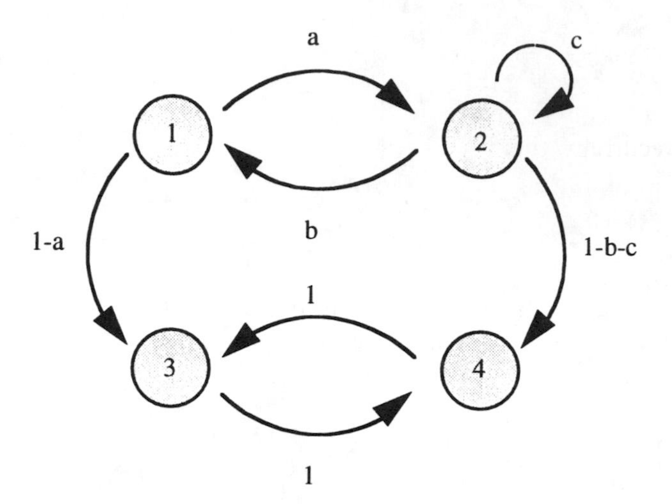

Figure 5.2 Markov diagram for Example 5.4.

5.2.4. Long-Run Behavior of Finite Markov Chains

We now provide some general theorems that are useful for applications. We first treat the finite-state case, for which long-run and stationary behavior can be neatly characterized in terms of the probability transition matrix and for which stationary and steady-state distributions are probability vectors. Specifically, according to Eq. 5.32, the vector

$$\boldsymbol{\pi} = (\pi_1, \pi_2, \ldots, \pi_N)' \tag{5.41}$$

provides the stationary distribution for the Markov chain if and only if $\mathbf{P}'\boldsymbol{\pi} = \boldsymbol{\pi}$, or in terms of the $N \times N$ matrix $\boldsymbol{\Pi}$ whose columns are each $\boldsymbol{\pi}$,

$$\mathbf{P}'\boldsymbol{\Pi} = \boldsymbol{\Pi} \tag{5.42}$$

Since $\boldsymbol{\pi}$ is a probability vector, $0 \leq \pi_n \leq 1$ for all n and $\pi_1 + \pi_2 + \cdots + \pi_N = 1$. Moreover, according to Eq. 5.30, the Markov chain has steady-state distribution $\boldsymbol{\pi}$ if and only if

$$\lim_{k \to \infty} \mathbf{P}^k = \boldsymbol{\Pi} \tag{5.43}$$

Relative to the probability vector $\boldsymbol{\pi}$, Eq. 5.31 takes the form

$$\lim_{k\to\infty} \mathbf{p}(k) = \boldsymbol{\pi} \tag{5.44}$$

Theorem 5.3. Let **P** be the transition probability matrix for an aperiodic, irreducible, homogeneous Markov chain possessing finite state space $S = \{1, 2, \ldots, N\}$. Then there exists a unique probability vector $\boldsymbol{\pi}$, with $0 < \pi_n < 1$, providing both the stationary and steady-state distributions. Convergence to $\boldsymbol{\pi}$ is *geometric* in the sense that there exist constants α and ε, $\alpha > 0$ and $0 < \varepsilon < 1$, such that for any states m and n,

$$\max \{|p_{m,n}(k) - \pi_n|\} < \alpha\varepsilon^k \quad \blacksquare \tag{5.45}$$

Should a homogeneous Markov chain possessing finite state space only be irreducible, and not necessarily aperiodic, then we can still assert that it has a stationary distribution, but not that it has a steady-state distribution. Indeed, this is precisely what happened for the two-state Markov chain {3,4} of Example 5.4.

Together with $\boldsymbol{\pi}$ being a probability vector, the relation $\mathbf{P}'\boldsymbol{\pi} = \boldsymbol{\pi}$ can be used to find the stationary (steady-state) distribution of a finite-state, irreducible, aperiodic Markov chain.

***Example* 5.5.** For the random walk with barriers discussed in Example 5.3, the stationary distribution is found by solving the 5-equation system composed of $\mathbf{P}'\boldsymbol{\pi} = \boldsymbol{\pi}$ and the total probability constraint. The resulting linear system is

$$q\pi_2 = \pi_1$$

$$\pi_1 + q\pi_3 = \pi_2$$

$$p\pi_2 + \pi_4 = \pi_3$$

$$p\pi_3 = \pi_4$$

$$\pi_1 + \pi_2 + \pi_3 + \pi_4 = 1$$

If $p = q = 1/2$, then the stationary distribution vector is

$$\boldsymbol{\pi} = (1/6, 1/3, 1/3, 1/6)' \quad \blacksquare$$

5.2.5. Long-Run Behavior of Markov Chains with Infinite State Spaces

We now turn to stationarity and steady-state behavior for homogeneous Markov chains possessing countably infinite state spaces. We will say that a sequence $\{x_n : n \in S\}$ is *absolutely convergent* if the infinite series with terms $|x_n|$ is convergent.

Theorem 5.4. For a discrete-time, irreducible, homogeneous Markov chain with state space $S = \{1, 2, ...\}$, the following conditions are equivalent:

i) The Markov chain possesses a stationary distribution.

ii) S is positive recurrent.

iii) There exists an absolutely convergent sequence $\{x_n\colon n \in S\}$, not identically 0, such that

$$x_n = \sum_{m \in S} x_m p_{m,n} \tag{5.46}$$

for all $n \in S$. In such a case, the stationary distribution $\{\pi_n\colon n \in S\}$ is given by

$$\pi_n = \frac{1}{\mu_{nn}} = \frac{x_n}{\sum_{m \in S} x_m} \quad \blacksquare \tag{5.47}$$

***Example* 5.6.** We consider a random walk with a single barrier at 0 for the state space $S = \{0, 1, 2, ...\}$; more specifically, we consider the *generalized random walk*, for which the transition probabilities depend on the state. To simplify notation, we let $r_k = p_{k,k}$ and $p_k = p_{k,k+1}$ for $k = 0, 1, 2, ...,$ and $q_k = p_{k,k-1}$ for $k = 1, 2, 3, ...$ The transition probabilities can be conveniently displayed as an infinite transition probability matrix

$$\mathbf{P} = \begin{pmatrix} r_0 & p_0 & 0 & 0 & \cdots \\ q_1 & r_1 & p_1 & 0 & \cdots \\ 0 & q_2 & r_2 & p_2 & \cdots \\ 0 & 0 & q_3 & r_3 & \cdots \\ \vdots & \vdots & \vdots & \vdots & \ddots \end{pmatrix}$$

The generalized random walk with a left barrier has application in modeling certain adaptive nonlinear image filters. Theorem 5.4 allows us to determine both positive recurrence and stationarity of the model. The system of Eq. 5.46 reduces to

$$x_0 = x_0 r_0 + x_1 q_1$$

$$x_i = x_{i-1} p_{i-1} + x_i r_i + x_{i+1} q_{i+1} \qquad (i = 1, 2, \ldots)$$

Since $r_0 + p_0 = 1$ and $p_i + q_i + r_i = 1$ for $i = 1, 2, \ldots$, the system can be expressed as

$$x_1 q_1 = x_0 p_0$$

$$x_{i+1} q_{i+1} = x_i p_i + x_i q_i - x_{i-1} p_{i-1} \qquad (i = 1, 2, 3, \ldots)$$

Solving the system for x_1 and x_2 yields

$$x_1 = x_0 \frac{p_0}{q_1}$$

$$x_2 = \frac{x_1 p_1 + x_1 q_1 - x_0 p_0}{q_2} = \frac{x_1 p_1}{q_2} = x_0 \frac{p_0 p_1}{q_1 q_2}$$

Proceeding recursively,

$$x_n = x_0 \frac{p_0 p_1 \cdots p_{n-1}}{q_1 q_2 \cdots q_n}$$

According to Theorem 5.4, the generalized random walk is positive recurrent and has a stationary distribution if and only if

$$\sum_{n=1}^{\infty} \frac{p_0 p_1 \cdots p_{n-1}}{q_1 q_2 \cdots q_n} < \infty$$

Depending on the complexity of the transition probabilities p_n and q_n, checking for convergence of the preceding infinite sum can be difficult. In the simple case where the right and left transitions are fixed at p and q, respectively, the series converges if and only if $p < q$. According to Eq. 5.47, if $p < q$, then, for $n = 0, 1, 2, \ldots$,

$$\pi_n = \frac{(p/q)^n}{\sum_{n=0}^{\infty} (p/q)^n} = \frac{q-p}{q}\left(\frac{p}{q}\right)^n$$

$$\mu_{nn} = \frac{q}{q-p}\left(\frac{q}{p}\right)^n$$ ■

Theorem 5.5. A discrete-time, irreducible, aperiodic, positive recurrent, homogeneous Markov chain with state space $S = \{1, 2,...\}$ has a steady-state distribution if and only if there exists an absolutely convergent sequence $\{x_n: n \in S\}$ for which Eq. 5.46 holds for all $n \in S$. The steady-state distribution $\{\pi_n: n \in S\}$ is the unique solution of the system

$$\pi_n = \sum_{m \in S} \pi_m p_{m,n} \qquad \text{(for all } n \in S) \tag{5.48}$$

$$\sum_{n=1}^{\infty} \pi_n = 1 \tag{5.49}$$

Moreover, the steady-state distribution is also the stationary distribution. ■

Since the Markov chain in Theorem 5.5 is assumed to be positive recurrent, as well as aperiodic, the fact that it has a stationary distribution follows from Theorem 5.4. A key difference in the theorems is the assumption of aperiodicity in Theorem 5.5. Note that the general random walk is aperiodic because $p_{n,n} > 0$ for all $n \in S$. Hence, when finding its stationary distribution, we *ipso facto* found the steady-state distribution. The equations for the steady-state distribution take the same form in both the finite and countably infinite cases; however, in the finite case they can be solved directly by matrix methods, whereas the infinite case is more problematic. In the infinite case, where the number of equations is infinite, one can sometimes apply the method of generating functions.

***Example* 5.7.** As in Example 5.6, consider the random walk with a barrier at 0. For $k = 1, 2,...$, fix $p_k = p > 0$, $q_k = q > 0$, and $r_k = r = 1 - p - q$. For $k = 0$, let $r_0 = 1 - p = q + r$ and $p_0 = p$. Then the infinite transition probability matrix is

$$\mathbf{P} = \begin{pmatrix} q+r & p & 0 & 0 & \cdots \\ q & r & p & 0 & \cdots \\ 0 & q & r & p & \cdots \\ 0 & 0 & q & r & \cdots \\ \vdots & \vdots & \vdots & \vdots & \ddots \end{pmatrix}$$

The system of Eq. 5.48 takes the form

$$\pi_0 = (q + r)\pi_0 + q\pi_1$$

$$\pi_n = p\pi_{n-1} + r\pi_n + q\pi_{n+1} \quad (n = 1, 2,\ldots)$$

Multiplying the equation for π_n by z^n yields the system

$$\pi_0 = q\pi_0 + r\pi_0 + q\pi_1$$

$$\pi_n z^n = p\pi_{n-1}z^n + r\pi_n z^n + q\pi_{n+1}z^n \quad (n = 1, 2,\ldots)$$

Let

$$A(z) = \sum_{n=0}^{\infty} \pi_n z^n$$

Summing over the system yields

$$A(z) = q\pi_0 + pzA(z) + rA(z) + z^{-1}(qA(z) - q\pi_0)$$

Solving,

$$A(z) = \frac{q\pi_0}{q - pz}$$

Owing to the probability condition of Eq. 5.49, $A(z) \to 1$ as $z \to 1$. Therefore, letting $z \to 1$ yields $\pi_0 = (q - p)/q$. Hence,

$$A(z) = \frac{q-p}{q-pz}$$

$$= \frac{q-p}{q}\frac{1}{1-(p/q)z}$$

$$= \frac{q-p}{q}\sum_{n=0}^{\infty}\left(\frac{p}{q}\right)^n z^n$$

Inversion of the transform gives the same solution for π_n as found in Example 5.6. ■

5.3. Steady-State Distributions for Continuous-Time Markov Chains

To this point we have concentrated on Markov chains over a discrete-time parameter. This section considers Markov chains defined over continuous time. We will first concentrate on the birth-death model and then show how the differential approach to finding the steady-state distribution for that model can be embedded in a more general methodology involving the Chapman-Kolmogorov equations.

5.3.1. Irreducible Continuous-Time Markov Chains

For continuous time, the Chapman-Kolmogorov equations take the form

$$p_{m,n}(t+s) = \sum_{l \in S} p_{m,l}(t) p_{l,n}(s) \tag{5.50}$$

for all t, $s > 0$. The following existence theorem applies for continuous-time processes.

Theorem 5.6. For an irreducible, continuous-time Markov chain, the limit as $t \to \infty$ of each state probability $p_n(t)$ exists independently of the initial probability. The limits are either 0 for all states or are all positive and form a steady-state distribution, in which case the distribution is also stationary. ■

If for an irreducible, continuous-time Markov chain the limits of the state probabilities are all positive, then the Markov chain is called *positive recurrent*.

A practical method for obtaining the steady-state distribution is to use the Chapman-Kolmogorov equations to express $p_n(t + \delta)$, form the difference quotient, and then let $\delta \to 0$ to find the derivative $p_n'(t)$ with respect to t, namely,

$$\frac{d}{dt} p_n(t) = \lim_{\delta \to 0} \frac{p_n(t+\delta) - p_n(t)}{\delta} \tag{5.51}$$

Given the existence of the steady-state distribution, $p_n(t) \to \pi_n$ as $t \to \infty$ and $p_n'(t) \to 0$ as $t \to \infty$. The Chapman-Kolmogorov equations thereby lead to a system of equations involving the steady-state probabilities. Our task is to solve this system in conjunction with the total-probability condition $\pi_1 + \pi_2 + \pi_3 + \cdots = 0$ to find the steady-state probabilities. We will demonstrate the method by applying it to the birth-death process.

5.3.2. Birth-Death Model — Queues

Whenever there are jobs to be serviced, variability in job arrivals and service times creates varying degrees of congestion, the result being a waiting line, or *queue*. In imaging, queuing problems arise in real-time image processing in regard to image data and algorithms awaiting space in memory or requiring CPU service. Various computational design constraints affect the manner in which imaging programs are serviced: for instance, architectural parallelism, operating system design, and multitasking disciplines. Queuing-type models are also employed for filter design and analysis.

The population of jobs awaiting service in a queue can be modeled in a general stochastic fashion by considering the generic problem of population increase and decrease. A birth is an event that increases the population; a death is an event that decreases the population. Births correspond to arrivals and deaths correspond to completions of service. To construct a stochastic model for births and deaths, let $N(t)$ denote the size of the population at time t. The *birth-death model* satisfies three axioms:

B1. (*birth*) If $N(t) = n$, then
- (a) the probability of exactly one birth in the time interval $(t, t + \delta]$ is $\lambda_n\delta + o(\delta)$,
- (b) the probability of no births during $(t, t + \delta]$ is $1 - \lambda_n\delta + o(\delta)$,
- (c) the probability of more than one birth during $(t, t + \delta]$ is $o(\delta)$,
- (d) births occurring in $(t, t + \delta]$ are independent of the time duration since the last event.

B2. (*death*) If $N(t) = n$, then
- (a) the probability of exactly one death during $(t, t + \delta]$ is $\mu_n\delta + o(\delta)$,
- (b) the probability of no deaths during $(t, t + \delta]$ is $1 - \mu_n\delta + o(\delta)$,
- (c) the probability of more than one death during $(t, t + \delta]$ is $o(\delta)$,
- (d) when $N(t) = 0$, the probability of at least one death occurring during $(t, t + \delta]$ is 0,
- (e) deaths occurring in $(t, t + \delta]$ are independent of the time duration since the last event.

B3. For $N(t) = n$, birth and death events are independent.

Application of the Chapman-Kolmogorov equations to the birth-death process is based on the properties of $o(\delta)$ and the birth-death axioms. For $n > 0$, the following transition probabilities result:

$$p_{n,n-1}(t, t+\delta) = (\mu_n\delta + o(\delta))(1 - \lambda_n\delta + o(\delta)) = \mu_n\delta + o(\delta) \tag{5.52}$$

$$p_{n,n}(t, t+\delta) = (1 - \lambda_n\delta + o(\delta))(1 - \mu_n\delta + o(\delta)) = 1 - \lambda_n\delta - \mu_n\delta + o(\delta) \tag{5.53}$$

$$p_{n,n+1}(t, t+\delta) = (\lambda_n\delta + o(\delta))(1 - \mu_n\delta + o(\delta)) = \lambda_n\delta + o(\delta) \tag{5.54}$$

Note that the probability of a birth and a death in the interval $(t, t + \delta]$ when $N(t) = n$ is $\lambda_n \mu_n \delta^2 + o(\delta) = o(\delta)$, so that there is no extra contribution to the probability $p_{n,n}(t, t + \delta)$ arising from a birth and death in the interval. For $n = 0$,

$$p_{0,0}(t, t+\delta) = 1 - \lambda_0 \delta + o(\delta) \tag{5.55}$$

$$p_{0,1}(t, t+\delta) = \lambda_0 \delta + o(\delta) \tag{5.56}$$

Finally,

$$\sum_{\substack{l=0 \\ l \notin \{n-1,n,n+1\}}}^{\infty} p_{n,l}(t, t+\delta) = o(\delta) \tag{5.57}$$

According to these equations, the Chapman-Kolmogorov equations yield

$$\begin{aligned} p_n(t+\delta) = {} & (1 - \lambda_n \delta - \mu_n \delta + o(\delta)) p_n(t) + (\lambda_{n-1}\delta + o(\delta)) p_{n-1}(t) \\ & + (\mu_{n+1}\delta + o(\delta)) p_{n+1}(t) + o(\delta) \end{aligned} \tag{5.58}$$

Hence,

$$\begin{aligned} \frac{p_n(t+\delta) - p_n(t)}{\delta} = {} & -\left(\lambda_n + \mu_n + \frac{o(\delta)}{\delta}\right) p_n(t) + \left(\lambda_{n-1} + \frac{o(\delta)}{\delta}\right) p_{n-1}(t) \\ & + \left(\mu_{n+1} + \frac{o(\delta)}{\delta}\right) p_{n+1}(t) + \frac{o(\delta)}{\delta} \end{aligned} \tag{5.59}$$

Letting $\delta \to 0$ yields the derivative with respect to t:

$$\frac{d}{dt} p_n(t) = -(\lambda_n + \mu_n) p_n(t) + \lambda_{n-1} p_{n-1}(t) + \mu_{n+1} p_{n+1}(t) \tag{5.60}$$

The steady-state solution is found by letting $t \to \infty$, letting

$$\pi_n = \lim_{t \to \infty} p_n(t) \tag{5.61}$$

and setting

$$\lim_{t\to\infty}\frac{d}{dt}p_n(t)=0 \tag{5.62}$$

Letting $t \to \infty$ in Eq. 5.60 yields the system

$$0=-(\lambda_n+\mu_n)\pi_n+\lambda_{n-1}\pi_{n-1}+\mu_{n+1}\pi_{n+1} \quad (n=1,2,3,\ldots) \tag{5.63}$$

$$0=-\lambda_0\pi_0+\mu_1\pi_1 \quad (n=0) \tag{5.64}$$

From the boundary condition for $n = 0$,

$$\pi_1=\frac{\lambda_0}{\mu_1}\pi_0 \tag{5.65}$$

Proceeding recursively according to Eq. 5.63 yields

$$\pi_n=\frac{\lambda_0\lambda_1\cdots\lambda_{n-1}}{\mu_1\mu_2\cdots\mu_n}\pi_0 \tag{5.66}$$

According to the probability condition, $\pi_0+\pi_1+\pi_2+\cdots=1$. Therefore

$$\pi_0=\frac{1}{1+\sum_{n=1}^{\infty}\frac{\lambda_0\lambda_1\cdots\lambda_{n-1}}{\mu_1\mu_2\cdots\mu_n}} \tag{5.67}$$

$\{\pi_0, \pi_1, \pi_2,\ldots\}$ is the steady-state distribution if and only if $\pi_0 > 0$, that is, if and only if

$$\sum_{n=1}^{\infty}\frac{\lambda_0\lambda_1\cdots\lambda_{n-1}}{\mu_1\mu_2\cdots\mu_n}<\infty \tag{5.68}$$

Suppose $\lambda_n = \lambda$ for $n = 0, 1, 2,\ldots$ and $\mu_n = \mu$ for $n = 1, 2,\ldots$. Then the steady-state condition $\pi_0 > 0$ becomes $\lambda/\mu < 1$ and, when $\lambda/\mu < 1$,

$$\pi_0=\frac{1}{\sum_{n=0}^{\infty}\left(\frac{\lambda}{\mu}\right)^n}=1-\frac{\lambda}{\mu} \tag{5.69}$$

Substitution into Eq. 5.66 yields the full steady-state distribution,

$$\pi_n = \left(1 - \frac{\lambda}{\mu}\right)\left(\frac{\lambda}{\mu}\right)^n \qquad (n = 0, 1, 2, \ldots) \tag{5.70}$$

***Example* 5.8.** (*M/M/*1 *queue*). Consider a single-server queue in which there is no bound on queue length. Arrivals to the system are assumed to be Poisson distributed with mean λ, and the service time for each job is assumed to be exponentially distributed with mean $1/\mu$. Jobs are serviced on a first-come-first-served (FIFO) basis. The system is known as an *M/M/*1 queue. Since jobs arrive in a Poisson stream, their interarrival time is exponentially distributed with mean $1/\lambda$, and since their service time is exponentially distributed, they leave the system in a Poisson departure stream with mean μ. Based on the properties of the Poisson process, the axioms of the birth-death model are satisfied with $\lambda_n = \lambda$ and $\mu_n = \mu$. The ratio $\rho = \lambda/\mu$ is called the *utilization factor* and, so long as $\rho < 1$, there exists a steady-state distribution, which by Eq. 5.70 is given by

$$\pi_n = (1 - \rho)\rho^n$$

This is a geometric distribution and therefore its mean (expected number of jobs in the system) and variance are given by

$$E[N] = \rho(1 - \rho)^{-1}$$

$$\mathrm{Var}[N] = \rho(1 - \rho)^{-2}$$

where we have let N denote the number of jobs in the system in the steady state. If L denotes the number of jobs waiting (not in service), then

$$E[L] = 0\pi_0 + 0\pi_1 + \sum_{n=1}^{\infty} n\pi_{n+1} = \rho^2(1 - \rho)^{-1}$$

An important characteristic of the system is the expected time an arriving job must wait for service. If a job arrives to enter the system, then the waiting time in the queue (prior to service) is the sum of the waiting times of jobs in the queue plus the remaining service time of the job currently in service. Owing to memorylessness of the exponential distribution, the remaining service time of the job in service remains exponentially distributed with mean service time $1/\mu$. Thus, if there are n jobs in the system, then the waiting time is a sum of independent exponential distributions with means $1/\mu$ and this sum is gamma distributed with parameters $\alpha = 1/\mu$ and $\beta = n$. Assuming the system is in the steady state, waiting time W is not a function of arrival time and the expected waiting time given there are $n > 0$ jobs in the system is

$$E[W \mid N = n] = \frac{n}{\mu}$$

Clearly, $E[W \mid N = 0] = 0$. According to Theorem 4.1,

$$E[W] = \sum_{n=1}^{\infty} \frac{n}{\mu} \pi_n = \frac{\rho(1-\rho)}{\mu} \sum_{n=1}^{\infty} n\rho^{n-1} = \frac{\rho}{\mu(1-\rho)}$$

■

***Example* 5.9.** Suppose the $M/M/1$ queue discipline is changed so that the maximum number of jobs allowed in the system is v. As with $n = 0$, there are now boundary conditions for $n = v$. The transition probabilities for v are

$$p_{v,v}(t, t+\delta) = 1 - \mu_v\delta + o(\delta)$$

$$p_{v-1,v}(t, t+\delta) = \lambda_{v-1}\delta + o(\delta)$$

The Chapman-Kolmogorov equation corresponding to Eq. 5.58 for v is

$$p_v(t+\delta) = (1 - \mu_v\delta + o(\delta))p_v(t) + (\lambda_{v-1}\delta + o(\delta))p_{v-1}(t)$$

Equations 5.63 and 5.64 still hold for $n = 1, 2, \ldots, v - 1$. The steady-state system is

$$0 = -(\lambda_n + \mu_n)\pi_n + \lambda_{n-1}\pi_{n-1} + \mu_{n+1}\pi_{n+1} \qquad (n = 1, 2, 3, \ldots, v-1)$$

$$0 = -\lambda_0\pi_0 + \mu_1\pi_1 \qquad (n = 0)$$

$$0 = \lambda_{v-1}\pi_{v-1} - \mu_v\pi_v \qquad (n = v)$$

Setting $\lambda_n = \lambda$ and $\mu_n = \mu$ and solving the system recursively yields $\pi_n = \pi_0\rho^n$. Applying the total probability condition $\pi_0 + \cdots + \pi_v = 1$ yields π_0. The steady-state distribution is

$$\pi_n = \left(\frac{1-\rho}{1-\rho^{v+1}}\right)\rho^n \qquad (n = 0, 1, \ldots, v)$$

■

5.3.3. Forward and Backward Kolmogorov Equations

The steady-state differential technique is seen to be related to the discrete-time theory by examining the manner in which differentiation affects the Chapman-Kolmogorov equations. There are two one-sided derivatives to consider for the transition probabilities:

$$\gamma_{nn} = \lim_{t\to 0} \frac{p_{n,n}(0) - p_{n,n}(t)}{t} = -p'_{n,n}(0) \tag{5.71}$$

$$\gamma_{mn} = \lim_{t\to 0} \frac{p_{m,n}(t) - p_{m,n}(0)}{t} = p'_{m,n}(0) \qquad (m \neq n) \tag{5.72}$$

For γ_{nn}, $p_{n,n}(0) = 1$; for γ_{mn}, $p_{m,n}(0) = 0$. We will assume the derivatives exist and are finite (although, in general, they need not be). γ_{nn} and γ_{mn} are called the *intensity of passage* for state n and the *intensity of transition* to state n from state m, respectively.

According to the Chapman-Kolmogorov equations,

$$p_{m,n}(t+\delta) = \sum_{l\in S} p_{m,l}(t) p_{l,n}(\delta) \tag{5.73}$$

Hence,

$$\frac{p_{m,n}(t+\delta) - p_{m,n}(t)}{\delta} = \frac{\sum_{l\neq n} p_{m,l}(t) p_{l,n}(\delta)}{\delta} + p_{m,n}(t) \frac{p_{n,n}(\delta) - 1}{\delta} \tag{5.74}$$

Letting $\delta \to 0$ yields the *forward Kolmogorov* equations:

$$p'_{m,n}(t) = -\gamma_{nn} p_{m,n}(t) + \sum_{l\neq n} \gamma_{l,n} p_{m,l}(t) \tag{5.75}$$

Rewriting the Chapman-Kolmogorov equations as

$$p_{m,n}(\delta + t) = \sum_{l\in S} p_{m,l}(\delta) p_{l,n}(t) \tag{5.76}$$

and proceeding similarly yields the *backward Kolmogorov equations*:

$$p'_{m,n}(t) = -\gamma_{mm} p_{m,n}(t) + \sum_{l\neq m} \gamma_{m,l} p_{l,n}(t) \tag{5.77}$$

Letting $t \to 0$ in the forward Kolmogorov equations under the assumption of a steady-state distribution $\{\pi_n\colon n \in S\}$ yields the steady-state system equation

$$0 = -\gamma_{nn}\pi_n + \sum_{l \neq n} \gamma_{l,n}\pi_l \qquad (n = 0, 1, 2, \ldots) \tag{5.78}$$

Together with the total probability condition, this system can be used to find the steady-state distribution.

For the birth-death process, the relevant derivatives are given by

$$\gamma_{n,n-1} = \lim_{\delta \to 0} \frac{\mu_n \delta + o(\delta)}{\delta} = \mu_n \tag{5.79}$$

$$\gamma_{nn} = \lim_{\delta \to 0} \frac{1 - (1 - \lambda_n \delta - \mu_n \delta + o(\delta))}{\delta} = \lambda_n + \mu_n \tag{5.80}$$

$$\gamma_{n,n+1} = \lim_{\delta \to 0} \frac{\lambda_n \delta + o(\delta)}{\delta} = \lambda_n \tag{5.81}$$

All others vanish. Substituting these into the steady-state system equation, Eq. 5.78, yields Eq. 5.63.

***Example* 5.10.** (*M/M/s queue*) The single-server system of Example 5.8 is now altered so there are multiple servers, s of them. Arrivals occur in a Poisson stream with parameter λ, and service times are independent and identically exponentially distributed with mean $1/\mu$. If a job arrives and there exists at least one free server, then the job randomly goes to a free server; if there are no free servers, then the job enters the unlimited queue. When a service is completed, the job at the front of the queue gets served. This system is known as an *M/M/s queue*. If the number n of jobs in the system is less than or equal to s, then the probability that a job leaves the system during the time interval $(t, t + \delta]$ is $n\mu + o(\delta)$. This follows from the probability formula for the union of n events and the fact that the probability of more than one job finishing service during the interval is $o(\delta)$. If there are at least s jobs in the system, then the probability that a job leaves the system during the time interval $(t, t + \delta]$ is $s\mu + o(\delta)$. Hence, the forward Kolmogorov equations are

$$p_0'(t) = -\lambda p_0(t) + \mu p_1(t) \qquad (n = 0)$$

$$p_n'(t) = -(\lambda + n\mu)p_n(t) + \lambda p_{n-1}(t) + (n + 1)\mu p_{n+1}(t) \qquad (n = 1, 2, \ldots, s - 1)$$

$$p_n'(t) = -(\lambda + s\mu)p_n(t) + \lambda p_{n-1}(t) + s\mu p_{n+1}(t) \qquad (n = s, s + 1, \ldots)$$

The resulting steady-state system is

$$0 = -\lambda\pi_0 + \mu\pi_1 \qquad (n = 0)$$

$$0 = -(\lambda + n\mu)\pi_n + \lambda\pi_{n-1} + (n+1)\mu\pi_{n+1} \quad (n = 1, 2, 3,..., v-1)$$

$$0 = -(\lambda + s\mu)\pi_n + \lambda\pi_{n-1} + s\mu\pi_{n+1} \quad (n = s, s+1,...)$$

Let $\rho = \lambda/s\mu$ be the system's utilization factor. If $\rho < 1$, then the steady-state distribution is

$$\pi_n = \begin{cases} \dfrac{(s\rho)^n}{n!}\pi_0, & \text{if } n \leq s \\ \dfrac{s^s\rho^n}{s!}\pi_0, & \text{if } n \geq s \end{cases}$$

where

$$\pi_0 = \frac{1}{\sum_{n=0}^{s-1}\dfrac{(s\rho)^n}{n!} + \dfrac{(s\rho)^s}{(1-\rho)s!}}$$

■

5.4. Markov Random Fields

As defined for time processes, the Markov property is one-dimensional because it involves order: conditioning by several past observations reduces to conditioning by the most recent. The Markov property can be generalized so that conditioning by several past observations reduces to conditioning by a subset consisting of the most recent observations. A further generalization involves adapting the Markov property to random fields. Because the domain of the random function is multidimensional, points are not ordered and reduction of conditioning cannot be treated in the same manner as for stochastic processes. We restrict our attention to the digitally important case where the domain is a finite subset of discrete d-dimensional space Z^d and the state space is finite. Markov random fields have numerous applications in image processing, including texture modeling, restoration, and segmentation.

5.4.1. Neighborhood Systems

Let $A = \{t_1, t_2,..., t_n\} \subset Z^d$ be a finite set, called the set of *sites*. Associate with A a family of pixel neighborhoods, $\mathcal{N} = \{N_1, N_2,..., N_n\}$, such that $N_i \subset A$, $t_i \notin N_i$, and $t_j \in N_k$ if and only if $t_k \in N_j$. $\mathcal{N}$ is called a *neighborhood system* for the set of sites and the points in N_i are called *neighbors* of t_i. The pair $(A, \mathcal{N})$ denotes a set A of sites endowed with a neighborhood system $\mathcal{N}$. A subset C

$\subset A$ is called a *clique* if C is a singleton subset of A or if every pair of distinct sites in C are neighbors. Let $\mathcal{C}$ denote the set of cliques.

***Example* 5.11.** Consider the set of sites $A = \{0, 1, 2, \ldots, m-1\}^2 \subset Z^2$, the Cartesian grid. Let the neighborhood system $\mathcal{N}_0$ be defined in the following manner: for each (i, j) lying in the interior of A, associate the neighborhood $N_{i,j}$ composed of its strong (4-connected) neighbors:

$$N_{i,j} = \begin{pmatrix} 0 & 1 & 0 \\ 1 & \mathbf{0} & 1 \\ 0 & 1 & 0 \end{pmatrix}_{(i,j)}$$

the subscript indicating that the neighborhood is translated to (i, j). Neighborhoods for boundary sites must be adjusted so that they are subsets of A. For instance,

$$N_{0,0} = \begin{pmatrix} 1 & 0 \\ \mathbf{0} & 1 \end{pmatrix}$$

$$N_{1,0} = \begin{pmatrix} 0 & 1 & 0 \\ 1 & \mathbf{0} & 1 \\ 0 & 0 & 0 \end{pmatrix}_{(1,0)}$$

There are three types of cliques: $\mathcal{C}_1$, $\mathcal{C}_2$, and $\mathcal{C}_3$ arc subsets of the clique set consisting of a single pixel, two horizontally adjacent pixels, and two vertically adjacent pixels, respectively. Schematically,

$$\mathcal{C}_1 = (1) \qquad \mathcal{C}_2 = (1 \quad 1) \qquad \mathcal{C}_3 = \begin{pmatrix} 1 \\ 1 \end{pmatrix}$$

The full clique set is $\mathcal{C} = \mathcal{C}_1 \cup \mathcal{C}_2 \cup \mathcal{C}_3$.

As a second illustration, define the neighborhood system $\mathcal{N}_1$ by associating with each (i, j) in the interior of A the neighborhood consisting of all of its (8-connected) neighbors:

$$N_{i,j} = \begin{pmatrix} 1 & 1 & 1 \\ 1 & \mathbf{0} & 1 \\ 1 & 1 & 1 \end{pmatrix}_{(i,j)}$$

As with the neighborhood system $\mathcal{N}_0$, boundary adjustments are required. For instance,

$$N_{0,0} = \begin{pmatrix} 1 & 1 \\ \mathbf{0} & 1 \end{pmatrix}$$

$$N_{1,0} = \begin{pmatrix} 1 & 1 & 1 \\ 1 & \mathbf{0} & 1 \\ 0 & 0 & 0 \end{pmatrix}_{(1,0)}$$

Besides the clique subsets C_1, C_2, and C_3 for $\mathcal{N}_0$, the neighborhood system $\mathcal{N}_1$ has the additional clique subsets

$$C_4 = \begin{pmatrix} 0 & 1 \\ 1 & 0 \end{pmatrix} \qquad C_5 = \begin{pmatrix} 1 & 0 \\ 0 & 1 \end{pmatrix} \qquad C_6 = \begin{pmatrix} 1 & 1 \\ 1 & 0 \end{pmatrix} \qquad C_7 = \begin{pmatrix} 1 & 0 \\ 1 & 1 \end{pmatrix}$$

$$C_8 = \begin{pmatrix} 1 & 1 \\ 0 & 1 \end{pmatrix} \qquad C_9 = \begin{pmatrix} 0 & 1 \\ 1 & 1 \end{pmatrix} \qquad C_{10} = \begin{pmatrix} 1 & 1 \\ 1 & 1 \end{pmatrix}$$

The full clique set is $C = C_1 \cup C_2 \cup \cdots \cup C_{10}$. ■

Since a site set is finite, a random field defined on it will be a random vector. Let $X(t)$ be a random function, $X: A \to S$, the state space being $S = \{0, 1,..., M-1\}$. $X(t)$ is a *Markov random field* relative to the neighborhood system $\mathcal{N}$ if

$$P(x_1, x_2,..., x_n; t_1, t_2,..., t_n) = P(X(t_1) = x_1, X(t_2) = x_2,..., X(t_n) = x_n) > 0 \tag{5.82}$$

for all $(x_1, x_2,..., x_n) \in S^n$ and

$$P(X(t_i) = x_i \mid X(t_j) = x_j \text{ for } j \neq i) = P(X(t_i) = x_i \mid X(t_j) = x_j \text{ for } t_j \in N_i) \tag{5.83}$$

for all $t_i \in A$ and for all $(x_1, x_2,..., x_n) \in S^n$. The conditions of Eqs. 5.82 and 5.83 are termed the *positivity* and *dependency* conditions, respectively. The dependency condition gives the Markovian character to the model since it reduces conditioning from the entire site set to the neighborhood of a pixel. The finiteness of A implies that the dependency condition can always be satisfied if the neighborhoods are sufficiently large; however, the usefulness of the model depends on a reduction of neighborhood sizes sufficient to make the model computationally tractable. To simplify notation, we will denote the probability of Eq. 5.82 by $P(x_1, x_2,..., x_n)$ or, more simply, by $P(\mathbf{x})$. In the context of Markov random fields, realizations of the process, $(x_1, x_2, \ldots, x_n) \in S^n$, are usually called *configurations*. A configuration will be denoted by $X(\mathbf{t})$, $\mathbf{t} = (t_1, t_2, \ldots, t_n) \in A$.

5.4.2. Determination by Conditional Probabilities

Regardless of whether or not $X(t)$ is a Markov random field, if its joint distribution satisfies the positivity condition, then the joint distribution is determined by the conditional probabilities. To see this, consider $\mathbf{x} = (x_1, x_2, \ldots, x_n)$, $\mathbf{y} = (y_1, y_2, \ldots, y_n) \in S^n$. Apply the multiplication principle to obtain

$$P(x_1, x_2, \ldots, x_n) = P(x_n \mid x_1, x_2, \ldots, x_{n-1})P(x_1, x_2, \ldots, x_{n-1}) \tag{5.84}$$

$$P(x_1, x_2, \ldots, x_{n-1}, y_n) = P(y_n \mid x_1, x_2, \ldots, x_{n-1})P(x_1, x_2, \ldots, x_{n-1}) \tag{5.85}$$

Solving for $P(x_1, x_2, \ldots, x_{n-1})$ in Eq. 5.85 and inserting the solution into Eq. 5.84 yields

$$P(x_1, x_2, \ldots, x_n) = \frac{P(x_n | x_1, x_2, \ldots, x_{n-1})}{P(y_n | x_1, x_2, \ldots, x_{n-1})} P(x_1, x_2, \ldots, x_{n-1}, y_n) \tag{5.86}$$

Applying the multiplication principle again yields

$$P(x_1, x_2, \ldots, x_{n-1}, y_n) = P(x_{n-1} \mid x_1, x_2, \ldots, x_{n-2}, y_n)P(x_1, x_2, \ldots, x_{n-2}, y_n) \tag{5.87}$$

$$P(x_1, x_2, \ldots, x_{n-2}, y_{n-1}, y_n) = P(y_{n-1} \mid x_1, x_2, \ldots, x_{n-2}, y_n)P(x_1, x_2, \ldots, x_{n-2}, y_n) \tag{5.88}$$

Solving for $P(x_1, x_2, \ldots, x_{n-2}, y_n)$ in Eq. 5.88 and inserting the solution into Eq. 5.87 yields

$$P(x_1, x_2, \ldots, x_{n-1}, y_n) = \frac{P(x_{n-1} | x_1, x_2, \ldots, x_{n-2}, y_n)}{P(y_{n-1} | x_1, x_2, \ldots, x_{n-2}, y_n)} P(x_1, x_2, \ldots, x_{n-2}, y_{n-1}, y_n) \tag{5.89}$$

Substitution of this result into Eq. 5.86 yields

$$P(x_1, x_2, \ldots, x_n) = \tag{5.90}$$

$$\frac{P(x_n | x_1, x_2, \ldots, x_{n-1})}{P(y_n | x_1, x_2, \ldots, x_{n-1})} \frac{P(x_{n-1} | x_1, x_2, \ldots, x_{n-2}, y_n)}{P(y_{n-1} | x_1, x_2, \ldots, x_{n-2}, y_n)} P(x_1, x_2, \ldots, x_{n-2}, y_{n-1}, y_n)$$

Proceeding inductively yields

$$\frac{P(x_1, x_2, \ldots, x_n)}{P(y_1, y_2, \ldots, y_n)} = \prod_{i=1}^{n} \frac{P(x_i | x_1, \ldots, x_{i-1}, y_{i+1}, \ldots, y_n)}{P(y_i | x_1, \ldots, x_{i-1}, y_{i+1}, \ldots, y_n)} \tag{5.91}$$

Hence, the conditional probabilities determine all quotients $P(\mathbf{x})/P(\mathbf{y})$. Enumerating the elements $\mathbf{x} \in S^n$ by $\mathbf{x}_1$, $\mathbf{x}_2, \ldots$, $\mathbf{x}_r$ and dividing the total probability constraint

$$P(\mathbf{x}_1) + P(\mathbf{x}_2) + \cdots + P(\mathbf{x}_r) = 1 \tag{5.92}$$

by $P(\mathbf{x}_r)$ yields

$$\frac{1}{P(\mathbf{x}_r)} = 1 + \sum_{i=1}^{r-1} \frac{P(\mathbf{x}_i)}{P(\mathbf{x}_r)} \tag{5.93}$$

Since the quotients are given by the conditional probabilities, $P(\mathbf{x}_r)$ can be expressed in terms of the conditional probabilities, as can all other unconditioned probabilities.

Now suppose $X(t)$ is a Markov random field relative to the neighborhood system $\mathcal{N}$. For any $N_i \in \mathcal{N}$, let $N_i^- = \{t_j \in N_i: j < i\}$ and $N_i^+ = \{t_j \in N_i: j > i\}$. According to the dependency condition, the quotient of Eq. 5.91 reduces to

$$\frac{P(x_1, x_2, \ldots, x_n)}{P(y_1, y_2, \ldots, y_n)} = \prod_{i=1}^{n} \frac{P(x_i | x_j \text{ for } t_j \in N_i^-, y_j \text{ for } t_j \in N_i^+)}{P(y_i | x_j \text{ for } t_j \in N_i^-, y_j \text{ for } t_j \in N_i^+)} \tag{5.94}$$

Thus, the probability distribution of $X(t)$ is fully determined by the conditional probabilities over the neighborhood system.

A key aspect of Markov chains is the manner in which specification of the conditional probabilities determines the finite-dimensional joint distributions of the process. Given an initial state vector and one-step transition probabilities, one can obtain the joint distributions. This is evidenced not only in the transition-probability matrices but also in Eq. 5.17, where by pulling out the product of conditional probabilities in the fifth equality and employing probability notation for the case of discrete densities, we arrive at the representation

$$P(x_1, x_2, \ldots, x_n) = P(x_1) \prod_{i=2}^{n} P(x_i | x_{i-1}) \tag{5.95}$$

Assignment of conditional probabilities is not totally free: they must be made in such a way that the preceding equation leads to a legitimate probability distribution. This creates little difficulty because we need only ensure that the one-step transition-probability matrices are stochastic. To see the important role played by conditional probability modeling, one need only recognize the roles played in applications by the transition-probability matrices and the Chapman-Kolmogorov equations.

Looking at Eq. 5.94, we see that it is much more difficult to assign conditional probabilities based on a neighborhood system while being assured that the resulting unconditioned probabilities form a legitimate probability distribution. Just as important, they must be assigned consistently: each factorization of Eq. 5.91 must result in the same unconditioned probability. Consequently, construction of useful Markov random fields poses a fundamental problem.

5.4.3. Gibbs Distributions

Key to construction are Gibbs distributions. A *Gibbs distribution* relative to the site-neighborhood pair $(A, \mathcal{N})$ is a probability measure ζ on S^n possessing the representation

$$\zeta(\mathbf{x}) = \frac{1}{Z}\exp\left(-\frac{U(\mathbf{x})}{T}\right) \tag{5.96}$$

where Z and T are constants and U is a function on S^n. U is called the *energy function* and is of the form

$$U(\mathbf{x}) = \sum_{C \in \mathcal{C}} V_C(\mathbf{x}) \tag{5.97}$$

where $\mathcal{C}$ is the set of cliques. For each $C \in \mathcal{C}$ and $\mathbf{x} \in S^n$, $V_C(\mathbf{x})$ depends only on the coordinates of $\mathbf{x}$ that lie in C. The collection $\{V_C\colon C \in \mathcal{C}\}$ is known as a *potential*. The constant Z normalizes the exponential so that $\zeta(\mathbf{x})$ is a probability measure:

$$Z = \sum_{\mathbf{x}} \exp\left(-\frac{U(\mathbf{x})}{T}\right) \tag{5.98}$$

The next theorem states the fundamental relationship between Gibbs distributions and Markov random fields.

Theorem 5.7. Let $\mathcal{N}$ be a neighborhood system for a set A of sites and $X\colon A \to S$ be a random field. $X\colon A \to S$ is a Markov random field relative to $(A, \mathcal{N})$ if and only if there exists a Gibbs distribution $\zeta(\mathbf{x})$ relative to $(A, \mathcal{N})$ such that $P(X(\mathbf{t}) = \mathbf{x}) = \zeta(\mathbf{x})$. ■

According to the theorem, if $\zeta(\mathbf{x})$ is a Gibbs distribution relative to $(A, \mathcal{N})$ and the distribution of a random field X: $A \to S$ is defined by $P(X(\mathbf{t}) = \mathbf{x}) = \zeta(\mathbf{x})$, then $X(t)$ is a Markov random field (called a *Gibbs random field*) with

$$P(x_1, x_2,..., x_n) = \frac{1}{Z}\exp\left(-\frac{1}{T}\sum_{C \in \mathcal{C}} V_C(x_{C,1}, x_{C,2}, \ldots, x_{C,n(C)})\right) \tag{5.99}$$

where $\{x_{C,1}, x_{C,2},..., x_{C,n(C)}\}$ is the subset of $\{x_1, x_2,..., x_n\}$ whose sites belong to the clique C. The conditional probabilities are given by

$$P(X(t_i) = x_i \mid X(t_j) = x_j, j \neq i) = \frac{P(x_1, x_2, \ldots, x_n)}{P(x_1, x_2, \ldots, x_{i-1}, x_{i+1}, \ldots, x_n)} \tag{5.100}$$

$$= \frac{1}{Z_i}\exp\left[-\frac{1}{T}\sum_{\{C \in \mathcal{C}: t_i \in C\}} V_C(x_{C,1}, x_{C,2}, \ldots, x_{C,n(C)})\right]$$

where the normalizing constant Z_i depends on i. The conditional probability only depends on x_i and components x_k for which $t_k \in N_i$ because if $t_i \in C$ then all other sites in C must be neighbors of t_i.

From the perspective of modeling, the salient part of the theorem is in the other direction, namely, that all Markov random fields are, in fact, Gibbs fields. The definition of a Gibbs random field is rather general, the only restriction besides its form being the restriction on V_C in terms of its dependence on the values associated with the clique C. There is no condition of consistency among different values of the potential. Since potential functions, *ipso facto*, give rise to the local dependency properties inherent in a Markov random field and therefore, in some sense, characterize our understanding of a model, their choice represents a basic decision in modeling via Markov random fields.

***Example* 5.12.** Let $A = \{0, 1, 2,..., m-1\}^2 \subset Z^2$ and let $\mathcal{N}_0$ be the nearest-neighbor system defined in Example 5.11 and possessing the clique set $\mathcal{C} = \mathcal{C}_1 \cup \mathcal{C}_2 \cup \mathcal{C}_3$. The potential consists of three types of functions, each associated with a clique type: $V_{\{(i,j)\}}$ with $\mathcal{C}_1$, $V_{\{(i,j),(i+1,j)\}}$ with $\mathcal{C}_2$, and $V_{\{(i,j),(i,j+1)\}}$ with $\mathcal{C}_3$. The energy function is of the form

$$U(\mathbf{x}) = \sum V_{\{(i,j)\}}(x_{i,j}) + \sum V_{\{(i,j),(i+1,j)\}}(x_{i,j}, x_{i+1,j}) + \sum V_{\{(i,j),(i,j+1)\}}(x_{i,j}, x_{i,j+1})$$

$\mathbf{x}$	$U(\mathbf{x})$	$\mathbf{x}$	$U(\mathbf{x})$
$\begin{pmatrix} 0 & 0 \\ 0 & 0 \end{pmatrix}$	0	$\begin{pmatrix} 0 & 1 \\ 0 & 1 \end{pmatrix}$	$2\alpha + \beta$
$\begin{pmatrix} 1 & 0 \\ 0 & 0 \end{pmatrix}$	α	$\begin{pmatrix} 0 & 1 \\ 1 & 0 \end{pmatrix}$	2α
$\begin{pmatrix} 0 & 1 \\ 0 & 0 \end{pmatrix}$	α	$\begin{pmatrix} 0 & 0 \\ 1 & 1 \end{pmatrix}$	$2\alpha + \beta$
$\begin{pmatrix} 0 & 0 \\ 0 & 1 \end{pmatrix}$	α	$\begin{pmatrix} 1 & 1 \\ 1 & 0 \end{pmatrix}$	$3\alpha + 2\beta$
$\begin{pmatrix} 0 & 0 \\ 1 & 0 \end{pmatrix}$	α	$\begin{pmatrix} 0 & 1 \\ 1 & 1 \end{pmatrix}$	$3\alpha + 2\beta$
$\begin{pmatrix} 1 & 1 \\ 0 & 0 \end{pmatrix}$	$2\alpha + \beta$	$\begin{pmatrix} 1 & 0 \\ 1 & 1 \end{pmatrix}$	$3\alpha + 2\beta$
$\begin{pmatrix} 1 & 0 \\ 0 & 1 \end{pmatrix}$	2α	$\begin{pmatrix} 1 & 1 \\ 0 & 1 \end{pmatrix}$	$3\alpha + 2\beta$
$\begin{pmatrix} 1 & 0 \\ 1 & 0 \end{pmatrix}$	$2\alpha + \beta$	$\begin{pmatrix} 1 & 1 \\ 1 & 1 \end{pmatrix}$	$4\alpha + 4\beta$

Figure 5.3 Configurations and corresponding energy function values for Example 5.12.

where the sums are over $(i, j) \in A$ for which the terms are defined. The *Ising model* results from a binary 0-1 state space and a certain choice of the potential. Define

$$V_{\{(i,j)\}} = \alpha x_{i,j}$$

$$V_{\{(i,j),(i+1,j)\}} = \beta x_{i,j} x_{i+1,j}$$

$$V_{\{(i,j),(i,j+1)\}} = \beta x_{i,j} x_{i,j+1}$$

To illustrate the model we consider only four sites, so that

$$A = \{(0, 0), (0, 1), (1, 0), (1, 1)\}$$

Since $S = \{0, 1\}$, there are 16 configurations. These are given in Fig. 5.3 along with their corresponding energy function values. As an example of how $U(\mathbf{x})$ is computed, consider the configuration

$$\mathbf{x} = \begin{pmatrix} 0 & 1 \\ 1 & 1 \end{pmatrix}$$

The corresponding potential values are

$$V_{\{(0,0)\}}(1) = V_{\{(1,0)\}}(1) = V_{\{(1,1)\}}(1) = \alpha$$

$$V_{\{(0,0),(1,0)}(1, 1) = V_{\{(1,1),(1,0)\}}(1, 1) = \beta$$

$$V_{\{(0,1)\}}(0) = V_{\{(0,1),(0,0)\}}(0, 1) = V_{\{(0,1),(1,1)\}}(0, 1) = 0$$

Thus,

$$U(\mathbf{x}) = 3\alpha + 2\beta$$

Letting $T = 1$,

$$P(\mathbf{x}) = Z^{-1}e^{-(3\alpha + 2\beta)}$$

As an illustration of Markov random field conditioning, note that for pixel (0, 1) full conditioning gives the probability

$$P(X(0, 1) = 0 \mid X(0, 0) = 1, X(1, 1) = 1, X(1, 0) = 0)$$

$$= \frac{P\begin{pmatrix} 0 & 1 \\ 1 & 0 \end{pmatrix}}{P\begin{pmatrix} 0 & 1 \\ 1 & 0 \end{pmatrix} + P\begin{pmatrix} 1 & 1 \\ 1 & 0 \end{pmatrix}}$$

$$= \frac{e^{-2\alpha}}{e^{-2\alpha} + e^{-3\alpha-2\beta}}$$

$$= (1 + e^{-\alpha-2\beta})^{-1}$$

On the other hand, using the dependency condition yields

$$P(X(0, 1) = 0 \mid X(0, 0) = 1, X(1, 1) = 1)$$

$$= \frac{P\begin{pmatrix} 0 & 1 \\ 1 & 0 \end{pmatrix} + P\begin{pmatrix} 0 & 1 \\ 1 & 1 \end{pmatrix}}{P\begin{pmatrix} 0 & 1 \\ 1 & 0 \end{pmatrix} + P\begin{pmatrix} 0 & 1 \\ 1 & 1 \end{pmatrix} + P\begin{pmatrix} 1 & 1 \\ 1 & 0 \end{pmatrix} + P\begin{pmatrix} 1 & 1 \\ 1 & 1 \end{pmatrix}}$$

$$= \frac{e^{-2\alpha} + e^{-3\alpha-2\beta}}{e^{-2\alpha} + e^{-3\alpha-2\beta} + e^{-3\alpha-2\beta} + e^{-4\alpha-4\beta}}$$

$$= (1 + e^{-\alpha-2\beta})^{-1}$$

which is the same result obtained with full conditioning. ■

5.5. Random Boolean Model

On occasion we have loosely employed the term *random set*. Although a seemingly intuitive notion, the theory of random sets is mathematically sophisticated and we will only briefly introduce that theory in Section 5.7. For the present, we discuss (without complete mathematical rigor) a random set model having applications in various domains, including binary image processing, microscopy, and queuing theory.

5.5.1. Germ-Grain Model

Consider a geometric shape (set) that is random owing to its specification in terms of random parameters. We assume the set is topologically closed, meaning it contains its boundary. Such a set can be denoted by S_{A}, where $\mathrm{A} = (A_1, A_2,..., A_n)'$ is a vector of random parameters that determine the realizations of S_{A}. For instance, $(A_1, A_2)'$ may give the width and height of a closed rectangle. If the rectangle is to have a random rotation, then the parameter vector will take the form $(A_1, A_2, A_3)'$, where A_3 specifies the angle of rotation. A numerical property of a random shape depends on the parameters and therefore takes the form $\Gamma = \Gamma(A_1, A_2,..., A_n)$. Γ is a random variable and its distribution depends on the distribution of the random vector A. In the case of the random rectangle, the area is $\mathrm{A} = A_1A_2$ and the perimeter is $\Pi = 2A_1 + 2A_2$. If the width and height are independent, then the density for Π is the convolution of the densities for $2A_1$ and $2A_2$. Except for simple geometric situations when the parameters are independent, it is difficult to obtain exact closed-form expressions for densities of geometric parameters.

The Boolean random set model is composed of a spatial Poisson process Ξ in $\Re^k$ (for imaging, $k = 2$) and a random shape process S. The random shape is translated to each Poisson point to generate the collection of sets $\mathcal{B} = \{S_\xi + \xi: \xi \in \Xi\}$, where the notation S_ξ is meant to indicate that there is a set process identically distributed to S at each point ξ of the Poisson process and where the translate $S_\xi + \xi$ is defined by

$$S_\xi + \xi = \{z + \xi: z \in S_\xi\} \tag{5.101}$$

The model assumes that the events S_ξ are independent of each other and independent of the Poisson process. The points ξ are called *germs*, the random set S the *primary grain*, the translated set $S_\xi + \xi$ a *grain*, and the collection of sets $\mathcal{B}$ the *random Boolean model*. The random image generated by the model is

$$G = \bigcup_{\xi \in \Xi} S_\xi + \xi \tag{5.102}$$

The union is called the *germ-grain model*. For instance, if the primary grain is a randomly rotated rectangle of random length and width, then each realization of the germ-grain model is a binary image formed as a union of such rectangles without any restriction on rectangle intersection. The Boolean model plays a major role in the theory of coverage and is employed in imaging for texture modeling and filter optimization.

Since the union of two independent Poisson processes possessing intensities λ and ν is a Poisson process with intensity $\lambda + \nu$, the union of two independent Boolean models, one with germ intensity λ and primary grain S^λ and the other with germ intensity ν and primary grain S^ν, is a Boolean model with germ intensity $\lambda + \nu$ and primary grain $S^{\lambda,\nu}$, where the distribution of $S^{\lambda,\nu}$ arises from the distributions of S^λ and S^ν by weighting them according to the intensities, the distribution of S^λ receiving weight $\lambda/(\lambda + \mu)$ and the distribution of S^ν receiving weight $\nu/(\lambda + \nu)$. More generally, the union of any finite number of independent Boolean models is a Boolean model.

5.5.2. Vacancy

If D is a (Borel) set in $\Re^k$ and G is the random set generated by a random Boolean model $\mathcal{B} = \{S_\xi + \xi: \xi \in \Xi\}$ according to Eq. 5.102, then the *vacancy* V_D of D relative to $\mathcal{B}$ is the area of D not covered by G. If, for each $z \in \Re^k$ we define $X(z) = 0$ if $z \in G$ and $X(z) = 1$ if $z \notin G$, then $X(z)$ is a binary random function and

$$V_D = \int_D X(z)\, dz \tag{5.103}$$

Taking the expectation and (legitimately) taking the expectation under the integral yields

$$E[V_D] = \int_D E[X(z)]\, dz \tag{5.104}$$

We need to evaluate $E[X(z)]$, the probability that z is not covered. Owing to stationarity,

$$\begin{aligned} E[X(z)] &= P(z \notin G) \\ &= P(z \notin S_\xi + \xi \text{ for all } \xi \in \Xi) \\ &= P(\xi \notin -S_\xi + z \text{ for all } \xi \in \Xi) \\ &= P(\xi \notin S_\xi \text{ for all } \xi \in \Xi) \end{aligned} \tag{5.105}$$

where the last equality follows from stationarity and where the germ process generating ξ is independent of the grain process generating S_ξ. To evaluate the latter probability, first suppose S_ξ is contained in the ball B of radius r centered at the origin. If S is the primary grain and if a point ξ is placed uniformly randomly into B, then

$$P(\xi \notin S_\xi) = \frac{\nu(B) - E[\nu(S)]}{\nu(B)} = 1 - \frac{E[\nu(S)]}{\nu(B)} \tag{5.106}$$

where ν denotes volume (measure). If n points $\xi_1, \xi_2, \ldots, \xi_n$ are independently and uniformly randomly placed into B, then

$$P(\xi_i \notin S_{\xi_i} \text{ for } i = 1, 2, \ldots, n) = \left(1 - \frac{E[\nu(S)]}{\nu(B)}\right)^n \tag{5.107}$$

If N is the number of Poisson points falling in B, then $E[N] = \lambda\nu(B)$ and, by Theorem 4.1,

$$\begin{aligned} P(\xi \notin S_\xi \text{ for all } \xi \in \Xi) &= \sum_{n=0}^{\infty} P(\xi \notin S_\xi \text{ for all } \xi \in \Xi | N = n) P(N = n) \\ &= \sum_{n=0}^{\infty} \left(1 - \frac{E[\nu(S)]}{\nu(B)}\right)^n e^{-\lambda\nu(B)} \frac{(\lambda\nu(B))^n}{n!} \\ &= e^{-\lambda\nu(B)} \sum_{n=0}^{\infty} \frac{\lambda^n (\nu(B) - E[\nu(S)])^n}{n!} \\ &= e^{-\lambda E[\nu(S)]} \end{aligned} \tag{5.108}$$

Although the preceding expression has been derived for bounded S, it can be shown to hold for unbounded S. Hence, from Eqs. 5.104 and 5.105,

$$E[V_D] = \nu(D)\, e^{-\lambda E[\nu(S)]} \tag{5.109}$$

The uncovered portions of D are known as the *pores* and the *porosity* is the expected proportion of D that is not covered, namely,

$$\frac{E[V_D]}{\nu(D)} = e^{-\lambda E[\nu(S)]} \tag{5.110}$$

Generally, porosity is the probability of an arbitrary point not being covered; equivalently, it is the proportion of the whole space not covered. Since Eqs. 5.105 and 5.108 make no reference to D, the views of porosity as a point-coverage probability and as the proportion of a set not covered are consistent and, owing to stationarity, porosity is expressed as

$$P(0 \notin G) = e^{-\lambda E[\nu(S)]} \tag{5.111}$$

where 0 is the origin in $\Re^d$.

Equation 5.108 can be used to obtain the variance of V_D. First, the second moment is

$$\begin{aligned} E[V_D^2] &= E\left[\left(\int_D X(z)\,dz\right)^2\right] \\ &= E\left[\int_D \int_D X(z)X(w)\,dz\,dw\right] \\ &= \int_D \int_D E[X(z)X(w)]\,dz\,dw \end{aligned} \tag{5.112}$$

But

$$\begin{aligned} E[X(z)X(w)] &= P(z \notin S_\xi + \xi,\ w \notin S_\xi + \xi \text{ for all } \xi \in \Xi) \\ &= P(\xi \notin -S_\xi + z,\ \xi \notin -S_\xi + w \text{ for all } \xi \in \Xi) \\ &= P(\xi \notin -S_\xi + z - w,\ \xi \notin -S_\xi \text{ for all } \xi \in \Xi) \\ &= P(\xi \notin (-S_\xi + z - w) \cup (-S_\xi) \text{ for all } \xi \in \Xi) \end{aligned}$$

$$= P(\xi \notin (S_\xi + w - z) \cup S_\xi \text{ for all } \xi \in \Xi)$$

$$= \exp[-\lambda E[\nu((w - z + S) \cup S)]]$$

$$= \exp[-\lambda E[\nu(w - z + S) + \nu(S) - \nu((w - z + S) \cap S)]]$$

$$= \exp[-2\lambda E[\nu(S)] + \lambda E[\nu((w - z + S) \cap S)]] \tag{5.113}$$

Hence,

$$\mathrm{Var}[V_D]$$

$$= \int_D \int_D \exp[-2\lambda E[\nu(S)] + \lambda E[\nu((w - z + S) \cap S)]]\, dz\, dw - \nu(D)^2 \exp[-2\lambda E[\nu(S)]]$$

$$= \exp[-2\lambda E[\nu(S)]] \int_D \int_D (\exp[\lambda E[\nu((w - z + S) \cap S)]] - 1)\, dz\, dw \tag{5.114}$$

From Eq. 5.110 we see that the expected vacancy depends only on the expected volume of S; from Eq. 5.114 we see that the variance of the vacancy depends on both the expected volume of S and the expected volumes of intersections of translates of S.

***Example* 5.13.** Let the primary grain S be a disk of random radius R possessing a gamma distribution with parameters α and β. Realizations of the germ-grain process for different radius distributions are shown in the two parts of Fig. 5.4. In part b of the figure, we have shown outlines of disk boundaries to illustrate the overlapping effect (although, as in part a, there are no such outlines in the actual realization). Since the second moment of the gamma distribution is $E[R^2] = \beta^2(\alpha + 1)\alpha$, the porosity and expected vacancy are

$$P(0 \notin G) = \exp[-\lambda\pi\beta^2(\alpha + 1)\alpha]$$

$$E[V_D] = \nu(D)\exp[-\lambda\pi\beta^2(\alpha + 1)\alpha]$$

The variance of the vacancy is more complicated, even in this relatively simple case of random disks. If S is a disk of radius R and $|z - w| < 2R$, then

$$\nu((w - z + S) \cap S) = R^2(\zeta - \sin \zeta)$$

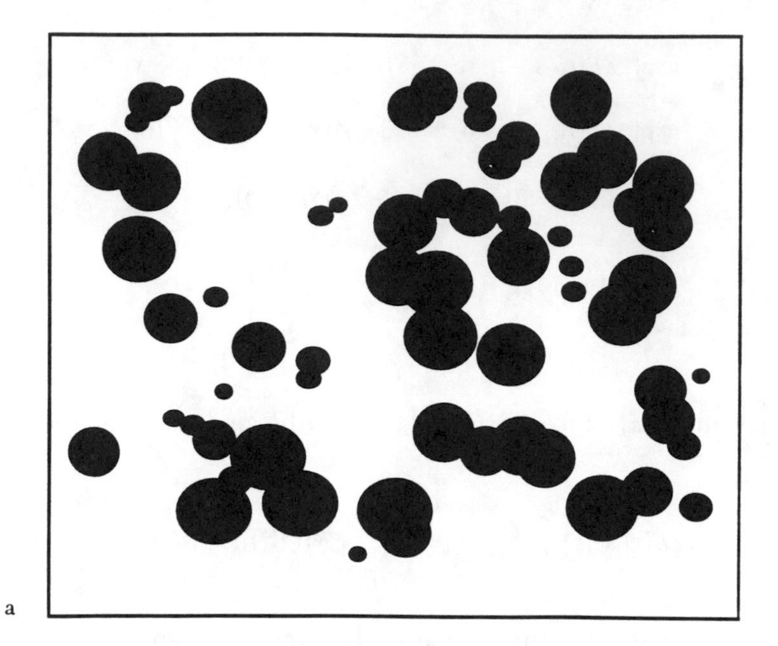

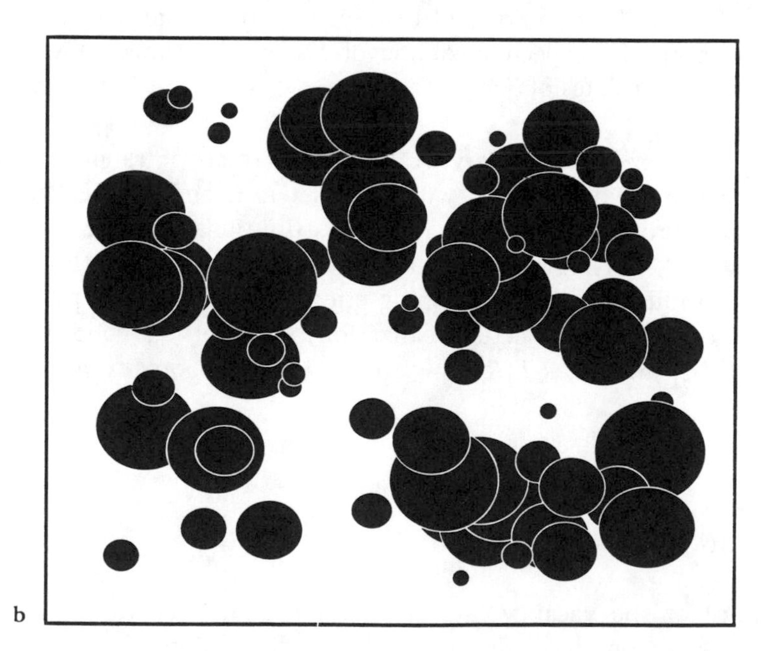

Figure 5.4 Realizations of random-disk Boolean model, where in part b the outlines of disk boundaries are shown to illustrate the overlapping effect.

where $\cos \zeta/2 = |z - w|/2R$; otherwise, the area of the intersection is 0. If we denote the intersection area by $g(|z - w|, R)$ and the gamma density by $f(r)$, then Eq. 5.114 gives

$$\mathrm{Var}[V_D] = \exp[-2\lambda\pi\beta^2(\alpha + 1)\alpha] \int_D\int_D \left(\exp\left[\lambda \int_0^\infty g(|z-w|, r) f(r)\, dr \right] - 1 \right) dz\, dw \qquad \blacksquare$$

Since $P(X(0) = 1)$ is the porosity, Eq. 5.111 shows that the mean of $X(z)$ is the constant $\mu_X = e^{-\lambda E[\nu(S)]}$. From Eq. 5.113, the autocorrelation function for X is

$$R_X(z, w) = \exp[-\lambda E[\nu((w - z + S) \cup S)]] \tag{5.115}$$

and therefore the covariance function is

$$K_X(z, w) = \exp[-\lambda E[\nu((z - w + S) \cup S)]] - e^{-2\lambda E[\nu(S)]} \tag{5.116}$$

This covariance is known as the *covariance of the pores*. Since $X(z)$ is SS stationary, it is WS stationary and its autocorrelation and covariance functions are functions of $|z - w|$. Letting $\tau = z - w$, $r_X(\tau)$ gives the probability that two points $|\tau|$ apart are white. Assuming S is bounded, $k_X(\tau) = 0$ for sufficiently large $|\tau|$ because $\nu((\tau + S) \cup S) = 2\nu(S)$ for sufficiently large $|\tau|$. Hence, the process $X(z)$ is mean-ergodic.

The *covariance of the grains* is the covariance of the complement of $X(z)$, defined by $X^c(z) = 1$ if and only if z is covered. The complement has the autocorrelation function

$$\begin{aligned} R_{X^c}(z, w) &= P(X(z) = 0, X(w) = 0) \\ &= P(1 - X(z) = 1, 1 - X(w) = 1) \qquad (5.117) \\ &= E[(1 - X(z))\,(1 - X(w))] \\ &= 1 - 2\mu_X + R_X(z, w) \end{aligned}$$

Since $E[X^c(z)] = 1 - \mu_X$, subtracting $(1 - \mu_X)^2$ from both sides yields

$$K_{X^c}(z, w) = K_X(z, w) \tag{5.118}$$

The *variogram*, which is also a function of $|\tau| = |z - w|$, is defined by

$$\gamma(z, w) = E[X(z)X^c(w)]$$
$$= P(X(z) = 1, X(w) = 0)$$
$$= P(X(z) = 1) - P(X(z) = 1, X(w) = 1) \tag{5.119}$$
$$= e^{-\lambda E[\nu(S)]} - R_X(z, w)$$

5.5.3. Hitting

We consider the number of grains in a random Boolean model that *hit* a given set in $\Re^2$. Given two sets B and D, the *dilation* (or *Minkowski sum*) of B and D is defined by

$$B \oplus D = \{z + w: z \in B, w \in D\} = \bigcup_{w \in D} B + w \tag{5.120}$$

Dilation is commutative, $B \oplus D = D \oplus B$, and has the hitting formulation

$$B \oplus D = \{z: (-B + z) \cap D \neq \varnothing\} \tag{5.121}$$

A random set is *isotropic* if it is uniformly randomly rotated about the origin. Letting ∂S denote the boundary of S and $\nu(\partial S)$ denote the length of the boundary, we can state the following theorem.

Theorem 5.8. Let $\mathcal{B}$ be a Boolean random set in $\Re^2$ with germ intensity λ and primary grain S, and let $D \subset \Re^2$ be nonrandom. Then the number N of grains in the Boolean model that intersect D is Poisson distributed with mean

$$E[N] = \lambda E[\nu((-S) \oplus D)] \tag{5.122}$$

If it is further assumed that S is isotropic and convex and D is convex, then

$$E[N] = \lambda(\nu(D) + E[\nu(S)] + (2\pi)^{-1}\nu(\partial D)E[\nu(\partial S)]) \quad \blacksquare \tag{5.123}$$

In the general case, because N is Poisson distributed, Eq. 5.122 shows that the probability of no grains intersecting a set D is

$$P((S_\xi + \xi) \cap D = \varnothing \text{ for all } \xi \in \Xi) = P(N = 0)$$
$$= \exp[-\lambda E[\nu((-S) \oplus D)]] \tag{5.124}$$

Let us focus our attention on the case where S is isotropic and convex and D is convex. Then the probability that no grains intersect D is

$$P(N=0) = \exp[-\lambda(v(D) + E[v(S)] + (2\pi)^{-1}v(\partial D)E[v(\partial S)])] \tag{5.125}$$

If $S_\xi + \xi$ is an arbitrary grain in the Boolean model, then the probability that no other grain in the model intersects $S_\xi + \xi$ ($S_\xi + \xi$ is isolated) can be found from $P(N = 0)$. Let

$$Y = \begin{cases} 1, & \text{if } (S_\xi + \xi) \cap (S_{\xi'} + \xi') = \varnothing \text{ for all } \xi' \in \Xi, \xi' \neq \xi \\ 0, & \text{otherwise} \end{cases} \tag{5.126}$$

By Theorem 5.8 (and Theorem 4.1),

$$P((S_\xi + \xi) \cap (S_{\xi'} + \xi') = \varnothing \text{ for all } \xi' \in \Xi, \xi' \neq \xi) = E[E[Y \mid S]] \tag{5.127}$$

$$= \exp[-\lambda E[(v(S)]]E[\exp[-\lambda(v(S) + (2\pi)^{-1}v(\partial S)E[v(\partial S)])]]$$

***Example* 5.14.** Let S be isotropic and convex and D be convex. If D is a single point and N the number of grains in the Boolean model covering the point, then

$$E[N] = \lambda E[v(S)]$$

If S is an isotropic (randomly rotated) rectangle with length and width being independent, gamma-distributed random variables having parameters α and β, then

$$E[N] = \lambda(v(D) + \alpha^2\beta^2 + (2\pi)^{-1}4\alpha\beta v(\partial D))$$

If S is a disk whose radius R is gamma distributed with parameters α and β, then

$$E[N] = \lambda(v(D) + \pi\beta^2(\alpha + 1)\alpha + \alpha\beta v(\partial D))$$

If S is a disk whose radius R is gamma distributed with parameters α and β, then the probability that the grain $S_\xi + \xi$ is isolated is

$$P((S_\xi + \xi) \cap (S_{\xi'} + \xi') = \varnothing \text{ for all } \xi' \in \Xi, \xi' \neq \xi)$$

$$= \exp[-\lambda\pi\beta^2(\alpha + 1)\alpha]E[\exp[-\lambda\pi(R^2 + 2\alpha\beta R)]] \qquad \blacksquare$$

***Example* 5.15.** Equation 5.125 can be used to estimate λ and the mean area and perimeter of the primary grain. Let $q = P(N = 0)$. Taking logarithms yields

$$-\log q = \lambda(\nu(D) + E[\nu(S)] + (2\pi)^{-1}\nu(\partial D)E[\nu(\partial S)])$$

Let D be a ball of radius r. Then this expression becomes

$$-\log q = \lambda(\pi r^2 + E[\nu(\partial S)]r + E[\nu(S)])$$

$$= b_2 r^2 + b_1 r + b_0$$

An estimate for q can be obtained for different values of r by scanning the image to estimate the probability of zero hits for different radii. Regression on r^2, r, and 1 provides estimates for b_2, b_1, and b_0, thereby providing estimates for λ, $E[\nu(\partial S)]$, and $E[\nu(S)]$. ■

5.5.4. Linear Boolean Model

On the real line, the random Boolean model consists of a Poisson point process Ξ on $\Re$ and grains that are subsets of $\Re$, each grain being identically distributed to a primary grain. It is referred to as the *linear Boolean model*. We focus on the *simple linear Boolean model*, for which the primary grain is a compact segment. The observed germ-grain process is comprised of alternating covered and uncovered intervals, called *clumps* and *gaps* (or *spacings*), respectively.

The Boolean model is determined by its driving Poisson germ process Ξ and its grain process. Therefore, by Theorem 2.5, if the Poisson points ξ_1, ξ_2,... are replaced by the points $\xi_1 + \eta_1$, $\xi_2 + \eta_2$,..., where η_1, η_2,... are independent, identically distributed, and independent of Ξ, then the new germ-grain model is identically distributed to the original germ-grain model. Thus, for the simple linear Boolean model, it makes no difference whether segment centers or endpoints are placed at germ points. Henceforth, we assume that segments are placed with their left endpoints at germs, so that they emanate to the right. This approach models an $M/G/\infty$ queue, in which germs and grains correspond to arrivals and service intervals, respectively, and where M, G, and ∞ refer to Poisson arrivals, a general service distribution, and an infinite number of servers. There is no waiting for service. Clumps and gaps correspond to busy and empty periods, respectively.

Our intent is to characterize clump and gap lengths. Arbitrarily select a clump, and let Y_1, Y_2,... and Z_1, Z_2,... denote the succeeding clump and gap lengths, respectively. Owing to the independence of numbers of Poisson points in nonoverlapping intervals and the independence of the grains, the random variables Y_1, Y_2,..., Z_1, Z_2,... are independent. Moreover, Y_1, Y_2,... are identically distributed, as are Z_1, Z_2,.... Since Z_i is the time from the end of a clump until the next Poisson point, according to Theorem 2.4 it possesses an exponential distribution with mean λ^{-1}, where λ is the intensity of the Poisson process.

To find the probability distribution function $C(x)$ of the clump lengths, let $F(x)$ be the probability distribution function of the segment lengths, assume that the mean segment length, μ_F, is finite, and assume that the first gap commences at the origin. Then the number of clumps or clump parts in the interval $[0, t]$ is given by

$$N(t) = \sum_{n=1}^{\infty} H_n(t) \tag{5.128}$$

where $H_n(t) = 1$ if

$$\sum_{i=1}^{n-1} Y_i + \sum_{i=1}^{n} Z_i \leq t \tag{5.129}$$

and $H_n(t) = 0$ otherwise. $N(t)$ has mean

$$\mu_N(t) = \sum_{n=1}^{\infty} P\left(\sum_{i=1}^{n-1} Y_i + \sum_{i=1}^{n} Z_i \leq t\right) \tag{5.130}$$

Let $c(s)$ and $g(s)$ be the Laplace-Stieltjes transforms of the distributions of the clump and gap lengths, respectively. Then the Laplace-Stieltjes transform of $\mu_N(t)$ is given by

$$m_N(s) = \int_0^{\infty} e^{-st}\, d\mu_N(t)$$

$$= \sum_{n=1}^{\infty} \int_0^{\infty} e^{-st}\, dP\left(\sum_{i=1}^{n-1} Y_i + \sum_{i=1}^{n} Z_i \leq t\right) \tag{5.131}$$

$$= \sum_{n=1}^{\infty} c^{n-1}(s) g^n(s)$$

Since the gap length is exponentially distributed with mean λ^{-1},

$$g(s) = (1 + (s/\lambda))^{-1} \tag{5.132}$$

Therefore,

$$m_N(s) = \sum_{n=1}^{\infty} \frac{c^{n-1}(s)}{(1+s/\lambda)^n}$$

$$= \frac{1}{1+s/\lambda} \sum_{n=0}^{\infty} \frac{c^n(s)}{(1+s/\lambda)^n}$$

$$= \frac{1}{1+s/\lambda} \frac{1}{1-(1+s/\lambda)^{-1}c(s)} \tag{5.133}$$

$$= \frac{1}{1+s/\lambda - c(s)}$$

We will find $c(s)$ by finding an analytic expression for $m_N(s)$. Suppose R Poisson points fall in $[0, t]$ and X is the length of the grain $S_{t'}$ commencing at the Poisson point t'. The probability that $S_{t'}$ does not cover t is given by

$$P(X < t - t') = F(t - t') \tag{5.134}$$

Since the Poisson points are independently and uniformly distributed over $[0, t]$, the conditional probability (conditioned on $R = r$) that any segment (out of the r segments) does not cover t is

$$q(t|r) = t^{-1} \int_0^t F(t-x)\, dx \tag{5.135}$$

Since r does not appear in the integral, we can simply write $q(t)$. The probability that no segment covers t is $q(t)^R$. The unconditional probability, $q_0(t)$, that no segment covers t is obtained by averaging over R (which satisfies a Poisson distribution with mean λt):

$$q_0(t) = E[q(t)^R]$$

$$= \sum_{r=0}^{\infty} q(t)^r e^{-\lambda t} \frac{(\lambda t)^r}{r!}$$

$$= \exp[\lambda t(q(t) - 1)]$$

$$= \exp\left[\lambda \int_0^t (F(t-x)-1)\,dx\right] \tag{5.136}$$

$$= \exp\left[-\lambda \int_0^t (1-F(x))\,dx\right]$$

We wish to express $\mu_N(t)$ in terms of $q_0(t)$. For small $h > 0$, the probability that a clump begins between t and $t + h$ is given by the product of the probability that t is not covered and the probability that a Poisson point falls in $[t, t + h]$. This product is given by

$$q_0(t)(\lambda h + o(h)) = q_0(t)\lambda h + o(h) \tag{5.137}$$

Hence, the expected number of clumps commencing between 0 at $t + h$ is

$$\mu_N(t+h) = \mu_N(t) + \lambda q_0(t)h + o(h) \tag{5.138}$$

Therefore,

$$\frac{d}{dt}\mu_N(t) = \lim_{h\to 0}\frac{\mu_N(t+h)-\mu_N(t)}{h} = \lambda q_0(t) \tag{5.139}$$

and $\mu_N(t)$ and its Laplace-Stieltjes transform are given by

$$\mu_N(t) = \lambda \int_0^t q_0(t)\,dt = \lambda \int_0^t \exp\left[-\lambda \int_0^u (1-F(x))\,dx\right] du \tag{5.140}$$

$$m_N(s) = \int_0^\infty e^{-st}\,d\mu_N(t) = \lambda \int_0^\infty \exp\left[-st-\lambda \int_0^t (1-F(x))\,dx\right] dt \tag{5.141}$$

respectively. Setting Eq. 5.133 equal to Eq. 5.141 yields the next theorem.

Theorem 5.9. In the simple linear Boolean model, if the Poisson germ process has intensity λ and the grain lengths have probability distribution function $F(x)$ with finite mean, then the clump-length distribution has Laplace-Stieltjes transform

$$c(s) = 1 + \frac{s}{\lambda} + \left(\lambda \int_0^{\infty} \exp\left[-st - \lambda \int_0^{t} (1 - F(x))\, dx \right] dt \right)^{-1} \quad \blacksquare \tag{5.142}$$

The problem with the representation of Eq. 5.142 is that, in general, it does not yield a closed-form representation for the clump-length distribution. Since we observe the clumps and gaps, these must be used to estimate the underlying model parameters, and a closed-form expression would be beneficial. There are special cases in which closed-form representations are known, for instance, when there is a constant segment length. Expressions are known for the mean and variance of the clump-length distribution in the general case. These can be obtained by differentiation of Eq. 5.142 and are given by

$$\mu_C = \lambda^{-1}(e^{\lambda\mu_F} - 1) \tag{5.143}$$

$$\sigma_C^2 = 2\lambda^{-1} e^{\lambda\mu_F} \int_0^{\infty} \left(\exp\left[\lambda t \int_t^{\infty} (1 - F(x))\, dx \right] - 1 \right) dt - \lambda^{-2}(e^{\lambda\mu_F} - 1)^2 \tag{5.144}$$

where the last expression holds when the segment-length variance is finite, and σ_C is finite if and only if the segment-length variance is finite.

Suppose we wish to use the method of maximum likelihood to estimate the parameters of a simple linear Boolean model based on an observation in $[0, t]$. Since we observe the alternating clumps and gaps of the germ-grain model, we must use their lengths. Let $\boldsymbol{\theta}$ denote the vector of parameters for the Boolean model. $\boldsymbol{\theta}$ includes the intensity λ of the Poisson process and the parameters of the distribution governing the primary grain. Let $X_i = Y_i + Z_i$ be the sum of the ith clump and gap lengths. Assume that the clump length possesses a continuous distribution, and let $f_X(x)$ and $f_Y(y)$ denote the densities of X_i and Y_i, respectively (keeping in mind that Z_i is exponentially distributed with mean λ^{-1} and that $Y_1, Y_2, \ldots, Z_1, Z_2, \ldots$ are independent). Estimation is based on an observation consisting of the random variables N (the number of complete clumps observed in $[0, t]$), $Y_1, Y_2, \ldots, Y_N, Z_1, Z_2, \ldots, Z_N$, and $R_t = t - (X_1 + X_2 \cdots + X_N)$, which is the length of the remainder of the interval after the final completely observed clump. Using the *Radon-Nikodym theorem*, it can be shown that the joint likelihood function for these random variables satisfies the proportionality relation

$$L(\boldsymbol{\theta}) \propto \prod_{i=1}^{N} \lambda e^{-\lambda Z_i} \prod_{i=1}^{N} f_Y(Y_i|\boldsymbol{\theta}) \left(1 - \int_0^{R_t} f_X(x|\boldsymbol{\theta})\, dx \right) \tag{5.145}$$

where the constant of proportionality is independent of $\boldsymbol{\theta}$ and can be ignored when minimizing the likelihood function to obtain a maximum-likelihood estimator. So long as the segment length has finite variance, the effect of R_t on the maximum-likelihood estimator asymptotically vanishes as $t \to 0$. Therefore, dropping the last factor in the preceding equation yields the following approximate proportionality relationship for the likelihood function:

$$L_1(\boldsymbol{\theta}) \propto \prod_{i=1}^{N} \lambda e^{-\lambda Z_i} \prod_{i=1}^{N} f_Y(Y_i|\boldsymbol{\theta})$$

$$= \lambda^N \exp\left[-\lambda \sum_{i=1}^{N} Z_i\right] \prod_{i=1}^{N} f_Y(Y_i|\boldsymbol{\theta}) \tag{5.146}$$

The difficulty with both the exact and approximate likelihood functions is that, in general, we do not have a closed-form expression for the density of the clump length; we have only its Laplace-Stieltjes transform.

As mentioned previously, there exists a closed-form representation for the clump-length distribution when the segment length is fixed. In that case (omitting the details), we can arrive at an approximate likelihood function corresponding to Eq. 5.146, in which the fixed segment length is a and there is only the single parameter λ to estimate:

$$L_1(\lambda) \propto \lambda^N (1 - e^{-a\lambda})^{N-M} \exp\left[-\lambda \sum_{i=1}^{N} Z_i - Ma\lambda\right] \prod_{i=1}^{N-M} f_Y(Y_i|\lambda) \tag{5.147}$$

where M is the number of single-segment clumps (which can be determined when the segment length is fixed) and $Y_1, Y_2, \ldots, Y_{N-M}$ are the lengths of the multisegment clumps.

Although we will not pursue the matter here, the discrete random Boolean model is defined by selecting germs according to a Bernoulli distribution and having the primary grain be a discrete random set. It is interesting to note that the choice of segment center is relevant in the discrete linear Boolean model, so that the right-emanating (*directional*) model cannot be assumed to generate an arbitrary germ-grain coverage process. The good point is that, in general, the clump-length distribution can be directly expressed in a recursive fashion for the directional discrete linear Boolean model.

5.6. Granulometries

The random Boolean model and other binary random-set models can be used to model grain and texture images. A certain class of filters defined in accordance with granular sieving has proven useful for both classification and grain selection for granular images. This section introduces these *granulometric filters* and shows how they can be employed for texture classification and removal of undesirable grains.

5.6.1. Openings

For sets S and B, the *opening* of S by B is defined by

$$\phi_B(S) = \bigcup_{B+z \subset S} B + z \tag{5.148}$$

The infix notation $S \circ B$ is also used to denote opening. $\phi_B(S)$ is the portion of S swept out by (*structuring element*) B as B is translated about inside S. For instance, if S is a square of edge length $2s$ and B is a disk of radius $r \leq s$, then $\phi_B(S)$ is the square with its corners rounded off, the amount of rounding depending on r; if $r > s$, then $\phi_B(S) = \varnothing$. If S is an isosceles triangle with horizontal base and height h, and B is a vertical line of length $l < h$, then $\phi_B(S)$ is an irregular pentagon formed as the original triangle with base corners removed.

As an operator, opening possesses four basic properties. It is

(O1) *translation invariant*: $\phi_B(S + x) = \phi_B(S) + x$

(O2) *increasing*: $S_1 \subset S_2$ implies $\phi_B(S_1) \subset \phi_B(S_2)$

(O3) *antiextensive*: $\phi_B(S) \subset S$

(O4) *idempotent*: $\phi_B(\phi_B(S)) = \phi_B(S)$

If $\phi_B(A) = A$, then $\phi_A(S) \subset \phi_B(S)$ for any set S. A much deeper proposition states that a compact set B is convex if and only if, for $r > s > 0$, $\phi_{sB}(rB) = rB$. Assuming B is compact and convex, if $r > s > 0$, then $\phi_{rB}(S) \subset \phi_{sB}(S)$. If we think of grains in image S falling through the holes sB and rB, more will fall through the hole rB, thereby yielding a more diminished filtered image.

Properties O1 through O4 are characteristic of grain *sieving*. If image components are sieved by a filter, translation invariance means that grain translation does not affect sieving, increasingness means that subgrains are more readily sieved than the grains of which they are subsets, antiextensivity means that grains are not adjoined by sieving, and idempotence means that once sieved, no more grains are filtered by using the sieve again.

A set operator that is translation invariant, increasing, antiextensive, and idempotent is called a *τ-opening*. Every τ-opening Ψ has a representation of the form

$$\Psi(S) = \bigcup_{B \in \mathcal{B}} \phi_B(S) \tag{5.149}$$

where $\mathcal{B}$ is a collection of sets called a *base* for Ψ. A set is *invariant* under Ψ, meaning $\Psi(S) = S$, if and only if S is a union of translates of sets in $\mathcal{B}$. If all base sets are connected, then Ψ is distributive in the sense that, if S_1 and S_2 are disjoint compact sets, then $\Psi(S_1 \cup S_2) = \Psi(S_1) \cup \Psi(S_2)$.

Given a class $\mathcal{G}$ of convex sets, a parameterized family $\{\Psi_t\}$, $t > 0$, of τ-openings is defined by

$$\Psi_t(S) = \bigcup_{B \in \mathcal{G}} \phi_{tB}(S) \tag{5.150}$$

$\{\Psi_t\}$ is called a *granulometry*, $\mathcal{G}$ is called a *generator* of the granulometry, and the family is extended to $t = 0$ by defining $\Psi_0(S) = S$. For fixed $t > 0$, $t\mathcal{G} = \{tB: B \in \mathcal{G}\}$ is a base for Ψ_t. (A more general notion of granulometry exists but we will not discuss it.) Because each generator set is convex, each is connected. We confine ourselves to generators composed of finite numbers of compact, convex sets. The simplest granulometry is composed of parameterized openings, namely, $\{\Psi_t\} = \{\phi_{tB}\}$, in which case $\mathcal{G} = \{B\}$. Strictly speaking, a granulometry is a class $\{\Psi_t\}$ of operators; however, when considering the outputs of $\{\Psi_t\}$ on a set S, we will call the collection $\{\Psi_t(S)\}$ of outputs a granulometry.

If $\Omega(t)$ is the volume (area) of $\Psi_t(S)$, then $\Omega(t)$ is a decreasing function of t, known as a *size distribution*. A normalized size distribution is defined by

$$\Phi(t) = 1 - \frac{\Omega(t)}{\Omega(0)} \tag{5.151}$$

If S is compact and B consists of more than a single point, then $\Phi(t)$ increases from 0 to 1 and is continuous from the left. Thus, it defines a probability distribution function. Its derivative, $\Phi'(t) = d\Phi(t)/dt$, is a probability density. Both $\Phi(t)$ and $\Phi'(t)$ are called the *pattern spectrum* of S relative to $\mathcal{G}$. The moments of $\Phi(t)$ are used for classification.

***Example* 5.16.** Figure 5.5 shows portions of three toner-particle images from an electrophotographic process. In part a, the toner particles are fairly well uniformly spread across the image, whereas in parts b and c, the particles suffer from increasing degrees of agglomeration. Single-structuring-element digital granulometries have been applied to binarized versions of the images using a digital-ball generator. The resulting pattern spectra are shown in Fig. 5.6. Notice how agglomeration has resulted in pattern spectra shifts to the right, especially with regard to skewing. This shifting can be expected to cause significant changes in the mean, variance, and skewness of the pattern spectra. The corresponding pattern-spectra moments can be used to detect significant agglomeration or, more generally, to classify electrophotographic images relative to toner agglomeration. ■

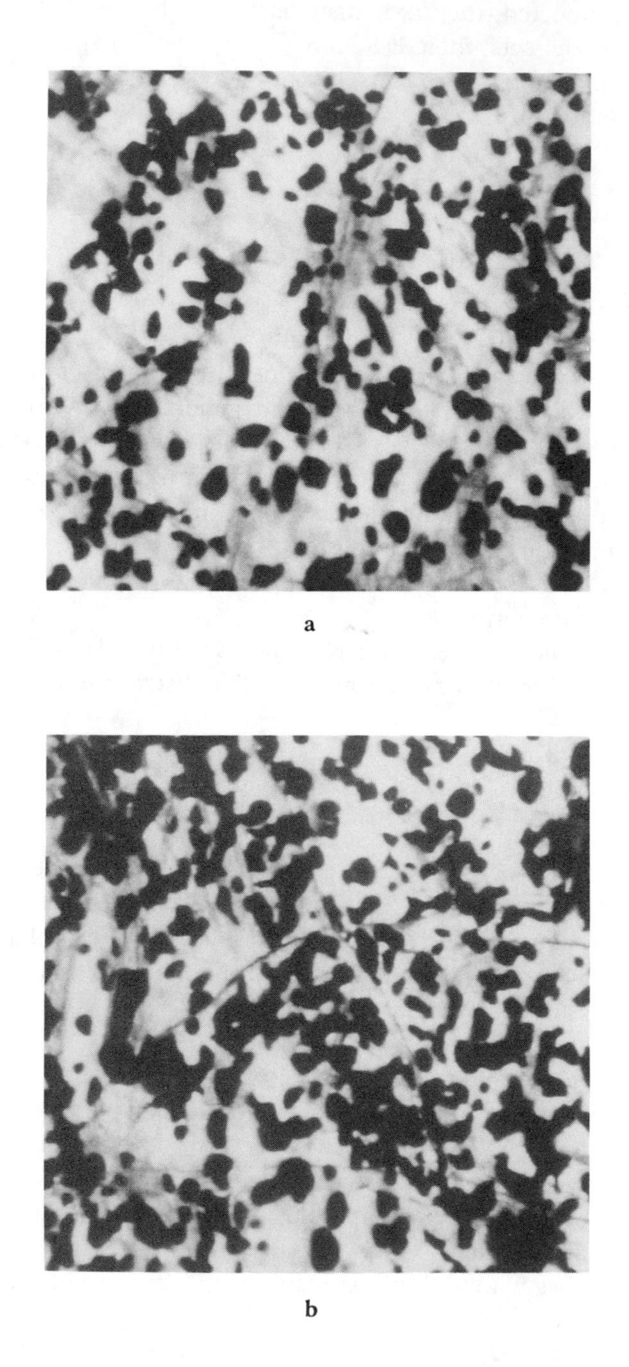

a

b

Figure 5.5 Toner-particle realizations: (a) little agglomeration; (b) modest agglomeration;

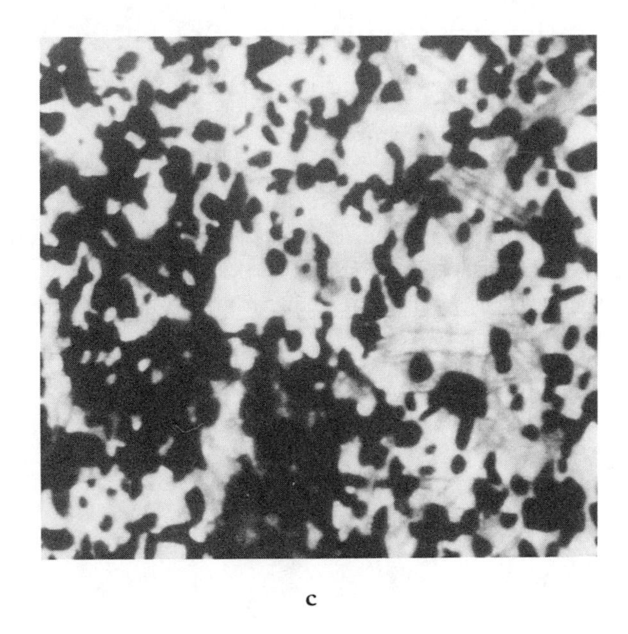

c

Figure 5.5 (cont): (c) excessive agglomeration.

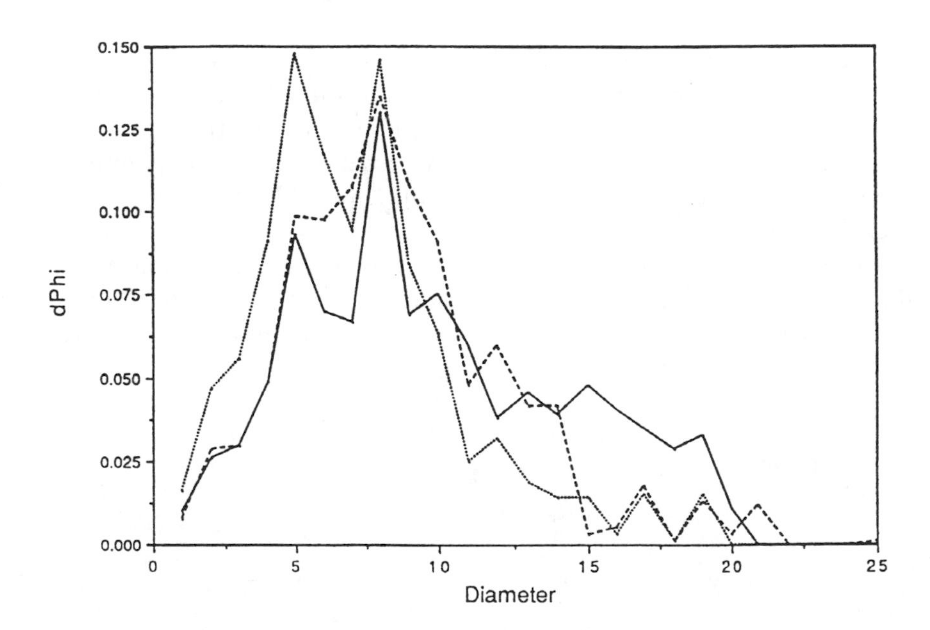

Figure 5.6 Pattern spectra for realizations of Fig. 5.5: dotted line for part a; dashed line for part b; solid line for part c.

5.6.2. Classification by Granulometric Moments

Given a binary image process (random set) S and a collection of convex, compact sets $B_1, B_2,..., B_w$, the pattern spectra resulting from the granulometries $\{\phi_{tB_j}(S)\}$, $j = 1, 2,..., w$, are random functions and the corresponding pattern-spectrum (granulometric) moments are random variables. Taking the first q moments of each granulometry generates $m = qw$ descriptors, $\mu^{(k)}(S; B_j)$, for the process S. Given a class $\{S_1, S_2,..., S_M\}$ of binary image processes to be classified, each has an associated random feature vector

$$\mathbf{F}_i = \begin{pmatrix} \mu^{(1)}(S_i; B_1) \\ \mu^{(2)}(S_i; B_1) \\ \vdots \\ \mu^{(q)}(S_i; B_1) \\ \mu^{(1)}(S_i; B_2) \\ \vdots \\ \mu^{(q)}(S_i; B_w) \end{pmatrix} \tag{5.152}$$

Classification can be achieved as it would be for any class of feature vectors, for instance, Gaussian maximum-likelihood, linear discriminants, or neural networks.

Since the moments of the random function $\Phi(t)$ are used for classification, three basic problems arise: (1) find expressions for the granulometric moments; (2) find expressions for the moments of the granulometric moments; and (3) describe the probability distributions of the granulometric moments. Generally, these problems are difficult. We consider a particular model.

In the Boolean random set model, there is no restriction placed on the manner in which grains intersect. Consider now a random image process S whose realizations are disjoint unions of scalar multiples of compact primitives $A_1, A_2,..., A_n$:

$$S = \bigcup_{i=1}^{n} \bigcup_{j=1}^{m_i} (r_{ij} A_i + x_{ij}) \tag{5.153}$$

S is random owing to the number of grains, $N = m_1 + m_2 + \cdots + m_n$, the randomness of the scalar multiples r_{ij}, and the locations x_{ij}. Although S does not result from the random Boolean model, there are random sets that satisfy the union representation and the nonintersection constraint. For this model, there exists a representation for the pattern-spectrum moments of S in terms of the pattern-spectrum moments of the primitives. We will prove the theorem after stating it and illustrating its use.

Theorem 5.10. If B is compact and convex, and S satisfies the model of Eq. 5.153, then the kth pattern-spectrum moment for the granulometry $\{\phi_{tB}(S)\}$ is given by

$$\mu^{(k)}(S) = \frac{\sum_{i=1}^{n} v(A_i)\mu^{(k)}(A_i)\sum_{j=1}^{m_i} r_{ij}^{k+2}}{\sum_{i=1}^{n} v(A_i)\sum_{j=1}^{m_i} r_{ij}^{2}} \tag{5.154}$$

where $\mu^{(k)}(A_i)$ is the kth pattern-spectrum moment for the granulometry $\{\phi_{tB}(A_i)\}$ and $v(A_i)$ is the volume of A_i. ■

An interesting reduction occurs for the case of a single primitive A. Then

$$S = \bigcup_{j=1}^{N} r_j A + x_j \tag{5.155}$$

For fixed N and p = 3, 4,..., define the random variable

$$Z_p = \left(\sum_{i=1}^{N} r_i^{\,p}\right)\left(\sum_{i=1}^{N} r_i^{2}\right)^{-1} \tag{5.156}$$

and assume that the scalar multiples r_j are independent and identically distributed, their common distribution being known as a *sizing distribution*. It can be shown that Z_p is asymptotically normal as $N \to \infty$, and there exist asymptotic expressions for the moments of Z_p. For instance, if μ_i denotes the ith central moment of the sizing distribution, then

$$E[Z_3] = \frac{\mu_3 + 3\mu_1\mu_2 + \mu_1^3}{\mu_2 + \mu_1^2} + O(N^{-1}) \tag{5.157}$$

For the single-primitive model of Eq. 5.155, the representation of Eq. 5.154 reduces to

$$\mu^{(k)}(S) = \mu^{(k)}(A)Z_{k+2} \tag{5.158}$$

The kth granulometric moment of S is a constant times Z_{k+2}. Thus, the granulometric moments are asymptotically normal (as $N \to \infty$). Since there exist asymptotic expressions for the moments of Z_{k+2}, there exist asymptotic expressions for the moments of the granulometric moments. Denoting the mean and variance of the pattern spectrum by M and V, respectively, for the model of Eq. 5.155,

$$E[M(S)] = \mu^{(1)}(A)E[Z_3] \tag{5.159}$$

$$\mathrm{Var}[M(S)] = \mu^{(1)}(A)^2\mathrm{Var}[Z_3] \tag{5.160}$$

$$E[V(S)] = E[\mu^{(2)}(S) - \mu^{(1)}(S)^2] = \mu^{(2)}(A)E[Z_4] - \mu^{(1)}(A)^2E[Z_3^{\,2}] \tag{5.161}$$

***Example* 5.17.** Suppose the sizing distribution is normally distributed with mean μ and variance σ^2. According to Eq. 5.157,

$$E[Z_3] = \mu\left(1+\frac{2}{1+\mu^2\sigma^{-2}}\right) + O(N^{-1})$$

It can also be shown that

$$\mathrm{Var}[Z_3] = \frac{\sigma^2(\mu^8+8\mu^6\sigma^2+12\mu^4\sigma^4+12\mu^2\sigma^6+15\sigma^8)}{N(\sigma^2+\mu^2)^4} + O(N^{-3/2})$$

Consider the image process consisting of disjoint randomly sized squares, each of which is missing the upper right corner. Suppose the sizing parameter is normally distributed with mean $\mu = 10$ and variance $\sigma^2 = 4$. Figure 5.7 shows a realization of the process. A granulometry is performed with a vertical linear structuring element. From Eq. 5.159 and the above expression for $E[Z_3]$ it is found that $E[M(S)] = 17.95$ (asymptotically). Thirty realizations of the image process have been generated and M computed for each. The average value of the computed values provides the estimate $\overline{M(S)} = 17.89$. ■

***Example* 5.18.** Suppose the sizing distribution is gamma with parameters α and β. According to Eq. 5.157,

$$E[Z_3] = (\alpha + 2)\beta + O(N^{-1})$$

It can also be shown that

$$\text{Var}[Z_3] = \frac{\beta^2(\alpha^4 + 6\alpha^3 + 39\alpha^2 + 74\alpha + 96)}{N\alpha(\alpha+1)^2} + O(N^{-3/2})$$ ■

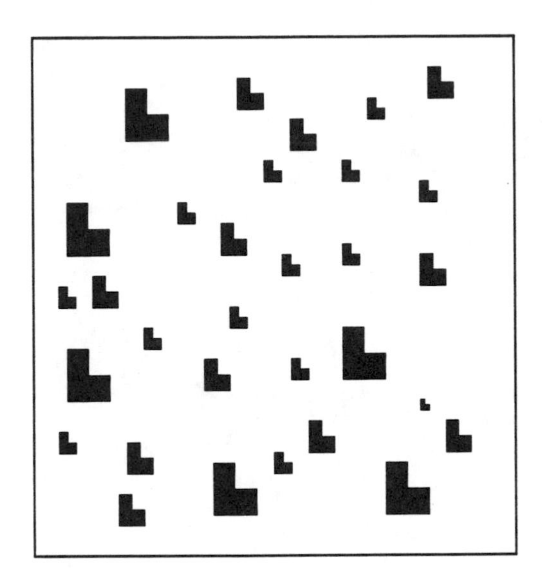

Figure 5.7 Realization of image process consisting of disjoint randomly sized squares missing upper right corners.

To prove Theorem 5.10, note that $\mu^{(k)}(S)$ is a random variable and $\mu^{(k)}(A_i)$ is a constant for $i = 1, 2, \ldots, m$. Let Ω, Ω_i, Φ, and Φ_i denote the size distribution for $\{\phi_{tB}(S)\}$, the size distribution for $\{\phi_{tB}(A_i)\}$, the normalized size distribution for Ω, and the normalized size distribution for Ω_i, respectively. Based on the opening relation $\phi_{tB}(S) = t\phi_B(S/t)$, which is not difficult to demonstrate, and referring to the model of Eq. 5.153,

$$\phi_{tB}(r_{ij}A_i) = r_{ij}\phi_{(t/r_{ij})B}(A_i) \tag{5.162}$$

Distributivity of the granulometry yields

$$\Omega(t) = \sum_{i=1}^{m}\sum_{j=1}^{m_i} \nu[\phi_{tB}(r_{ij}A_i)]$$

$$= \sum_{i=1}^{m}\sum_{j=1}^{m_i} \nu[r_{ij}\phi_{(t/r_{ij})B}(A_i)]$$

$$= \sum_{i=1}^{m} \sum_{j=1}^{m_i} r_{ij}^2 \Omega_i(t/r_{ij}) \tag{5.163}$$

Normalization yields

$$\Phi(t) = 1 - \frac{\sum_{i=1}^{m} \sum_{j=1}^{m_i} r_{ij}^2 \Omega_i(t/r_{ij})}{\sum_{i=1}^{m} v[A_i] \sum_{j=1}^{m_i} r_{ij}^2} \tag{5.164}$$

Since

$$\Omega_i'(t/r_{ij}) = -v[A_i]\Phi_i'(t/r_{ij}) \tag{5.165}$$

differentiation gives

$$\Phi'(t) = \frac{\sum_{i=1}^{m} v[A_i] \sum_{j=1}^{m_i} r_{ij} \Phi_i'(t/r_{ij})}{\sum_{i=1}^{m} v[A_i] \sum_{j=1}^{m_i} r_{ij}^2} \tag{5.166}$$

Hence,

$$\mu^{(k)}(S) = \frac{\sum_{i=1}^{m} v[A_i] \sum_{j=1}^{m_i} r_{ij} \int_0^\infty t^k \Phi_i'(t/r_{ij})\,dt}{\sum_{i=1}^{m} v[A_i] \sum_{j=1}^{m_i} r_{ij}^2}$$

$$= \frac{\sum_{i=1}^{m} v[A_i] \sum_{j=1}^{m_i} r_{ij}^{k+2} \int_0^\infty u^k \Phi_i'(u)\,du}{\sum_{i=1}^{m} v[A_i] \sum_{j=1}^{m_i} r_{ij}^2} \tag{5.167}$$

which reduces to Eq. 5.154 since the integral in the numerator equals $\mu^{(k)}(A_i)$.

5.6.3. Adaptive Reconstructive Openings

Adaptive design of linear filters and neural networks is used to circumvent the mathematical and estimation difficulties arising in direct optimization. Adaptation schemes are also employed in the design of other filters for similar reasons. We show how openings can be designed adaptively for the purpose of restoring binary images degraded by union noise and how characterization of filter adaptation falls naturally into the context of Markov chains.

If a binary image S is composed of a disjoint collection of grains (compact, connected components), then $\{\Psi_t\}$ acts as a parameterized family of sieving filters: if G is a grain and no translate of tB fits inside (is a subset of) G, then G is totally eliminated by Ψ_t; if, on the other hand, at least one translate of tB fits inside G, then part (perhaps all) of G is passed. The *reconstructive τ-opening* Λ_t associated with Ψ_t is defined in terms of image grains by $\Lambda_t(G) = G$ if $\Psi_t(G) \neq \varnothing$; $\Lambda_t(G) = \varnothing$ if $\Psi_t(G) = \varnothing$. The reconstructive filter is a true sieve in that it either fully passes or fully eliminates each grain.

Consider the random set

$$S = \bigcup_{i=1}^{I} C_i + x_i \tag{5.168}$$

where I is a random natural number giving the number of grains in the image, $C_1, C_2, \ldots, C_I$ are connected, compact random sets, and $x_1, x_2, \ldots, x_I$ are random points constrained by the requirement that the translates $C_1 + x_1, C_2 + x_2, \ldots, C_I + x_I$ form a disjoint collection. Then, for the τ-opening defined by Eq. 5.150,

$$\Psi_t(S) = \bigcup_{i=1}^{I} \Psi_t(C_i) + x_i \tag{5.169}$$

where, for any i, $\Psi_t(C_i) \subset C_i$. The reconstructive τ-opening associated with Ψ_t is given by

$$\Lambda_t(S) = \bigcup_{i \in I_t[S]} C_i + x_i \tag{5.170}$$

where $I_t[S] = \{i\text{: } \Psi_t(C_i) \neq \varnothing\}$. The set model of Eq. 5.168 is not restrictive since every set is a union of its maximally connected components.

If C is any compact random set and $\{\Psi_t\}$ is a granulometry, then the $\{\Psi_t\}$-measure (*granulometric measure*) of C is the random variable defined by

$$M_C = \sup\{t: \Psi_t(C) \neq \varnothing\} \tag{5.171}$$

Note that $\Psi_t(C) \neq \varnothing$ if and only if some translate of tB fits inside C. If C is connected, then M_C is given in terms of the associated reconstructive filter by

$$M_C = \sup\{t: \Lambda_t(C) = C\} \tag{5.172}$$

Our intent is to adaptively design a reconstructive τ-opening for the *signal-union-noise restoration model* by letting the filter parameter adapt. For the model, suppose there are two disjoint random sets S and N, S taking the form of Eq. 5.168 and

$$N = \bigcup_{j=1}^{J} D_j + y_j \tag{5.173}$$

where I and J are random natural numbers, $C_1, C_2, \ldots, C_I$ and $D_1, D_2, \ldots, D_J$ are collections of compact, connected random sets identically distributed to the primary grains C and D, respectively, and $x_1, x_2, \ldots, x_I, y_1, y_2, \ldots, y_J$ are random points constrained by the mutual disjointness of $C_1 + x_1$, $C_2 + x_2, \ldots, C_I + x_I$, $D_1 + y_1$, $D_2 + y_2, \ldots, D_J + y_J$. S is the ideal image, N the noise, and $S \cup N$ the observed process. The model could result from overlapping signal and noise grains that are segmented to produce the disjoint representations of Eqs. 5.168 and 5.173. If Λ_t is a reconstructive τ-opening, then each grain in the union $S \cup N$ is either fully passed or fully eliminated. Error results from signal grains erroneously removed and noise grains erroneously passed. There is a reconstructive τ-opening optimization theory for the $S \cup N$ model but the error equations are cumbersome.

For adaptive design, realizations of $S \cup N$ are formed from realizations of S and N, a scanning procedure is used to randomly select grains from $S \cup N$, the reconstructive filter Λ_t is applied to each grain encountered during scanning, and the filter parameter is adapted according to whether the filter correctly or incorrectly passes the grain. We let $r(n)$ denote the value of the parameter, where n counts the number of grains encountered.

When a grain G arrives, the following possibilities exist:

$$\begin{aligned} &\text{(a) } G \text{ is a noise grain and } \Lambda_{r(n)}(G) = G, \\ &\text{(b) } G \text{ is a signal grain and } \Lambda_{r(n)}(G) = \varnothing, \\ &\text{(c) } G \text{ is a noise grain and } \Lambda_{r(n)}(G) = \varnothing, \\ &\text{(d) } G \text{ is a signal grain and } \Lambda_{r(n)}(G) = G. \end{aligned} \tag{5.174}$$

In cases (c) and (d), the filter has behaved as desired; in cases (a) and (b), it has not. The following adaptation procedure can be employed:

$$\begin{array}{lll} \text{(i) } r \to r + \mu & \text{if condition (a) occurs,} & \\ \text{(ii) } r \to r - \mu & \text{if condition (b) occurs,} & (5.175) \\ \text{(iii) } r \to r & \text{if conditions (c) or (d) occur,} & \end{array}$$

where μ is a fixed increment. For computer implementation, we let $r(n)$ be a nonnegative integer and $\mu = 1$. The random parameter determines a discrete-time Markov chain.

We must compute the one-step transition probabilities $p_{r,r+1}$, $p_{r,r-1}$, and $p_{r,r}$. These depend on the probabilities $P(S)$ and $P(N)$ of selecting a signal or noise grain, respectively, in the scanning procedure. Various scanning protocols can be employed, so we leave these probabilities generic. In terms of granulometric measure, conditions (a) through (d) above can be rewritten in the following manner:

$$\begin{array}{ll} \text{(a) } G \text{ is a noise grain and } M_G \geq r(n), & \\ \text{(b) } G \text{ is a signal grain and } M_G < r(n), & \\ \text{(c) } G \text{ is a noise grain and } M_G < r(n), & (5.176) \\ \text{(d) } G \text{ is a signal grain and } M_G \geq r(n). & \end{array}$$

Letting M_S and M_N be the granulometric measures of the signal and noise primary grains, respectively, if grain selection and passing the filter are independent, then the transition probabilities are given by

$$\begin{array}{ll} \text{(i) } p_{r,r+1} = P(N)P(M_N \geq r), & \\ \text{(ii) } p_{r,r-1} = P(S)P(M_S < r), & (5.177) \\ \text{(iii) } p_{r,r} = P(S)P(M_S \geq r) + P(N)P(M_N < r). & \end{array}$$

We assume that the distribution supports for M_S and M_N are intervals with endpoints $a_S < b_S$ and $a_N < b_N$, respectively, where $0 \leq a_S$, $0 \leq a_N$, and it may be that $b_S = \infty$ or $b_N = \infty$. We also assume that $a_N \leq a_S < b_N \leq b_S$. Nonnull intersection of the distribution supports ensures that the adaptive filter does not trivially converge to a filter that totally restores S.

There are four cases regarding state communication. Suppose $a_S < 1$ and $b_N = \infty$. Then the Markov chain is irreducible since all states communicate. Suppose $a_S \geq 1$ and $b_N = \infty$. Then, for each state $r \leq a_S$, r is accessible from state s if $s < r$, but s is not accessible from r; on the other hand, all states $r \geq a_S$ communicate and form a single equivalence class. Suppose $a_S < 1$ and $b_N < \infty$. Then, for each state $r \geq b_N$, r is accessible from state s if $s > r$, but s is not accessible from r; on the other hand, all states $r \leq b_N$ communicate and form a single equivalence class. Suppose $1 \leq a_S < b_N < \infty$. Then the states below a_S are accessible from states below themselves but not conversely, states above b_N are accessible from states above themselves but not conversely, and all states r such that $a_S \leq r \leq b_N$ communicate and form a single equivalence class. In sum, the states between a_S and b_N form an irreducible equivalence class C of the state space and each state outside C is nonrecurrent. With probability 1, the chain will eventually enter C and, once inside C, will not leave. Thus, we focus our attention on C. Within C, the chain is irreducible and aperiodic. If $b_N < \infty$, then the state space is finite and, according to Theorem 5.3, the Markov chain possesses a common steady-state and stationary distribution.

Suppose $b_N = \infty$ and (without loss of generality) $a_S < 1$, so that the Markov chain is irreducible and aperiodic over the state space $S = \{0, 1, 2,...\}$. In fact, it is a generalized random walk and fits the model of Example 5.6. According to the results of that example, the Markov chain is positive recurrent and has a stationary distribution if and only if

$$\sum_{n=1}^{\infty} \frac{p_{0,1}p_{1,2}\cdots p_{n-1,n}}{p_{1,0}p_{2,1}\cdots p_{n,n-1}} < \infty \tag{5.178}$$

It can be shown that the series is convergent. Then (as in Example 5.6), Eq. 5.46 results in

$$x_n = \frac{p_{0,1}p_{1,2}\cdots p_{n-1,n}}{p_{1,0}p_{2,1}\cdots p_{n,n-1}} x_0 \tag{5.179}$$

According to Eq. 5.47, the stationary distribution is given by

$$\pi_0 = \frac{1}{1+\sum_{n=1}^{\infty} \frac{p_{0,1}p_{1,2}\cdots p_{n-1,n}}{p_{1,0}p_{2,1}\cdots p_{n,n-1}}} \tag{5.180}$$

$$\pi_n = \frac{\frac{p_{0,1}p_{1,2}\cdots p_{n-1,n}}{p_{1,0}p_{2,1}\cdots p_{n,n-1}}}{1+\sum_{n=1}^{\infty} \frac{p_{0,1}p_{1,2}\cdots p_{n-1,n}}{p_{1,0}p_{2,1}\cdots p_{n,n-1}}} \quad (n = 1, 2,...) \tag{5.181}$$

According to Theorem 5.4, the stationary distribution is also the steady-state distribution.

***Example* 5.19.** Even though we are assured under the preceding mild assumptions that the parameter chain has a stationary distribution, the complicated forms of Eqs. 5.180 and 5.181 make closed-form analytic formulation of the stationary distribution next to impossible and numerical simulation techniques must be used to estimate the distribution. However, for the uniform model, the state space is finite and a closed-form solution is achievable. Suppose M_N and M_S are uniformly distributed over $[a, b]$ and $[c, d]$, respectively, with $a < c < b < d$ and, for computational convenience, a, b, c, and d are integers. The effective state space for the parameter is $[c, b]$ because all other states are nonrecurrent. Let $m_S = (d - c)^{-1}$ and $m_N = (b - a)^{-1}$. Owing to the uniformity of the signal and noise granulometric measures, Eq. 5.177 yields

$$p_{r,r+1} = P(N)m_N(b - r)$$

$$p_{r,r-1} = P(S)m_S(r - c)$$

The denominator of Eq. 5.181 becomes

$$1 + \sum_{r=c+1}^{b} \prod_{k=c+1}^{r} \frac{p_{k-1,k}}{p_{k,k-1}} = 1 + \sum_{r=c+1}^{b} \prod_{k=c+1}^{r} \frac{P(N)m_N(b-k+1)}{P(S)m_S(k-c)}$$

$$= 1 + \sum_{i=1}^{b-c} \left[\prod_{j=1}^{i} \left(\frac{P(N)m_N}{P(S)m_S} \right) \frac{(b-c)-(j-1)}{j} \right]$$

$$= 1 + \sum_{i=1}^{b-c} \binom{b-c}{i} \left(\frac{P(N)m_N}{P(S)m_S} \right)^i$$

$$= \left(1 + \frac{P(N)m_N}{P(S)m_S} \right)^{b-c}$$

the second equality following from the substitutions $i = r - c$ and $j = k - c$, and the third following because the second is a binomial expansion. Note that the numerator of Eq. 5.181 has been evaluated in the process of the preceding derivation. From Eqs. 5.180 and 5.181, the stationary (steady-state) distribution is given by

$$\pi_c = \left(\frac{m_S P(S)}{m_S P(S) + m_N P(N)} \right)^{b-c}$$

$$\pi_{c+i} = \left(\frac{m_S P(S)}{m_S P(S) + m_N P(N)} \right)^{b-c} \binom{b-c}{i} \left(\frac{m_N P(N)}{m_S P(S)} \right)^i, \quad 1 \le i \le b - c \qquad ■$$

***Example* 5.20.** In this example we empirically apply the adaptive procedure. Figure 5.8 shows a realization of a union process. There are 200 signal grains (black), each being a randomly rotated ellipse with major axis A_S, minor axis B_S, and $A_S/B_S = 3$, where A_S is normally distributed with mean 20 and variance 2. There are 200 noise grains (gray), each being a randomly rotated ellipse with major axis A_N, minor axis B_N, and $A_N/B_N = 2$, where A_N is normally distributed with mean 18 and variance 2. To separate the grain types, we use a granulometry whose generator consists of four unit-length straight lines: vertical, horizontal, 45°, and –45°. The granulometric measure is the maximum of the grain measurements in the four directions. Adaptation is accomplished by horizontally scanning each realization and adapting upon encountering a grain. Figure 5.9 shows the empirically derived signal and noise granulometric-measure distributions obtained from five realizations. It also shows the corresponding empirical steady-state distribution for the parameter. The empirical mean and variance of the parameter in the steady state are 17.18 and 2.08, respectively. If we set $t = 18$ and apply the filter to the realization of Fig. 5.8, then the output of Fig. 5.10 results. ■

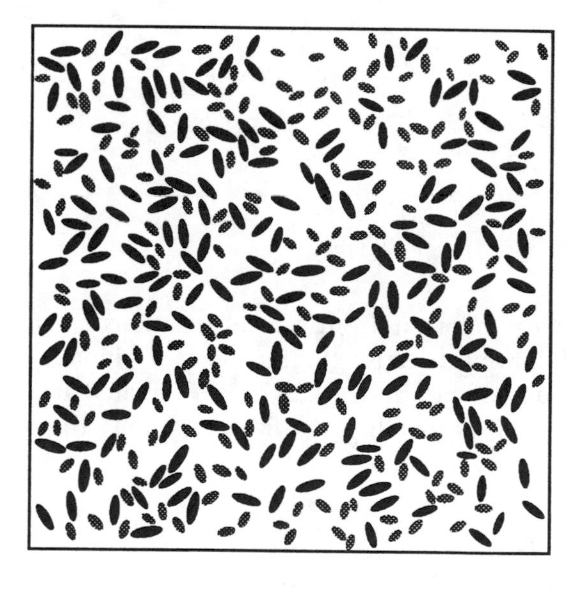

Figure 5.8 Realization of signal-union-noise process for Example 5.20.

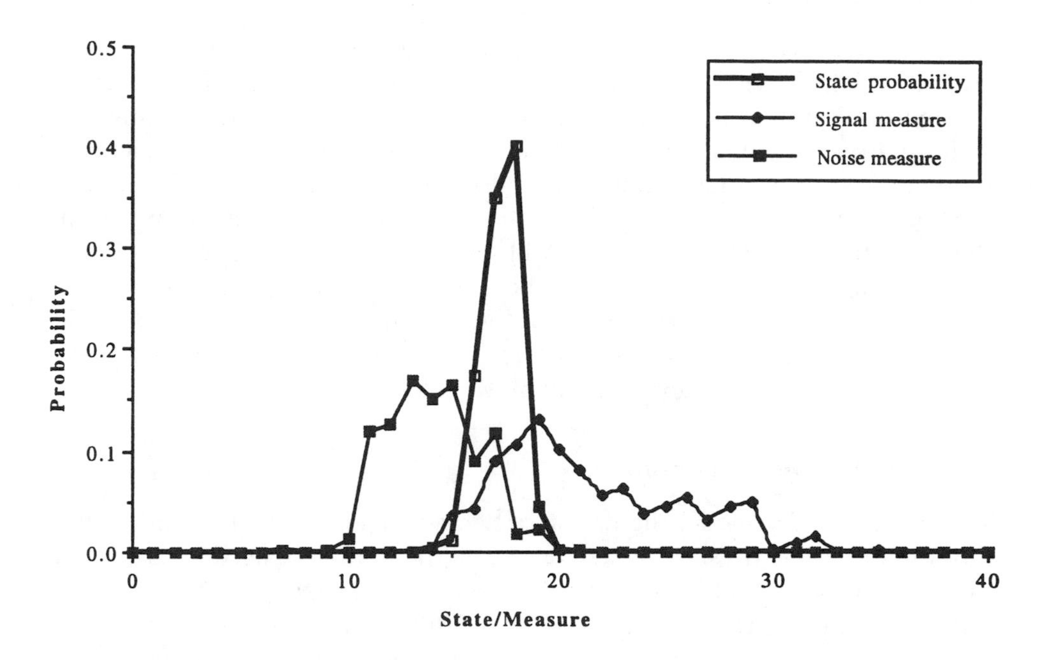

Figure 5.9 Emprically derived signal and noise granulometric-measure distributions and empirical steady-state distribution for the parameter.

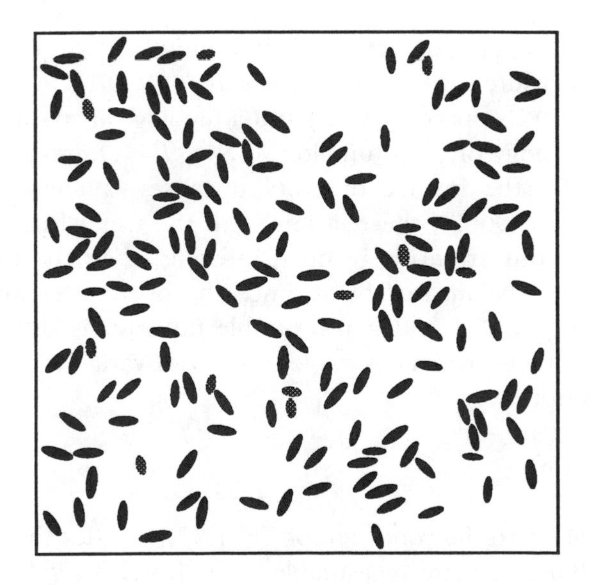

Figure 5.10 Output of filter applied to realization of Fig. 5.8.

5.7. Random Sets

On numerous occasions in the text we have discussed random sets. While one might wish to view a random set as simply a 0-1 valued random function, the morphological applications of random sets provide sufficient motivation to have a distinct theory of random sets. This is especially so for Euclidean random sets, an example being the random Boolean model. In this section our aim is to present the topological and probabilistic foundation for the theory of random closed sets. This will require basic concepts from point-set topology and measure theory. For those without a background in measure theory, we suggest a cursory reading of the section just to get the flavor of the subject and to see the kind of issues that distinguish the theory of random sets from the classical theory of random variables.

5.7.1. Hit-or-Miss Topology

To mathematically model the notion of a random set, we first consider what types of probabilistic statements concerning random sets need to be asked. In the case of a random variable X, since each realization of X is a real number, one needs to ask the probability of X lying in an interval I, $P(X \in I)$. The definition of a random variable is motivated by this requirement. X is assumed to be a real-valued function defined on a probability space $(S, \mathfrak{E}, P)$. Then $P(X \in I)$ can be defined via the inverse mapping:

$$P(X \in I) = P(X^{-1}(I)) = P(\{\omega \in S\colon X(\omega) \in I\}) \tag{5.182}$$

This requires $X^{-1}(I) \in \mathfrak{E}$, that is, $X^{-1}(I)$ must be an event or, in the terminology of measure theory, a *measurable set*.

Given a probability space $(S, \mathfrak{E}, P)$ and a topological space $(T, \mathfrak{I})$, T being a set and $\mathfrak{I}$ a topology on T, a function $X\colon S \to T$ is *measurable* if $X^{-1}(V) \in \mathfrak{E}$ for every $V \in \mathfrak{I}$ [the inverse of every open set is measurable]. If the topological space is $(\mathfrak{R}, \mathfrak{I})$, the real line with the usual Euclidean topology, and if we restrict our attention to open intervals, then the inverse of every open interval must be measurable. Since the open intervals generate the Euclidean topology on $\mathfrak{R}$ and a measurable function is one for which the inverse of every open set is measurable, a random variable is defined to be a measurable function

$$X\colon (S, \mathfrak{E}, P) \to (\mathfrak{R}, \mathfrak{I}) \tag{5.183}$$

(which is equivalent to the condition of Eq. 1.30). Under this definition, the inverses of all Borel sets are measurable. The Borel σ-algebra, which is the smallest σ-algebra containing $\mathfrak{I}$, is closed under countable unions, countable intersections, and complementation. Hence, closed sets, countable unions of closed sets, and countable intersections of closed sets are Borel sets and the inverses of any of these kinds of sets resulting from a random variable are measurable. In particular, $X^{-1}(I) \in \mathcal{E}$ for any interval I and the probability of Eq. 5.182 is defined.

Every random variable X: $(S, \mathcal{E}, P) \rightarrow (\mathfrak{R}, \mathfrak{I})$ induces a probability measure P_X on $\mathcal{B}$, the Borel σ-algebra in $\mathfrak{R}$, by

$$P_X(B) = P(X^{-1}(B)) = P(\{\omega \in S: X(\omega) \in B\}) \tag{5.184}$$

for any Borel set B. $P_X(B)$ gives the probability $P(X \in B)$. Because $\mathfrak{R}$ is a locally compact Hausdorff space in which open sets are σ-compact, as a positive Borel measure, P_X is *regular*. Hence, for every Borel set B,

$$\begin{aligned} P_X(B) &= \inf \{P_X(G): B \subset G, G \text{ open}\} \\ &= \sup \{P_X(K): K \subset B, K \text{ compact}\} \end{aligned} \tag{5.185}$$

There are similarities between the measure-theoretic foundations of random closed sets and random variables. The theory of random closed sets is grounded on probability statements concerning whether or not a random closed set X hits or misses a fixed compact set or a fixed open set: (1) Given a fixed compact set K, what is the probability of X missing K, $P(X \cap K = \varnothing)$ or, equivalently, the probability of X hitting K, $P(X \cap K \neq \varnothing)$? (2) Given a fixed open set G, what is the probability of X hitting G, $P(X \cap G \neq \varnothing)$ or, equivalently, of missing G, $P(X \cap G = \varnothing)$? The probability statements need to be answered relative to a probability P on a probability space $(S, \mathcal{E}, P)$, namely,

$$P(X \cap K = \varnothing) = P(\{\omega \in S: X(\omega) \cap K = \varnothing\}) \tag{5.186}$$

$$P(X \cap G \neq \varnothing) = P(\{\omega \in S: X(\omega) \cap G \neq \varnothing\}) \tag{5.187}$$

where $X(\omega)$ is the realization of the random set corresponding to $\omega \in S$. As in the case of a random variable, the random set X needs to be defined on a probability space and needs to be a measurable function. But, whereas for random variables the topology on the range space $\mathfrak{R}$ was generated by the open intervals, now the topology will be on the space of closed sets and be generated by the closed-set collections $\{F: F \cap K = \varnothing\}$, K compact, and $\{F: F \cap G \neq \varnothing\}$, G open. The "points" in the resulting topological space will be closed Euclidean sets and the open sets in the topological space will be certain collections of closed Euclidean sets. Once this topology of closed sets has been formed, random closed sets can be defined in a manner not unlike random variables.

Although our main interest is in two-dimensional binary images, no extra difficulty arises from considering subsets of $\mathfrak{R}^d$, $d \geq 1$. Hence, we consider an image to be a closed subset of $\Re^d$, and we let $\mathcal{F}$, $\mathcal{K}$, and $\mathcal{G}$ denote the collections of closed, compact, and open subsets of $\Re^d$, respectively. For any subset $H \subset \Re^d$, the classes of closed sets hitting and missing H are respectively defined by

$$\mathcal{F}_H = \{F \in \mathcal{F}: F \cap H \neq \varnothing\} \tag{5.188}$$

$$\mathcal{F}^H = \{F \in \mathcal{F}: F \cap H = \varnothing\} \tag{5.189}$$

Since we say F *hits* H if $F \cap H \neq \varnothing$ and F *misses* H if $F \cap H = \varnothing$, $\mathcal{F}_H$ and $\mathcal{F}^H$ are the collections of closed sets that hit and miss H, respectively.

A topology $\Im_{\mathcal{F}}$ for the space $\mathcal{F}$ of closed sets is generated by two collections of closed sets: $\{\mathcal{F}^K: K \in \mathcal{K}\}$ and $\{\mathcal{F}_G: G \in \mathcal{G}\}$. The sets (of closed sets) $\mathcal{F}^K$ and $\mathcal{F}_G$ form a subbase for the topology $\Im_{\mathcal{F}}$, so that the base B for $\Im_{\mathcal{F}}$ is the set of all finite intersections of sets within $\{\mathcal{F}^K: K \in \mathcal{K}\}$ and $\{\mathcal{F}_G: G \in \mathcal{G}\}$:

$$\mathrm{B} = \{\mathcal{F}^{K_1} \cap \mathcal{F}^{K_2} \cap \cdots \cap \mathcal{F}^{K_n} \cap \mathcal{F}_{G_1} \cap \mathcal{F}_{G_2} \cap \cdots \cap \mathcal{F}_{G_m} : K_1, \ldots, K_n \in \mathcal{K}, G_1, \ldots, G_m \in \mathcal{G}\} \tag{5.190}$$

where n and m are arbitrary. Since each $\{\mathcal{F}^K: K \in \mathcal{K}\}$ and $\{\mathcal{F}_G: G \in \mathcal{G}\}$ are sets of sets, the base is a family of sets of sets. A typical set in B consists of all closed sets that miss $K_1, K_2, \ldots, K_n$ and hit $G_1, G_2, \ldots, G_m$. Since missing $K_1, K_2, \ldots, K_n$ is equivalent to missing the union of $K_1, K_2, \ldots, K_n$, which is still a compact set, the base can be reformulated as

$$\mathrm{B} = \{\mathcal{F}^K \cap \mathcal{F}_{G_1} \cap \mathcal{F}_{G_2} \cap \cdots \cap \mathcal{F}_{G_m} : K \in \mathcal{K}, G_1, \ldots, G_m \in \mathcal{G}\} \tag{5.191}$$

The sets $\mathcal{F}^K$ and $\mathcal{F}_G$ belong to the base and therefore are open in the resulting topology, known as the *hit-or-miss topology* for the space $\mathcal{F}$ of closed sets. Because B is a base for $\Im_{\mathcal{F}}$, open sets in $\Im_{\mathcal{F}}$ are unions of sets of closed sets of the form given in Eq. 5.191. The topological space $(\mathcal{F}, \Im_{\mathcal{F}})$ is compact, Hausdorff, and has a countable base.

5.7.2. Convergence and Continuity

A key aspect of any topology is convergence within the topology and the manner in which convergence characterizes continuity. Since $(\mathcal{F}, \Im_{\mathcal{F}})$ has a countable base, continuity is characterized by sequential convergence: if $\{F_n\}$ is a sequence of closed sets and F is a closed set, then F_n *converges* to the *limit* F if and only if for any set in the base for $\Im_{\mathcal{F}}$ containing F, F_n lies within the base set for sufficiently large n. We write either $F_n \to F$ or $\lim F_n = F$. From the definition of the base, this means that $F_n \to F$ if and only if the following two conditions hold:

C1. If open set G hits F, then G hits F_n for sufficiently large n.

C2. If compact set K misses F, then K misses F_n for sufficiently large n.

If $\{F_n\}$ is a decreasing sequence ($F_n \supset F_{n+1}$) and

$$F = \bigcap_{n=1}^{\infty} F_n \tag{5.192}$$

then we write $F_n \downarrow F$. If $\{F_n\}$ is an increasing sequence ($F_n \subset F_{n+1}$) and

$$A = \bigcup_{n=1}^{\infty} F_n \tag{5.193}$$

then we write $F_n \uparrow A$. If $F_n \downarrow F$ then $\lim F_n = F$ and if $F_n \uparrow A$ then $\lim F_n = \overline{A}$, the closure of A. (Note that $F_n \downarrow F$ implies $F \in \mathcal{F}$ but $F_n \uparrow A$ does not imply $A \in \mathcal{F}$.) To demonstrate that $F_n \downarrow F$ implies $\lim F_n = F$, we need to check conditions C1 and C2. Suppose open set G hits F. Since $F = \cap_n F_n$, G hits every F_n, so that C1 holds. Suppose compact set K misses F. Then there exists F_N such that K misses F_N. Since $\{F_n\}$ is a decreasing sequence, K misses F_n for $n \geq N$, so that C2 is verified. Hence, $\lim F_n = F$. A similar argument applies to $F_n \uparrow A$.

With convergence defined, definition of continuity proceeds customarily. A mapping $F \to \Psi(F)$ is *continuous* from $\mathcal{F} \to \mathcal{F}$ if and only if $F_n \to F$ implies $\Psi(F_n) \to \Psi(F)$. A mapping $(F, H) \to \Psi(F, H)$ is continuous from the product space $\mathcal{F} \times \mathcal{F} \to \mathcal{F}$ if and only if $F_n \to F$ and $H_n \to H$ implies $\Psi(F_n, H_n) \to \Psi(F, H)$.

An important continuous mapping on $\mathcal{F} \times \mathcal{F}$ is the union, $(F, H) \to F \cup H$. To show that union is continuous (and to give a flavor of hit-or-miss proofs), we verify conditions C1 and C2. Suppose $F_n \to F$, $H_n \to H$, and open set G hits $F \cup H$. Then G hits F or G hits H. Without loss of generality, assume G hits F. Since $F_n \to F$, there exists N such that F_n hits G for $n \geq N$. Therefore, $F_n \cup H_n$ hits G for $n \geq N$. As for condition C2, suppose compact K misses $F \cup H$. Then K misses F and K misses H. Since $F_n \to F$, there exists N_1 such that F_n misses K for $n \geq N_1$. Since $H_n \to H$, there exists N_2 such that H_n misses K for $n \geq N_2$. Hence, for $n \geq \max\{N_1, N_2\}$, $F_n \cup H_n$ misses K.

Many key mappings are not continuous but are semicontinuous and for these we need the following definitions. If $\{F_n\}$ is a sequence in $\mathcal{F}$, then the *limit inferior* of $\{F_n\}$, denoted $\underline{\lim}\ F_n$, is the largest closed set satisfying condition C1 and the *limit superior* of $\{F_n\}$, denoted $\overline{\lim}\, F_n$, is the smallest closed set satisfying condition C2. It is immediate from the definitions that $\lim F_n = F$ if and only if

$$\underline{\lim}\, F_n = \overline{\lim}\, F_n = F \tag{5.194}$$

If T is a topological space and mapping Ψ: $T \to \mathcal{F}$, then Ψ is *upper semicontinuous* (*u.s.c.*) if, for any compact set $K \in \mathcal{K}$, $\Psi^{-1}(\mathcal{F}^K)$ is open in T. Ψ is *lower semicontinuous* (*l.s.c.*) if, for any open set $G \in \mathcal{G}$, $\Psi^{-1}(\mathcal{F}_G)$ is open in T. If T possesses a countable base (which it does if $T = \mathcal{F}$), then upper semicontinuity and lower semicontinuity are characterized in terms of limits superior and inferior, respectively: Ψ is u.s.c. if and only if $x_n \to x$ in T implies

$$\Psi(x) \supset \overline{\lim}\, \Psi(x_n) \tag{5.195}$$

in $\mathcal{F}$; Ψ is l.s.c. if and only if $x_n \to x$ in T implies

$$\Psi(x) \subset \underline{\lim}\, \Psi(x_n) \tag{5.196}$$

in $\mathcal{F}$.

When we restrict our attention to compact sets, we are concerned with the space $\mathcal{K}$. A topology $\mathfrak{I}_{\mathcal{K}}$ for $\mathcal{K}$ is generated by the collections $\mathcal{K}^F$ for $F \in \mathcal{F}$ and $\mathcal{K}_G$ for $G \in \mathcal{G}$. $\mathfrak{I}_{\mathcal{K}}$ is called the *myope topology* and when we speak of the topological space $\mathcal{K}$, we mean the space $(\mathcal{K}, \mathfrak{I}_{\mathcal{K}})$. The myope topology is not equivalent to the relative topology induced on $\mathcal{K}$ as a subset of $\mathcal{F}$ endowed with $\mathfrak{I}_{\mathcal{F}}$. The myope topology is strictly finer than the induced topology. A set $\mathcal{C}$ of compact sets is compact in the myope topology if and only if $\mathcal{C}$ is closed in $\mathcal{F}$ and there exists a compact set K_0 such that $K \subset K_0$ for all $K \in \mathcal{C}$. The myope topology and the hit-or-miss topology on $\mathcal{F}$ are equivalent on subsets of $\mathcal{K}$ that are compact in the myope topology. A sequence $\{K_n\} \subset \mathcal{K}$ is convergent in $(\mathcal{K}, \mathfrak{I}_{\mathcal{K}})$ if and only if there exists $K_0 \in \mathcal{K}$ such that $K_n \subset K_0$ for all n and $\{K_n\}$ converges in $(\mathcal{F}, \mathfrak{I}_{\mathcal{F}})$. If ψ: $\mathcal{K} \to [-\infty, \infty]$ is an increasing function [$K_1 \subset K_2$ implies $\psi(K_1) \leq \psi(K_2)$], then ψ is u.s.c. if and only if $K_n \downarrow K$ in $\mathcal{K}$ implies $\psi(K_n) \downarrow \psi(K)$.

Removing the null set from $\mathcal{K}$ yields the space $\mathcal{K}_0 = \mathcal{K} - \{\varnothing\}$, which we endow with the relative myope topology. This topology is equivalent to the metric-space topology derived from the *Hausdorff metric*, which is defined by

$$\rho(K, K') = \max \{ \sup_{x \in K} d(x, K'), \sup_{x \in K'} d(x, K) \} \tag{5.197}$$

where

$$d(x, K) = \inf_{y \in K} |x - y| \tag{5.198}$$

The Hausdorff metric $\rho(K, K')$ is obtained by finding the supremum of all distances of points in K from K', finding the supremum of all distances of points in K' from K, and then taking the maximum of these suprema. The Hausdorff metric is used in both morphological image processing and the theory of fractals to provide a metric for convergence and approximation.

Many key imaging operations are either continuous or semicontinuous. Although we have introduced morphological operations in various places in the text, for unity we will define the four fundamental structural operations here as they apply in Euclidean space. We provide both their operator and infix notations. For two sets $A, B \in \Re^d$, *dilation*, *erosion*, *opening*, and *closing* are defined by

$$\begin{aligned}
\delta_B(A) &= A \oplus B = \{a + b: a \in A, b \in B\} \\
\varepsilon_B(A) &= A \ominus B = \{z: B_z \subset A\} \\
\phi_B(A) &= A \circ B = (A \ominus B) \oplus B \\
\gamma_B(A) &= A \bullet B = (A \oplus (-B)) \ominus (-B)
\end{aligned} \tag{5.199}$$

respectively. Operator notation is used when B is considered as a structuring element determining the operation; infix notation is used when the operation is viewed as a binary operation. In the latter case, only dilation is commutative. Owing to various identities, the definitions could be given in other forms. For instance, Eqs. 5.121 and 5.148 provide hitting and fitting formulations of dilation and opening, respectively. The following mappings are continuous:

$$\begin{aligned}
&\mathcal{F} \times \mathcal{K} \to \mathcal{F} \text{ by } (F, K) \to F \oplus K \\
&\mathcal{F} \times \mathcal{F} \to \mathcal{F} \text{ by } (F_1, F_2) \to F_1 \cup F_2 \\
&\mathcal{K} \to \mathcal{K} \text{ by } K \to \mathrm{H}(K) \quad (\textit{convex hull}) \\
&\Re^+ \times \mathcal{F} \to \mathcal{F} \text{ by } (r, F) \to rF \; (\textit{scalar multiplication})
\end{aligned} \tag{5.200}$$

The following mappings are u.s.c.:

$$\begin{aligned}
&\mathcal{F} \times \mathcal{K}_0 \to \mathcal{F} \text{ by } (F, K) \to F \ominus K \\
&\mathcal{F} \times \mathcal{K}_0 \to \mathcal{F} \text{ by } (F, K) \to F \circ K \\
&\mathcal{F} \times \mathcal{K}_0 \to \mathcal{F} \text{ by } (F, K) \to F \bullet K \\
&\mathcal{F} \times \mathcal{F} \to \mathcal{F} \text{ by } (F_1, F_2) \to F_1 \cap F_2
\end{aligned} \tag{5.201}$$

The following mappings are l.s.c.:

$$\begin{aligned}
&\mathcal{F} \to \mathcal{F} \text{ by } F \to \overline{F^c} \\
&\mathcal{F} \to \mathcal{F} \text{ by } F \to \partial F \quad (\textit{boundary})
\end{aligned} \tag{5.202}$$

$\mathcal{F} \to \mathcal{F}$ by $F \to \mathrm{H}(F)$

5.7.3. Random Closed Sets

Having introduced the topological space $(\mathcal{F}, \mathfrak{I}_{\mathcal{F}})$, we can proceed to the definition of a random closed set. Let $\Sigma_{\mathcal{F}}$ be the Borel σ-algebra generated by the hit-or-miss topology $\mathfrak{I}_{\mathcal{F}}$. The Borel sets are sets in $\Sigma_{\mathcal{F}}$ and each Borel set is a set of closed Euclidean sets. Let (S, Σ, μ) be a probability space. A *random closed set* (*RACS*) is a measurable function

$$X: (S, \Sigma, \mu) \to (\mathcal{F}, \Sigma_{\mathcal{F}}) \tag{5.203}$$

If X is a RACS, then by definition it is a measurable function. Hence, for any Borel set $\mathcal{B} \in \Sigma_{\mathcal{F}}$, $X^{-1}(\mathcal{B})$ is a measurable set, that is, $X^{-1}(\mathcal{B}) \in \Sigma$. A probability measure is induced on $\Sigma_{\mathcal{F}}$ by

$$P_X(\mathcal{B}) = \mu(X^{-1}(\mathcal{B})) = \mu(\{\omega \in S: X(\omega) \in \mathcal{B}\}) \tag{5.204}$$

where one must keep in mind that, for each $\omega \in S$, $X(\omega)$ is a closed Euclidean set. As in the case of ordinary random variables, the induced probability $P_X(\mathcal{B})$ gives the inclusion probability $P(X \in \mathcal{B})$, the difference here being that the Borel set $\mathcal{B}$ is a collection of closed Euclidean sets, not a single set. In particular, for compact K and open G,

$$P_X(\mathcal{F}^K) = \mu(X^{-1}(\mathcal{F}^K)) = \mu(\{\omega \in S: X(\omega) \cap K = \varnothing\}) \tag{5.205}$$

$$P_X(\mathcal{F}_G) = \mu(X^{-1}(\mathcal{F}_G)) = \mu(\{\omega \in S: X(\omega) \cap G \neq \varnothing\}) \tag{5.206}$$

$P_X(\mathcal{F}^K)$ and $P_X(\mathcal{F}_G)$ give the respective probabilities of X missing K and hitting G. Since $(\mathcal{F}, \mathfrak{I}_{\mathcal{F}})$ is a compact Hausdorff space and P_X is a positive measure, P_X is regular and, for any Borel set $\mathcal{B} \in \Sigma_{\mathcal{F}}$,

$$\begin{aligned} P_X(\mathcal{B}) &= \inf\{P_X(\mathcal{V}): \mathcal{B} \subset \mathcal{V}, \mathcal{V} \text{ open in } (\mathcal{F}, \mathfrak{I}_{\mathcal{F}})\} \\ &= \sup\{P_X(\mathcal{C}): \mathcal{C} \subset \mathcal{B}, \mathcal{C} \text{ compact in } (\mathcal{F}, \mathfrak{I}_{\mathcal{F}})\} \end{aligned} \tag{5.207}$$

Equation 5.207 corresponds to Eq. 5.185 for random variables.

Owing to the definition of a RACS as a measurable function, if X is a RACS and Ψ is a measurable function, then $\Psi(X)$ is a RACS. Since a semicontinuous function is measurable, if Ψ is semicontinuous, then $\Psi(X)$ is a RACS. Hence, if we apply any of the mappings of Eqs. 5.200 through 5.202 to a RACS, then the output is a RACS.

Example 5.21. Let X be the closed disk centered at the origin whose radius possesses a gamma distribution with parameters α and β. For any arbitrary compact set K let m_K be the minimum absolute value attained by points in K,

$$m_K = \min\{|z|: z \in K\}$$

Assuming $\Re^+ = [0, \infty)$ is the sample space on which X is defined and μ is the Borel measure induced by the gamma distribution, the probability that X misses K is given by

$$P_X(\mathcal{F}^K) = \mu(\{x \in \Re^+ : X(x) \cap K = \varnothing\})$$

$$= \frac{\beta^{-\alpha}}{\Gamma(\alpha)} \int_0^{m_K} x^{\alpha-1} e^{-x/\beta}\, dx \qquad \blacksquare$$

5.7.4. Capacity Functional

For any RACS X, its *capacity functional* T_X is defined on $\mathcal{K}$ by

$$T_X(K) = P_X(\mathcal{F}_K) = P(X \in \mathcal{F}_K) = P(X \cap K \neq \varnothing) \tag{5.208}$$

$T_X(K)$ is the probability that the RACS X hits K. The complementary *generating functional* Q_X is defined on $\mathcal{K}$ by

$$Q_X(K) = 1 - T_X(K) \tag{5.209}$$

It follows at once that

$$Q_X(K) = P_X(\mathcal{F}^K) = P(X \in \mathcal{F}^K) = P(X \cap K = \varnothing) \tag{5.210}$$

$Q_X(K)$ is the probability that the RACS X hits K.

Three properties of the capacity functional are immediate: $0 \le T_X(K) \le 1$, $T_X(\varnothing) = 0$, and T_X is increasing [$K_1 \subset K_2$ implies $T_X(K_1) \le T_X(K_2)$]. In addition, the capacity functional is u.s.c. Indeed, as an increasing functional, T_X is u.s.c. if and only if $K_n \downarrow K$ in $\mathcal{K}$ implies $T_X(K_n) \downarrow T_X(K)$. But $K_n \downarrow K$ implies $\mathcal{F}_{K_n} \downarrow \mathcal{F}_K$, which, because P_X is a probability measure, implies $P_X(\mathcal{F}_{K_n}) \downarrow P_X(\mathcal{F}_K)$, which says precisely that $T_X(K_n) \downarrow T_X(K)$.

Let $K, K_1, K_2, \ldots$ be compact sets and define the functions

$$Q_{X,1}(K; K_1) = T_X(K \cup K_1) - T_X(K)$$

$$Q_{X,2}(K; K_1, K_2) = Q_{X,1}(K; K_1) - Q_{X,1}(K \cup K_2; K_1) \tag{5.211}$$

$$\vdots$$

$$Q_{X,n}(K; K_1, K_2,..., K_n) = Q_{X,n-1}(K; K_1, K_2,..., K_{n-1}) - Q_{X,n-1}(K \cup K_n; K_1, K_2,..., K_{n-1})$$

Then

$$\begin{aligned} Q_{X,1}(K; K_1) &= P(X \cap (K \cup K_1) \neq \varnothing) - P(X \cap K \neq \varnothing) \\ &= P(X \cap K_1 \neq \varnothing) - P(X \cap K \neq \varnothing, X \cap K_1 \neq \varnothing) \\ &= P(X \cap K_1 \neq \varnothing, X \cap K = \varnothing) \end{aligned} \tag{5.212}$$

which is the probability that X hits K_1 and misses K. In general, for $n = 1, 2,...,$

$$Q_{X,n}(K; K_1, K_2,..., K_n) = P(X \cap K_1 \neq \varnothing, X \cap K_2 \neq \varnothing,..., X \cap K_n \neq \varnothing, X \cap K = \varnothing) \tag{5.213}$$

which is the probability that X hits $K_1, K_2,..., K_n$ and misses K. As a probability,

$$Q_{X,n}(K; K_1, K_2,..., K_n) \geq 0 \tag{5.214}$$

The functionals $Q_{X,n}$ play a key role in characterizing the RACS X.

Suppose a functional $\psi: \mathcal{K} \to [-\infty, \infty]$ and, for $n = 1, 2,..., Q_{\psi,n}$ is defined by the system of Eq. 5.211 with ψ in place of T_X. If ψ is u.s.c. as a mapping on $\mathcal{K}$ and if $Q_{\psi,n} \geq 0$ for all $K, K_1, K_2,... \in \mathcal{K}$, then ψ is called a *Choquet capacity*. A Choquet capacity is necessarily increasing. Since T_X is u.s.c. and $Q_{X,n}$ satisfies the system, T_X is a Choquet capacity. This leads us to one of the most celebrated theorems in random set theory (binary image processing).

Theorem 5.11. (*Choquet-Matheron-Kendall*) Suppose $\psi: \mathcal{K} \to [-\infty, \infty]$. Then there exists a unique probability P defined on $\Sigma_{\mathcal{F}}$ such that $P(\mathcal{F}_K) = \psi(K)$ for all $K \in \mathcal{K}$ if and only if ψ is a Choquet capacity such that $\psi(\varnothing) = 0$ and $0 \leq \psi(K) \leq 1$. ■

Since, for any random set X, the corresponding capacity functional is a Choquet capacity for which $T_X(\varnothing) = 0$ and $0 \leq T_X(K) \leq 1$, the probability $P_X(\mathcal{F}_K) = P(X \cap K \neq \varnothing)$ is determined [equal to $T_X(K)$] for all compact sets K. Since $\Sigma_{\mathcal{F}}$ is generated by the classes $\mathcal{F}^K$, $K \in \mathcal{K}$, and

$$P_X(\mathcal{F}^K) = 1 - P_X(\mathcal{F}_K) \tag{5.215}$$

the induced probability P_X is fully determined by T_X.

The situation is akin to the relationship between a random variable X and its probability distribution function F_X: there the induced probability P_X is on the Borel sets in $\Re$, P_X is fully determined by F_X, and we say that F_X is the *law* of the random variable. Thus, it is commonplace to model random variables by known parametric distributions and then statistically estimate the parameters from realizations. For random sets, this approach is much more difficult. A probability distribution function involves probabilities of the half-infinite intervals $(-\infty, b]$ for $b \in \Re$, whereas the capacity involves probabilities for all compact sets in $\Re^d$. One needs both useful models and ways to estimate the parameters of those models.

Example 5.22. According to Theorem 5.8, if the RACS X is the germ-grain process of a random Boolean set model with primary grain S and germ intensity λ, then

$$Q_X(K) = \exp[-\lambda E[\nu((-S) \oplus K)]]$$

where ν denotes volume. Even if the model is appropriate, the problem is estimating both the germ intensity λ and the distribution governing the primary grain S (for instance, see Example 5.15). We have seen in the previous section the difficulty of maximum-likelihood estimation even in the simple linear Boolean model. ■

Example 5.23. For a RACS X and a set $K \in \mathcal{K}$, it is often important to know the probability $P(K \subset X)$. Typically, it is extremely difficult (and, for the most part, not known how) to relate $P(K \subset X)$ to the generating functional or, equivalently, to the capacity functional for X. For a finite set, the problem can be solved in a straightforward manner using the probability addition theorem (Eq. 1.10). If $K = \{z_1, z_2,..., z_m\}$ is a finite set, then

$$\begin{aligned} Q_X(K) &= 1 - P(X \cap \{z_1, z_2,..., z_m\} \neq \varnothing\}) \\ &= 1 - P\left(\bigcup_{j=1}^{m} (X \cap \{z_j\} \neq \varnothing)\right) \\ &= 1 - \sum_{j=1}^{m} (-1)^{j+1} \sum_{1 \le i_1 < i_2 < \cdots < i_j \le m} P\left(\bigcap_{k=1}^{j} (X \cap \{z_{i_k}\} \neq \varnothing)\right) \\ &= 1 - \sum_{j=1}^{m} (-1)^{j+1} \sum_{1 \le i_1 < i_2 < \cdots < i_j \le m} P(\{z_{i_1}, z_{i_2}, \ldots, z_{i_j}\} \subset X) \end{aligned}$$

In the other direction,

$$
\begin{aligned}
P(K \subset X) &= P(\{z_1, z_2, \ldots, z_m\} \subset X) \\
&= P\left(\bigcap_{j=1}^{m} (X \cap \{z_j\} \neq \varnothing)\right) \\
&= 1 - P\left(\bigcup_{j=1}^{m} (X \cap \{z_j\} = \varnothing)\right) \\
&= 1 - \sum_{j=1}^{m} (-1)^{j+1} \sum_{1 \le i_1 < i_2 < \cdots < i_j \le m} P\left(\bigcap_{k=1}^{j} (X \cap \{z_{i_k}\} = \varnothing)\right) \\
&= 1 - \sum_{j=1}^{m} (-1)^{j+1} \sum_{1 \le i_1 < i_2 < \cdots < i_j \le m} Q_X(\{z_{i_1}, z_{i_2}, \ldots, z_{i_j}\})
\end{aligned}
$$

$P(z_1 \in X, z_2 \in X)$ gives the *autocorrelation* of X. Based upon the preceding equation,

$$P(z_1 \in X, z_2 \in X) = 1 - Q_X(\{z_1\}) - Q_X(\{z_2\}) + Q_X(\{z_1, z_2\}) \qquad \blacksquare$$

The capacity theory for a RACS has been presented in terms of the capacity functional being defined on compact sets $K \in \mathcal{K}$; however, it could have been presented in terms of the capacity functional being defined on open sets $G \in \mathcal{G}$. Suppose we define the capacity functional on $\mathcal{G}$, namely, for $G \in \mathcal{G}$,

$$T_X(G) = P_X(\mathcal{F}_G) = P(X \in \mathcal{F}_G) = P(X \cap G \neq \varnothing) \tag{5.216}$$

If $\{K_n\} \subset \mathcal{K}$ and $K_n \uparrow G \in \mathcal{G}$, then $\mathcal{F}_{K_n} \uparrow \mathcal{F}_G$, implying $P_X(\mathcal{F}_{K_n}) \uparrow P_X(\mathcal{F}_G)$, which in turn implies $T_X(K_n) \uparrow T_X(G)$. Thus, for any open Euclidean set G,

$$T_X(G) = \sup \{T_X(K) \colon K \in \mathcal{K}, K \subset G\} \tag{5.217}$$

Going in the other direction, if $\{G_n\} \subset \mathcal{G}$ and $G_n \downarrow K \in \mathcal{K}$, then $\mathcal{F}_{G_n} \downarrow \mathcal{F}_K$ and therefore $T_X(G_n) \downarrow T_X(K)$. Thus, for any compact Euclidean set K,

$$T_X(K) = \inf \{T_X(G): G \in \mathcal{G}, K \subset G\} \tag{5.218}$$

The capacity functional is characterized on compact and open sets by defining it on either.

As defined on open sets, T_X is increasing, $0 \le T_X(G) \le 1$, and $T_X(\emptyset) = 0$. Rather than being u.s.c. as on compact sets, T_X is l.s.c. on open sets, meaning that $G_n \uparrow G$ in $\mathcal{G}$ implies $T_X(G_n) \uparrow T_X(G)$. The functions $Q_{X,n}$ can be defined by using open sets $G, G_1, G_2,...$ and again we would have $Q_{X,n} \ge 0$. A function $\psi: \mathcal{G} \to [-\infty, \infty]$ is also a Choquet capacity if it is l.s.c. and the functions $Q_{\psi,n}$ are nonnegative. Theorem 5.11 has an alternate form that applies in the current circumstances: if $\psi: \mathcal{G} \to [-\infty, \infty]$, then there exists a unique probability P defined on $\Sigma_{\mathcal{F}}$ such that $P(\mathcal{F}_G) = \psi(G)$ for all $G \in \mathcal{G}$ if and only if ψ is a Choquet capacity such that $\psi(\emptyset) = 0$ and $0 \le \psi(G) \le 1$.

Exercises for Chapter 5

1. Consider a two-state Markov chain whose one-step transition matrix is given in Eq. 5.18 and for which $|1 - u| < 1$. Given that the chain is in state i ($i = 0$ or 1), let α_i be the number of new steps that it remains in state i until it moves out of state i. Show that

$$E[\alpha_0] = \frac{a}{1-a} \qquad E[\alpha_1] = \frac{d}{1-d}$$

$$\text{Var}[\alpha_0] = \frac{a}{(1-a)^2} \qquad \text{Var}[\alpha_1] = \frac{d}{(1-d)^2}$$

   Hint: Use properties of the geometric distribution.
2. Suppose a Markov chain has the following one-step transition matrix:

$$\mathbf{P} = \begin{pmatrix} 0.2 & 0.5 & 0.3 & 0 & 0 \\ 0.3 & 0.4 & 0.3 & 0 & 0 \\ 0.1 & 0.4 & 0.3 & 0.1 & 0.1 \\ 0 & 0 & 0 & 0.5 & 0.5 \\ 0 & 0 & 0 & 0.5 & 0.5 \end{pmatrix}$$

   Find the equivalence classes for the chain.
3. Reconsider the random walk with barriers of Example 5.3. This time assume that if the particle is in the 0 position, it can either stay there with probability q or it can go right with probability p, and if it is in the 3 position, it can either stay there with probability p or go left with probability q. Find the one-step transition probability matrix and compute the state vector $\mathbf{p}(2)$.
4. A letter is to be chosen randomly from among the letter set $\{b, c, u, r, p\}$ and then replaced. The process is repeated with the intent being to spell the word *up*. Model the recognition scheme using three states, realizing that once the word *up* is successfully spelled, the process stays in the "fully spelled" state. Find the one-step transition probability matrix and then find the state vector $\mathbf{p}(3)$.
5. State n is said to be *absorbing* if $p_{n,n} = 1$. Show that the mean recurrence time of an absorbing state is 1, thereby showing it to be positive recurrent.
6. Consider a modified random walk with barriers having state space $S = \{0, 1, 2, 3, 4\}$. With the exception of state 2, right and left transition probabilities are given by p and $q = 1 - p$, respectively, with states 0 and

4 acting as barriers. State 2 is absorbing, so that once in state 2, the Markov chain remains there forever. Write down the one-step transition matrix and find the state vector $\mathbf{p}(2)$. Prove that all states except state 2 are nonrecurrent.

7. Consider the following one-step transition matrix:

$$\mathbf{P} = \begin{pmatrix} 0.5 & 0.2 & 0.2 & 0 & 0 & 0 & 0.1 \\ 0.2 & 0.6 & 0.2 & 0 & 0 & 0 & 0 \\ 0 & 0.5 & 0.5 & 0 & 0 & 0 & 0 \\ 0 & 0 & 0 & 0.4 & 0.3 & 0 & 0.3 \\ 0 & 0 & 0 & 0.2 & 0.7 & 0.1 & 0 \\ 0 & 0 & 0 & 0 & 0 & 0.5 & 0.5 \\ 0 & 0 & 0 & 0 & 0 & 0.2 & 0.8 \end{pmatrix}$$

Identify the recurrent and nonrecurrent equivalence classes.

8. Let the recurrent and nonrecurrent classes of a homogeneous Markov chain be $C_1, C_2, \ldots, C_m$ and $C_{m+1}, C_{m+2}, \ldots, C_\nu$, respectively, where the ordering places recurrent classes prior to nonrecurrent classes. Let the corresponding transition probability matrices of the equivalence classes be $\mathbf{P}_1, \mathbf{P}_2, \ldots, \mathbf{P}_m, \mathbf{P}_{m+1}, \ldots, \mathbf{P}_\nu$ and, for equivalence classes n and r, let $\mathbf{Q}_{n,r}$ be the matrix of transition probabilities from states in class n to states in class r. Then the transition probability matrix for the overall chain takes the form

$$\mathbf{P} = \begin{pmatrix} \mathbf{P}_1 & & & & & & 0 \\ & \mathbf{P}_2 & & & & & \\ 0 & & \ddots & & & & \\ & & & \mathbf{P}_m & & & \\ \mathbf{Q}_{m+1,1} & \mathbf{Q}_{m+1,2} & \cdots & \mathbf{Q}_{m+1,m} & \mathbf{P}_{m+1} & & \\ \vdots & \vdots & \ddots & \vdots & \vdots & \ddots & \\ \mathbf{Q}_{\nu,1} & \mathbf{Q}_{\nu,2} & \cdots & \mathbf{Q}_{\nu,m} & \mathbf{Q}_{\nu.m+1} & \cdots & \mathbf{P}_\nu \end{pmatrix}$$

where the 0s indicate that the entire unlabeled regions are filled with 0s. In this form, the one-step transition probability matrix is said to be in *canonical* form. Note that there need not be a unique canonical form because it depends on the listing of equivalence classes; however, one can always assume the matrix to be in canonical form. Find canonical forms for the transition probability matrices of Exercises 5.2 and 5.7.

9. The following theorem can be proved: In a finite Markov chain, if n is a nonrecurrent state, then

$$\lim_{j \to \infty} p_n(j) = 0$$

regardless of the initial state. Deduce from the theorem that, in a finite Markov chain, not all states can be nonrecurrent. For a challenge, prove the theorem.

10. Suppose an object randomly transitions among three locations in accordance with the following transition probability matrix:

$$\mathbf{P} = \begin{pmatrix} 0.2 & 0.8 & 0 \\ 0 & 0.5 & 0.5 \\ 0.4 & 0 & 0.6 \end{pmatrix}$$

Find the stationary distribution vector.

11. For the *M/M/s* queue, show that the probability of a particular server being busy at a point in time in the steady state is equal to ρ. Hint: If there are at least s jobs in the system, then all servers are busy; if there are $k < s$ jobs in the system, then the probability that the server is busy is k/s.
12. *Balking* occurs in a queuing system if the arrival rate diminishes with increasing queue length because not all jobs coming to the system actually arrive (in the sense that they enter the system and do not leave because the queue is too long). For an *M/M/*1 queue with balking, the state equations of Eqs. 5.63 and 5.64 still apply, as do the solutions of Eqs. 5.66 and 5.67. Show that if balking occurs with $\lambda_n = \lambda/(n+1)$ for $n = 0, 1, 2, \ldots$, then $\{\pi_n: n = 0, 1, \ldots\}$ satisfies a Poisson distribution with mean λ/μ.
13. Suppose balking is service-dependent with $\lambda_n = \lambda e^{-cn/\mu}$, for $n = 0, 1, 2, \ldots$ and some constant $c > 0$. Let $\gamma = e^{-c/2\mu}$ and show that

$$\pi_0 = \left(\sum_{k=0}^{\infty} \rho^k \gamma^{k(k-1)} \right)^{-1}$$

$$\pi_n = \rho^n \gamma^{n(n-1)} \pi_0$$

14. In an *M/M/*1 queue with *state-dependent service*, the server reacts to the number of jobs in the system. Suppose $\lambda_n = \lambda$ for $n = 0, 1, 2, \ldots$ and $\mu_n = n^c \mu$ for $n = 1, 2, 3, \ldots$, where c is a *pressure coefficient*. Show that

$$\pi_0 = \left(\sum_{k=1}^{\infty} \frac{\rho^k}{(k!)^c} \right)^{-1}$$

$$\pi_n = \frac{\rho^n}{(n!)^c}\pi_0$$

15. Suppose the *M/M/*1 model is adapted so that there is an additional server whenever the number of jobs in the system exceeds N (> N) and the additional server departs whenever the number of jobs does not exceed N. Find the steady-state equations and by solving these recursively show that the steady-state distribution is given by

$$\pi_0 = \frac{(1-\rho)(2-\rho)}{2-\rho-\rho^{N+1}}$$

$$\pi_n = \rho^n \pi_0 \qquad (n = 1,\ 2,\ldots,\ N)$$

$$\pi_n = \frac{\rho^n}{2^{n-N}}\pi_0 \qquad (n = N+1, N+2, \ldots)$$

16. For an *M/M/*1 queue model, let $X(n)$, $n \geq 1$, be the number of jobs in the queue at the moment when the nth person served has just completed service. Show that $X(n)$ is a Markov chain. Hint: Let $Y(n)$ be the number of jobs arriving during the time the nth job is being serviced, define $\zeta(x) = 1$ if $x \neq 0$ and $\zeta(x) = 0$ if $x = 0$, and use the fact that

$$X(n+1) = X(n) - \zeta[X(n)] + Y(n+1)$$

X(n) is known as an *embedded Markov chain.*

17. Referring to the embedded Markov chain of Exercise 5.16, let a_k denote the probability that k customers arrive during the time a job is being serviced and find the infinite-dimensional one-step transition probability matrix in terms of a_0, a_1, a_2,....
18. Starting from the appropriate form of Eq. 5.32, show that the generating function of the stationary distribution of the embedded Markov chain of Exercise 5.16 is given by

$$A(z) = \frac{(z-1)\sum_{j=0}^{\infty} a_j z^j}{z - \sum_{j=0}^{\infty} a_j z^j}\pi_0$$

By letting $z \to 1$, show that

$$\pi_0 = 1 - \sum_{k=0}^{\infty} k a_k$$

which is the mean number of jobs arriving during the service time of a job. Throughout, assume this mean is strictly less than 1.

19. Referring to Exercise 5.17 and assuming that the Poisson arrivals are governed by parameter λ and the exponential service time has mean μ, show that the probability of k arrivals during the service time of a single job is given by

$$a_k = \int_0^{\infty} \frac{(\lambda t)^k}{k!\mu} \exp\left[-\left(\lambda + \frac{1}{\mu}\right)t\right] dt$$

21. Equation 5.24 defines a first recurrence probability for a state n. In fact, for any states m and n we can define the *first-passage probability* $f_{mn}(i)$ to be the probability that, starting from state m, the process reaches state n for the first time in i steps. Find all first-passage probabilities for a two-state Markov chain with states 0 and 1.
22. Prove that the first-passage probabilities (Exercise 5.21) are related to the transition probabilities of a homogeneous Markov chain by

$$p_{m,n}(k) = \sum_{i=1}^{k} f_{mn}(i) p_{n,n}(k-i)$$

where we define $p_{m,n}(0) = 1$ if $m = n$ and $p_{m,n}(0) = 0$ if $m \neq n$.

23. From the first-passage probabilities, in analogy to Eq. 5.25 we can define

$$f_{mn} = \sum_{i=1}^{\infty} f_{mn}(i)$$

for any two states m and n. This gives the probability that, starting in state m, the process ever visits state n. Now, for any state n, the *occupation time* of the state in the first i transitions, $N_n(i)$, is the number of times the chain is in state n during the first i transitions. Rigorously define the random process $Z_n(k)$ by

$$Z_n(k) = \begin{cases} 1, & \text{if } X(k) = n \\ 0, & \text{if } X(k) \neq n \end{cases}$$

Then we define

$$N_n(i) = \sum_{k=1}^{i} Z_n(k)$$

By letting the sum be to ∞, we define $N_n(\infty)$. Show that

$$f_{mn} = P(N_n(\infty) - N_n(j) > 0 | X(j) = m)$$

for any time j for which the conditional probability is defined, meaning that the probability of the conditioning event is nonzero.

24. The probability that there are infinitely many visits to state n when initially starting in state m is defined by

$$g_{mn} = P(N_n(\infty) - N_n(j) = \infty | X(j) = m)$$

It can be proven that $g_{mn} = f_{mn} g_{nn}$ and, for any states m and n,

$$g_{nn} = \lim_{k \to \infty} f_{nn}^k$$

From this limit deduce that, for any state n, either $g_{nn} = 1$ or $g_{nn} = 0$.

25. Let Q be the set of nonrecurrent states of a Markov chain with state space S. The *time before absorption* is the random variable T giving the time the process spends in the class of nonrecurrent states before passing into the class of recurrent states. Rigorously,

$$T = 1 + \sum_{n \in Q} N_n(\infty)$$

The *mean time to absorption*, given that the process starts in state m, is given by

$$\mu_m = E[T | X(0) = m]$$

Prove that

$$\mu_m = \sum_{n \in Q} \sum_{i=0}^{\infty} p_{m,n}(i)$$

and that the mean times to absorption satisfy the following system of equations:

$$\mu_m = 1 + \sum_{n \in Q} p_{m,n} \mu_n \qquad (m \in Q)$$

Note that, for a finite Markov chain, the mean times to absorption are finite and the preceding system has a unique solution. If a Markov chain has an infinite number of nonrecurrent states, then mean times to absorption may be infinite.

26. Let Q be the set of nonrecurrent states of a finite Markov chain with state space S and let $\mathbf{Q}$ be the one-step transition probability matrix for states in Q. Let $\mathbf{m}$ be the column vector whose components are the mean times to absorption of the states in Q. Then the system of equations in Exercise 5.25 takes the form

$$\mathbf{m} = \mathbf{u} + \mathbf{Qm}$$

where $\mathbf{u}$ is the column vector whose entries are all 1. Hence,

$$\mathbf{m} = (\mathbf{I} - \mathbf{Q})^{-1}\mathbf{u}$$

Apply this vector-matrix equation to find the mean times to absorption (the vector $\mathbf{m}$) for nonrecurrent states of the Markov chain of Exercise 5.2.

27. For the Ising model of Example 5.12, show that $U(\mathbf{x}) = 4\alpha + 4\beta$ for

$$\mathbf{x} = \begin{pmatrix} 1 & 1 \\ 1 & 1 \end{pmatrix}$$

28. For the Ising model of Example 5.12, evaluate

$$P(X(1,1) = 0 \mid X(0,0) = 0, X(1,0) = 1, X(0,1) = 1)$$

by both full conditioning and application of the dependency condition.

29. Write down the form of the energy function for the neighborhood system $\mathcal{N}_1$ of Example 5.11. For the Ising model of Example 5.12, each function composing the potential is formed as a scalar times the product of the variables corresponding to the associated clique type — the constant α for a single variable and the constant β for two variables. Extend the model to the present setting involving $\mathcal{N}_1$ by using the scalars δ and ε for the cases of three and four variables, respectively. Using the same set of sites as in Example 5.12, find $U(\mathbf{x})$ for the configuration

$$\mathbf{x} = \begin{pmatrix} 1 & 1 \\ 1 & 0 \end{pmatrix}$$

30. Consider the site set A of Example 5.11 and define the neighborhood system $\mathcal{N}_2$ in the following manner: for each (i, j) lying in the interior of A, associate the neighborhood

$$N_{i,j} = \begin{pmatrix} 1 & 0 & 1 \\ 0 & \mathbf{0} & 0 \\ 1 & 0 & 1 \end{pmatrix}_{(i,j)}$$

As for $\mathcal{N}_0$, neighborhoods for boundary sites must be adjusted so they are subsets of A. Find the three types of cliques that compose the clique set and write down the form of the energy function. As in the Ising model, define each function in the potential as a scalar times the product of the variables corresponding to the associated clique type — the constant α for a single variable and the constant β for two variables. Find $U(\mathbf{x})$ for

$$\mathbf{x} = \begin{pmatrix} 0 & 1 \\ 1 & 0 \end{pmatrix}$$

31. Suppose the primary grain S in a random Boolean model is an ellipse with the axes possessing independent normally distributed lengths, one with mean μ and variance σ^2 and the other with mean ξ and variance η^2. Find the porosity of the process. Note that we are assuming the means are sufficiently large with respect to the variances that we can neglect the possibility of negative lengths.
32. Find the covariance of the pores and the variogram for the random Boolean model of Example 5.13.
33. Let S be an isotropic random rectangle whose length and width possess independent gamma distributions with parameter pairs α, β and ρ, υ, respectively, and let D be convex. Find the expected number of grains that intersect D.
34. For the Boolean model of Exercise 5.33, find the probability that a given grain is isolated.
35. Simulate a realization of a random Boolean model with $\lambda = 0.1$ and primary grain S being a disk with random radius R possessing an exponential distribution with mean $\mu = 2$. Estimate the mean and covariance of the grains using both mean and covariance ergodicity. Also find them theoretically.
36. Show the following properties of dilation:

$$B \oplus D = D \oplus B$$

$$(B \oplus C) \oplus D = B \oplus (C \oplus D)$$

$$B \oplus (D + x) = (B \oplus D) + x$$

$$C \oplus (B \cup D) = (C \oplus B) \cup (C \oplus D)$$

$$t(B \oplus D) = tB \oplus tD, \quad t > 0$$

In turn, these properties provide commutativity, associativity, translation invariance, distributivity over union, and distributivity of scalar multiplication over dilation.

37. Show that the dilation of a ball of radius r with itself is a ball of radius $2r$. By associativity, the n-fold self-dilation of a ball of radius r with itself is a ball of radius nr.
38. Prove that opening is antiextensive and idempotent.
39. A deterministic binary image consists of a square of edge length 1 for which there is an angle of $\pi/6$ between one of its edges and the x axis. Find the pattern spectrum for a vertical linear structuring element of unit length.
40. Suppose that the sizing distribution for the single-primitive model of Eq. 5.155 is uniform over the interval $[a, b]$, $0 < a < b < \infty$, the structuring element is a vertical line, and the image primitive is a square of edge length 1 having a horizontal base. Find the expected value of the pattern-spectrum mean.
41. Run 30 realizations of the random binary image process of Exercise 5.40 with $a = 1$ and $b = 3$ (keeping in mind that the grains must be disjoint) and estimate the pattern spectrum mean. Rerun the 30 realizations under the assumption that the process is a random Boolean model with $\lambda = 0.1$ and estimate the pattern spectrum mean. Discard squares hitting the edge of the frame.
42. Suppose G is a random nonrotated ellipse with horizontal and vertical axes possessing independent exponential distributions having means μ_1 and μ_2, respectively, and suppose the generator of the granulometry $\{\Psi_t\}$ contains a single disk of unit diameter. Find the probability distribution of the granulometric measure M_G.
43. The steady-state distribution of Eqs. 5.180 and 5.181 was obtained by recognizing that the Markov chain is a random walk. Proceed directly and find the appropriate Chapman-Kolmogorov equations. From these, deduce the steady-state distribution.
44. Construct signal and noise models according to Eqs. 5.168 and 5.173 so that both M_S and M_N are uniformly distributed. Then apply the results of Example 5.19 assuming $P(N) = P(S)$.

45. In the text it states that $F_n \to F$ in $(\mathcal{F}, \mathfrak{I}_{\mathcal{F}})$ if and only if conditions C1 and C2 hold. Give a detailed argument to demonstrate this equivalence.
46. Show that conditions C1 and C2 are equivalent to conditions C1' and C2', respectively:
    C1': For any $x \in F$ there exists a sequence of points $x_n \in F_n$ such that $x_n \to x$ in $\mathfrak{R}^d$.
    C2': If $\{F_{n_k}\}$ is a subsequence of $\{F_n\}$, $x_{n_k} \in F_{n_k}$, and $x_{n_k} \to x$ in $\mathfrak{R}^d$, then $x \in F$.
    These conditions can often be used in place of C1 and C2 to prove convergence.
47. Prove that $F_n \downarrow F$ implies $\lim F_n = F$ in $(\mathcal{F}, \mathfrak{I}_{\mathcal{F}})$.
48. Prove that $F_n \uparrow A$ implies $\lim F_n = \overline{A}$ in $(\mathcal{F}, \mathfrak{I}_{\mathcal{F}})$.
49. Prove that $F_n \downarrow F$ and $H_n \downarrow H$ implies the following limits in $(\mathcal{F}, \mathfrak{I}_{\mathcal{F}})$:

$$\lim F_n \cap H_n = F \cap H$$

$$\lim F_n \cup H_n = F \cup H$$

50. Prove that $F_n \uparrow A$ and $H_n \uparrow B$ implies

$$\lim F_n \cap H_n = \overline{A \cap B}$$

$$\lim F_n \cup H_n = \overline{A \cup B}$$

51. Prove that the mapping $(F, H) \to F \cap H$ from $\mathcal{F} \times \mathcal{F} \to \mathcal{F}$ is u.s.c.
52. Prove that the formulations of dilation in Eqs. 5.121 and 5.199 are equivalent.
53. Prove the duality relation for erosion and dilation:

$$A \oplus B = [A^c \ominus (-B)]^c$$

54. Show that erosion has the alternative formulation

$$A \ominus B = \bigcap_{b \in B} A - b$$

55. Show that the opening definitions of Eqs. 5.148 and 5.199 are equivalent.
56. Prove the duality relation for opening and closing:

$$A \bullet B = (A^c \circ B)^c$$

57. Suppose K is a nonempty compact set and F is a closed set. Show that the following sets are closed: $F \oplus K$, $F \ominus K$, $F \circ K$, and $F \bullet K$.
58. Suppose K is a nonempty compact set and G is an open set. Show that the following sets are open: $G \oplus K$, $G \ominus K$, $G \circ K$, and $G \bullet K$.
59. Suppose K is a nonempty compact set and C is a compact set. Show that the following sets are compact: $C \oplus K$, $C \ominus K$, $C \circ K$, and $C \bullet K$.
60. Prove that dilation is continuous as a mapping from $\mathcal{F} \times \mathcal{K} \rightarrow \mathcal{F}$.
61. Prove that erosion is u.s.c. as a mapping from $\mathcal{F} \times \mathcal{K}_0 \rightarrow \mathcal{F}$.
62. Prove that, for a RACS X, a closed set F, and a compact set K,

$$T_{X \cap F}(K) = T_X(K \cap F)$$

63. Prove that, for a RACS X, $\rho > 0$, and a compact set K,

$$T_{\rho X}(K) = T_X\left(\frac{1}{\rho}K\right)$$

64. Prove that, for a RACS X and a compact set K,

$$T_{-X}(K) = T_X(-K)$$

65. Prove that, for a RACS X, a nonempty compact set K, and a point $x \in \Re$,

$$T_{X \ominus K}(\{x\}) = P(K + x \subset X)$$

66. Prove that, for a RACS X and nonempty compact sets K and C,

$$T_{X \oplus C}(K) = P(K \oplus (-C) \subset X)$$

67. A RACS X is said to be *stationary* if, for every set $\mathcal{A}$ of closed sets in the σ-algebra generated by $\mathcal{F}$ and for any $x \in \Re^d$,

$$P(X \oplus \{x\} \in \mathcal{A}) = P(X \in \mathcal{A})$$

Show that X is stationary if and only if

$$T_X(K \oplus \{x\}) = T_X(K)$$

for any compact set K and $x \in \Re^d$.
68. Two compact sets K_1 and K_2 are said to be *separated* by compact set K if, for any points $x_1 \in K_1$ and $x_2 \in K_2$, there is nonnull intersection between

K and the line segment from x_1 to x_2. Show that, if closed convex set F hits K_1 and K_2, then F hits K.

69. A RACS X is said to be a *semi-Markov RACS* if, for any compact sets K_1, K_2, and K such that K_1 and K_2 are separated by K,

$$P(X \cap K_1 \in \mathcal{A}_1, X \cap K_2 \in \mathcal{A}_2 | X \cap K = \varnothing)$$

$$= P(X \cap K_1 \in \mathcal{A}_1 | X \cap K = \varnothing)\ P(X \cap K_2 \in \mathcal{A}_2 | X \cap K = \varnothing)$$

where $\mathcal{A}_1$ and $\mathcal{A}_2$ are sets in the σ-algebra generated by $\mathcal{F}$. Show that X is semi-Markovian if and only if

$$Q_X(K_1 \cup K_2 \cup K) Q_X(K) = Q_X(K_1 \cup K) Q_X(K_2 \cup K)$$

for any compacts set K_1, K_2, and K such that K_1 and K_2 are separated by K.

70. Use Theorem 5.11 to prove Theorem 2.5.

Bibliography

1. Abramson, N. M., *Information Theory and Coding*, McGraw-Hill, New York, 1963.
2. Alexander, S. T., *Adaptive Signal Processing: Theory and Applications*, Springer-Verlag, New York, 1986.
3. Andrews, H. C., and B. R. Hunt, *Digital Image Restoration*, Prentice-Hall, Englewood Cliffs, NJ, 1977.
4. Asmussen, S., *Applied Probability and Queues*, Wiley, New York, 1987.
5. Bartlett, M. S., *An Introduction to Stochastic Processes with Special Reference to Methods and Applications*, 3rd ed., Cambridge University Press, London, 1978.
6. Bailey, N. T. J., *The Elements of Stochastic Processes with Applications to the Natural Sciences*, Wiley, New York, 1964.
7. Bellanger, M., *Adaptive Digital Filters and Signal Analysis*, Marcel Dekker, New York, 1987.
8. Besag, J., Spatial interaction and the statistical analysis of lattice systems, *J. Royal Statistical Society B*, **36**(1974).
9. Bharucha-Reid, A. T., *Elements of the Theory of Markov Processes and Their Applications*, McGraw-Hill, New York, 1960.
10. Bhat, U. N., *Elements of Applied Stochastic Processes*, Wiley, New York, 1972.
11. Bhattacharya, R. N., *Stochastic Processes with Applications*, Wiley, New York, 1990.
12. Bishop, C. M., *Neural Networks for Pattern Recognition*, Clarendon Press, Oxford, UK, 1995.
13. Bouleau, N., and D. Lepingle, *Numerical Methods for Stochastic Processes*, Wiley, New York, 1993.
14. Boullion, T. L., and P. L. Odell, *Generalized Inverse Matrices*, Wiley-Interscience, New York, 1971.
15. Bow, S.-T., *Pattern Recognition and Image Preprocessing*, Marcel Dekker, New York, 1992.
16. Bracewell, R. N., *Two Dimensional Imaging*, Prentice-Hall, Englewood Cliffs, NJ, 1995.
17. Bucy, R. S., and P. D. Joseph, *Filtering for Stochastic Processes with Applications to Guidance*, Interscience Publishers, New York, 1968.
18. Catlin, D. E., *Estimation, Control, and the Discrete Kalman Filter*, Springer-Verlag, Berlin, 1989.
19. Chellappa, R., *Digital Image Processing*, rev. ed., IEEE Computer Society Press, Los Alamitos, California, 1992.
20. Chui, C. K., and G. Chen, *Kalman Filtering*, Springer-Verlag, New York, 1987.

21. Choquet, G., "Theory of capacities," *Annals Institute Fourier*, **V**(1953–54).
22. Clarke, R. J., *Transform Coding of Images*, Academic Press, New York, 1985.
23. Cooper, G. R., and C. D. McGillem, *Probabilistic Methods of Signal and System Analysis*, Holt, Rinehart and Winston, New York, 1971.
24. Cowan, C. F., and P. M. Grant, *Adaptive Filters*, Prentice-Hall, Englewood Cliffs, NJ, 1985.
25. Cox, D. R., and H. D. Miller, *The Theory of Stochastic Processes*, Methuen, London, 1968.
26. Cramér, H., *Mathematical Methods of Statistics*, Princeton University Press, Princeton, NJ, 1946.
27. Cramér, H., and M. R. Leadbetter, *Stationary and Related Stochastic Processes; Sample Function Properties and Their Applications*, Wiley, New York, 1967.
28. Cramér, H., *Random Variables and Probability Distributions*, 3rd ed., Cambridge University Press, London, 1970.
29. Cramér, H., "Contributions to the theory of statistical estimation," *Skandinavisk Aktuarietidskrift*, 1946.
30. Cressie, N., *Statistics for Spatial Data*, Wiley, New York, 1991.
31. Cressie, N., and G. M. Lasslett, "Random set theory and problems of modeling, *SIAM Review*, **29**, No. 4(1987).
32. Davenport, W. B., Jr., and W. L. Root, *An Introduction to the Theory of Random Signals and Noise*, IEEE Press, New York, 1987.
33. Davenport, W. B., Jr., *Probability and Random Processes; an Introduction for Applied Scientists and Engineers,* McGraw-Hill, New York, 1970.
34. David, H. A., *Order Statistics*, Wiley, New York, 1970.
35. DeGroot, M. H., *Optimal Statistical Decisions*, McGraw-Hill, New York, 1970.
36. Desai, U. B., *Modeling and Applications of Stochastic Processes*, Kluwer Academic Publishers, Boston, 1986.
37. Dobrushin, P. L., "The description of a random field by means of conditional probabilities and conditions of its regularity," *Theory of Probability and Its Applications*, **13**, No. 2(1968).
38. Doob, L. J., *Stochastic Processes*, Wiley, New York, 1953.
39. Dougherty, E. R., ed., *Mathematical Morphology in Image Processing*, Marcel Dekker, New York, 1993.
40. Dougherty, E. R. *Probability and Statistics for the Engineering, Computing, and Physical Sciences,* Prentice-Hall, Englewood Cliffs, NJ, 1990.
41. Dougherty, E. R., and J. Astola, eds., *Nonlinear Image Filters*, SPIE Press and IEEE Press, Bellingham, WA, 1998.
42. Dougherty, E. R., and C. R. Giardina, *Image Processing—Continuous to Discrete*, Prentice-Hall, Englewood Cliffs, NJ, 1987.
43. Duda, R. O., and P. E. Hart, *Pattern Classification and Scene Analysis*, Wiley, New York, 1973.

44. Dunford, N., and J. T. Schwartz, *Linear Operators: Part I: General Theory*, Wiley, New York, 1976.
45. Einstein, A., *Investigations on the Theory of the Brownian Movement*, Dover, New York, 1956 (contains translations of Einstein's 1905 papers).
46. Feller, W., *An Introduction to Probability Theory and Its Applications*, Vol. 1, 3rd ed., Wiley, New York, 1968.
47. Feller, W., *An Introduction to Probability Theory and Its Applications*, Vol. 2, 2nd ed., Wiley, New York, 1971.
48. Freund, J. E., and R. E. Walpole, *Mathematical Statistics*, 4th ed., Prentice-Hall, Englewood Cliffs, NJ, 1987.
49. Fukunaga, K., *Introduction to Statistical Pattern Recognition*, Academic Press, New York, 1972.
50. Gelb, A., ed., *Applied Optimal Estimation*, MIT Press, Cambridge, MA, 1974.
51. Gelfand, I. M., and G. E. Shilov, *Generalized Functions*, Academic Press, New York, 1965.
52. Gnedenko, B. V., *The Theory of Probability*, 4th ed., Chelsea, New York, 1967.
53. Gonzalez, R. K., and R. E. Woods, *Digital Image Processing*, Addison-Wesley, Reading, PA, 1992.
54. Goffman, C., and G. Pedrick, *First Course in Functional Analysis*, Prentice-Hall, Englewood Cliffs, NJ, 1965.
55. Goldberg, R. R., *Fourier Transforms*, Cambridge University Press, Cambridge, MA, 1965.
56. Goutsias, J., "On the morphological analysis of random shapes," *Mathematical Imaging and Vision*, **2**, No. 2/3(1992).
57. Hall, P., *Introduction to the Theory of Coverage Processes*, Wiley, New York, 1988.
58. Halmos, P. R., *Measure Theory*, Van Nostrand, Princeton, 1974.
59. Halmos, P. R., *Finite Dimensional Vector Spaces*, Van Nostrand, Princeton, 1958.
60. Haralick, R. M., and L. G. Shapiro, *Computer and Robot Vision*, Vols. 1 & 2, Addison-Wesley, Reading, PA, 1992.
61. Haykin, S., *Adaptive Filter Theory*, 2nd ed., Prentice-Hall, Englewood Cliffs, NJ. 1991.
62. Haykin, S., *Neural Networks*, Macmillan, New York, 1994.
63. Helstrom, C. W., *Statistical Theory of Signal Detection*, 2nd ed., Pergamon Press, New York, 1968.
64. Hoel, P. G., *Introduction to Mathematical Statistics*, 4th ed., Wiley, New York, 1971.
65. Hoel, P. G., S. C. Port, and C. J. Stone, *Introduction to Probability Theory*, Houghton Mifflin, Boston, MA, 1971.
66. Hoel, P. G., S. C. Port, and C. J. Stone, *Introduction to Statistical Theory*, Houghton Mifflin, Boston, MA, 1971.

67. Hoel, P. G., S. C .Port, and C. J. Stone, *Introduction to Stochastic Theory*, Houghton Mifflin, Boston, MA, 1972.
68. Hoffman, K., and R. Kunze, *Linear Algebra*, Prentice-Hall, Englewood Cliffs, NJ, 1971.
69. Hogg, R. V., and A.T. Craig, *Introduction to Mathematical Statistics*, 4th ed., Macmillan, New York, 1978.
70. Isaacson, D. L., and R. W. Madsen, *Markov Chains: Theory and Applications*, Wiley, New York, 1976.
71. Jain, A. K., *Fundamentals of Digital Image Processing*, Prentice-Hall, Englewood Cliffs, NJ, 1989.
72. Jenkins, G. M., and D. G. Watts, *Spectral Analysis and its Applications*, Holden-Day, San Francisco, CA, 1968.
73. Jeulin, D., ed., *Advances in Theory and Applications of Random Sets*, World Scientific, London, 1997.
74. Kalman, R. E., A New Approach to Linear Filtering and Prediction Problems, *Basic Engineering (ASME)*, **82D**, (1960).
75. Kalman, R. E., and R. Bucy, New Results in Linear Filtering and Prediction, *Basic Engineering (ASME)*, **83D**, (1961).
76. Karhunen, K., Uber lineare Methoden in der Wahrscheinlichkeitsrechnung, *Ann. Acad. Sci. Fennicae*, A I, No. 37(1947).
77. Kemeny, J., L. J. Snell and A. Knapp, *Denumerable Markov Chains*, 2nd ed., Springer-Verlag, New York, 1976.
78. Kemeny, J., and L. J. Snell, *Finite Markov Chains*, Springer-Verlag, New York, 1960.
79. Kendall, M. G., and P. A. P. Moran, *Geometrical Probability*, Griffin, London, 1963.
80. Kendall, M., and A. Stuart, *The Advanced Theory of Statistics,* 4th ed., Macmillan, New York, 1977-1983.
81. Khinchin, A. I., "Theory of correlation of stationary stochastic processes," *Uspekh. mat. nauk*, 5 (1938).
82. Khinchin, A. I., *Mathematical Foundations of Information Theory,* Dover, New York, 1957.
83. Kleinrock, L., *Queuing Systems,* Vol. I: *Theory*, Wiley, New York, 1975.
84. Kleinrock, L., *Queuing Systems,* Vol. 2: *Computer Applications,* Wiley, New York, 1976.
85. Kolmogorov, A., "Stationary sequences in Hilbert space," *Bulletin Math. University Moscow* 2(1941).
86. Kolmogorov, A., "The interpolation and extrapolation of stationary random sequences," *Akad. nauk SSSR, ser. mat.*, 5, No. 1(1941).
87. Kolmogorov, A., *Foundations of the Theory of Probability*, 2nd ed., Chelsea, New York, 1956.
88. Kolmogorov, A., "Interpolation and extrapolation of stochastic random sequences," in *Linear Least-Squares Estimation*, T. Kailath, ed., Hutchinson and Ross, 1977.

89. Kreider, D. L., R. G. Kuller, D. R. Ostberg and F. W. Perkins, *An Introduction to Linear Analysis*, Addison-Wesley, Reading, PA, 1966.
90. Kuznetsov, P. I., V. I. Tikhonov, and R. L. Stratonovich, *Non-Linear Transformations of Stochastic Processes*, Pergamon Press, New York, 1965.
91. Lewis, F. L., *Optimal Estimation: With an Introduction to Stochastic Control Theory,* Wiley, New York, 1986.
92. Lewis, T. O., and P. L. Odell, *Estimation in Linear Models*, Prentice-Hall, Englewood Cliffs, NJ, 1971.
93. Lim, J. S., *Two-Dimensional Signal and Image Processing*, Prentice-Hall, Englewood Cliffs, NJ, 1990.
94. Loéve, M., *Probability Theory I*, 4th ed., Springer-Verlag, New York, 1977.
95. Loéve, M., *Probability Theory II*, 4th ed., Springer-Verlag, New York, 1978.
96. Loce, R. P., and E. R. Dougherty, *Enhancement and Restoration of Digital Documents: Statistical Design of Nonlinear Algorithms,* SPIE Press, Bellingham, WA, 1997.
97. Mardia, K. V., and G. K. Kanji, eds., *Statistics and Images*, Carfax Pub., Abingdon, Oxfordshire, UK, 1993.
98. Matheron, G., *Random Sets and Integral Geometry*, Wiley, New York, 1975.
99. Medhi, J., *Stochastic Processes*, 2nd ed., Wiley, New York, 1994.
100. Miller, K. S., *Multidimensional Gaussian Distributions*, Wiley, New York, 1964.
101. Molchanov, I. A., *Statistics of the Boolean Model for Practitioners and Mathematicians,* Wiley, Chichester, UK, 1997.
102. Morse, P., *Queues, Inventories and Maintenance: the Analysis of Operational Systems with Variable Demand and Supply,* Wiley, New York, 1958.
103. Netravali, A. N., and B. G. Haskell, *Digital Pictures: Representation and Compression*, Plenum Press, New York, 1988.
104. Oppenheim, A. V., and R. W. Schafer, *Digital Signal Processing*, Prentice-Hall, Englewood Cliffs, NJ, 1975.
105. Oppenheim, A. V., A. S. Willsky, and I. T. Young, *Signals and Systems*, Prentice-Hall, Englewood Cliffs, NJ, 1983.
106. Papoulis, A., *Probability, Random Variables, and Stochastic Processes,* 2nd ed., McGraw-Hill, New York, 1984.
107. Parzen, E., *Modern Probability Theory and Its Applications,* Wiley, New York, 1960.
108. Parzen, E., *Stochastic Processes*, Holden-Day, San Francisco, CA, 1962.
109. Pitas, I., *Digital Image Processing Algorithms*, Prentice-Hall, New York, 1993.
110. Poor, H. V., *An Introduction to Signal Detection and Estimation*, 2nd, Springer-Verlag, Berlin, 1994.
111. Pratt, W. K., *Digital Image Processing*, 2nd, Wiley, New York, 1991.
112. Preston, C. J., *Random Fields*, Springer-Verlag, New York, 1976.

113. Prokhorov, V. S., *Probability Theory: Basic Concepts, Limit Theorems, Random Processes,* Springer-Verlag, New York, 1969.
114. Pugachev, V. S., *Theory of Random Functions and Its Applications to Control Problems*, Pergamon Press, Oxford, (U.S. ed. distributed by Addison-Wesley, Reading, MA), 1965.
115. Pugachev, V. S., "The application of canonical expansions of random functions to the determination of the optimal linear system," *Avtomat. i telemekh.*, **27**, No. 6(1956).
116. Pugachev, V. S., "Integral canonical representations of random functions and their application to the determination of optimal linear systems," *Avtomat. i telemekh.*, **28**, No. 11(1957).
117. Rabbani, M., and P. W. Jones, *Digital Image Compression Techniques*, SPIE Press, Bellingham, WA, 1991.
118. Rao, C. R., *Linear Statistical Inference and Its Applications*, 2nd ed., Wiley, New York, 1973.
119. Rosenblatt, M., *Random Processes*, 2nd ed., Springer-Verlag, New York, 1974.
120. Rosenfeld, A., and A. C. Kak, *Digital Picture Processing*, Vols. 1 & 2, Academic Press, New York, 1982.
121. Ross, S. M., *A First Course in Probability*, 4th ed., Macmillan, New York, 1994.
122. Ross, S. M., *Stochastic Processes*, Wiley, New York, 1983.
123. Rozanov, Y. A., *Introductory Probability Theory*, Prentice-Hall, Englewood Cliffs, NJ, 1969.
124. Rozanov, Y. A., *Innovation Processes*, Wiley, New York, 1977.
125. Rubinstein, R., *Simulation and the Monte Carlo Method*, Wiley, New York, 1981.
126. Rudin, W., *Real and Complex Analysis*, McGraw-Hill, New York, 1966.
127. Rudin, W., *Functional Analysis*, McGraw-Hill, New York, 1973.
128. Ruiz-Pala, E., C. Avila-Beloso and W. W. Hines, *Waiting-Line Models; an Introduction to Their Theory and Applications*, Reinhold, New York, 1967.
129. Rumelhart, D. E., and J. L. McClelland, eds, *Parallel Distributed Processing: Explorations in the Microstructure of Cognition*, Vol. 1, MIT Press, Cambridge, MA, 1986.
130. Ruymgaart, P. A., and T. T. Soong, *Mathematics of Kalman-Bucy Filtering*, Springer-Verlang, Berlin, 1985.
131. Santalo, L. A., *Integral Geometry and Geometric Probability*, Addison-Wesley, Reading, PA, 1978.
132. Semenov, V. M., "A contribution to the extrapolation of random time series," *Sborn. VVIA im. Zhukovskogo*, **1** (1954).
133. Serra, J., *Image Analysis and Mathematical Morphology*, Academic Press, London, 1982.
134. Serra, J., ed., *Image Analysis and Mathematical Morphology*, Vol. 2, Academic Press, New York, 1988.

135. Shannon, C. E., "A mathematical theory of communication," *Bell Systems Technical Journal*, **27** (1948).
136. Sklansky, J., and G. N. Wassel, *Pattern Classifiers and Trainable Machines*, Springer-Verlag, New York, 1981.
137. Stark, H., and J. W. Woods, *Probability, Random Processes, and Estimation Theory for Engineers*, Prentice-Hall, Englewood Cliffs, NJ, 1986.
138. Stewart, J. W., ed., *Numerical Solutions of Markov Chains*, Marcel Dekker, New York, 1991.
139. Stoyan, D., and W. S. Kendall, *Stochastic Geometry and Its Applications*, Wiley, New York, 1987.
140. Taylor, H. M., and K. Samuel, *An Introduction to Stochastic Modeling*, Academic Press, Cambridge, MA, 1994.
141. Wald, A., *Sequential Analysis*, Dover Publications, New York, 1973.
142. Weeks, A. R., *Fundamentals of Electronic Image Processing*, SPIE Press and IEEE Press, Bellingham, WA, 1996.
143. Wegman, E. J., and D. J. DePriest, *Statistical Image Processing and Graphics*, Marcel Dekker, New York, 1986.
144. White, H., *Artificial Neural Networks: Approximation and Learning Theory,* Blackwell, Cambridge, UK, 1992.
145. Widrow, B., and S. D. Stearns, *Adaptive Signal Processing*, Prentice-Hall, Englewood Cliffs, NJ, 1985.
146. Wiener, N., "Generalized Harmonic Analysis," Acta Math., **55**(1930).
147. Wiener, N., *Nonlinear Problems in Random Theory*, Wiley, New York, 1958.
148. Wiener, N., *Extrapolation, Interpolation and Smoothing of Stationary Time Series,* MIT Press, Cambridge, MA, 1964.
149. Wong, E., and B. Hajek, *Stochastic Processes in Engineering Systems*, Springer-Verlag, New York, 1985.
150. Yaglom, A. M., *Stationary Random Functions*, Prentice-Hall, Englewood Cliffs, NJ, 1962.
151. Yosida, K., *Functional Analysis*, Springer-Verlag, New York, 1968.

Index

Edward R. Dougherty is director of the Imaging Division of the Texas Center for Applied Technology and professor of Electrical Engineering at Texas A&M University. He holds an MS in computer science from Stevens Institute of Technology and a PhD in mathematics from Rutgers University. He is author of ten books, an editor of two books on image processing, and has written numerous papers on nonlinear image processing. He currently is Editor of the SPIE/IS&T *Journal of Electronic Imaging* and the SPIE/IEEE Series on Imaging Science and Engineering.